A Perfect Storm in the Amazon Wilderness

Success and Failure in the Fight to Save an Ecosystem of Critical Importance to the Planet

Volume 1

The Conventional Economy and the Drivers of Change

Timothy J. Killeen

To follow in Volumes 2 and 3:

THE WHITE HORSE PRESS
The Old Vicarage, Winwick, Cambridgeshire, PE28 5PN, UK
www.whpress.co.uk

A Perfect Storm in the Amazon Wilderness

The research for this book was supported by a grant from the Andes and Amazon Program of the Gordon and Betty Moore Foundation. The author is indebted to his colleagues at the Museo de Historia Natural Noel Kempff Mercado for their collaboration, particularly Lisette Correa for the selection of photographs.

A catalogue record for this book is available from the British Library

ISBN (combined volumes) 978-1-912186-22-8. doi: 10.3197/9781912186228

Volume 1. ISBN 978-1-912186-23-5
Volume 2. ISBN 978-1-912186-24-2
Volume 3. ISBN 978-1-912186-59-4

A Perfect Storm in the Amazon Wilderness

Contents

Chapter 1: The State of the Amazon

Chapter 2: Infrastructure Defines the Future

Chapter 3: Agriculture

Chapter 4: Land: The Ultimate Commodity

Figures, Tables and Text Boxes

Chapter 1

The State of the Amazon

The Amazon, home to the largest tropical forest on the planet, is an irreplaceable natural asset with enormous biodiversity and a critically important component in global carbon and water cycles. The Pan Amazon, which includes the full watershed and the rainforests of the Guiana Shield, is a geopolitical territory spanning nine nations that have been entrusted with the stewardship of its natural resources (Figure 1.1).

Fifteen years ago, the prospects for conserving this globally important natural asset were very much in doubt. Rampant deforestation driven by multiple social and economic phenomena threatened to transform its landscapes, degrade its aquatic resources and overwhelm its indigenous communities. Governments pursued construction of large-scale infrastructure projects as they sought to leverage unprecedented demand for global commodities with increased access to international financial markets. The resultant boom in economic activity motivated individuals and corporations to invest in business opportunities in the Amazon that progressively expanded the footprint of modern society. Climate scientists showed how a warmer planet would impact ecosystem function, as well as how a deforested landscape might disrupt moisture flows over the continent. The panorama was grim, and the combination of threats was referred to as *A Perfect Storm in the Amazon Wilderness*,[1] borrowing a phrase from popular culture that described the destructive synergies between multiple forces of change.

Fortunately, the citizens of the Amazonian nations were aware of the risk from uncontrolled development and demanded that their governments intervene to halt, or at the very least slow, the destruction. Concerned individuals from across the planet, in support of public and private conservation initiatives, joined them. The Pan Amazonian nations now boast the most extensive network of protected areas of any geographic region on Earth and have recognised the legal rights of indigenous communities by formalising their claims to ancestral lands. These two parallel efforts were

doi: 10.3197/9781912186228.ch01

implemented in a remarkably short span of time, reflecting the support of the area's constituent populations and the capacity of global society to mobilise financial resources for environmental action and social justice. Simultaneously, a dramatic reduction in deforestation rates gave hope to advocates seeking systemic changes in development paradigms, particularly in Brazil, where the agribusiness sector reformed its production systems after recognising that its commercial interests were best served by improving its environmental performance.

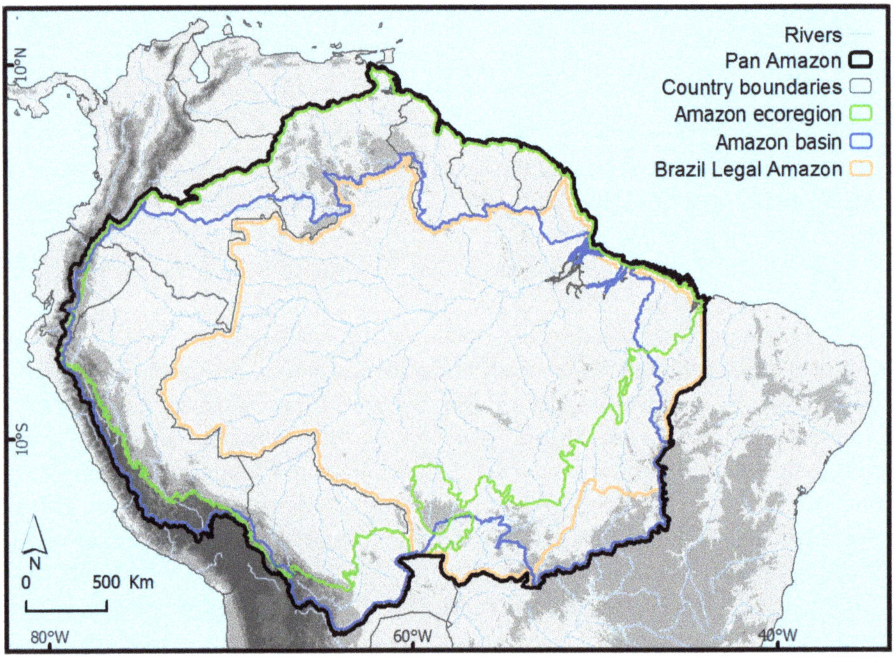

Figure 1.1: The Pan Amazon includes both the Amazon Ecoregion (green) and the Amazon Basin (blue). The southern boundary reflects Brazilian legislation governing fiscal and regulatory policy referred to as the Legal Amazon (tan).

Data source: Amazon Cooperation Treaty Organization.

The success of conservation initiatives and the decline in deforestation are essential for the long-term survival of the Amazon, but they have not changed the long-term trajectory of the Pan Amazon. Fully sixty per cent of the region remains open to non-sustainable activities, including logging, artisanal gold mining and settlement by small-scale farmers. Deforestation rates have crept upward across the region and registered historical highs in Colombia, Peru and Bolivia. Worse still, the predicted impacts of climate

Photograph © Rhett Butler, courtesy of Mongabay.

The Amazon is the world's largest intact tropical forest wilderness. The Río Javari on the border between Peru and Brazil.

change have manifested, in part due to increasing temperatures but, more ominously, by modifying precipitation regimes that threaten to tip the region – or at least its southern half – into a cataclysmic shift in ecosystem function that could lead to widespread forest dieback.[2]

Recent events, particularly the increase in forest fires and an election in Brazil, have placed Amazonian conservation once again in the forefront of the global media, which is now dominated by social networks that have succeeded in dramatising the issue at the local, national and international levels. Societies are demanding solutions, but these will be neither easy nor simple because the causes of environmental degradation in the Amazon are complex and span infrastructure, agriculture, minerals, finance and governance. Meaningful reform is impeded by the predominance of conventional business models, reinforced by deeply ingrained cultural attitudes, corruption, and inequality. The response to the COVID-19 pandemic exposed the inability of governments to safeguard their populations, particularly indigenous communities whose fear of disease is rooted in centuries of experience, as well as the rural and urban poor whose endemic exposure to infectious diseases and parasites increases the risk of mortality and morbidity.

Changing the development pathway of the Pan Amazon is like turning an ocean liner; steady pressure must be applied to the rudder of state over a long period in order to drive incremental change across multiple sectors of the regional economy. Regulation and market incentives that influence human behaviour and corporate decisions must be aligned with conservation outcomes so that sustainable development is less aspirational and more operational. This will require profound reforms in financial and commercial markets, as well as real change in regulatory systems and enhanced law enforcement. With few exceptions, sustainable models in forest and fisheries management have not yielded the economic returns needed to make them competitive with conventional extractive models. Even worse, the monetisation of ecosystem services has generated a mere fraction of the resources required to change human behaviour on the forest frontier, much less to subsidise the reforestation efforts that climate scientists view as essential for stabilising the hydrological regime of the Southern Amazon.

This second edition of *A Perfect Storm in the Amazon Wilderness* provides an overview of the topics most relevant to the conservation of the region's biodiversity, ecosystem services and indigenous cultures, as well as a description of the conventional and sustainable development models that are vying for space within the regional economy. Events of the last ten years are discussed in detail because future events will have to build upon – or modify – the cultural and economic forces driving events in the Pan Amazon. The text provides a longer historical perspective to show how policies create legacies that reverberate over decades, long after they have been recognised as being fundamentally flawed.

Drivers of Environmental Degradation

Why do people clear forest? To anybody who has lived on the forest frontier, the answer is as simple as it is obvious: it is essential to the livelihoods of the region's inhabitants. In some cases, it may be to grow food to feed a family but, more often, people clear forest to generate wealth by selling timber, cultivating a crop or raising livestock. The flow of goods between rural and urban societies is as old as civilisation, but in today's global economy the connection between the producer and the consumer is mediated by a supply chain that is complex and not particularly transparent. For the last several decades, increases in demand for food and fibre have been met by the expansion of agricultural supply chains into tropical forest wilderness. Producers operating on these landscapes are responding to global demand for the goods they produce; they are acting in their own self-interest to create wealth for their families and jobs for their communities (Chapter 3). Many are fully aware that deforestation is a global problem but maintain

Photograph © Rhett Butler, courtesy of Mongabay.

Santa Cruz, Bolivia

© Paula Vilela / shutterstock.com.

Mato Grosso

The production of beef cattle is the largest driver of deforestation and a mainstay of the rural economy in the Southern Amazon. Landholders clear forest on the forest frontier where land can be acquired via illegal or legal transactions. Over time, ranchers increase pasture area by clearing remnant forest, while improving productivity through grazing management, feed technology and animal genetics.

that they should not bear the cost of conserving biodiversity or fighting global warming, especially when wealthy nations have sacrificed their own forest and polluted the atmosphere in pursuit of economic growth. A common refrain, voiced across economic spectra, is that wealthy nations should assume the cost of forest conservation.

Many consumers in wealthy nations are concerned about tropical deforestation, and some express a willingness to pay for conservation measures. Nonetheless, many are unaware that the food they eat or the shampoo they use has been manufactured using products that originate from recently deforested landscapes. Consumers are isolated from producers by a host of middlemen, manufacturers and retailers, all of whom benefit from, and contribute to, the commercial forces that drive deforestation. Similarly, consumption of mineral commodities, such as iron ore, aluminum and petroleum, have contributed to the degradation of the forests and waters of the Amazon. The extractive industries impact the forest directly by operating mines and oil fields and indirectly when those facilities create access that opens the gateway for migration and the expansion of the agricultural frontier (Chapter 5). Financial institutions and individual investors who lend money or own shares in mining or petroleum corporations or agribusiness firms also share the responsibility for environmental degradation.

Deforestation is influenced enormously by the dynamics of rural real estate markets. Primary forest has value because its timber can be harvested for cash income but, once logged, land is referred to as 'unproductive' or as having 'productive potential'. In contrast, pastures, plantations and cropland are considered to be 'productive land' because they generate revenue. This may seem perverse to an ecologist who understands the potential long-term productivity of a forest, but clearing land generates cash flow over the short term and adds value to real estate assets over the medium- and long-term (Chapter 4).

Public policies also foster deforestation. Most Amazonian countries have legal mechanisms for transferring public land to private individuals that explicitly allow – or even require – deforestation. Similarly, small-scale deforestation and the use of fire is either allowed or openly tolerated in almost all Amazonian jurisdictions. Local and regional governments, with the support of multilateral development agencies, build roads in wilderness landscapes where it is implicitly understood that land speculation will invariably lead to deforestation (Chapter 2). These policies remain in place because they enjoy the support of the economic interests of construction companies, landholders and agribusinesses, as well as the electoral power of landless peasants seeking a pathway out of poverty. Rhetoric supporting forest conservation is widespread, but acting to curtail deforestation is politically perilous. The potential for a political movement to drive deforesta-

© Dr. Morley Read / shutterstock.com.

Highways are deforestation vectors that improve access to previously remote areas, attracting settlers who clear land to establish farms and ranches. Road improvement is broadly supported by rural inhabitants because it facilitates commerce and increases land value. Aerial shot of an Amazonian highway in Ecuador and a clearing made by a subsistence farmer.

tion is exemplified by the election of Jair Bolsonaro, who campaigned on a promise to reverse the conservation policies of the previous three decades.

Culture also plays a role. Consider the pioneer who created a successful farm over a lifetime of hard work and is understandably proud of that accomplishment. His or her children and grandchildren are likely to have similar views – even if they now also hold views supporting forest conservation. Frontier societies are populated with individuals who believe conventional development is beneficial, a life-view reinforced by educational systems and spiritual leaders (Chapter 6).

Corruption is another obvious accelerant to the forces driving environmental degradation. The *Lava Jato* scandals associated with the construction of public infrastructure projects revealed how graft[*] distorted economic feasibility studies for projects that might have been rejected based solely on financial criteria.[†] It is often assumed that improved governance will empower environmental advocates and slow deforestation, but initiatives to decentralise administrative processes place decisions in the hands of local politicians who tend to favour conventional business models (Chapter 7).

Geographers and economists have created a classification system and lexicon to facilitate the discussion about the drivers of deforestation.[3] The term 'proximate causes' refers to those phenomena and actors directly responsible for deforestation; typically, these are on or near the landscape

* In the US English sense of obtaining advantage through dishonest use of power.

† *Lava Jato* (car wash) refers to a corruption scandal involving large construction companies and politicians in Brazil, which eventually spread to involve contracts and political leaders in the Andean Republics (Ch. 6).

Table 1.1a: Proximate causes of environmental degradation

• **Infrastructure Development** ° Trunk highways ° Secondary road networks ° Railroads ° Dams/waterways • **Agriculture** ° Beef ° Food crops ° Cash crops ° Illicit drugs • **Forest Production** ° Timber ° Wildfire ° Logging ° Hunting	• **Mineral Extraction** ° Industrial minerals ° Oil and gas ° Gold • **Land Speculation** ° Legal ° Illegal • **Governance** ° Corruption ° Administrative inefficiency ° Decentralisation ° Civil unrest • **Demographics** ° Population growth ° Migration

Table 1.1b: Indirect drivers and actors of environmental degradation

• **Global demand for agricultural commodities** ° Industrial crops: soy, beef, palm oil ° Cash crops: coffee, cocoa, sugar ° Timber, pulp, biofuels ° Illicit drugs • **National demand for basic food-stuffs** ° Protein: beef, dairy, poultry, ° Staples: rice, manioc, fruit • **Global demand for mineral commodities** ° Industrial metals ° Oil and gas ° Gold • **Market Intermediaries** ° Multinational corporations ° State-owned monopolies ° Domestic conglomerates	• **National regulatory framework** ° Food security and biofuel policies ° Land-use regulations ° Decentralisation processes ° Environmental review systems ° Inoperative judicial systems ° Regional development strategies ° **Financial institutions** ° Multilateral development agencies ° National development funds ° Direct foreign investment ° Domestic banks ° **Technical assistance and extension** ° Public extension services ° Grower associations ° Alternative development (illicit drugs)

being impacted by development (Table 1.1a). In many cases, there is an obvious link: cattle ranchers, for example, clear forest to expand pastures to raise beef. Sometimes there is a strong correlation but not a direct link: a new road cuts through a forest landscape, which opens access to settlers who spread out to establish homesteads.

Conversely, economists use the term 'indirect drivers' to describe factors that occur at considerable distance from the forest frontier but create an economic force that motivates the behaviour of individuals in frontier society. These include markets where commodities are traded, regulatory agencies that oversee land use and policies intended to promote economic growth (Table 1.1b). Because the influence is both indirect and distant, it may be difficult to establish the connection between cause and effect. For example, certain varieties of elite coffee that originate in specific montane tropical localities have become popular in North America and Europe. Demand for these coffees creates a premium that is paid to growers from these regions, which motivates them to expand their production by clearing the patch of forest next to their coffee grove.

The use of these terms and the stratification of the causes of deforestation may seem like an academic exercise with limited practical application, but it has been used to identify leverage points that can make supply chains more transparent and force intermediaries to modify their business practices.[4] Similarly, an evaluation of the drivers linked to regulatory frameworks or financial systems can identify the existence of perverse incentives that reward individuals or corporations that deforest land. Governments have recognised the need to reform laws, while multilateral agencies have embraced social and environmental standards that seek to avoid or mitigate the environmental impacts associated with their investments.[5]

The interactions between markets, regulatory systems, technology and culture are by definition complex, which is why it has been so difficult to stop deforestation.[6] Recent successes in reducing deforestation have been based on integrated approaches, which embrace the concept of incremental change and the need to involve all stakeholders with legitimate interests in the activities that cause, either directly or indirectly, the environmental degradation that threatens the Pan Amazon (Chapter 8).

The Geography of Environmental Degradation

The Pan Amazon spans approximately 825 million hectares, of which approximately ninety million hectares have been lost to deforestation; this corresponds to ~13% of the original forest cover (Table 1.2). At first glance, this percentage value might not seem alarming, particularly in the context of the total extant forest cover. However, cleared land is embedded within

Table 1.2: Original forest cover and deforestation in the Amazonian jurisdictions[25]

Country	Total Original Forest Cover (km²)	Total Deforestation	Historical Deforestation (% of Original Forest Cover)	Deforestation 2010–2018 (km²)	Deforestation 2010–2018 (% of Total Deforestation)
Bolivia	483,550	67,125	13.9	14,612	21.8
Brazil	4,243,362	670,861	15.8	50,035	7.5
Colombia	419,450	25,585	6.1	5,931	23.2
Ecuador	123,330	11,025	8.9	742	6.7
F. Guiana	75,000	650	0.9	360	55.4
Guyana	151,690	6,891	0.6	653	9.5
Peru	725,110	72,624	10.0	10,779	14.8
Suriname	146,931	6,683	0.7	723	10.8
Venezuela	396,335	13,063	3.3	1,740	13.3
Total	**6,764,758**	**874,507**	**12.8**	**85,575**	**9.8**

Text Box 1.1: Human-Modified Landscapes

Forest frontiers are highway or river corridors that penetrate or are adjacent to wilderness landscapes; residents and immigrant pioneers exploit timber resources and appropriate public lands by clearing forest to install low technology agriculture production systems.

Agricultural frontiers emerge after the improvement of trunk highways ensure year-round access and secondary roads are extended into surrounding forest. The rate of deforestation increases; forest remnants are progressively fragmented and isolated. Timber exploitation remains important, but agriculture increases in area and technological sophistication.

Consolidated frontiers are landscapes where pastures, fields, or plantations exceed forest cover; absolute rates of deforestation fall, but relative rates remain high as landholders clear forest remnants. Private sector investment in industrial infrastructure adds value to agricultural production, while the state improves basic infrastructure in urban centers.

Gold rush frontiers are similar to forest frontiers, but deforestation is caused by illegal gold miners who transform and destroy floodplain habitats. Access may be by road, river or light plane.

Coca frontiers are agricultural frontiers populated by settlers who clear small patches in the forest to cultivate coca leaf for illicit drug markets.

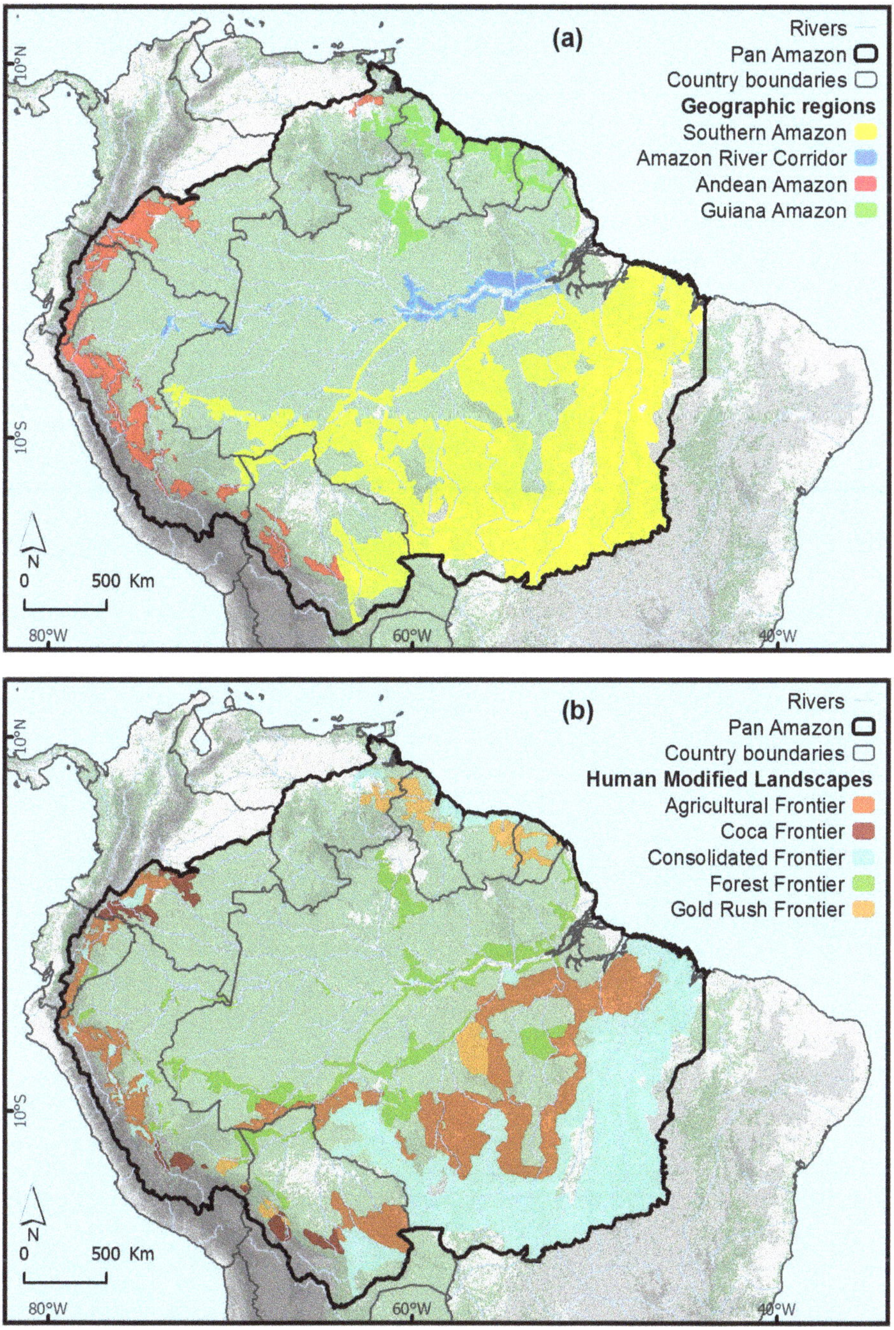

Figure 1.2: The human-modified landscapes of the Pan Amazon can be stratified according to geographic regions (a) and level of economic development (b).

landscape mosaics that include an approximately equivalent area of fragmented forest, both of which are surrounded by extensive areas of degraded forest damaged by illegal logging, wildfire, and overhunting. A conservative estimate of the total area of all these 'human-modified landscapes' is approximately 250 million hectares, which represents about thirty per cent of the geographic area of the Pan Amazon (Figure 1.2).

The human-modified landscapes of the Amazon can be stratified into four macro-regions, based on biophysical and cultural attributes that have determined their recent development. Each landscape has a unique development trajectory, but there are identifiable stages that reflect levels of infrastructure investment, agricultural production system, levels of technology, and social capital (see Text Box 1.1 and Figure 1.3).

Amazon River corridor

The upland terraces that flank the main stem of the Amazon River* are the oldest human-modified landscapes in the Pan Amazon. People have occupied these lands for millennia,† but their ethnic composition has changed over time as immigrants and escaped slaves intermarried with indigenous people to forge unique cultural groups referred to as *cabloco* (Brazil) or *ribereños* (Peru) (Chapter 6). The river connects rural communities with about two dozen small towns and six major urban centres: Belem, Macapá, Santarem, Manaus, Tabatinga, Leticia (Colombia), and Iquitos (Peru). The most densely populated stretch of the river is located between the top of the delta and Manaus, but isolated communities extend upstream to the junction of the Ucayali and Marañon rivers in Eastern Peru.

The foundation of the rural economy is based on the wild fish catch and non-timber forest products, particularly the super-fruit *açaí*, a palm that dominates floodplain habitats of the lower Amazon (Chapter 8). Most families cultivate a garden plot as part of their subsistence economy, while a limited number of secondary roads provide access to forest resources beyond the immediate confines of the river corridor. As a waterway, the river functions as a bulk transport system for bauxite mines located on upland landscapes adjacent to the floodplain (Chapter 5), as well as for commodities trans-shipped at ports that service the agricultural industry of the Southern Amazon (Chapter 2). Fluvial transport also serves the oil and gas sector while providing cost-effective transport for timber harvested

---

* Brazilians divide the Amazon into two sections: The name 'Amazon' is used for the section below the junction of the Solimoes and Rio Negro at Manaus; upstream they refer to the river as the 'Solimões', which extends to the junction of the Marañon and Ucayali rivers in Peru. Peruvians refer to the Solimões as the Río Amazonas.

† There is abundant archaeological evidence in the form of black earth soils documenting the extent of Pre-Colombian societies (Ch. 6).

from remote regions accessed by dozens of tributaries. Manaus is a travel destination marketed as the gateway to the Amazon, while Leticia and Iquitos have developed nature and cultural tourism linked to the domestic and international markets (Chapter 8).

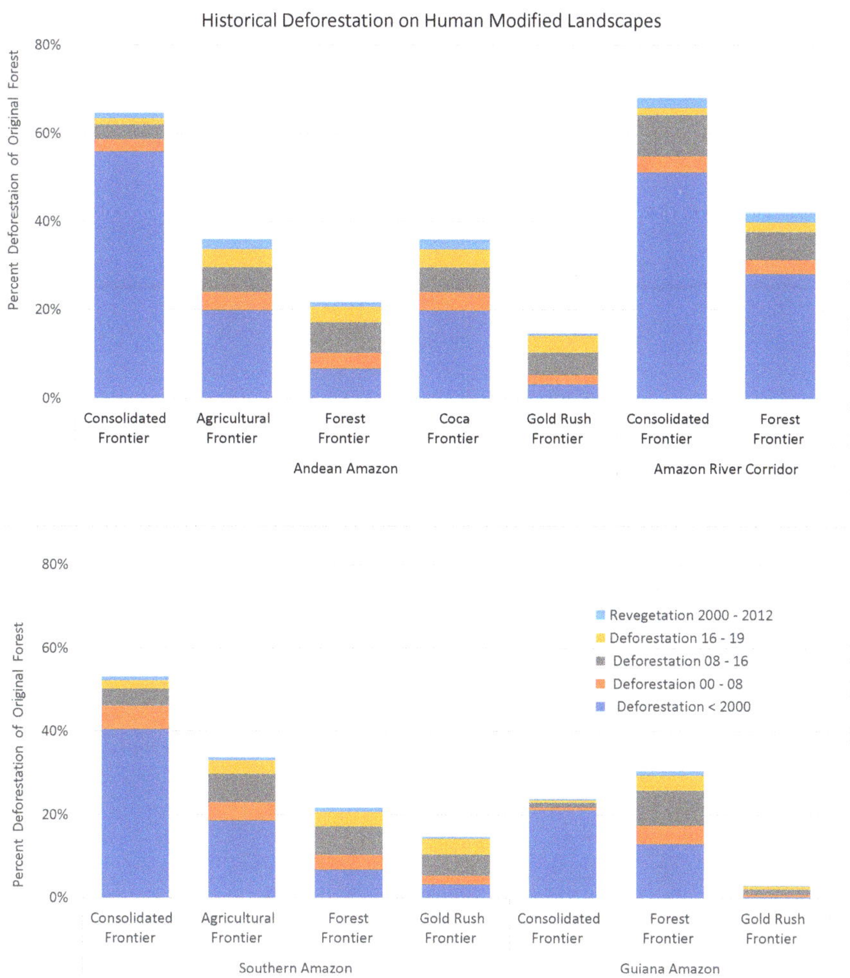

Figure 1.3: The human-modified landscapes of the Pan Amazon have different deforestation histories depending upon the phase of their development.

Source of land-use change data: Global Land Analysis and Discovery (GLAD).

The natural habitats of the Amazon River Corridor are remarkably well conserved, particularly above Manaus, where a wilderness riverscape extends more than 2,000 kilometres. Near Manaus, both sides of the river have evolved into forest frontiers, while the upland landscape between Oriximiná and Prainha in northern Pará has been deforested by the small farms and ranches established in the late 1960s and 1970s (Figure 1.2). Across the river in Santarem, grain silos and port facilities at the terminus of BR-163 are driving the conversion of pasture into cropland, which has caused this landscape in the heart of the Amazon to transition into an agricultural frontier dominated by the cultivation of soy.

Southern Amazon

The national development strategies epitomised by a highway network carved out of forest wilderness in the 1970s and 1980s set in motion a development trajectory that caused this region to be known as the Arc of Deforestation (Chapters 2 and 6). More than seventy million hectares of tropical forests have been sacrificed to create an agricultural economy that stretches from Eastern Pará, Brazil to Santa Cruz, Bolivia.

Low rates of deforestation within forest frontiers are a function of their remoteness and the poor state of their transportation infrastructure; landscapes closer to markets have transitioned into agricultural or consolidated frontiers. Rapid development has occurred on landscapes deemed to be geopolitically important (Rondônia) or where the soils were particularly apt for field crops (Mato Grosso and Santa Cruz). Public lands have been distributed to more than a million pioneer families and several thousand private companies via a variety of legal and extra-legal mechanisms (Chapter 6). The resulting land tenure mosaic reflects both the diversity of landholdings (small, medium, large, and massive), as well as the ongoing appropriation of state lands that continues to drive deforestation on forest frontiers (Chapter 4).

Most smallholders dedicate a portion of their land to produce basic foodstuffs, but the production of beef cattle is the major driver of deforestation on both forest and agricultural frontiers (Chapter 3). The global demand for soy in the 1990s stimulated the expansion of agribusiness and enabled business models dependent upon overseas export markets. The cultivation of maize in rotation with soy catalysed the explosive growth of the poultry and swine sectors, while the cultivation of oil palm diversified the rural economy in Northeast Pará (Chapter 3). Global competition is driving investment in bulk transport systems that link the production landscapes of the Southern Amazon to the grain terminals on the Amazon River (Chapter 2).

© PARALAXIS / shutterstock.com.

© PARALAXIS / shutterstock.com.

The cultivation of field crops is the preferred production strategy where topography and soils permit allow farmers to harvest two crops per year. The soy-maize production model contributed $US 25 billion to the Mato Grosso economy in 2020. Corporate farms predominate on some landscapes (a), but they share the export market with thousands of family farms. All rely on logistical infrastructure, such as these silos on BR-163 highway (b), which connects to global markets via ports on the Amazon River.

The development strategies of the 1970s included programmes for the exploitation of the region's mineral resources. Its recent history has been marked by chaotic gold rushes that ebb and flow with the price of gold (Chapter 5). More long-lasting development has resulted from the exploitation of world-class deposits of industrial metals at the Serra de Carajás in Pará, which stimulated investments in railroads, industrial mills and hydropower facilities. Infrastructure development represents a significant part of the regional economy, in part because the Brazilian government has built more than fifty hydropower facilities in the region, including several mega-scale facilities (> 1,000 MW) on the Madeira, Tocantins and Xingu rivers (Chapter 2).

The Southern Amazon can be stratified into five forest frontiers, eight agricultural frontiers, five consolidated frontiers and one active gold-rush frontier (Figure 1.3). If current trends continue, the forest frontiers will transition into agricultural frontiers and agricultural frontiers into consolidated frontiers. These human-modified landscapes, which are organized around a network of trunk highways, are separated by forest remnants that are increasingly isolated from each other and from the continuous forest landscapes of the West, Central, and Northern Amazon.

Andean Amazon

The cultural traditions and the national development strategies pursued by the Andean republics created human-modified landscapes that are fundamentally different from the Southern Amazon. Each country built a series of individual highways that connect a specific sector of the Andean highlands with an adjacent region of the Amazon lowlands. All roads traverse the Andean foothills, where tens of thousands of settlers established homesteads on steep slopes inappropriate for agriculture (Chapter 2). Soil erosion limits the productive capacity of farmsteads while threatening the integrity of the roads built on landscapes with extraordinarily high levels of rainfall.

Colonisation has been driven by migration from indigenous highland communities who have laid claim to small landholdings using forest-fallow production systems to produce basic foodstuffs for domestic markets (Chapter 6). Over time, settlers have diversified their production systems to include oil palm, coffee and cacao, some of which is destined for overseas markets. Pastures are abundant, and many families own cattle; however, the beef industry is not a driver of deforestation but an artefact of the forest-fallow production system. On selected landscapes, farmers cultivate coca leaf, sometimes legally for domestic consumption but more often as an illicit crop for the global cocaine market. A limited number of corporations

Deforestation in the Andean foothills is particularly damaging due to high rainfall and soil erosion on steep slopes that limits the productive life span of cleared fields; settlers tend to be small farmers producing basic foodstuffs for the domestic market.

have invested in large-scale oil palm plantations and associated industrial facilities in Peru and Ecuador (Chapter 3).

The mineral wealth of the region is concentrated in the High Andes, where global corporations operate industrial mines in polymetallic ore bodies that have been exploited for centuries. Gold that originated from these geological formations has been transported to the alluvial sediments on the piedmont that are exploited by small-scale miners in selected landscapes experiencing a gold rush. The Western Amazon has significant oil and gas reserves located beneath the alluvial landscapes east of the Andes, and the revenues derived from their exploitation have been a pillar of national economies and state budgets since the 1960s (Chapter 5).

Investment in infrastructure has accelerated in the last two decades thanks to an international effort to coordinate national development strategies and promote economic integration among countries via the Initiative for the Integration of the Regional Infrastructure of South America (IIRSA). Among the highest IIRSA priorities have been transportation corridors that transect wilderness landscapes, linking the Pacific Coast with the Amazon River and the agricultural landscapes of the Southern Amazon. The massive elevational drop in rivers that originate in the Andes has favoured the development of dozens of medium- and large-scale hydropower facilities. With one notable exception, efforts to promote mega-dams have not been successful because the proposed investments have not withstood financial due diligence (Chapter 2).

The human-modified landscapes of the Andean Amazon include four forest frontiers and six agricultural frontiers, seven coca frontiers and two gold-rush frontiers (Figure 1.3). Migration and forest loss occur altitudinally, as agricultural frontiers expand upslope, and laterally, as pioneers settle adjacent valleys or emerge onto nearby lowland landscapes. Unlike the Southern Amazon, where forest frontiers are associated with roads, the most dynamic forest frontiers in the Andes occur along river corridors (Chapter 6). Individual lowland landscapes have remained isolated from each other for decades; only Ecuador has built a road that connects all of the lowland settlement zones with an integrated trunk highway. The decision (or delay in deciding) to integrate settlement zones along the piedmont has maintained a limited number of biological corridors that connect the lowland forests of the Western Amazon and the montane forests of the Andean foothills (Figure 1.2a).

Guianan Amazon

The human-modified landscapes of the Guiana Shield have avoided the settlement and colonisation phenomena that caused the widespread deforestation in the Southern Amazon and Andean Amazon. Historical settlements are clustered along the coast of Guyana, Suriname and French Guiana, while Venezuela has pursued a national development strategy based on petroleum development and urbanisation (Chapter 6). Only the Brazilian state of Roraima was the target of resettlement schemes in the 1970s when a paved highway was built to connect Manaus with Boa Vista and, eventually, with the Venezuelan and Guyana national road networks. Currently, Brazil is building a modern highway in Amapá to connect the Amazon port city of Macapá with the development corridor of the Guiana Coast (Chapter 2).

Suriname and Guyana were once major producers of bauxite, but those mines have shut down after exhausting the easily exploitable deposits, leaving behind degraded landscapes and toxic tailing ponds. Northern Pará and Amapá have significant reserves of ferrous minerals, copper and gold, but most are located within a strategic mineral reserve that is closed temporarily to exploitation (Chapter 5). Venezuela has economically attractive bauxite deposits and world-class reserves of iron ore; nonetheless, its industrial mines and processing facilities are operating at minimal levels due to economic mismanagement.

The very considerable gold resources of the Guianan Amazon are associated with a geological formation known as a greenstone belt that has sustained a decades-long gold rush by small-scale miners, including native-born maroons and temporary migrants from Brazil. Most of the deforestation in the Guianan Amazon is caused by these miners; consequently, it is dispersed across landscapes as small patches that are difficult to detect using commonly available satellite images. Periodic gold rushes have occurred in Bolivar state in Venezuela and along the border with the Brazilian state of Roraima. Large-scale corporate gold mines are operating or under development in Guyana, Suriname, French Guiana, and Amapá (Chapter 5).

The human-modified landscapes of the Guiana Shield include two gold-rush frontiers, two forest frontiers and two consolidated frontiers. The recent discovery of large reserves of oil and gas offshore from Guyana and Suriname will transform their economies over the short term.

Hydrological degradation

The Amazon River system is the world's largest freshwater ecosystem, with nearly twenty per cent of the Earth's freshwater discharge. It is a megadiverse aquatic biome and an enormously productive ecosystem that provides livelihoods and essential protein resources for its resident populations. It is also a strategic economic asset due to its hydropower capacity and as a waterway in a region that lacks roads (Chapter 2). The massive volume of water that flows through the Amazon River system provides an inherent level of resiliency, which is augmented by intact floodplain habitats that buffer seasonal fluctuations in waterflow and absorb the effluents of human society. Nonetheless, development phenomena have impacted the rivers and streams within human-modified landscapes.

Illegal gold mining has impacted the biogeochemistry of the rivers that drain the gold-rush landscapes where small-scale gold miners are active (Figure 1.2b). The spatial footprint from individual mining operations is [relatively] small, but the environmental impact is magnified by the tendency of miners to operate within floodplains. The widespread use of mercury, a heavy metal with well-documented deleterious effects on human health, is creating an environmental legacy that will plague the region for decades (Chapter 5). Virtually all small-scale miners violate environmental laws and evade taxes; many are guilty of human rights violations (Chapter 7). Illegal gold mining is likely to increase over the near term due to the volatility of financial markets, which motivates global investors to buy gold as a hedge against uncertainty.

Dams and reservoirs cause permanent and long-term impacts by modifying hydraulic regimes, capturing sediment, and interrupting fish migration (Figure 1.4). Multilateral financial agencies have provided key financing to both public and private entities; companies from China have acquired distressed hydropower assets in Peru and Brazil while providing turn-key solutions to Bolivia and Ecuador. All the Pan Amazon nations have pursued hydropower development as part of their national energy strategies, and Brazil is leveraging the construction of hydropower projects to develop waterways as bulk transport systems.

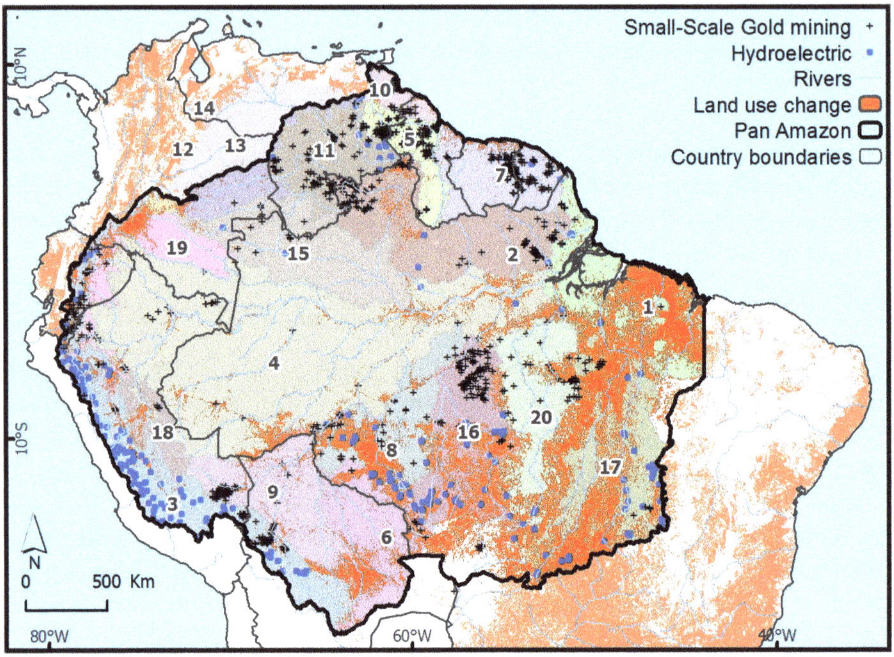

Figure 1.4: There are three major sources of hydrological degradation: (1) dams that trap sediments, interrupt hydraulic cycles and block fish migration (blue box); (2) illegal gold mines that destroy floodplains and release mercury (black x); and (3) run-off from land-use change (red areas). The coloured polygons are 'hydrosheds', which stratify river basins based on biogeochemical attributes.[26]

Sources: gold mine landscapes – Google Earth & RAISG (*Rede Amazônica de Informação Socioambiental Georreferenciada*); dams – RAISG; Hydrosheds – B. Lehner and G. Grill. 2013. *Hydrological Processes* **27** (15): 2171–2186; Land cover data – ESA. Land Cover CCI Product User Guide Version 2. Tech. Rep. 2017.

Land-use change and agricultural runoff have impacted the geochemistry of most of the headwaters of the rivers of the Southern Amazon, where the conversion of ~50% of the original forest cover has increased sediment loads and altered the nutrient status of the region's unique 'clear-water' rivers. These impacts should be mitigated by environmental regulations that obligate landholders to conserve forest in riparian corridors; however, noncompliance is widespread, and efforts to motivate landholders to reforest riparian corridors are constrained by cost and apathy (Chapter 7). Deforestation has impacted precipitation regimes, which will decrease runoff volumes, a phenomenon that will become greater over time as farmers adopt irrigation technology to mitigate the threat of seasonal drought (Chapter 4).

The Political Economy of the Pan Amazon

Politics is idiosyncratic to each country, but trends in economic policies span borders. The political economy of the Pan Amazon is the legacy of strategic development plans that began in the mid decades of the twentieth century when the Amazonian nations turned their attention to the development of the Amazon (Chapter 6).

The nationalist period (1960–1985) was characterised by authoritarian governments that sought to use the power of the state to harness the natural resources of the Amazon. The push into the region was seen as a way to generate economic growth, mitigate poverty and avoid political unrest driven by socialist ideology. Governments were insensitive to environmental and cultural impacts and viewed the Amazon as an unpopulated region that could absorb a growing population. Highways were extended into the wilderness, and public lands were distributed to landless peasants and corporate investors. Mining ventures and oil exploitation were subsidised by the state. Development was synonymous with deforestation.

The neoliberal period (1985–2005) began with a wave of market-oriented reforms designed to foster economic growth via the private sector. Referred to as the 'Washington Consensus', each country enacted a suite of policies to privatise state-owned entities, eliminate budget deficits, reinforce property rights, promote free trade and facilitate foreign investment. The state's role was to provide essential services, such as law enforcement and the administration of a streamlined regulatory apparatus. The Amazon was integrated into an increasingly globalised economy; meanwhile, environmental and social advocates drew attention to the magnitude of deforestation and the plight of indigenous communities. The concept of sustainable development emerged from academia and soon dominated policy forums. Commodity exports were synonymous with development.

The populist period (2005–2019) was a reaction to the austerity associated with neoliberal policies and a cultural aversion to foreign influence. Governments enjoyed robust electoral majorities by promising to address social inequality while embracing a form of democratic socialism that enhanced the role of the state in the national economy. Unlike previous socialist movements, however, these governments protected private sector actors who were generating the economic growth and export revenues essential for financing their political agendas.* Like their neoliberal precursors, populist governments made public commitments to sustainability, supported conservation initiatives and acted to protect the rights of indigenous communities, all while investing in infrastructure and conventional development paradigms. Sustainability was a synonym for the *status quo*.

---

* An exception to this was the government of Venezuela, which has pursued an extreme form of socialism combined with authoritarian rule.

Each of these phases created 'facts on the ground' that would constrain or enhance development and conservation options in subsequent decades. For example, the trunk roads carved out of the forest between 1970 and 1980 created the framework for future highway investments that are a recurrent feature in the annual budget of the Brazilian state. The development of oil and gas fields in the Andean republics during the 1960s now generates revenue streams essential for macroeconomic stability while creating infrastructure assets, such as pipelines, that are used to expand operations into wilderness landscapes (Chapter 5). The export-driven agribusiness production model consolidated during the late 1990s is now an indispensable component of the national economy and has endowed agribusiness with the financial capacity to invest in bulk transport systems (Chapter 2). The beef industry in Brazil and Bolivia has tens of thousands of constituents with sufficient political power to allow them to ignore land-use regulations intended to slow deforestation (Chapter 7). The creation of protected areas and the recognition of the territorial rights of indigenous people removed almost fifty per cent of the surface area of the Amazon from the reach of conventional development (Chapter 10).

The political nature of the next period is uncertain. In 2019, Jair Bolsonaro was elected president of Brazil with an avowed agenda of reversing the conservation policies of the past two decades and returning to the unbridled development that characterised the 1970s. In the Andean republics, there is widespread dissatisfaction with political elites tainted by corruption, but newly elected governments continue to pursue conventional development models, tolerating deforestation and the destruction of aquatic ecosystems caused by small-scale gold miners. Venezuela is a basket case, while the republics of the Guyana Coast are in the process of embracing an economy reliant on fossil fuels.

Understanding the conventional economy

The economies of nations are traditionally evaluated by their gross domestic product (GDP), a metric that measures the total economic output of a nation. The GDP of the Pan Amazon was approximately $US 270 billion in 2017 (Figure 1.5),* a modest number in a global economy valued at $US 80 trillion in the same year. By way of comparison, this is approximately equal to the total income of the online retail company that has appropriated the

* This value was compiled from reports published by national statistical agencies that stratify information by sub-national jurisdiction and sub-sector; Guyana and Suriname report only national data; Venezuela does not report GDP data stratified by region and is excluded. The GDP metric reported here is based on 'current value' and compared among jurisdictions using mean annual exchange rates in 2017.

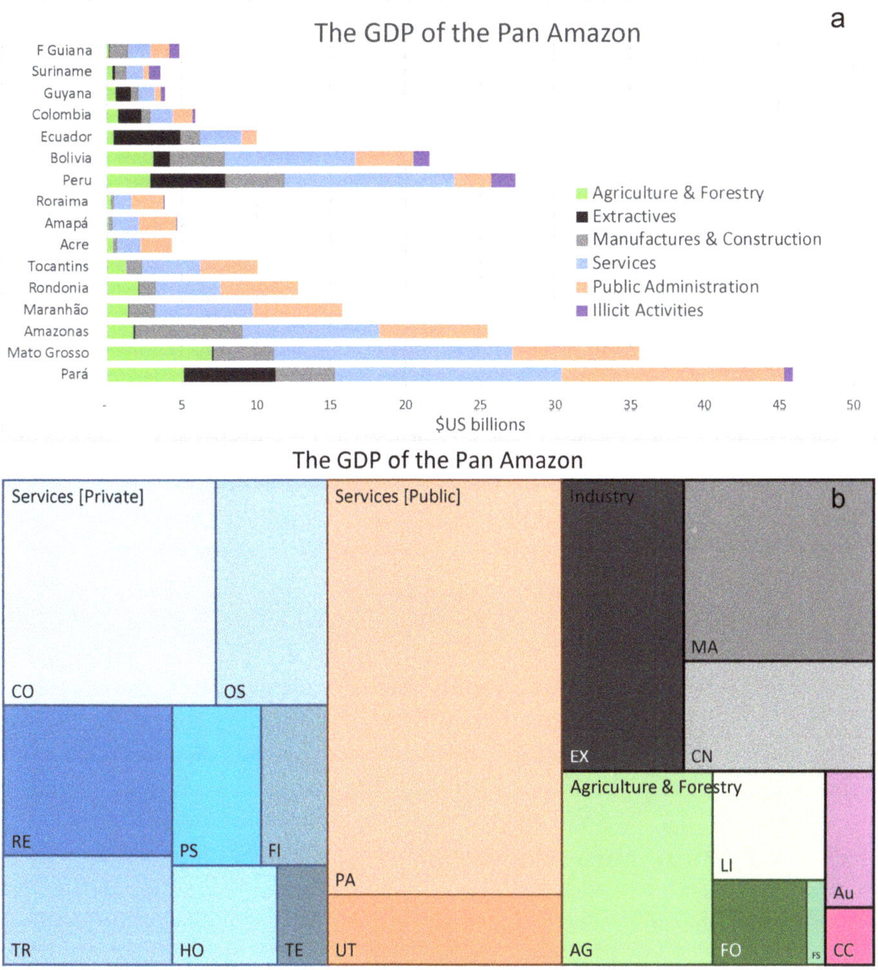

Figure 1.5: The GDP of the Pan Amazon stratified by political jurisdiction and sector (a) and subsector (b). Agriculture & Forestry: Agriculture (AG), Fisheries & Aquaculture (FI), Forestry (FO), Livestock (LI); Industry: Extractives (EX), Manufacturing (MA), Construction (CN); Services (Private Sector): Real Estate (RE), Commerce (CO), Transportation (TR), Hospitality (HO), Telecommunications (TE), Finance (FI), Professional Services (PS), Other services (OS); Services (Public Sector): Public Administration (PA), Utilities (UT); Illicit Activities: Coca/Cocaine (CC), Artisanal Gold (Au).[27]

name of the world's largest tropical forest.* The contribution to national GDP from the jurisdictions located within the Pan Amazon range from a high of 100% (Guyana and Suriname) to a low of 0.2% (French Guiana). The contribution of Amazonian regions to national GDP is small but significant in Brazil (8%), Peru (13%) and Ecuador (10%), less in Colombia (2%), and considerably more in Bolivia (59%), where three major urban centres lie within the Amazon basin.†

There are numerous problems with using GDP as an analytical metric (see below); nonetheless, it is the most commonly used statistic for evaluating the economy of a nation. The first priority of most governments is to promote economic growth, and their motivation is as simple as it is obvious: an increase in GDP reflects increased wealth, which can be used to reduce poverty; a decrease connotes a recession, which usually means an increase in poverty. Governments, and their advisors in multilateral development institutions, use GDP and its underlying metrics to identify how fiscal policies, such as taxes, subsidies and public investment impact the conventional economy. GDP is particularly informative when it is disaggregated into component metrics that measure economic production for economic [sub] sectors and organised by subnational jurisdiction (see Annexes 1.1 and 1.2).

The GDP metric has limited utility when evaluating the economic health of a society.[7] Its detractors point out five major limitations, all of which are germane to the Amazon: (1) it provides no information on inequality; (2) it underestimates the contribution of the informal sector;‡ (3) it makes no attempt to measure the economic value of subsistence activities; (4) it does not distinguish between sustainable activities, such as the harvest of renewable resources, and non-sustainable business models, such as the exploitation of non-renewable resources; (5) it fails to account for negative outcomes that create a long-term economic liability, such as an oil spill§ or the loss of a key ecosystem service. In spite of these limitations, or because they are so obvious, a review of the conventional economy using GDP metrics highlights the challenges and opportunities facing the pursuit of a sustainable economy.

---

* Amazon.com reported $US 233 billion in total sales in 2018 with an annual growth rate of ~30%; in contrast, GDP for the Pan Amazon increased by about 2% between 2017 and 2018.

† Santa Cruz, La Paz and Cochabamba together comprise about 55% of the population.

‡ When lacking data, economists estimate metrics using models based on (i) differences in expenditures and income or (ii) currency demand, or (iii) by tracking the consumption of a commodity correlated with economic activity such as electricity; source: Federal Reserve Bank of St. Louis.

§ Ironically, expenditures to remediate an oil spill will register as a positive contribution to GDP.

Perhaps the most revealing number in the sectoral GDP statistics is the minuscule contribution of the forest sector (< 2%), a paltry sum when considering the intrinsic value* of the Amazon's vast renewable natural resources (Figure 1.5). The most obvious explanation for this low number is the failure to assign value to subsistence activities. Indigenous people and traditional communities harvest food and fibre from natural ecosystems; most forest families grow food for their own consumption. These activities have tangible economic value and are central to the livelihoods of forest families, but they are ignored by measurements of GDP. Additionally, most households complement subsistence activities by harvesting timber, non-timber forest products and wildlife. Some of this production is captured by the statistics, particularly for the commercialisation of forest goods with strong export markets, such as Brazil nuts and palm fruits; however, other valuable products, such as timber and fish, are sold to middlemen who operate within the informal sector of the domestic economy (Chapter 8). Nonetheless, if the real contribution of the forest economy was twice the value of the official statistics, it would still lag the sectors of the economy driving deforestation and other forms of environmental degradation. The low valuation of forest products shines a spotlight on the challenge of using the forest economy as an alternative development strategy to displace agriculture and livestock.

The informal economy in Latin American nations accounts for approximately thirty per cent of total economic activity;[8] that proportion is greater in frontier communities, where 'cash is king' and the institutions of the state are weak or absent. This is even more true for smallholder landscapes where subsistence farming is combined with the cultivation of foodstuffs commercialised in domestic markets (Chapter 3). Approximately ten per cent of the previously deforested lands in Brazil and Bolivia have been settled by small farmers; although their spatial footprint is limited, they constitute about seventy per cent of rural families and are an important source of basic foodstuffs, such as manioc, rice, beans and a variety of tropical fruits.[9]

In Peru and Ecuador, the predominance of smallholders is much greater, representing about 98 per cent of all landholdings and occupying more than ninety per cent of agricultural landscapes (Chapter 4). The production model pursued by the majority of smallholders on frontier

---

* Investors use the term intrinsic value to describe the potential value of an asset based on an objective evaluation that considers the long-term potential return; it is used in juxtaposition to the term market value, which reflects the earnings generated by the asset. Philosophers and environmental economists use intrinsic value to describe a point of view that resists efforts to ascribe a monetary value to nature, arguing that it has value in and of itself. Both usages are valid for the Amazon.

Coffee cultivation can be a driver of deforestation when new fields are established by clearing natural forest, or a sustainable form of tropical agriculture if producers expand by recovering abandoned fields and secondary forest. Photograph from Ecuador.

landscape is based on slash-and-burn technology, which is used to establish and maintain a forest-fallow production system. Most farmers invest in perennial production systems over time as they diversify their crops and plantations, but they expand cultivation at the expense of remnant forests within their properties. The full value of their production is not incorporated into GDP, which causes the official statistics to underestimate their contribution to the regional economy, as well as the economic forces that drive deforestation by smallholders.

A more significant factor in the underappreciation of agriculture and livestock production is the methodological framework designed to avoid double accounting when compiling the GDP metric. Unlike the underreporting caused by the informal economy or subsistence farmers, this is not a bug (flaw) but a feature of the GDP bookkeeping methodology. The value of the production for any sector is measured only once, and in the case of agriculture, that data is captured at the 'farmgate' a term used to describe the price paid to the producer. All subsequent transactions 'add value' to the commodity and are accrued to a supply chain participant; for example, the increased value of dressed beef and soy oil are accrued to the manufacturing sector, while the cost of hauling grains to export terminals is allocated to the transportation sector. Similarly, expenditures for inputs made by farmers and ranchers prior to the harvest or the sale of livestock are subtracted from farmgate revenues and assigned to their respective service sectors, which includes veterinarians, seed companies, appliance dealers, fuel companies, and agrochemical venders. A comparison of the total gross value of farm production in Mato Grosso compared to the value-added metric used to compile sectoral GDP reveals that 45 per cent of the total proceeds are allocated to service providers or manufacturers in the commodity supply chain.[10]

The service sector is the largest component of GDP in seven jurisdictions and the second most important sector in the remaining ten (see Annexes 1.1 and 1.2). The predominance of the service sector is not uncommon among nations because it is a basket of many different economic activities. The growth of the service sector is also the consequence of the ongoing urbanisation of Amazonian society (Chapter 6). More than fifty per cent of the region's inhabitants reside in cities with populations greater than 100,000, and the overwhelming majority work in the service sector. Many of the services in large cities are environmentally benign and could be easily accommodated within a 'green' economy, including telecommunications, information management, health care, hospitality, and finance. Only Manaus has a strong manufacturing sector, an anomalous situation maintained by subsidies and tariff barriers. The other large cities (Belem, Santa Cruz, Cuiabá, Santarem, Porto Velho) are economically diverse, but their manufacturing and service companies rely directly or indirectly on revenues from the extractive industry or the agricultural and livestock sectors. The dependence of mid-sized cities (10,000 to 100,000 residents) on the rural economy is even more pronounced because they are the economic gateway for private sector services to farms, ranches and rural communities.

Mid-size cities and towns are also where rural inhabitants access public services, most importantly health care and secondary education, but also technical assistance and financial credit. The poor quality of rural schools motivates many families to maintain a residence in nearby small towns, one of several factors contributing to rural-urban migration. Urban inhabitants also enjoy access to basic services taken for granted in advanced economies, including sanitation, electricity, access to the internet and higher education. All of these are absent in the rural Amazon. Government expenditures are relatively large in Brazil and are the leading sector in Acre, Amapá, Rondônia and Roraima (see Annex 1.2), which reflects that nation's willingness to subsidise its frontier jurisdictions via revenue transfers from federal to state and local budgets. This includes operating budgets for law enforcement and agricultural research and extension, as well as support for a large public university system (Chapter 9), environmental oversight (Chapter 7) and management of protected areas (Chapter 11).

Brazil's generosity contrasts with the nations of the Andes, where small public budgets in Amazonian jurisdictions are a legacy of their centralised governance systems (Chapter 7). The somewhat greater contribution in Bolivia is due to the inclusion of its capital city (La Paz) within the Pan Amazon, while Colombia's is the consequence of the budget allocated to its security forces (Chapter 6 and 7). Guyana and Suriname have budgets that were historically similar to the Andean republics, but public expenditures will surge after 2021, when offshore oil fields start producing oil and natural gas (Chapter 5).

Public budgets provide one of the easiest avenues for channeling financial resources to shift the Amazonian economy away from non-sustainable production paradigms, which is why the jurisdictional approach is gaining popularity as a way to organise the payment-for-ecosystem services.* The challenge will be to translate an increase in state expenditures into a modification of behaviour by private sector actors. Brazil pursued a version of this strategy from 2004 to 2018 when it successfully reduced deforestation within its Amazonian states by eighty per cent (Chapter 10); however, this effort has caused a political backlash by landowners opposed to the regulatory measures imposed by the state.

All the Pan Amazonian nations suffer from a deficit in basic infrastructure, a consequence of decades of underinvestment caused by political instability, poor governance and financial austerity imposed by multilateral financial institutions.[11] A surge in construction activity occurred between 2005 and 2015 when the global commodity boom provided national governments with revenues that allowed them to radically increase investments in basic infrastructure (Chapter 2). Urban areas benefited most because that was where the need was greatest; however, investments in transportation networks and energy systems were a priority as governments sought to increase economic growth by integrating frontier landscapes into the national economy and harnessing the natural resources of the Amazon.

The contribution of the construction industry to regional GDP is large across all jurisdictions, placing just behind agriculture as a component of the conventional economy (see Figure 1.6). Financing for infrastructure comes from a combination of annual budgets, debt issued from national development banks, government-backed bonds and multilateral development agencies. State-backed entities from China have become a prominent participant in large-scale hydropower projects, while private investors have assumed a leading role in the development of railroads (Chapter 2).

* See the section below, 'Monetising the Value of Ecosystem Services – or Not' on page 55.

Large-scale construction projects in Amazonian jurisdictions have been harshly criticised for their environmental and social impacts (Chapter 7); nonetheless, they enjoy the support of elected officials from successive governments. The construction sector is an unabashed proponent of investment in transportation and energy infrastructure and views deforestation and hydrological degradation as acceptable environmental impacts.

There is an inherent synergy between expenditures in construction and the value of real estate. Investments add value to an asset, while property values increase following improvements in public infrastructure. The reported contribution of real estate transactions to GDP is approximately the same as that of construction and, likewise, is largely the consequence of investments in urban centres. The declared value of real estate transactions, however, is often under-reported by buyers and sellers in order to evade taxes, a practice more prevalent on frontier landscapes where contracts are executed without the intermediation of banks (Chapter 4).* This common practice is another example of how the informal economy gives rise to corrupt practices, and its contribution to GDP is underestimated. Real estate markets are further distorted by the highly lucrative activity of land grabbing and, in the Andean republics, money laundering linked to illicit drugs (Chapter 3). Reining in land grabbing is impeded by the agencies that administer titles, which are plagued by administrative inefficiencies, a backlog of work spanning decades and functionaries complicit in criminal activity (Chapter 4).

The extractive industries in Pan Amazonian jurisdictions are massively important for the national economies of Colombia, Ecuador, Peru, Guyana, Suriname and, to a slightly lesser extent, Bolivia and the state of Pará (Annexes 1.1 and 1.2). All mineral resources in the Pan Amazon are the property of the state, which exploits them via a state-owned enterprise or some type of joint venture with corporations that specialise in mining or the production of hydrocarbons (Chapter 5). Revenues accrue to a region's GDP, even though they do not flow through the local economy; instead, they are deposited directly into the national treasury. This bookkeeping procedure distorts the value of the GDP-per-capita, which is often [mis-] reported by the mass media as a measure of human wellbeing. Forty-five per cent of GDP in Amazonian Ecuador is contributed by oil exports and, if you exclude that revenue from the regional GDP, the per capita value in 2017 would fall from US$ 11,500 to US$ 6,400.

Governments return mineral rents to producing regions as royalties that are included within GDP values reported for public administration.

---

* In Bolivia, transactions are actually formalised with two sets of parallel documents: one with the real value, which is kept private, and one with a 'cadastral' value, which is reported to the authorities. In spite of an obvious fraud, courts will accept the real value version in litigation.

© Dr. Morley Read / shutterstock.com.

The Trans Andean oil pipeline in Ecuador has caused numerous oil spills during its 40-year life span; the right-of-way is used by locals to access small farms cleared from the forest. Petroleum production is an important source of government revenues.

Royalties vary among jurisdictions and across commodities: Peru has the most generous revenue sharing mechanism, which combines royalties with corporate income tax in a system referred to as a Canon. Brazil has the most frugal royalty system but transfers a larger sum of money via the general budgetary system (Chapter 5). Like agriculture, the extractive industries generate benefits to regions via the service sector, which vary depending upon the mineral commodity: hydrocarbons pay higher royalties but consume fewer services, while miners pay lower royalties but consume more services.* In Ecuador and Colombia, the exploitation of petroleum was a major driver of Amazonian settlement and deforestation, while the development of the iron ore deposits in the Carajás highlands was part of a multi-sectoral development strategy (see Text Box 1.2).[12]

The mining sector has an illegal component that is one of the most lucrative activities in the Pan Amazon. Artisanal and small-scale gold mining generated an estimated seven billion dollars in 2017, of which about half was exported via channels invisible to authorities. Most small-scale miners use placer mining technologies that destroy floodplains while polluting watersheds with mercury. The most seriously impacted basins are the tributaries to Madeira (Madre de Dios and Beni) and Tapajós (Crepori), the Caroni River in Venezuela, the Essequibo in Guyana and the Courantyne on the border between Suriname and French Guiana (Chapter 5).

* In Ecuador, value added GDP was 50% of the gross oil production, while in Pará, value added GDP was 40% of gross revenues. Since production costs are fixed, these percentages vary depending on the price of these global commodities. Source: *Instituto Brasileiro de Geografia e Estatistica (IBGE) and Dirección Nacional de Síntesis Macroeconómica, Banco Central de Ecuador.*

Text Box 1.2: The Carajás – São Luis Development Corridor

Pará has developed a metallurgical industry, the result of a deliberate national strategy to add value to mineral exports; this has led to investment in industrial mills that transform bauxite into aluminum and iron ore into steel. Aluminum smelting is an energy-intensive process and that has motivated the construction of large-scale hydropower facilities on the Tocantins and Xingu Rivers. Steel mills and pig-iron foundries consume vast quantities of vegetable charcoal provided by landholders who established cattle ranches along the rail line between the mines at the Serra de Carajás and the port facilities near São Luis do Maranhão.

State governments in Pará have a tradition of being 'pro-mining', which reflects the economic benefits derived from the mines, including royalties and tax revenues, but more importantly from the economic growth spawned by the goods and services sold to the mining companies. The development corridor between the Serra de Carajás and São Luis do Maranhão has lost more than eighty per cent of its original forest cover.

© Christian Inga/shutterstock.com.

Placer gold mines on a tributary to the Río Madre de Dios in Southern Peru; these strip-mines are particularly pernicious because their impacts extend far beyond deforestation of the mine site, which typically spans the entire floodplain from river terrace to river terrace for dozens of kilometers. Watersheds are polluted by mercury and sediments, while fish migration is impeded both above and below the mine site. This highly lucrative activity is pursued by small to medium-sized miners who operate outside law.

Another illegal activity not captured by GDP is the coca-cocaine supply chain that originates in Bolivia, Colombia and Peru. The quasi-legal (or tolerated) cultivation of coca is associated with blatantly illegal laboratories that process coca leaf into cocaine. The coca-cocaine supply chain generates about 1.5 billion dollars annually within Amazonian jurisdictions, an amount multiplied several times over as money is laundered in the commerce, construction and real estate sectors. Coca cultivation is an important source of deforestation in the Andean foothills, where it occurs on landscapes surrounding protected areas and indigenous territories (Chapters 3 and 10).

The Natural History of the Amazon Rainforest

In the nineteenth and first half of the twentieth centuries, scientists affiliated with institutions in the Northern Hemisphere organised most expeditions into the Amazon. These botanists, zoologists and anthropologists were accompanied by native-born and immigrant scientists who pioneered the establishment of natural history museums in the nations of the Pan Amazon. Investment in research and education expanded dramatically as governments pursued development strategies predicated on the exploitation of the region's natural resources. Institutional capacity grew to encompass more than fifty universities and a half dozen research institutions within the Pan Amazon. Collections, publications and students grew exponentially and, by the end of the millennium, more specimens were accessioned annually in the regional museums than all of the legacy collections housed in foreign museums (Chapter 9).

The collections in European and North American museums are the foundation of the taxonomic classification systems at the centre of biodiversity science; unfortunately, that information was unavailable to local biologists, who struggled to identify the plants and animals they encountered during their fieldwork. Starting in the 1980s and accelerating in the 1990s and 2000s, innovation in information management and the creation of the internet revolutionised biodiversity science. Online information resources, such as taxonomic databases and digital image archives, have vastly improved the quantity and quality of floristic and faunistic inventories. Every country can now boast a relatively complete catalogue of all vertebrate groups and a robust checklist of vascular plants, which continue to improve as cadres of young biologists explore their countries. Biogeographers and population ecologists can now accurately map the distribution and abundance of species, which has improved the identification of endemic species unique to a specific region or locality, as well as objectively evaluate the risk of extinction for individual species.* Regional biologists routinely participate in the global effort sponsored by the International Union for Conservation of Nature (IUCN) to identify species at risk of extinction while leading the same effort within their own countries. Having these studies executed by native-born biologists laboring within domestic institutions has increased the legitimacy of that information in the eyes of government and society.

Forest ecology and carbon dynamics

A tropical rain forest is composed of thousands of species of long-lived trees, each with a natural history characterised by unique morphological, physiological and reproductive attributes. Very little was known about the ecological processes within these plant communities in 1980 when ecologists began to establish long-term studies to document their composition, structure and function. In one project near Manaus, scientists created an experiment to evaluate the impact of deforestation and forest fragmentation on plant and animal communities.† In another, dozens (eventually, hundreds) of botanists affiliated with local universities and research institutions created a network‡ of hundreds (eventually, thousands) of permanent one-hectare plots scattered across the entire region. They have used these plots to study Amazonian tree biodiversity, identifying which species are extraordinarily abundant (hyperdominant) and which are exceedingly rare, as well as documenting how tree communities vary across latitudinal, elevational and climatic gradients, information essential for understanding how the Amazon might change in the future due to global warming (Chapter 9).

---

* Governments also invested in research in the applied sciences that contributed to the economic development of the region and, intentionally or otherwise, accelerated deforestation and environmental degradation. Among the most devastating consequences of this research and development was the evaluation and introduction of cultivated grasses of African origin combined with the genetic improvement of cattle breeds from South Asia. The Brazilian agricultural research service (EMBRAPA) transformed tropical agriculture by developing varieties of soybeans adapted to the humid tropics (Chapter 3). Research by plant pathologists and entomologists in Colombia and Ecuador overcame inherent limitations to plantation production systems to support the expansion of oil palm (Colombia) and cacao (Ecuador).

† Ecologists affiliated with the *Instituto Nacional de Pesquisas da Amazônia* (INPA) in collaboration with the Smithsonian Institution initiated *The Biological Dynamics of Forest Fragments Project* (BDFFP), a long-term experiment begun in 1979 to document how deforestation and fragmentation affect the composition, structure, and function of a forest ecosystem:

‡ *Red Amazónica de Inventarios Forestales* (RAINFOR) has published more than 200 peer-reviewed papers authored by 200 contributing authors; see: http://www.rainfor.org/en

Because ecologists count trees and measure their dimensions, they have revealed the dimensions of the Amazon's massive carbon reserves; more importantly, because they repeat their measurements periodically, they have discovered that the Amazon ecosystem has functioned as a net carbon sink over the last several decades. Surprisingly, they discovered that the photosynthetic capacity of intact primary forest was so great that wilderness landscapes have been sequestering more carbon on an annual basis than was being lost via deforestation and forest degradation on the forest frontier. Unfortunately, these same studies identified a trend of increasing tree mortality and a shift in species composition that may cause intact primary forest to become a net source of carbon emissions over the next decade. The shift in ecosystem function is due, in part, to alterations in the physiological processes in leaves stressed by periodic drought and high temperatures but also to a shift in the composition of tree species caused by increased tree mortality (Chapter 9).

Investments in forest ecology were matched by parallel efforts to observe and evaluate forest landscapes using remote sensing technology. Brazil led the way in the early 1980s with a commitment to annually quantify deforestation using satellite images, a decision that would have an enormous impact on the public debate about the expansion of the agricultural frontier (Chapters 2 and 9), as well as the decisions made in international forums dealing with climate change (Chapter 10). The protocols and technology developed by the Brazilian space agency have been adopted as a global standard and are now used by all the Pan Amazonian nations to monitor their own forest frontiers. Over time, innovation in remote sensing technology led to the deployment of satellite-borne sensors that could monitor deforestation in real time, as well as identify forest degradation from logging and wildfire (Figure 1.6). These studies complemented field research by forest ecologists, which allowed researchers to spatially map the distribution of forest biomass, as well as detect how seasonal and interannual variation in weather was impacting ecosystem function at the landscape, regional, and continental scales (Chapter 10).

Climate change and moisture recycling

Starting in the 1990s, ecosystem ecologists and atmospheric scientists in the Brazilian space agency embarked on a sophisticated collaboration with NASA and other international research institutions; their goal was to understand and model the interactions between the atmosphere, the ocean and the forest ecosystem. They collected data using instruments mounted on satellites, airborne platforms and canopy towers, a collaboration that discovered how cyclical anomalies in ocean temperatures, such as El Niño/La Niña, drive the decadal-scale droughts and floods that impact the Am-

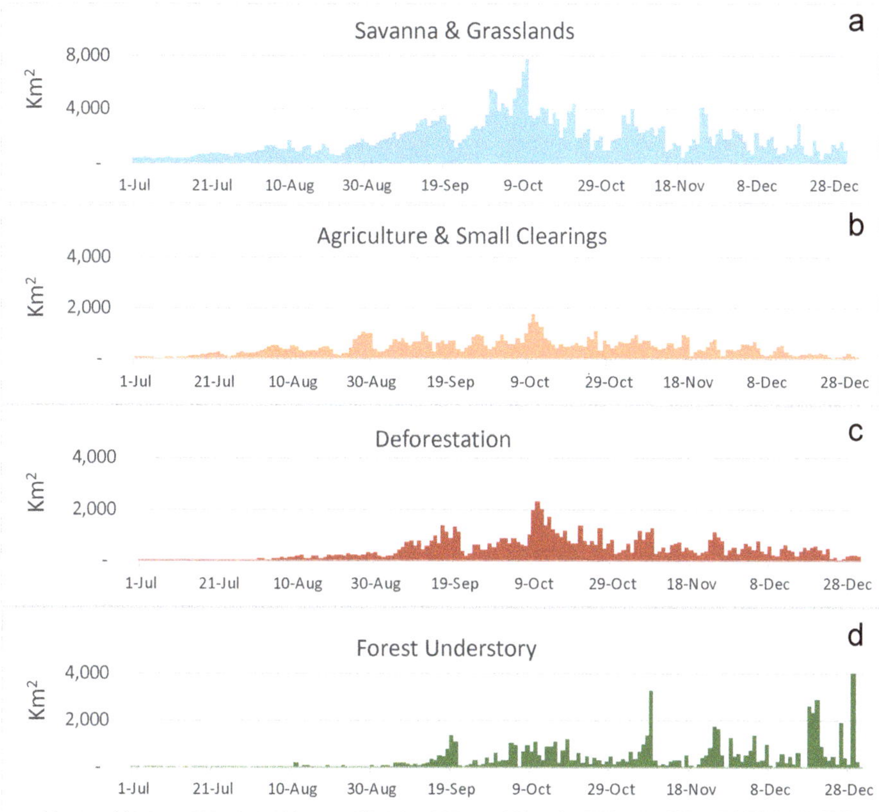

Figure 1.6: The 2020 fire season in the Amazon south of the Equator was typical: The burning season starts in July as landholders use fire to renovate pastureland (a) and clear fields on existing farms (b). The use of fire increases at the end of the dry season (September to October) when landholders burn recently deforested land to expand production (c). Managed fire escapes to create wildfire that impacts the understory of natural forest (d). The y-axis shows the area (km^2) burned as measured by the MODIS instrument on NASA satellites.

Data source: Global Fire Emissions Database, Amazon Dashboard, https://globalfire-data.org/pages/amazon-dashboard/

azon. Their most important finding was to elucidate how tropical forests recycle water between the surface and the upper troposphere via a process known as 'deep convection'; sometimes referred to as a 'biotic pump'; this natural system maintains the high precipitation that defines the Amazon rainforest (Chapter 10).

The impacts of deforestation, forest fragmentation and forest degradation are all weakening water recycling, and this is enhancing the intensity and frequency of seasonal and interannual drought. The risk is particularly acute in the Southern Amazon, a climatic transition zone where subtle shifts in ecological succession can determine whether a landscape is dominated by forest or savanna species. When and if a forest community is established, feedback mechanisms will reinforce the biotic pump, which favours an equilibrium state that supports the maintenance of rainforest. A rapid transition to a non-forest equilibrium can occur if a key environmental factor, such as drought, wildfire or logging, passes a threshold that alters the microclimate that favours forest species. When that occurs, rain forest trees suffer high rates of mortality and are replaced by species adapted to open savanna-like conditions (Chapter 10).

Climate models show that drought in the Amazon will become both more frequent and intense, while higher temperatures increase stress on tropical trees. There is increasing concern that the Southern Amazon could suffer from two or more consecutive years of drought, which could trigger a large-scale forest dieback of cataclysmic proportions. Known as the 'tipping point hypothesis', it is a clarion call of the dangers from uncontrolled deforestation, illegal logging and the indiscriminate use of fire by small farmers and ranchers. The impact of a collapsing forest ecosystem would extend well beyond the loss of biodiversity in the remnant forests of the Southern Amazon because it would signal a dramatic reduction in rainfall across the region – and beyond.

The most economically significant discovery of recent years grew out of a collaboration between meteorological agencies to integrate the subregional manifestations of the annual wet and dry seasons into a common continental-scale climate system. Christened the South American Monsoon (SAM), it mediates the flow of water from the Atlantic Ocean westward across the Amazon, south along the base of the Andes and, eventually, southeast into the subtropical landscapes of Bolivia, Paraguay, Central Brazil and Northern Argentina (Figure 1.7). The combination of the water recycling driven by deep convection within the Amazon and the distribution of water across the continent by the SAM directly links the productivity of one of the planet's most important bread baskets – and the economic health of four nations – with the conservation of the Amazon rain forest (Chapter 9).

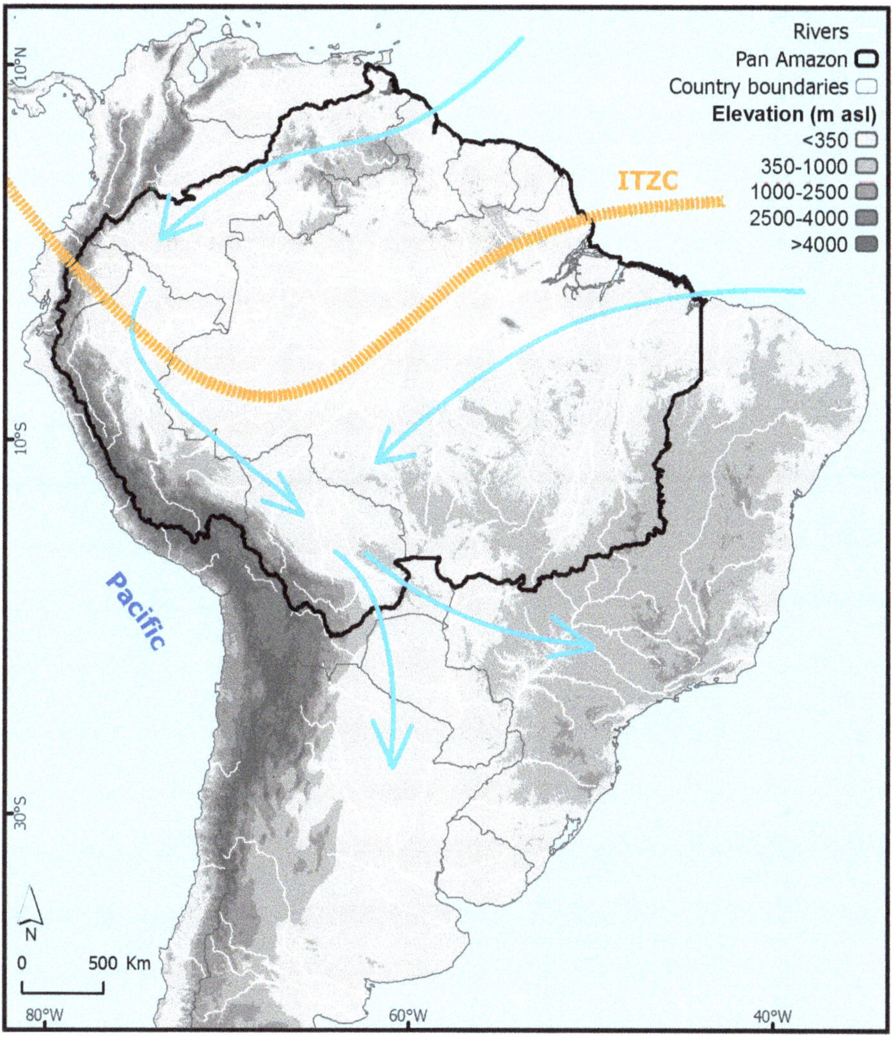

Figure 1.7: The South American Monsoon is established and controlled by the rotation of the Earth, the seasonal shift of the Inter Tropical Convergence Zone (ITCZ), the height of the Andes, and atmospheric pressure gradients driven by temperature differences across the South American land mass.

Data source: Marwan, N. and J. Kurths. 2015. 'Complex network based techniques to identify extreme events and (sudden) transitions in spatio-temporal systems'. *Chaos: An Interdisciplinary Journal of Nonlinear Science* **25** (9): 097609.

NASA (Public Domain)

Wildfires typically start when landholders burn pasture to renew forage or clear forest to establish new fields; managed fire often spreads into adjacent natural forest and, during drought years, can damage millions of hectares of primary forest habitat. This photograph of wildfires along the Río Xingu was taken by astronauts from the International Space Station in 2011.

NASA Earth Observatory, https://earthobservatory.nasa.gov/images/71256/fires-along-the-rio-xingu-brazil

Environmental Policy and Action on the Ground

From the outset, academics realised that the development programmes being deployed across the Amazon in the 1960s and 1970s would bring irreversible harm. Attempts to promote conservation were a natural outgrowth of their efforts to document the region's biodiversity and ecological complexity. They were not the first defenders of the Amazon, however; that distinction belongs to indigenous people who had been fighting to protect their way of life for centuries. These two groups, sometimes working together and sometimes independently, motivated citizens across the world to organise civil society groups, now known as 'non-governmental organisations' (NGOs), to advocate for biodiversity conservation, the protection of indigenous rights and eventually the transformation of the region's economy. Governments and the private sector have responded by implementing policies and adopting business practices that seek to staunch the deterioration of the environment.

It remains to be seen, however, whether these actions will be sufficient to change human behaviour on frontier landscapes or slow global warming within the timeframe needed to save the Amazon. These policies are now being called into question by newly elected governments in Brazil and Bolivia that have embraced conventional development models, by chaos in Venezuela, and equivocation in Ecuador, Peru, Guyana and Suriname.

Environmental governance

The Pan Amazon nations have constitutions that were reformed or rewritten in the last decades of the twentieth century. Previous versions either ignored nature or incorporated a simple clause assigning the state the 'duty' to protect [or improve] the environment. Brazil's constitution of 1988 was radically different: ten separate articles address nature conservation or environmental management by declaring that access to a healthy environment is a basic human right. The reformed national charters of Colombia (1991), Peru (1993), Bolivia (1994, 2009), Venezuela (1999) and Ecuador (2008) all include mandates to enact environmental legislation and create environmental ministries.* Ecuador's is the most emphatic, stating that Mother Nature (*Pachamama*) has rights that must be honoured by human society (Chapter 7).

Environmental ministries are responsible for developing environmental policy and administering agencies that regulate the public and private use of natural resources. Most share management responsibilities, however, with other agencies that reflect competing domestic agendas. For example, the national forest service is part of the Ministry of Agriculture in Brazil and Peru, while the regulation of water resources is shared with agriculture (Peru) or energy agencies (Brazil) or enshrined as a separate agency (Ecuador). Environmental ministries have well-defined responsibilities to oversee the management of the long-term environmental (brown) liabilities linked to the extractive sector (Chapter 5) and to review the potential harm from industrial development and infrastructure investments.

The stated goal of modern environmental management is to 'avoid, mitigate or compensate' the negative effects of modern development. The Environmental Impact Analysis (EIA) framework was introduced into Latin America in the 1990s, and EIAs are now routinely conducted as part of the due diligence and feasibility planning processes for major industrial or infrastructure investments. Like their counterparts in the advanced economies, these technical documents reflect the conflicting societal pressures between promoters of economic growth and advocates for nature conservation. Their methodologies are far from perfect, but their utilisation has

---

* The current constitutions of Guyana (1980) and Suriname (1987), adopted at their independence, have only a single clause dealing with the environment.

vastly diminished harm when compared to the *status quo ante*. The IIRSA initiative has sought to harmonise the EIA processes across the continent, which has improved the technical capacity of environmental ministries and facilitated private sector investment in much needed public infrastructure (Chapter 2).

Many environmental advocates view them as a greenwash, however, and there is ample objective evidence that the EIA service industry is biased toward the approval and completion of individual projects (Chapter 7). There are few examples of projects that have been canceled due to the discoveries identified in an EIA; instead, developers use the evaluation process to identify and modify specific aspects while ensuring projects move forward to completion.*

The dam and power plant at Jirau on the Río Madeira in Rondônia was built between 2008 and 2016 as part of the Programa de Aceleração do Crescimento – PAC (Growth Acceleration Program) over the objections of aquatic ecologists who advised it would alter natural flood cycles, impede the flow of sediments to downstream habitats and obstruct the migration of commercial catfish species essential to the livelihoods of thousands of river communities.

* The rare example of a cancelled project is the proposed dam at São Luis de Tapajós, which was canceled in 2017; in contrast, serial observations by both official and independent technical panels criticising the proposed dams on the Xingu River led to a reconfiguration of their components and the eventual construction of the Belo Monte facility between 2000 and 2019 (Ch. 2 and 7).

The limitations of the EIA system led to the development of a complementary approach known as a Strategic Environmental Analysis (SEA) designed to identify alternative development options far out into the future. The SEA is viewed as a way to avoid environmental degradation because it is predicated on the participation of civil society and local communities that would foresee and veto non-sustainable development pathways. There have been three high-profile attempts to conduct an SEA along highway corridors sponsored by the IIRSA initiative.* In each case, they contributed to the creation of one or more protected areas, but none substantially changed the economic and social forces driving deforestation in the landscapes impacted by those highways (Chapter 7).

Another strategic planning methodology embraced by governments is land-use planning that combines information on soil, water and biodiversity with social and economic data to make zoning recommendations for political jurisdictions. This approach was conceived in the 1970s to identify lands appropriate for agriculture but evolved into a more holistic system in the 1990s as governments reacted to the deforestation crisis. Referred to as *Zonificación Ecológica Econômica* (ZEE), these studies typically recognise the *status quo* of existing settlements and highways while discouraging development on fragile lands or areas of exceptional ecological or cultural value (Chapter 5). The legal standing of these technical documents varies among countries, but most are aspirational rather than mandatory. Settlers, local elites and entrepreneurs use them to identify land with arable soils but all too often ignore recommendations based on environmental criteria. ZEEs have been used in Brazil and Bolivia to identify areas that were incorporated into protected area systems and to bolster the claims of indigenous or traditional communities (see Chapter 11).

One of the most conflictive governance issues in the Pan Amazon revolves around forest management. The constitutions of all eight nations establish sovereign (state) control over forest resources. This obviously covers all aspects of the management of public forests, but it also includes the forest resources located within private properties. To observers from cultures that recognise the priority of property rights, this would seem an aberration; nonetheless, it is a fundamental aspect of these nations' legal systems.† Moreover, virtually all of the landholdings within the Amazon were issued (albeit provisional) deeds that recognise this legal principle;

* *Corridor Bioceanico (Santa Cruz–Puerto Saurez); Corridor Interoceanico del Sur and the Madre de Dios–Acre– Pando (MAP); Plano de Desenvolvimento Regional Sustentável la paraa Área de Influência da Rodovia BR0163 Cuiabá–Santarem.*

† This aspect of jurisprudence is linked to their colonial history when the land and its resources belonged to the crown, which granted concessions and use-rights to individuals and companies. It also underpins the legal basis for the state to claim dominion over all underground mineral resources (Ch. 5 and 7).

consequently, their property rights are clearly circumscribed by precedent and law. Regardless, human nature often supersedes legal tenets, and there is enormous resistance by property owners to comply with rules and regulations emanating from government agencies.

The conflict between private property and forest governance is most evident in the Brazilian government's attempts to enforce the Forest Code, a landmark regulatory framework first promulgated in 1936 and most recently revised in 2012. The 1965 version established that landholders in the Amazon could clear only twenty per cent of their land for agriculture or ranching and must retain eighty per cent in its natural state (Chapter 7). Landholders have the right to exploit forest resources, but they must comply with a variety of management criteria.* The Forest Code was blatantly ignored throughout most of the 1970s, 1980s and 1990s, when landholders cleared land in excess of the legal limit, as well as along river corridors that merit special protection.

In 2005, the government of Brazil launched the *Plano de Ação para Prevenção e Controle do Desmatamento na Amazônia Legal* (Action Plan for Prevention and Control of Deforestation in the Legal Amazon; PPCDAm), an ambitious campaign to bring the combined resources of the Brazilian state, using 'carrot and stick' policies, to the reduction of deforestation. Directed from the president's office, it imposed coordination across ten ministries and a dozen autonomous agencies. Satellites captured accurate and precise data in real time, which law enforcement teams – led by the public prosecutor's office, with support from police, tax authorities, and the environmental protection agency – used to identify illegal activities (Chapter 7). Simultaneously, the federal government coordinated its actions with state governments to implement a land registration and reporting system that provides landholders with a flexible pathway to remediate past infractions of the Forest Code. Consumer boycotts reinforced these actions, forcing multinational corporations to implement supply chain certification systems eliminating production sourced from properties engaged in illegal activities. Coercive measures were combined with incentives such as technical assistance and financial credit from both public and private entities, measures that increased yields and revenues from previously deforested landscapes via the 'intensification' of production instead of the 'extensification' of land use (Chapter 4).

These PPCDAm and related actions led to an eighty per cent drop in deforestation in the Brazilian Amazon between 2005 and 2012, an astonishing turn of events that led optimists to proclaim that deforestation

---

* In the case of timber, landholders must conduct an inventory and then adopt a cutting cycle that ensures the survival of viable populations of each species; since this is not known for most species, a 30-year harvest cycle is recommended as a default.

had been decoupled from agricultural production. The 'carrot and stick' incentive system was designed primarily for agribusiness, particularly farmers producing soy and beef for export markets. The most effective coercive policies excluded smallholders and, consequently, did little to change land-use practices on the most active forest frontiers. Ranchers realised that the government lacked the political will to collect environmental fines and developed workarounds to trade cattle from properties engaged in illegal deforestation. Deforestation increased from a low of 457,000 hectares in 2012 to 976,000 hectares in 2019, a value far below the 2.7 million hectares of deforestation registered in 2004, but analysts forecast the trend will continue in the regulatory environment following the elections of 2019.

The anti-deforestation campaigns in the Andean republics never contemplated an all-of-government effort to reduce deforestation, although, in Peru, a multi-agency task force has used a newly implemented monitoring system to intervene in a few highly publicised incidents.[13] Perhaps more importantly, there are no corresponding global commodity supply chains that might motivate companies, governments, and landholders to change their business practices (Chapter 3).*

The largest land set-aside in history

The commitment to environmental governance in the 1990s was preceded by a civil society movement to create national parks and wildlife refuges. In the first half of the twentieth century, the Amazonian nations created a dozen or so national parks, thanks largely to the efforts of farsighted and passionate individuals with political influence. In most cases, these efforts were both isolated and unsustainable because governments failed to allocate financial resources for their management and protection. Serious efforts to create national protected area systems began in the 1970s as part of the global effort to protect nature and wildlife. International NGOs and the United Nations played prominent roles because they had public relation skills and the legitimacy to garner the attention of national governments. More importantly, domestic opinion supported their creation (Chapter 10).

The first cohort of protected areas was characterised by spectacular examples of biodiversity and scenic beauty: the *tepuis* of Venezuela, the isolated table-mountains of Colombia and the snow-covered peaks of the High Andes. Academics used their knowledge, which was still rudimentary, to advocate for selected lowland landscapes that were exceptionally diverse or ecologically unique (Chapter 9). Brazil established reserves in different parts of the Amazon, operating on the assumption that they would be dif-

* Coffee and cacao have decentralised supply chains that lack critical points sensitive to boycotts, while the palm oil producers in Peru and Ecuador largely produce for domestic markets. Ch. 3.

ferent and, therefore, complementary. Ecuador was the first nation to create a coherent national system in the early 1980s when it designated areas in all of that country's major biogeographical formations. Brazil, Colombia and Venezuela created the core components of their national system between 1988 and 1991, and Bolivia followed suit in the mid-1990s. The growth in the protected area network proceeded at phenomenal speed, catalysed by donations from the advanced economies and multilateral development agencies. Designations increased by ~10% annually between 1965 and 1995 and by ~5% annually until 2015; as of 2019, approximately 28% of the surface area of the Pan Amazon had been set aside into some sort of state-sponsored protected area (Figure 1.8).

Many of these new protected landscapes were home to communities whose livelihoods depended on the natural resources within reserves, which created scenarios for potential conflict and failure. The Man & Biosphere Program of the United Nations provided a philosophical framework for managing the relations between resident populations and park managers while sponsoring pilot projects in high-profile protected areas. These experiences informed a rapidly emerging consensus that protected area systems must reflect the diversity of management challenges characteristic of developing countries. Each country responded accordingly and created a plethora of categories that reflect its respective social realities and the political compromises required to set aside large areas within their Amazonian jurisdictions (Chapter 12).

The IUCN* provides a classification system that offers a logical framework for comparing the multiple different categories in six groupings, which include those afforded 'strict protection,' such as national parks and monuments, and others to be managed for 'sustainable use,' such as forest reserves and buffer zones around national parks (Figure 1.8). The concept of sustainable use reserves was pioneered by Chico Mendez, a human rights activist who led the rubber tapper workers union in Rondônia and Acre in the 1970s and 1980s. His murder by land grabbers in 1988 motivated the Brazilian government to create a new category of protected area, *Reserva Extrativista* (RESEX), which recognised the rights of families whose livelihoods were dependent on forest and wildlife resources (Chapter 12).

The proportion of Amazonian territories set aside as nature reserves varies between countries, ranging from a high of 75 per cent in Venezuela to a low of five per cent in Guyana. The differences are smaller than they seem, however, depending upon what is considered a protected area. For example, forest reserves that allow logging have been incorporated into protected area systems in Brazil and Venezuela but not in Bolivia, Colombia, and Peru. The two countries with the lowest designated protected area,

* International Union for the Conservation of Nature (IUCN).

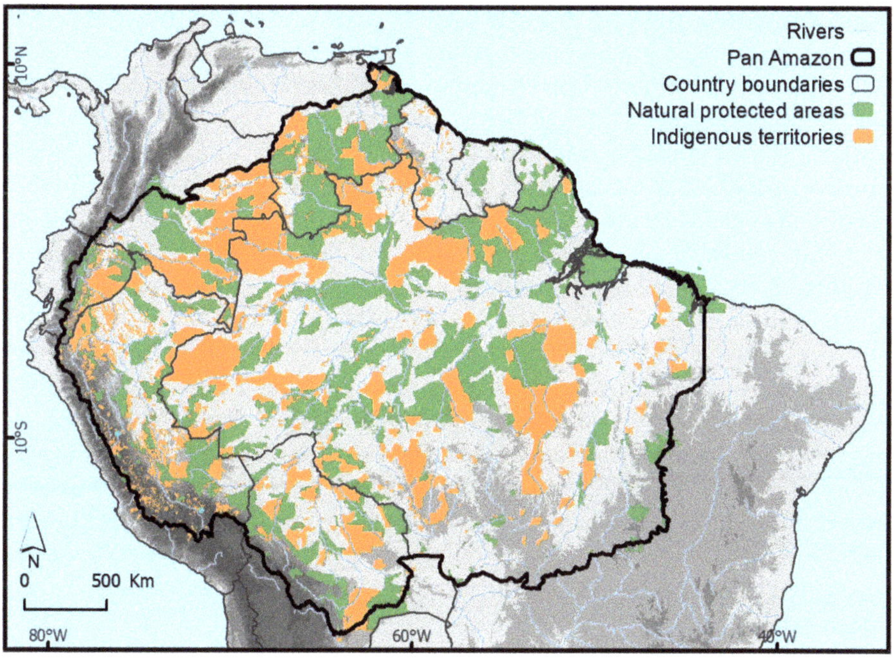

Figure 1.8: The protected area network can be roughly stratified into strict protection areas and sustainable-use areas; indigenous territories are broadly organised into community-based landholdings, which tend to be smaller in area and territorial based reserves that include multiple communities.

Data source: *Rede Amazônica de Informação Socioambiental Georreferenciada* (RAISG), https://www.amazoniasocioambiental.org/

Guyana (5%) and Suriname (15%) retain extensive areas within the forest estate (~50%) that are being held in reserve for future timber exploitation (Chapter 8). Some indigenous territories have been incorporated into protected area systems, while others have not (see below). Brazil and Peru also have large areas that have not been allocated to any management category or tenure regime.

The consolidation of the protected area system is an ongoing process. All but a few parks and reserves are understaffed, and all lack budgets adequate to the task of managing tens of millions of hectares of forest landscape. Some protected areas were created with pre-existing social conflicts linked to illegal mining,[*] the cultivation of coca[†] or cattle ranching.[‡] The goal of sustainable management is largely aspirational and will depend on the willingness of inhabitants to pursue livelihoods compatible with management guidelines, which in turn will depend on their ability to obtain a quality of life commensurate with their personal aspirations. Recent trends in several RESEX reserves show that families have opted to clear land and establish small agriculture operations because the economic benefits from forest-based production models do not meet their needs for cash income.[§] In some jurisdictions, opposition to protected areas by influential constituencies threatens to overturn or downgrade their status.[¶]

In spite of these challenges, or perhaps because of them, civil society groups continue their campaigns to expand protected area networks. The designation of national protected areas has slowed, but efforts to create regional and local parks, reserves, and recreational areas have increased, especially in the Andean republics where administrative authority has only recently been devolved to subnational jurisdictions (Chapter 7). Although local business elites support conventional development models, creating protected areas under local control is popular with broad sectors of the electorate.

An Indigenous revival

The explosive growth of protected area networks was accompanied by the simultaneous recognition of the territorial rights of indigenous people, an ongoing process that has yet to conclude. There are approximately 2.5 million indigenous people living in the Pan Amazon; about two-thirds live

---

* Brazil: Floresta Nacional (FLONA) Jamanxim, Crepori, Itaituba-I; Area Protección Ambiental (APA) Tapajós; Parque Nacional (PARNA) Jamanxim, Río Novo. Venezuela: Reserva Forestal (RF) Imataca, La Paraguay, Sipapo; Parque Nacional (PN) Canaima, Caura, Parima-Tapirapeco; Reserva de la Biósfera Alto Orinoco-Casiquiare; Peru: Reserva Nacional Tambopata; Bolivia; Parque Nacional Madidi; French Guiana: Parc Amazonien de Guyane

† Bolivia: Parque Nacional (PN) Isiboro-Securé, Carrasco; Area de Manejo Integral (AMNI) Amboró; Colombia: Distrito de Manejo Integrado (DMI) Macarena Norte.

‡ Brazil: FLONA Jamanxim, Reserva Biológica (RESEB) Nascentes da Serra do Cachimbo, APA Triunfo do Xingú.

§ Almost all resident families in oldest extractive reserve in Brazil, RESEX Chico Mendes, have cleared small forest patches ranging in size from a few to more than a hundred hectares and are increasingly dedicated to cattle ranching, while RESEX reserves in Rondônia resemble colonisation landscapes with linear deforestation similar to colonisation zones. Ch. 8.

¶ The state legislature of Rondônia revoked the status of ten protected areas in 2018 as a backlash when a governor created a protected area on the last day of his term by executive decree rather than legislative action. Ch. 10.

© Caio Pederneiras/shutterstock.com.

Photograph courtesy of Hermes Justiniano.

The nations of the Pan Amazonian have created the world's largest network of nature reserves to protect the biodiversity and scenic beauty of places such as the Anavilhanas archipelago on the Río Negro (Amazonas, Brazil) and the Serranía de Caparú in Noel Kempff National Park (Santa Cruz, Bolivia).

within their territories, which total between 170 and 220 hundred million hectares (Figure 1.8 and Table 1.3). They have occupied and defended these territories for centuries; historically, they used armed resistance, but they now rely on civil disobedience and political activism. Being granted legal title or explicit use-rights to their homeland, however, does not signify an end to conflict. Indigenous communities must physically protect their land and its resources from timber thieves, gold miners and land grabbers, as well as wage regulatory battles to stop highway construction, petroleum exploration or the development of hydropower infrastructure (Chapter 11).

Indigenous people are the Amazon's fiercest and most effective conservation advocates because the struggle for their territories is existential: if they lose their land, they will lose their identity and cease to exist as a people. They know this because they are the survivors of a holocaust.

The Native Americans of the Pan Amazon have survived wave after wave of genocidal events that started with the colonisation of the Western Hemisphere by the European powers in the fifteenth and sixteenth centuries.[14] Slavery, war and epidemic disease reduced their populations by an estimated ninety per cent by the mid decades of the nineteenth century. The prosperous communities that once populated the main stem of the Amazon River were not entirely annihilated; their languages disappeared, but survivors were absorbed, along with their knowledge, into the mestizo culture that now occupies the banks of the river.*

The ethnic groups that survived with their culture and languages intact did so by retreating to remote territories on upstream tributaries that limited their contact with the agents of Western civilisation. Some tribes interacted with missionaries and frontier merchants, participating in the Amazonian economy by trading a diversity of forest products, including gums, resins, fibre, fruits, nuts, wildlife and fish. They were unprepared, however, for the avalanche of invaders who arrived with the onset of the first rubber boom in the last half of the nineteenth century (Chapter 6). The number of dead has never been compiled, but tens of thousands perished from another round of disease and slavery. The surviving tribes moved deeper into the wilderness.

The period between the end of the first rubber boom and the nationalist policies of the 1970s was a time of relative calm. Outsiders continued to seek out indigenous communities, but they now came with benevolent intentions. Catholic and Protestant missionaries renewed their efforts to bring salvation to so-called heathen populations; their most consequential action was to educate young men, and sometimes women, as part of a deliberate strategy to assimilate ethnic groups into Western society. The

* Along the mainstem of the Amazon River, they interbred and merged with immigrants who gave rise to the *Cabloco, Quilambolo and Ribereño* traditional communities. See Ch. 6.

Table 1.3: The ethnic populations of the Pan Amazon and their territorial claims in 2017[28]

	Groups Extant/ Extinct	Indigenous Population	Territorial Units	Formalised + Legalised (hectares)	Claimed + Pending (hectares)
Brazil	174/9	708,474	428	106,906,612	8,789,287
Bolivia	33/4	244,386	94	11,711,762	7,164,531
Colombia	49/4	253,711	185	26,217,159	
Ecuador	8/1	349,935	+/-200	4,095,851	3,667.168
Peru	73/4	615,555	1,500	14,568,446	5,360,988
Guyana	8/0	82,561	128	3,167,297	~3,000.000
Suriname	10/1	25,792	50		3,424,063
Venezuela	29/4	271,996	10	1,186,539	30,580,859
F. Guiana	7/0	25,333	22	715,223	
Total	**340/39**	**2,577,743**	**2,461**	**168,568,889**	**56,138,396**

Brazilian government created the *Serviço de Proteção ao Índio* (SPI), which sought to 'pacify' indigenous groups and integrate them as distinct cultures into Brazilian society.* Expeditions and trading stations were manned by *sertanistas*, many of whom admired indigenous cultures and sought to protect them from modern society.† A third group of individuals combined both missionary and anthropological approaches: evangelical Christians organised a highly effective effort to preserve indigenous languages, recognising their essential role in cultural survival. Ironically, they used that knowledge to translate the bible into native languages and, in the process, attack the spiritual elements at the core of indigenous culture (Chapter 11).

All three of these interlopers acted as mentors to indigenous people as their societies adapted to a changing world in the first half of the twentieth century. Each contributed to the survival of the indigenous people with whom they interacted; unfortunately, many exchanges triggered another round of epidemic disease. Indigenous people were not passive recipients of their ministrations, however; they absorbed some lessons while ignoring others. More importantly, they retained their own leadership traditions, which they would harness to navigate the next existential threat to their

* The SPI was established under the leadership of Cândido Rondon, an army officer of indigenous descent, famous for his Amazonian expeditions and his philosophy of how to contact and interact with indigenous tribes.

† *Sertanistas* were known for their backwoods skills and ability to live among indigenous communities; some were autodidact anthropologists and contributed information to academic studies that documented ethnic cultures; others were scoundrels and took advantage of their positions within the SPI to engage in corrupt acts.

Photographs courtesy of Eric Stoner.

Megaron Txucarramae is a leader of the Kayapó, an indigenous nation (~8,000 individuals) that has pioneered self-governance within their territories in Pará, seen here addressing a seminar where forest rangers received training to combat wildfire, illegal logging and other types of encroachment on their territory in Mato Grosso.

way of life. The already considerable pace of change was hyper-charged by the tumult of the 1970s and 1980s as immigrants streamed into their territories with government assistance and the explicit intention of stealing their land. Fleeing further into the forest wilderness was no longer an option. They had to organise and fight, or they would perish. Typically, a group of individuals would form an association united by ethnicity and language; not infrequently, they were led by a charismatic individual who had been indoctrinated by a Western mentor. In a remarkably short period of time, individual ethnic associations united to form national federations to represent their interests to government. This happened independently in Brazil, Bolivia, Colombia, Ecuador and Peru; simultaneously, they organised internationally to create an Amazonian coalition of indigenous organisations.* By design or good fortune, their movement coalesced at a time when these nations were undergoing democratic renewal and constitutional reform. By the end of the 1980s, the Andean republics and Brazil explicitly recognised the rights of indigenous people to their ancestral territories and some form of autonomous government.

The first indigenous territory was actually conceived as a national park: *Parque Nacional do Xingu* was created in 1961 and, at the time, was the largest protected area in Brazil. Its proponents, a trio of brothers who were famous *sertanistas*,† based their ambitious proposal on their observations that indigenous cultures are inextricably linked to their livelihoods, which are entirely dependent on access to forest and aquatic resources (Chapter 11). To protect indigenous culture, it is necessary to conserve the landscape that supports the livelihood of the entire tribe. The park was declassified as a protected area in 1991 and recognised solely as an indigenous territory: *Parque Indígena do Xingu* (PIX).

Prior to its formation, the amount of land deeded to a village or community was calculated based on the area required to support a family using slash-and-burn technology, rather than the territory needed to pursue a forest livelihood. Another key decision was to include multiple communities in a single reserve, which also increased the size of the protected area. This provision obligated different tribes, many of whom were historically hostile, to collaborate in the administration of their shared territory (Chapter 10). The PIX set a precedent in Brazil, but very few indigenous territories were created during the military government between 1964 and 1985. The recognition of indigenous territories began in earnest with the constitutional reform of 1988.

---

* *COICA: Coordinadora de las Organizaciones Indígenas de la Cuenca Amazónica.*

† The Villas-Bôas brothers lived among the indigenous people on the upper Xingu River in the state of Mato Grosso; their proposal, made originally in 1952, had the support of Cândido Rondon and other key military officers. See Ch. 10.

The recognition of the territorial rights of indigenous communities in the Andean republics occurred first in Peru, when a military government initiated an agrarian reform policy in the early 1970s. The Peruvians adopted a community-based model that deeded land to individual villages, rather than a territory-based model that pooled the lands of multiple communities. This policy has created a fragmented land tenure map, which has facilitated the development of energy infrastructure while impeding indigenous organisations' efforts to limit the expansion of the oil industry in northeast Peru (Chapters 5 and 6).

In Bolivia, the government of Evo Morales changed the nature of its territorial system, which was originally established using ethnic affinities, to allow new settlements by migrants from indigenous highland communities. Guyana adopted a community-based approach that limited the forest area ceded to indigenous tribes while maximising the forest estate under government control. Suriname has yet to act to create any indigenous territories, in spite of a decision by the Interamerican Court of Human Rights in favour of the two largest groups,* who petitioned the court to seek restitution for environmental impacts linked to the bauxite industry (Chapter 5). Venezuela has recognised the rights of indigenous people in some protected areas but has yet to act on the territorial claims of its very sizable indigenous population (Table 1.3).

Although Peru relies heavily on the community-based model for allocating land to ethnic groups engaged with modern society, it has deployed the territory-based model to protect indigenous groups who are in voluntary isolation. Previously referred to as 'uncontacted' indigenous groups, these small bands are known to exist in Brazil, Bolivia, Colombia, Ecuador and Venezuela. Estimates vary, but there are probably less than 10,000 individuals living in about sixty bands in the most remote corners of the region (Figure 1.9).[15] These are among the most vulnerable groups within the Amazon because they are susceptible to common diseases and have not acquired the social skills to protect themselves from the vagaries of life that are an integral part of modern society.

The people living in voluntary isolation are not the only vulnerable cultures in the Amazon. Approximately ten per cent of the ethnic groups in the Amazon have become extinct since anthropologists compiled a (more or less complete) list in the first half of the twentieth century.[16] Cultural extinction is a forgone conclusion for forty groups with populations of less than 100 individuals, and the future is only marginally better for another 82 tribes with fewer than 500 souls, particularly if they have not retained the use of their language.[17] Overall, however, indigenous populations have approximately quadrupled since the 1970s, a positive sign that their health

* *Lokono (Carib)* and *Kalina (Arawak)*.

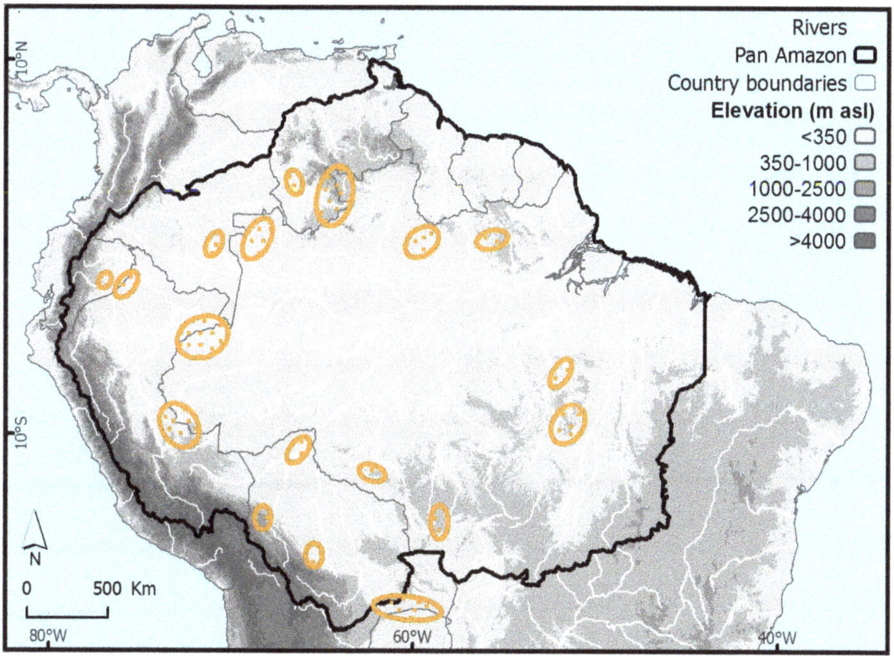

Figure 1.9: Most indigenous groups living in voluntary isolation live in remote areas on the border between Amazonian nations.

Map courtesy of International Work Group for Indigenous Affairs. Shelton et al. 2013.

and welfare has improved in parallel with efforts to secure their lands. Most of that growth has occurred within the fifty largest ethnic groups, who have been most successful in protecting their rights and prerogatives (see Chapter 11).

These numbers underestimate the actual indigenous population because of migration to urban centres by youth. Urban populations are both an opportunity and risk for indigenous people. If they retain their indigenous identity, urban populations can act as a conduit for information, technology, education and financial resources. Unfortunately, history has shown that it is more likely they will lose their language and adopt the cultural identity of the far more numerous mestizo population.

The revitalisation of indigenous communities has paid monumental dividends for Amazonian society (Text Box 1.3). Their commitment to biodiversity conservation ensures their territories will be managed as sustainable use reserves, and most adhere to management criteria similar to the most restrictive type of protected area. Their commitment is hard-wired into their culture, reinforced by recent history and the bitter struggle to defend their lands (Chapter 10). They are quite literally conservation warriors.

Text Box 1.3: Conservation Warriors

Kayapó: Opposed the development of multiple hydropower facilities on the Xingu river, confronting land grabbers and timber thieves. They have organised a self-defence system that serves as a model for other indigenous groups.

Mundurukú: Have protected their lands from gold miners and maintained a strategic presence at the falls at São Luis do Tapajós, where they mounted a successful campaign in opposition to a dam that would have opened the Tapajós as a waterway for grain exports.

Yanomami: Mobilised public opinion, which forced the state to use the army to combat illegal gold mining, while alerting the world to the plight of uncontacted communities in the face of epidemic diseases.

Amaerikaeri: Displaced from their lands by the illegal gold miners in the Madre de Dios, but continue to fight to protect their lands in the foothills.

Asheninka: Fought a long and hard battle with the Sendero Luminoso terrorist organisation and their allies among the coca growers in the foothills of the central Andes.

Moxeños: Defend their lands against a government intent on building a highway through the middle of their territories that would expand the settlement footprint for coca cultivation.

Waorani: Led the battle to stop the expansion of oil production in Yasuní National Park by taking their case to the Interamerican Court of Human Rights

Ashuar: Oppose new oil drilling by the state, while suing the government in court to hold them accountable for the damage from oil spills caused by negligence in the management of a forty-year old pipeline.

Photograph courtesy of Eric Stoner.

Warriors celebrating the Kuarup, an intertribal ceremony honoring the dead practiced by the inhabitants of Parque Indígena do Xingu (PIX), Mato Grosso. The PIX was established in 1961 as the Amazon's first multi-ethnic indigenous territory; today it is home to more than 8,000 Native Americans, belonging to 16 ethnic groups pertaining to five distinct linguistic lineages.

Source: *Povos Indigenas no Brasil, Instituto Sociamiental* (https://pib.socioambiental.org/).

Monetising the Value of Ecosystem Services – or Not

Environmental economics views the environment as a form of natural capital: land, water and biodiversity are viewed as assets that mediate the flow of goods and services from ecosystems toward human society. One of that discipline's most important innovations was the concept of payment for ecosystem services (PES), based on the observation that, under certain circumstances, societies are willing to pay for the provision of goods and services that originate from nature. The Pan Amazon is the most biologically diverse tropical forest on Earth, its forests and soils contain the largest stock of terrestrial carbon and it is home to the world's largest freshwater resource. Nonetheless, it has been difficult to discover and implement PES schemes capable of monetising the value of this enormous natural capital.

With regard to biodiversity, its value remains hypothetical until discovered; consequently, the market potential of the vast majority of Amazonian species is unknown. The foundation of the modern pharmaceutical industry is based on chemicals derived from plants, fungi and animals, but the discovery process is long, difficult and laden with risk. Similar obstacles impede investment by agribusiness: research to discover the economic potential of agrobiodiversity is largely carried out by public institutions. A PES scheme for biodiversity is not an option, because the potential beneficiaries of an undiscovered species have not been identified.

The value of water is easier to estimate because we consume it on a daily basis; nonetheless, developing PES schemes is difficult because Amazon has a surplus of water. It is implausible to ask consumers to pay for conservation in the name of water, except in a limited number of situations characterised by local scarcity. For example, urban water districts in the Andes often incorporate the cost of nature conservation in monthly water bills; similarly, rural communities dependent on irrigation in dry Andean valleys have developed PES schemes to compensate neighbors in cloud forest habitats that act as water towers. In all instances, the connection between provision and consumption is circumscribed to a local watershed. At larger scales, water is viewed as a free resource.

Two recent scientific discoveries provide an opportunity for a novel, continental-scale, PES scheme based on water: (1) Deep convection, which maintains rainfall over the Amazon, is threatened by deforestation and forest fragmentation; and (2) the SAM transports water from the Amazon to the agricultural landscapes of the subtropics (Chapter 10). The combined agricultural economy of the Paraná-Paraguay Basin was reported as approximately $US 200 billion in 2018; productivity losses due to drought stress are common and translate into billions of dollars of loss in revenue. Unfortunately, a PES scheme is unlikely because it would require the transfer of money to Brazil from a less wealthy nation (Paraguay) and a geopolitical competitor renowned for poor governance (Argentina).

Fortunately, Brazil has the capacity and institutional infrastructure to implement a domestic rainfall-based PES programme. The federal government already supports Amazonian states via revenue transfers embedded within the annual budget process; public expenditures are the largest component of the economy in Acre, Amapá, Rondônia and Roraima (Annex 1.1). The system could be expanded into a *de facto* PES system by increasing funding for programmes known to support water recycling on frontier landscapes, including forest conservation, reforestation, agroforestry and low carbon agriculture (Chapter 8).

The potential value of forest carbon has motivated all the nations on the planet to create a global PES scheme based on carbon offsets.* In 2009, the signatories to the United Nations Framework Convention on Climate Change (UNFCCC) agreed to implement a system referred to as 'Reducing Emissions from Deforestation and Forest Degradation' (REDD+). At the time, it was assumed the advanced economies, and possibly China, would agree to mandatory reductions in greenhouse gas (GHG) emissions and the adoption of a cap-and-trade carbon market to generate demand for forest carbon offsets. Unfortunately, the United States failed to adopt a coherent climate change strategy, and the anticipated cap-and-trade compliance market has yet to materialise.

In the intervening period, the countries of the Pan Amazon have invested in a variety of REDD+ initiatives, motivated by the possibility that the US would eventually adopt supportive policies. They were encouraged by multilateral development agencies and civil society groups that provided financial resources to create the necessary institutional infrastructure and to test modalities for investing the revenues from an eventual REDD+ system. This preliminary system has been operating since its inception by generating carbon offsets that have been monetised within a voluntary carbon market or via *ad hoc* agreements negotiated by multilateral or binational development agencies (Chapter 12).

The Brazilian government integrated its REDD+ policies within its national climate change strategy and elected to monetise emission reductions via the Amazon Fund (*Fundo Amazônia*). According to the rules established by the REDD+ process, the decline in deforestation between 2005 and 2017 reduced CO_2 emissions from the Brazilian Amazon by ~1.5 gigatons.[18] The Brazilian government contends those reductions should be worth ~$US 22 billion based on a projected price of about ~$US 15 per ton of CO_2. Donations to the Amazon Fund have totaled $US 1.3 billion, which translates

* A carbon offset is a reduction in emissions of CO2 or other greenhouse gases (GHG) that is traded to compensate for emissions made elsewhere.

into a carbon offset price of $US 0.86 per ton of CO_2.[19] Brazil has pursued policies to reduce deforestation for multiple reasons, including to protect its export markets and respond to a domestic constituency concerned about the Amazon (Chapter 10), but when viewed as investment in REDD+, those policies have not been particularly lucrative.*

Brazil's commitment to the REDD+ system is currently under review by the newly elected government, which campaigned on a platform of promoting conventional development in the Brazilian Amazon. The two major contributors to the Amazon Fund, the governments of Norway and Germany, suspended their contributions in late 2019 to protest against the change in policies by the newly elected administration of Jair Bolsonaro. This conflict has revealed different interpretations of the REDD+ agreement: Brazil contends that it should be compensated for past performance, while the donor countries believe their ongoing contribution is based on future emission reductions, or at the very least, a commitment to maintain policies to combat illegal deforestation.[20]

The Andean countries, Guyana and Suriname all participate in REDD+ initiatives organised by the United Nations and The World Bank, as well as in several binational programmes sponsored by individual donor countries. Numerous REDD+ initiatives have been arranged and financed by civil society organisations that have yet to monetise their carbon offsets; presumably, they are holding certificates in anticipation of a future cap-and-trade market. As of 2020, there was no published estimate of their market value, although they report a total surface area of 44 million hectares.[21]

Perhaps the most interesting REDD+ initiative is the Governors' Climate and Forest Task Force, a coalition of subnational jurisdictions working to create a framework through which emissions reductions can be monetised independently of national policies.† The jurisdictional approach simplifies the challenges and mitigates many of the risks associated with monitoring deforestation, while providing a politically expedient framework for distributing revenues generated by emission reductions. In addition, advocates propose to add value to REDD+ by incorporating mechanisms for counting carbon removals that might occur due to changes in soil management or other opportunities created by low emission development strategies (LEDS).

* Other REDD+ initiatives involving the Brazilian state include a project financed by the Green Climate Fund (GCF) to support the conservation and restoration of forest remnants on private properties ($96.4 million) and a programme implemented by the state environmental agency in Acre to promote forest conservation ($US 163 million).

† The initiative is led from California and Illinois in the USA, which seek to implement climate change policies in spite of the lack of action on the part of the federal government. Bridge financing for the GCF-TF is being provided by the governments of Norway, Germany and the United Kingdom.

Participating jurisdictions include all eight state governments in the Brazilian Amazon, Peru, Ecuador and Colombia (Chapter 12).

REDD+ has failed to provide the resources necessary to halt deforestation, much less transform the regional economy. This failure is ascribed to the inability of the advanced economies to put a significant price on carbon emissions; however, even if REDD+ revenues were increased, they might not be sufficient to overcome the multiple complex factors that drive deforestation. Even less likely is the prospect for money to be allocated for the reforestation of tens of millions of hectares of degraded pastures that climate modellers assert is needed to stabilise the precipitation regime of the South American continent.

The Challenge of the Future (and Lessons from the Recent Past)

The current development trajectory of the Pan Amazon is uncertain. The ongoing investment in protected areas and indigenous territories has created a solid foundation for conservation of the region's biodiversity. The dramatic reduction in deforestation in Brazil probably avoided an ecological catastrophe and identified key public policies with the potential for bending the arc of Amazonian development. Nonetheless, the momentum of fifty years of chaotic economic growth, disregard for the law and the economic power of vested interests continue to impede efforts to halt the environmental degradation that threatens the long-term integrity of the Pan Amazon.

Looking forward, multiple interrelated phenomena will determine the future of the region: some would support the development of a sustainable economy, while others would reinforce the behaviors linked to conventional business models. Quite a few are neutral in nature and have impacts that could be mitigated in a well-managed and diversified regional economy. They can be organised in the following four categories, based on their probability of occurrence and their potential to contribute to a sustainable future (Chapter 13).

Things that will definitely happen

Highway networks will continue to expand; existing roads will be upgraded; it is just a matter of time. Agricultural enterprises with overseas export markets will expand; this will displace some producers toward the forest frontier (ranchers) and motivate others to expand existing production models (smallholders). The extractive industries will dominate the economies of jurisdictions rich in mineral resources; their environmental performance will improve, but they will still create long-term environmental liabilities. Sustainable production technologies, such as aquaculture, will provide new economic opportunities, while selected forest products, such as *açaí*,

© PARALAXIS/shutterstock.com.

Río Branco, the capital of Acre, is home to approximately 50% of the state's citizens, a rural–urban distribution replicated throughout the Southern Amazon where three of every four inhabitants reside in towns with a population greater than 10,000. Many urban residents support nature conservation, but their economies are dependent on production systems linked to deforestation and the extractive industries.

will become new export commodities. Initiatives to decentralise the administrative functions of the state will empower both local elites who want to expand conventional production models and grassroots activists who advocate for environmental conservation and social justice. Urbanisation will continue to expand the opportunities and improve the living conditions of the region's inhabitants.

Things that might possibly happen

The management of protected areas should improve as their operations are incorporated into state budgets; nonetheless, political opposition may cause some to be downgraded or degazetted from the system. Indigenous communities will face challenges to protect their land; some may be tempted to sell access to natural resources in exchange for money. Widespread non-compliance with the land-use zoning regulations (e.g., Forest Code) and the loss of forest remnants may diminish the convective systems that maintain high precipitation regimes. Consumer demand for deforestation-free commodities should lead to the intensification of production on agricultural landscapes, while community-based business models could improve the management of wild fisheries and decrease levels of informality (illegality) in the timber industry. Nature and cultural tourism could improve the livelihoods of traditional forest communities but would require significant investment in infrastructure, human capacity and marketing. Democratic reform and regulatory oversight can improve the quality and objectivity of environmental and social review but are unlikely to eliminate ill-advised investments in infrastructure or the extractive sector. Financial resources from PES schemes could provide critical support for the operational budgets for protected areas and indigenous lands but may not provide sufficient investment capital to reform non-sustainable business models.

Things that should never happen

There should be no new trunk highways constructed through any wilderness area; there are no justifications based on economic criteria within the transportation sector. Large-scale hydropower facilities on major rivers should be eliminated from consideration due to irremediable impacts on sediment flows and fish migration; they are rarely economically viable using standard financial criteria. Global warming should not exceed 2° Celsius due to impacts on plant physiology, forest carbon dynamics and continental-scale modifications of precipitation regimes.

Things that absolutely need to happen

The most pernicious and destructive activities in the Pan Amazon are all blatantly illegal, and governments must take action to bring them under control. Some are obvious by-products of a cultural tolerance of corruption: illegal logging and land grabbing. Others are the product of inequality and the lucrative nature of the illicit activity: artisanal gold mining and the cultivation of coca. Eradicating the former will require institutional reform and sustained law enforcement action; the latter will require a more integrated approach because of the number of people involved and their willingness to confront the state with violence.

Current Policy Approaches: Certainly Necessary, but Are They Sufficient?

The consensus strategy for saving the Amazon is based on a pentad of self-reinforcing policies: (1) create protected areas and recognise indigenous reserves; (2) improve governance to combat illegal activities; (3) increase market demand for deforestation-free commodities; (4) enhance the economic value of forest livelihoods; and (5) generate financial revenues from PES schemes to underwrite the implementation of the first four strategic pillars. Two of these policies (3 and 5) rely on macroeconomic incentives, and two (1 and 2) on top-down initiatives emanating from central governments. Only one (4) seeks to change the microeconomics on the forest frontier, and it focuses on a part of the Amazonian economy that does not directly cause deforestation. All of these policies are reliant on REDD+ and, consequently, the development of a robust global carbon market.

This is not an unreasonable expectation. REDD+ provides an opportunity to mitigate global warming as well as alleviate the biodiversity crisis by subsidising policies that are cost-effective, timely and humane. REDD+ has always been viewed as an interim solution that would reduce emissions while the advanced economies, led by the United States, transition to a green economy. Unfortunately, the twenty-year delay in implementing a coherent climate change strategy has made REDD+ a less relevant policy tool.

It is increasingly likely that the Biden administration elected in 2020 will adopt a more aggressive climate change strategy that will reverberate across the global economy. Policies favoured by progressives call for a multi-sector investment and regulatory program that will rapidly transform the domestic economy.[22] In contrast, mainstream economists prefer a fiscally neutral carbon tax that will allow the market to determine which energy systems and technologies prevail.[23] Neither policy approach envisions a mandatory cap-and-trade system.[*] Moreover, many climate activists view REDD+ as 'greenwash' that would delay fundamental changes needed to create a zero-carbon economy, a view reinforced by the uncertainty in the permanence and additionality of forest carbon offsets originating in societies characterised by poor governance.[24]

Considering the large investment made by multilateral development agencies and tropical forest nations, some type of REDD+ mechanism will be implemented. Unfortunately, that system is more likely to remain – as it currently is – a mechanism for channeling overseas development assistance to developing nations combined with a voluntary carbon market used by corporations to offset emissions from fossil fuels (Chapter 10). That type of REDD+ system would provide essential support for protected area management and assist indigenous communities but it will not be sufficient to transform the economy of the Pan Amazon.

---

* In June 2020, there were eight different legislative proposals before the US Congress to adopt policies to put a price on carbon; all but one are based on a carbon tax; the cap-and-trade bill is referred to as a cap-and-dividend system where revenues are returned to taxpayers and does not include a provision for carbon offsets created via the REDD+ mechanism. See: Price On Carbon, https://priceoncarbon.org/business-society/history-of-federal-legislation-2/

The need for a Plan B: Tree-based production systems

To be truly successful, the transformation of the Amazonian economy will require the wholesale support of the ranchers and smallholders who occupy eighty million hectares of previously deforested land and an approximately equivalent area at risk to future deforestation. Most landholders pursue business models based on deforestation, none of which are particularly productive when viewed from the perspective of energy (carbohydrates) or nutrition (protein). They are, however, economically advantageous when viewed within the timeframes that constrain investment decisions, even when those decisions lead to the eventual degradation of soil resources (Chapter 3). If the economic logic that drives deforestation were reversed – if planting trees were more lucrative than cutting them down – most landholders would happily change their production systems. This supposition is essentially the business model for agroforestry, a tree-based production system targeted at smallholders, and for plantation forestry, a tree-based production system appropriate for larger landholdings (Chapter 8).

Both production models have been promoted by extension agents and foresters for decades. The most notable examples of agroforest crops are coffee and cacao, while plantation systems include palm oil and wood fibre (cellulose, charcoal, biofuel or timber). Agroforestry systems are relatively popular among environmental advocates and the smallholders that adopt them. Plantation systems, however, typically attract the wrath of environmental activists because they have been associated historically with large-scale deforestation and are usually predicated on the monoculture of non-native species (Chapter 3).

The preferred option of most conservation scientists is the restoration of forest landscapes using native species to recreate natural habitat that would produce, eventually, high-quality hardwood timber species. Unfortunately, that business model has a pay-back time measured in decades, rather than years, and is nonviable without very large subsidies, which presumably would be provided by global carbon markets. In Brazil, environmental advocates contend that landholders will eventually be forced to restore native habitat in order to come into compliance with the Forest Code (Chapter 7). Perhaps – but recent history has revealed the social friction associated with that policy pathway, and it certainly will not happen during the administration of Jair Bolsonaro.

Tree-based production systems have the advantage of restoring atmospheric water-recycling functionality to highly fragmented, deforested landscapes (Chapter 10). Amazonian producers can be persuaded to adopt tree-based production models, but only if there is a genuine demand for the commodities they produce. Several million hectares of new coffee and cacao plantings would almost surely swamp global commodity markets and destroy the economic incentive for cultivating them. In contrast, the production of wood fibre has a much larger growth pathway, particularly in light of the recent global awakening of the impacts caused by our plastic-based consumer economy. If fossil-fuel derived plastics were replaced by plant fibre, the potential economic opportunity for the Pan Amazon and other tropical forest regions would be enormous.

Tree-based production systems are not without risk, including the potential to displace cattle ranches to the forest frontier and the introduction of exotic species into natural habitats. Displacement risks could be mitigated using the same pentad of policy options described above, while the risk from invasive species could be avoided by using mixed plantings of native species, of which there are literally thousands of prospective candidates (Chapter 9). The pursuit of tree-based production systems based on a commodity with a global market represents yet another macroeconomic solution, but one using a business model that responds to the microeconomic challenges and social obstacles that have stymied current policies.

© JBK / shutterstock.com.

Açaí is a super fruit that has exploded in popularity over the last decade due to its nutritional benefits. It has been a staple of Amazonian commerce for centuries where it is harvested from a palm (Euterpe oleracea) that dominates floodplain forest in the lower Amazon. Landholders have started growing the açaí palm in plantations in order to meet global demand for exports that exceeded one billion dollars in 2019.

The Pan Amazon is a big, complicated region, and no single policy package will resolve the conundrum of reconciling nature conservation and economic development. Many, perhaps all, solutions will need to be local or regional in nature. Even the macroeconomic and top-down models favoured by policy specialists will need to be implemented in the context of geographically specific circumstances. This book attempts to lay these issues out in a systematic and logical narrative to facilitate the discovery of a pathway through a perfect storm of environmental mayhem to a sustainable future for the Pan Amazon and all of its inhabitants.

Annex 1.1: The relative contribution of the various sectors and subsectors to to the GDP of the Amazonian jurisdictions of the Andean republics (Left) and the Guiana coast (Right)

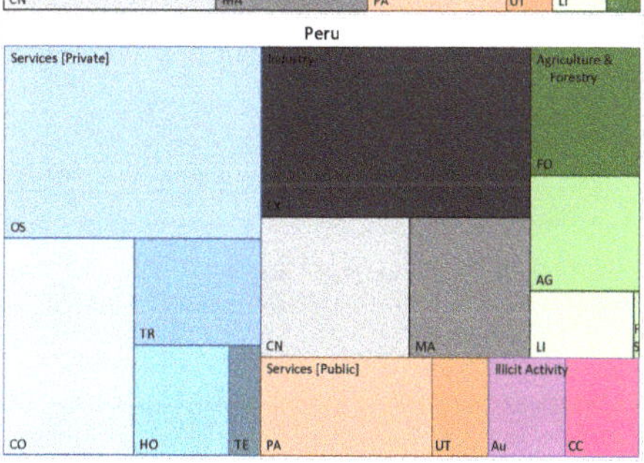

<u>Agriculture & Forestry</u>: Agriculture (AG), Fisheries & Aquaculture (FA), Forestry (FO), Livestock (LI);
<u>Industry</u>: Extractives (EX), Manufacturing (MA), Construction (CN);
<u>Services (Private Sector)</u>: Real Estate (RE), Commerce (CO), Transportation (TR), Hospitality (HO), Telecommunications (TE), Finance (FI), Professional Services (PS), Other services (OS);
<u>Services (Public sector):</u> Public Administration (PA), Utilities (UT); Space Center (SP)
<u>Illicit Activities</u>: Coca/Cocaine (CC), Artisanal Gold (Au)

Annex 1.2: The relative contribution of the various sectors and subsectors to the GDP of the states of the Brazilian Amazon

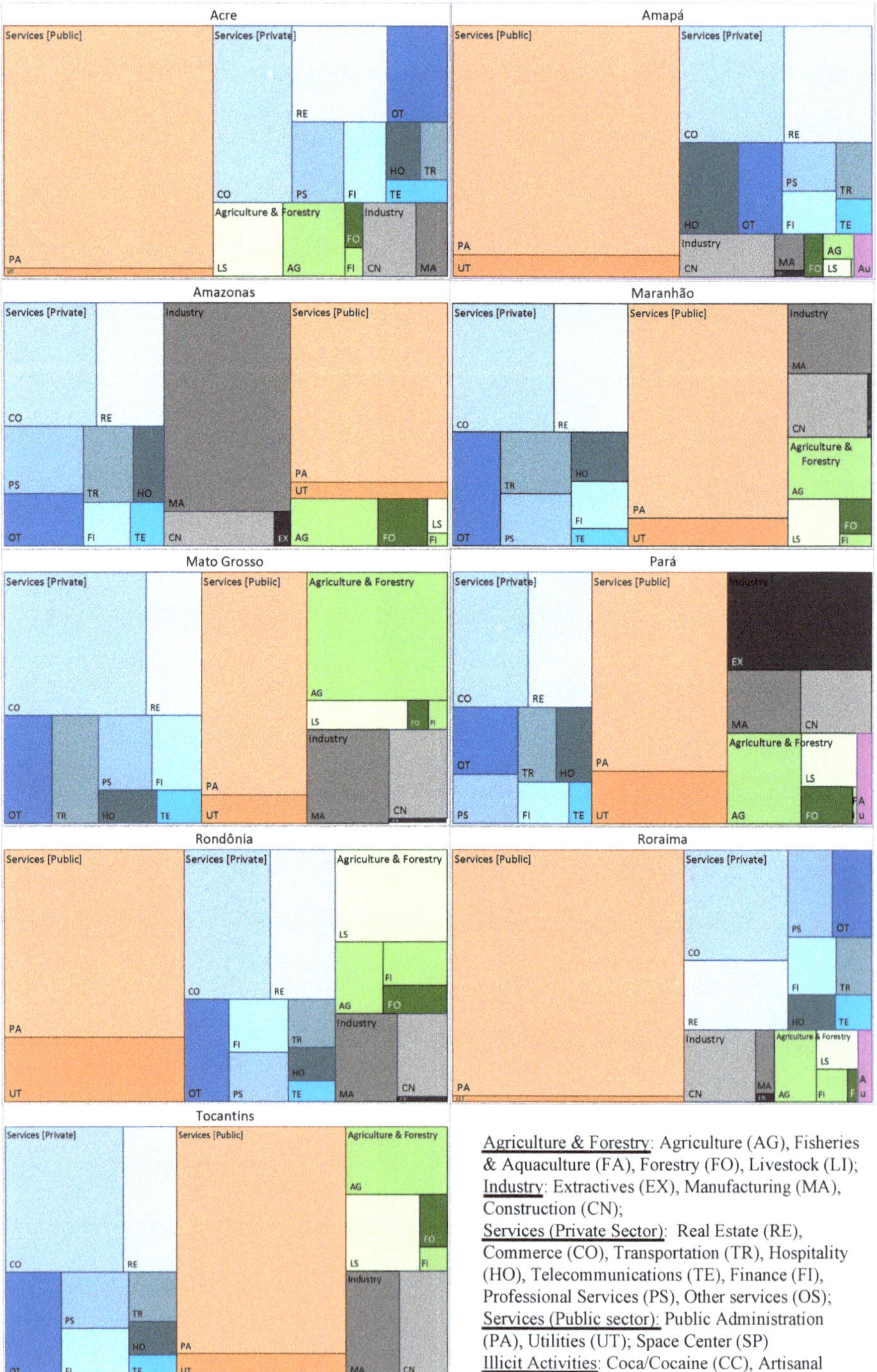

Agriculture & Forestry: Agriculture (AG), Fisheries & Aquaculture (FA), Forestry (FO), Livestock (LI); Industry: Extractives (EX), Manufacturing (MA), Construction (CN); Services (Private Sector): Real Estate (RE), Commerce (CO), Transportation (TR), Hospitality (HO), Telecommunications (TE), Finance (FI), Professional Services (PS), Other services (OS); Services (Public sector): Public Administration (PA), Utilities (UT); Space Center (SP) Illicit Activities: Coca/Cocaine (CC), Artisanal Gold (Au)

Bibliography

Angelsen, A., C. Martius, V. De Sy, A.E. Duchelle, A.M. Larson and T.T. Pham. 2018. *Transforming REDD+: Lessons and New Directions*. Bogor, Indonesia: Center for International Forestry Research (CIFOR).

Boucher, D., P. Elias, C. May-Tobin, K. Lininger, S. Roquemore and E. Saxon. 2011. *Root of the Problem: What's Driving Tropical Deforestation Today?* Cambridge, MA: Union of Concerned Scientists.

Busch, J. and K. Ferretti-Gallon. 2017. 'What drives deforestation and what stops it? A meta-analysis'. *Review of Environmental Economics and Policy* **11** (1): 3–23.

Fay, M., L. Alberto Andres, C. Fox, U. Narloch, S. Straub and M. Slawson. 2017. *Rethinking Infrastructure in Latin America and the Caribbean: Spending Better to Achieve More*. The World Bank.

Geist, H.J. and E.F. Lambin. 2001. What drives tropical deforestation. *LUCC Report series*, 4.

Hemming, J. 1987. *Amazon Frontier: The Defeat of the Brazilian Indians*. Cambridge, MA: Harvard University Press.

HSCCC – House Select Committee on the Climate Crisis (1 Jul. 2020) *The Solving the Climate Crisis: The Congressional Action Plan for Clean Economy and a Healthy, Resilient and Just America*.

IFC – International Finance Corporation. 2012. *IFC Performance Standards on Environmental and Social Sustainability*. International Finance Corporation, The World Bank Group

Killeen T.J., 2007. *A Perfect Storm in Amazon Wilderness, Conservation and Development in the Context of the Initiative for the Integration of the Regional Infrastructure of South America* (IIRSA). Washington, D.C.: Conservation International.

Lovejoy, Thomas E. and Carlos Nobre. 2018. 'Amazon tipping point'. *Science Advances* **4** (2): eaat2340.

Shelton, D., A. Vaz, B. Huertas-Castillo, C. Camacho Nassar, L.J. Bello C. Colleoni, J. Proaño and D. Mahecha (eds). 2013. *Indigenous Peoples in Voluntary Isolation and Initial Contact*. Copenhagen and Madrid: International Work Group for Indigenous Affairs (IWGIA) and Instituto de Promoción Estudios Sociales (IPES).

Sierra Praeli, Y. (27 Oct. 2020) 'Menonitas en Perú: fiscalías de Loreto y Ucayali investigan deforestación de 2500 hectáreas en la Amazonía', *Mongabay LATAM* https://es.mongabay.com/2020/10/menonitas-peru-investigacion-deforestacion-amazonia/

Simonet, G., A. Agrawal, S. Atmadja, F. Bénédet, M. Cromberg, C. de Perthuis, D. Haggard, N. Jansen, A. Karsenty, W. Liang, A. Morel, P. Newton, A-M. Sales, A. Satwika, B. Schaap, C. Seyller ,V. Silviana and G. Vaillant. 2020. *ID-RECCO, International Database on REDD+ Projects and Programs: Linking Economics, Carbon and Communities*. Version 4.0. http://www.reddprojectsdatabase.org

Sonter, L.J., D. Herrera, D.J.Barrett, G.L. Galford, C.J. Moran and B.S. Soares-Filho. 2017. 'Mining drives extensive deforestation in the Brazilian Amazon'. *Nature Communications* **8** (1): 1–7.

Steward, J.H. (ed.). 1946–1949. Handbook of South American Indians, *Smithsonian Institution, Bureau of American Ethnology,* Bulletin 143, Washington DC, US Government Printing Office.

Vuletin, G.J. 2008. *Measuring the Informal Economy in Latin America and the Caribbean,* International Monetary Fund, IMF Working Paper WP/08/102.

West, T.A., J. Börner and P.M. Fearnside. 2019. 'Climatic benefits from the 2006–2017 avoided deforestation in Amazonian Brazil'. *Frontiers in Forests and Global Change* 2: 1–11.

Notes to Chapter 1

1. Killeen T.J., 2007. *A Perfect Storm in Amazon Wilderness, Conservation and Development in the Context of the Initiative for the Integration of the Regional Infrastructure of South America* (IIRSA). Washington, D.C.: Conservation International.
2. Lovejoy, Thomas E. and Carlos Nobre. 2018. 'Amazon tipping point'. *Science Advances* **4** (2): eaat2340.
3. Geist, H.J. and E.F. Lambin. 2001. What drives tropical deforestation. *LUCC Report series*, 4, p. 116.
4. Boucher, D., P. Elias, C. May-Tobin, K. Lininger, S. Roquemore and E. Saxon. 2011. *Root of the Problem: What's Driving Tropical Deforestation Today?* Cambridge, MA: Union of Concerned Scientists.
5. IFC – International Finance Corporation. 2012. *IFC Performance Standards on Environmental and Social Sustainability*. International Finance Corporation, The World Bank Group. https://www.ifc.org/wps/wcm/connect/topics_ext_content/ifc_external_corporate_site/sustainability-at-ifc/publications/publications_handbook_pps
6. Busch, J. and K. Ferretti-Gallon. 2017. 'What drives deforestation and what stops it? A meta-analysis'. *Review of Environmental Economics and Policy* **11** (1): 3–23.
7. Khan Academy (9 June 2020) The limitations of GDP, AP®/College Macroeconomics, https://www.khanacademy.org/economics-finance-domain/ap-macroeconomics
8. Vuletin, G.J. 2008. *Measuring the Informal Economy in Latin America and the Caribbean*, International Monetary Fund, IMF Working Paper WP/08/102.
9. IBGE – Instituto Brasileiro de Geografia e Estatistica. 2017. Censo Agropecuario 2017, Tabela 6780 https://sidra.ibge.gov.br/tabela/6780
10. IBGE – Instituto Brasileiro de Geografia e Estatistica, Sistema de Contas Regionais – SCR. 2018. https://www.ibge.gov.br/estatisticas/economicas/contas-nacionais/2021-np-contas-regionais-do-brasil/9054-contas-regionais-do-brasil.html?=&t=resultados
11. Fay, M., L. Alberto Andres, C. Fox, U. Narloch, S. Straub and M. Slawson. 2017. *Rethinking Infrastructure in Latin America and the Caribbean: Spending Better to Achieve More.* The World Bank.
12. Sonter, L.J., D. Herrera, D.J.Barrett, G.L. Galford, C.J. Moran and B.S. Soares-Filho. 2017. 'Mining drives extensive deforestation in the Brazilian Amazon'. *Nature Communications* **8** (1): 1–7.
13. Sierra Praeli, Y. (27 Oct. 2020) 'Menonitas en Perú: fiscalías de Loreto y Ucayali investigan deforestación de 2500 hectáreas en la Amazonía', Mongabay.com
14. Hemming, J. 1987. *Amazon Frontier: The Defeat of the Brazilian Indians*. Cambridge, MA: Harvard University Press.
15. Shelton, D., A. Vaz, B. Huertas-Castillo, C. Camacho Nassar, L.J. Bello C. Colleoni, J. Proaño and D. Mahecha (eds). 2013. *Indigenous Peoples in Voluntary Isolation and Initial Contact*. Copenhagen and Madrid: International Work Group for Indigenous Affairs (IWGIA) and Instituto de Promoción Estudios Sociales (IPES).

16. Steward, J.H. (ed.). 1946–1949. *Handbook of South American Indians, Smithsonian Institution, Bureau of American Ethnology,* Bulletin 143. Washington DC: US Government Printing Office

17. Ethnologue.com. Languages of the World. https://www.ethnologue.com/country/GY/maps (28 Dec 2019).

18. West, T.A., J. Börner and P.M. Fearnside. 2019. 'Climatic benefits from the 2006–2017 avoided deforestation in Amazonian Brazil'. *Frontiers in Forests and Global Change* 2: 1–11.

19. Angelsen, A., C. Martius, V. De Sy, A.E. Duchelle, A.M. Larson and T.T. Pham. 2018. *Transforming REDD+: Lessons and New Directions.* Bogor, Indonesia: Center for International Forestry Research (CIFOR).

20. *The Guardian* (16 Aug. 2019) 'Norway halts Amazon fund donation in dispute with Brazil', https://www.theguardian.com/world/2019/aug/16/norway-halts-amazon-fund-donation-dispute-brazil-deforestation-jair-bolsonaro

21. Simonet, G., A. Agrawal, S. Atmadja, F. Bénédet, M. Cromberg, C. de Perthuis, D. Haggard, N. Jansen, A. Karsenty, W. Liang, A. Morel, P. Newton, A-M. Sales, A. Satwika, B. Schaap, C. Seyller ,V. Silviana and G. Vaillant. 2020. *ID-RECCO, International Database on REDD+ Projects and Programs: Linking Economics, Carbon and Communities.* Version 4.0. http://www.reddprojectsdatabase.org

22. HSCCC – House Select Committee on the Climate Crisis (1 July 2020) *The Solving the Climate Crisis: The Congressional Action Plan for Clean Economy and a Healthy, Resilient and Just America.* https://climatecrisis.house.gov/report

23. CLC – Climate Leadership Council (17 Jan. 2019) Economists' Statement on Carbon Dividends, as appeared in The Wall Street journal https://clcouncil.org/economists-statement/

24. ProPublica (22 May 2019) An even more Inconvenient Truth. https://features.propublica.org/brazil-carbon-offsets/inconvenient-truth-carbon-credits-dont-work-deforestation-redd-acre-cambodia/

25. Information sources for Table 1.2. Original forest cover and deforestation in the Amazonian jurisdictions:

Bolivia: Deforestación en el Estado Plurinacional de Bolivia. 2016-2017, (10 Jan 2021) Autoridad de Fiscalización y Control Social de Bosques y Tierra, Ministro de Medio Ambiente y Agua, http://www.abt.gob.bo/images/stories/Transparencia/InformesAnuales/memorias-2016-2017/Memoria_Deforestacion_2016_2017_opt.pdf

Brazil: Instituto Nacional de Pesquisas Espaciais. (10 Jan 2021) Coordenação Geral de Observação da Terra. Programa de Monitoramento de Amazônia e demais Biomas. Desmatamento – Amazônia Legal – Disponível em: http://terrabrasilis.dpi.inpe.br/downloads/

Colombia: Sistema de Monitoreo de Bosqiues y Carbonio (10 Jan 2021) Reportes de Bosque Natural http://smbyc.ideam.gov.co/Monitoreo-BC-WEB/reg/indexLogOn.jsp

Ecuador: Ministerio del Ambiente. 2012. Línea Base de Deforestación del Ecuador Continental, Quito-Ecuador, http://sociobosque.ambiente.gob.ec/files/Folleto%20mapa-parte1.pdf

Guyana: Guyana Forestry Commission 2017. Guyana REDD+ Monitoring Reporting & Verification System (MRVS), Year 7 Summary Report, https://

www.forestry.gov.gy/wp-content/uploads/2018/11/MRVS-Summary-Report-Year-7_November-2018_Final.pdf

Peru: Geobosques (10 Jan 2021) Plataforma de Monitoreo de Cambios sobre la Cobertura de los Bosques, http://geobosques.minam.gob.pe/geobosque/view/index.php

Suriname: GONINI (10 Jan 2021) , National Land Monitoring System of Suriname http://gonini.org/

Global Land Analysis and Discovery – GLAD (10 Jan 2021) Laboratory in the Department of Geographical Sciences, https://glad.umd.edu/gladmaps/globalmap.php

RAISG, 2015. Deforestation in the Amazonia (1970-2013). 48 pages, https://www.raisg.socioambiental.org

26. The hydrosheds numbered in Figure 1.4 are as following: (1) Amazonas Estuary & Coastal Drainages, (2) Amazon as Guiana Shield, (3) Amazonas High Andes, (5) Essequibo, (6) Guapore – Itenez, (7) Guianas, (8) Madeira Brazilian Shield, (9) Mamore – Madre de Dios Piedmont, (10) Orinoco Delta & Coastal Drainages, (14) Orinoco High Andes, (13) Orinoco Llanos, (12) Orinoco Piedmont, (15) Rio Negro, (16) Tapajos – Juruena, (17) Tocantins – Araguaia, (18) Ucayali – Urubamba Piedmont, (19) Western Amazon Piedmont, (20) Xingu; source., B. Lehner and G. Grill. 2013. Hydrological Processes 27 (15): 2171–2186; Land cover data – ESA. Land Cover CCI Product User Guide Version 2. Tech. Rep. 2017.

27. Information Sources for Figure 1.5 The GDP of the Pan Amazon stratified by political jurisdiction and sector.

Bolivia: Ministerio de Desarrollo Productivo y Economía Plural (MDPyEP) (10 Jan 2021) Bolivia: Producto Interno Bruto, Según Actividad Económica, https://siip.produccion.gob.bo/repSIIP2/formPib.php#!

Brazil: IBGE, Instituto Brasileiro de Geografia e Estatistica (10 Jan 2021) Sistema de Contas Regionais – SCR,. https://www.ibge.gov.br/estatisticas/economicas/contas-nacionais/2021-np-contas-regionais-do-brasil/

Colombia: DANE - Dirección Nacional de Estadística (10 Jan 2021) Estadísticas por tema, Cuentas nacionales, Cuentas nacionales departamentales: PIB por departamento, https://www.dane.gov.co/index.php/estadisticas-por-tema/cuentas-nacionales/cuentas-nacionales-departamentales

Ecuador, Banco Central del Ecuador (10 Jan 2021) Producto Interno Bruto, Cuentas Provinciales 2007–2019, https://contenido.bce.fin.ec/documentos/Estadisticas/SectorReal/CuentasProvinciales/Indice.htm

French Guiana: IDEOm Institut d'Émission des Départements d'Outre-Mer (10 Jan 2021), https://www.iedom.fr/IMG/pdf/tableau_de_bord_des_outre-mer___edition_2019_vf.pdf

Guyana: PSC, Private Sector Commission of Guyana (10 Jan 2021) http://psc.org.gy/resources.html#

Peru: Instituto Nacional de Estadística e Informática – INEI, (10 Jan 2021) Producto Bruto Interno Por Departamentos, https://www.inei.gob.pe/estadisticas/indice-tematico/producto-bruto-interno-por-departamentos-9089/

Suriname: Genetral Bureau of Statistics (10 Jan 2021) Gross domestic product, https://statistics-suriname.org/en/gross-domestic-product/

28. Information sources for Table 1.3 The ethnic populations of the Pan Amazon
Bolivia: *Centro de Documentación e Información Bolivia* – CEDIB (10 Jan. 2021); https://www.cedib.org/wp-content/uploads/2013/08/Tabla-Poblacion-Indigena1.pdf ;
Brazil: FUNAI, *Distribuição Espacial da População Indígena*. 2018. Distribuição Espacial da População Indígena, http://www.funai.gov.br/arquivos/conteudo/ascom/2013/img/12-Dez/encarte_censo_indigena_02%20B.pdf; *Instituto Socioambiental, Quadro geral dos povos* (10 Jan 2021) Quadro geral dos povos, https://pib.socioambiental.org/pt/Quadro_Geral_dos_Povos
Colombia: Organización Naciomal de Indígena de Colombia – ONIC (10 Jan. 2021) Pueblos Indígenas de Colombia, https://www.onic.org.co/pueblos
Ecuador: El Instituto Nacional de Estadística y Censos – INEC (10 Jan. 2021) Población que se autoidentificó indígena, según provincia de empadronamiento, nacionalidad o pueblo indígena al que pertenece y sexo, CENSO DE POBLACIÓN Y VIVIENDA (CPV-2010, https://www.ecuadorencifras.gob.ec/?s=indigena
Guyana. Ministry of Indigenous Affairs (10 Jan. 2021) Amerindian Land Titling (Alt) Project, https://moipa.gov.gy/indigenous-villages/
Peru: Ministerio de Cultura, (10 Jan. 2021) Base de datos de Pueblos Indígenas u Originarios – BDPI, Lista de pueblos indígenas u originarios; https://bdpi.cultura.gob.pe/pueblos-indigenas
Venezuela: *Censo Nacional 2001: Pueblos Indígenas según Declaración censos nacionales (from Allais, M.L. 2004. Población Indígena De Venezuela Según Los Censos Oficiales , II Encuentro Nacional de Demógrafos*, https://www.guao.org/sites/default/files/biblioteca/La%20poblaci%C3%B3n%20ind%C3%ADgena%20de%20Venezuela%20seg%C3%BAn%20los%20censos%20nacionales.pdf
Ethnologue.com. Languages of the World, 23rd Edition; https://www.ethnologue.com/
INWGIA -International Work Group for Indigenous Affairs, https://www.iwgia.org/en/indigenous-world.html
RAISG - *Rede Amazônica de Informação Socioambiental*, Cartographic Data, https://www.amazoniasocioambiental.org/en/maps/#!/download_

Chapter 2.

Infrastructure Defines the Future

The Merriam-Webster dictionary defines infrastructure as the 'underlying structure' of a country – specifically, the physical installations needed to ensure that its economy functions for the benefit of society. Modern infrastructure is made of steel and concrete and is ubiquitous in an advanced economy; typically, it is taken for granted by the people who depend upon it for their livelihoods. People living in emerging economies and developing countries do not suffer from this underappreciation of the value of infrastructure and, typically, are strong proponents of investing in it.

Conspicuous infrastructure assets include roads, bridges, railways, airports, ports, dams, power plants, energy grids, information networks, and water and sewer systems. Equally important are the physical assets that support key social services, such as schools, clinics, hospitals and recreational facilities. Most are built by the state, although some may be operated by private companies granted concessions by governments; quite a few are privately owned. Infrastructure assets are a perfect example of a long-term investment: they require a large initial investment in financial capital and pay dividends in the form of revenues and increased economic activity over decades or even centuries.*

Most infrastructure assets in the Pan Amazon are the product of long-term investment strategies formulated by governments at five-to-ten-year intervals. Regardless of the periodic shifts reflecting societal consensus and electoral cycles that have occurred over the last several decades, two themes have featured prominently in all the plans, programmes and projects: economic development and regional integration.

* Modern examples of infrastructure assets that have operated for more than a century include the Suez and Panama canals, the rail networks of North America and Europe; historical examples include the roads of the Roman Empire, the irrigation systems of Mesopotamia and the flood control levees of the Yellow River in China.

doi: 10.3197/9781912186228.ch02

Highway E-20 adjacent to the Río Napo in Amazonian Ecuador.

In Brazil, the modern era of infrastructure investment began in the 1970s with the *Programa de Integração Nacional* (PIN) that kicked off the construction of the highway network that has transformed the Southern Amazon. This was followed in the 1990s by the federal government diversifying its investment portfolio to include hydropower, waterways and railroads within priority geographies known as *Eixos Nacionais de Integração e Desenvolvimento* (ENID).* In the 2000s, infrastructure investments were at the core of the *Programa de Aceleração do Crescimento* (Growth Acceleration Program; PAC), which focused on the energy sector and included several mega-scale hydropower projects in the Amazon (see below).[1]

All of the Andean republics organised similar programmes that made highway systems a national priority, but some of their most important investments were in pipelines essential for the exploitation of hydrocarbon reserves that had been discovered in their Amazonian provinces. Historical aspirations and a shared cultural heritage motivated these nations' governments to create the *Comunidad Andina de Naciones* (CAN), a trading block that included within its founding principles investment in trans-frontier infrastructure assets. One of the ambitious early proposals was the Carretera Marginal de la Selva, a highway similar to the Pan Amazonian highway that would integrate their Amazonian provinces.[2] This concept was operationalised and expanded in the early 2000s when all the nations of South America came together to create the *Iniciativa para la Integración de la Infraestructura Regional Suramericana* (IIRSA)† (Figure 2.1).

* There were nine *Eixo de Integração* (Axes of Integration), of which four were located within the Brazilian Amazon: (1) Madeira – Amazonas, (2) Oeste, (3) Araguaia – Tocantins, (4) Arco Norte. The investment framework was part of Avança Brasil, which was the centre piece of the administration of Fernando Henrique Cardoso.

† IIRSA (Initiative for the Integration of the Regional Infrastructure of South America) was originally coordinated by the Interamerican Development Bank

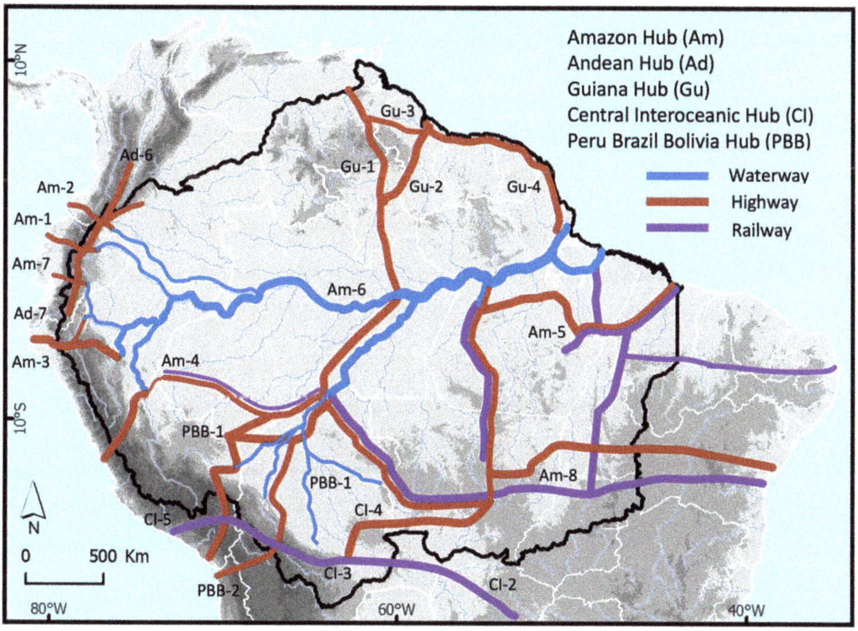

Figure 2.1: The Integration of the Regional Infrastructure of South America (IIRSA) initiative is a master plan for priority investments that are organised into hubs (4), groups (22) and projects (187). A major goal is to create multi-modal transportation corridors based on waterways, highways and railroads. The total projected IIRSA-sponsored investment in the Pan Amazon sums to $US 84.4 billion. See Annex 2.1.

Multilateral financial institutions, such as the World Bank, IDB and CAF* have played an essential role in financing the infrastructure that has transformed the human-modified landscapes of the Amazon (see Chapter 1). Although the resources they have deployed are limited by their pool of available capital, their participation has motivated governments to allocate greater capital to infrastructure and, more importantly, established a framework to leverage public resources with private capital. In addition,

(IDB), but after 2010 that responsibility was passed to the newly created *Unión de Naciones Suramericanas* (UNASUR), which governs the initiative via the *Consejo Suramericano de Infraestructura y Planeamiento* (South American Council for Infrastructure and Planning; COSIPLAN). The two acronyms (IIRSA and COSIPLAN) are synonyms: http://www.iirsa.org/Page/Detail?menuItemId=45

* IDB is the Interamerican Development Bank; the CAF is the Development Bank of Latin America, but retains its acronym from its previous name, *the Corporación Andina de Fomento.*

multilateral agencies finance – and influence the content of – strategy documents that guide long-term infrastructure investment; consequently, they share responsibility for both positive and negative outcomes associated with the infrastructure systems that have transformed the Pan Amazon.

Starting in about 2005, financial institutions from China began to play an important role in infrastructure development, typically by subsidising Chinese companies in the construction sector and, more recently, as underwriters for the acquisition of assets auctioned off by governments and corporations following the corruption scandals of the mid-2010s (see Chapter 6). Financial support is now organised under the banner of China's global policy initiative known as the Road and Belt Initiative.

The other major players in the field of infrastructure finance were the national development banks, semi-autonomous entities that leverage state resources with private capital to promote the participation of corporate actors and facilitate investment by sub-national jurisdictions. The most prominent of these is the *Banco de Desarrollo de Brazil* (BNDES), which has a long history of financing domestic infrastructure investments but expanded its activities to subsidise the operations of Brazilian construction companies competing for contracts tendered by the Andean republics. Investment in infrastructure reached a historical peak during the commodity export boom between 2005 and 2015, a period that provided the Pan American nations with unprecedented financial resources.

Infrastructure in the Pan Amazon has a bad reputation.[3] The largest projects have led to massive deforestation and hydrological degradation; many – if not most – have been beset by accusations of corruption. Nonetheless, investment in infrastructure has benefited millions of Amazonian citizens, particularly in urban areas that now house more than fifty per cent of the region's population. The region is set to begin another cycle of investment and development, in part because governments are once again seeking to expand the reach of conventional economic activities into the region but also because external events, particularly the COVID pandemic, are creating momentum within financial agencies to stimulate infrastructure investments as a means of restoring economic growth following the recession of 2020.[4]

Roads: Primary Vectors of Deforestation

It all starts with a road. In Brazil, the federal government commissioned the construction of mule trails and telegraph lines to link their coastal cities with long-established settlements on remote Amazonian tributaries.[*] In

* The most famous Amazonian explorer of the 20th century was Candido Rondon who established telegraph lines to Cuiabá and Porto Velho between 1900 and

the Andes, communication between the highlands and the lowlands has occurred over millennia via trails that traverse the foothills using routes dictated by topography. Most of these early roads had little impact on settlement and were associated with only a limited amount of deforestation, but most modern highways trace these routes into the wilderness.

Deforestation occurs when a government sends a clear signal that it is investing in a major trunk highway; objectives vary, but have included strategies to integrate an isolated population centre, open a region to settlement and create access to a valuable mineral asset. Settlers move into a region only when there is a commitment to keeping the road accessible on a permanent basis. There have been exceptions, particularly in Brazil in the 1980s, when the promise of future pavement did not materialise, but in general people will invest time and money in clearing the forest only when they believe they can transport their production to market. The examples are abundant and self-evident. All but six of the sixty human-modified landscapes defined in Chapter 1 are organised around a major trunk highway (Figure 2.2 and Annex 2.2).*

The existence and quality of secondary road networks is more important than trunk highways for determining the spatial extent and intensity of land use within a frontier landscape.[5] Access roads change a linear deforestation vector into a two-dimensional front that can trigger a geometric expansion in the deforestation rate. It is not sufficient just to have a secondary road network, however; it too must be open year-round in order to ensure that crops do not rot in the field. A functional secondary road network mandates investment in bridges, culverts and embankments; even more important is a budget to maintain these fragile assets in a region characterised by high seasonal rainfall.

The economic output of deforested lands is very much dependent on the quality of secondary roads because they connect ranches, farms, and plantations with the industrial infrastructure essential to agricultural supply chains, particularly grain silos, rice mills, palm-oil extraction refineries and beef-packing plants (see Chapter 3).[6] An overemphasis on trunk highways at the expense of investment in secondary roads in consolidated frontiers is a misallocation of public investment that results in sub-optimal economic growth.[7]

1915; he was followed by the Villa Boas brothers who cut trails across the Serra de Roncador to establish remote air strips across the Planalto de Mato Grosso during the 1940s (Hemming 2003).

* Those that are not closely linked to a highway use the Amazon River as a transportation corridor (HML #1 & 3) or are coca production landscapes where settlers are deliberately seeking isolation to avoid conflicts with the authorities (HML 35a & 35b).

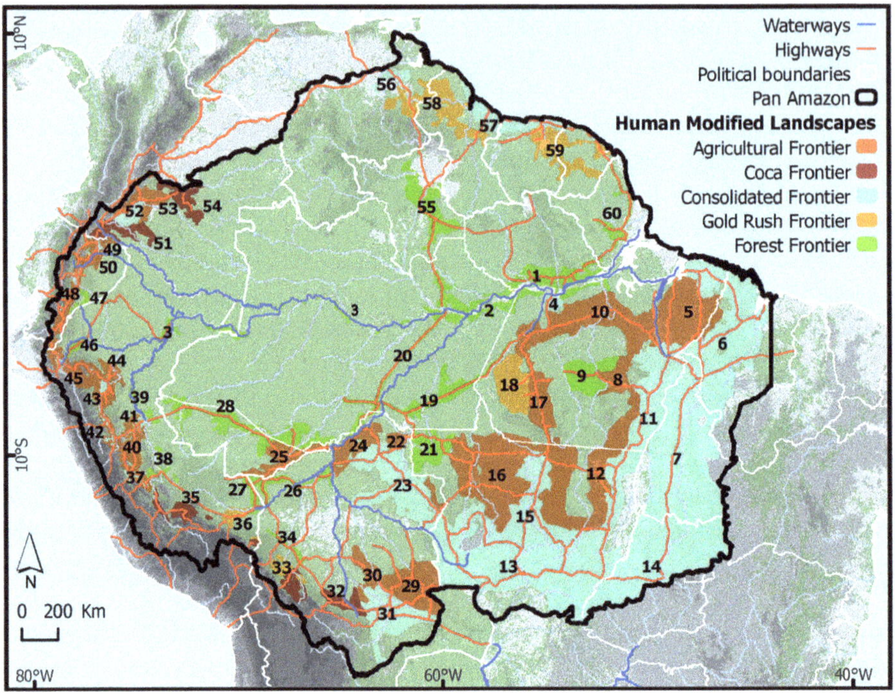

Figure 2.2: Human-modified landscapes are usually associated with one or more highway corridors (see Annex 2.2 for details).

Amazon River Corridor: (1) North Bank Amazon River, (2) South Bank Amazon River, (3) Solimoes, (4) Urban: Macapá, Manaus, Santarem.

Southern Amazon: (5) Belem – Marabá, (6) Carajás – São Luis do Maranhão, (7) Transbrasiliana, (8) São Felix do Xingu East, (9) São Felix do Xingu West, (10) Eastern Tranzamazônica, (11) Vale do Araguaia, (12) Upper Xingu, (13) Cuiabá – Roncador, (14) Alto Araguaia, (15) Sinop – Zapasal, (16) North Mato Grosso, (17) Novo Progresso, (18) Tapajos Crepori, (19) Western Tranzamazônica, (20) Novo Progresso / BR-319, (21) Northwest Mato Grosso, (22) East Rondônia, (23) Rondônia, (24) Interoceanico P Velho, (25) Interoceanico Acre, (26) Interoceanico Pando, (27) Interoceanico Made de Dios, (28) Cruzeiro do Sul, (29) Chiquitania, (30) Guayaros – Beni, (31) Santa Cruz.

Andean Amazon: (32) Chapare, (33) La Paz Yungas, (34) Yucumo – Ixiamas, (35) Peruvian Yungas , (36) Interoceanico Hueyepetu, (37) Selva Central, (38) Gran Pajonal – Atalaya, (39) Ucayali tributaries, (40) Ucayali – Huánuco – Pasco Piedmont, (41) Pucallpa, (42) Upper Huallaga, (43) Lower Huallaga, (44) Yurimaguas piedmont, (45) Marañon – Chachapoyas, (46) Saramiriza Piedmont, (47) Morona – Santiago, (48) Ecuador foothills, (49) Succumbios, (50) Orellana, (51) Putumayo – Caquetá, (52) Central Caquetá, (53) Caquetá – Macarena, (54) Macarena – Guaviare.

Guianan Amazon: (55) Roraima, (56) Arco Minero, (57) Coastal Guiana, (58) Greenstone Venezuela / Guyana, (59) Greenstone Suriname / French Guiana, (60) Amapá.

In aggregate, secondary roads require a greater investment than trunk highway systems, but the responsibility for building and maintaining these key transportation assets is almost always with under-funded local governments. Central governments have access to national budgets and international investment capital, which finance the construction of the trunk highway, but local governments must depend on limited revenues derived from local taxes or revenue transfers from the central government. Not infrequently, secondary roads are built by settlers out of necessity and constructed outside the framework of the environmental review and licensing system.

The Human-Modified Landscapes (HML) and the Brazilian highway network

The Southern Amazon has experienced massive deforestation, coupled with the degradation of soil and water resources.[8] The forest frontiers at the remote corners of the Brazilian highway system have remained isolated and impoverished, but the agricultural frontiers and consolidated landscapes of Maranhão, Mato Grosso, Pará, Rondônia and Tocantins are relatively prosperous. Their rural economies generate approximately $US 125 billion annually, representing about five per cent of the Brazilian economy.[9] That economic output is dependent upon the national and regional highway network and has created a strong constituency for highway development.

The oldest of the trunk highways in the Pan Amazon is the Rodovia Transbrasiliana (BR-010/BR-153), which was initiated in the 1960s simultaneously with establishment of the new federal capital at Brasilia. This north-south transportation corridor transects the upland landscapes between the Araguaía and Tocantins rivers and was the first permanent terrestrial link between Belem and southern Brazil (see Figure 2.3 and Figure 2.4). Its construction facilitated the expansion of the beef cattle industry from Minas Gerais into Goiás and Tocantins* (HML #7, #14) and was linked to northeastern Brazil by a pair of east-west highways (BR-222 and BR-226) that fostered the mass migration of rural poor into eastern Pará (see Chapter 6).

The steady improvement of these highways and their associated secondary and tertiary road networks coincided with the development of the hydropower facilities at Tucuruí on the Tocantins River (see below), the mining complex at Carajás and the metallurgical foundries in Marabá and São Luis do Maranhão (Figure 2.4). Simultaneously, the federal government created SUDAM,† an institution that managed a system of subsidies designed to promote agricultural development and the monetisation of the region's mineral resources (see Chapters 5 and 6). These policies succeeded in cre-

* Tocantins was separated from Goiás in 1988.

† *Superintendência do Desenvolvimento da Amazônia.*

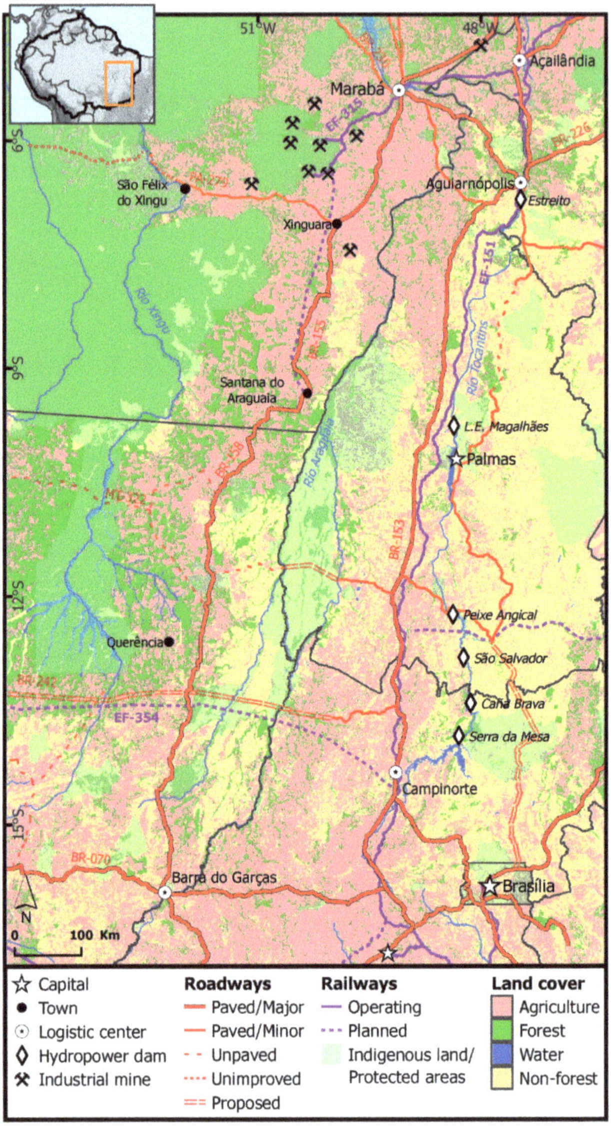

Figure 2.3: Two parallel highways built in the 1960s (BR-153) and 1990s (BR-158/155) opened the southeast Amazon for settlement. This area is now served by a railroad (EF-151) that was completed in 2015. Future planned development includes an extension of rail service into southeast Pará and the extension of the Tocantins Waterway from Marabá to near the border with Goiás.

ating wealth and the transformation of the regional landscape. By 2020, the Carajás -São Luis-Belem corridor had less than eighteen per cent remnant forest cover, the lowest proportional amount in the Pan Amazon (HML #6).* The landscape located south of Belem (HML #5) retained a greater forest area but is also home to the country's expanding palm oil industry and a numerous colonisation projects, where smallholder families pursue a combination of subsistence and market agriculture.

In the 1980s and 1990s, another north-south highway corridor (BR-155/BR-158) was constructed ~300 km to the west on the other side of the Araguaía River (Figure 2.3), which eventually connected the municipalities of Northeast Mato Grosso with their counterparts in Southeast Pará (HML: #11, #12). Deforestation has declined significantly since 2010 but remains relatively active on the frontier landscapes adjacent to the indigenous territories along the Río Xingu (see Chapter 11). The cultivation of soy and maize are displacing cattle ranching in Mato Grosso but still predominate in Pará (see Chapter 3).

Settlers moved into the landscapes of Central Pará following the construction of the PA-279, a regional highway that links Xinguara on BR-155 to São Felix do Xingu, once a small village on the river established during the rubber boom of the late nineteenth century (HML #8, #9). The landscapes west of the Rio Xingu are crisscrossed by unpaved roads that service a large area (occupied by large to medium-scale landholdings that have been incorporated into a multiple-use protected area (Área de Proteção Ambiental [APA] Triunfo do Xingu). Most of these properties, established during the land rush of the 1980s, have been characterised by slow but steady deforestation (Figure 2.5). The municipality of São Felix do Xingu has consistently ranked among the five Brazilian municipalities with the highest annual rate of deforestation in the Brazilian Amazon.[10,11]

One of the most economically dynamic regions in the Brazilian Amazon is synonymous with another highway project: The Cuiabá – Santarem Corridor. The social and economic forces that transformed the highway corridors east of the Rio Xingu are being replicated along BR-163, which links the prosperous farming landscapes of central Mato Grosso with the grain terminals and ports on the Tapajós and Amazon rivers (Figure 2.5). This highway was established in the 1970s during the *Programa de Integração Nacional* (PIN), but the northern sector soon fell into a state of disrepair. For approximately 25 years, it was a typical frontier landscape dominated by logging companies that could organise their operations by transporting timber during the dry season when the road was passable (HML #17).

---

* The reconstruction of BR-222 was included within the IIRSA highway portfolio: IIRSA, Amazon Hub, Group 5, - Conexión entre la Cuenca Amazónica y el Nordeste Septentrional de Brasil: AMA84 ($180 million): http://www.iirsa.org/proyectos/detalle_proyecto.aspx?h=1387

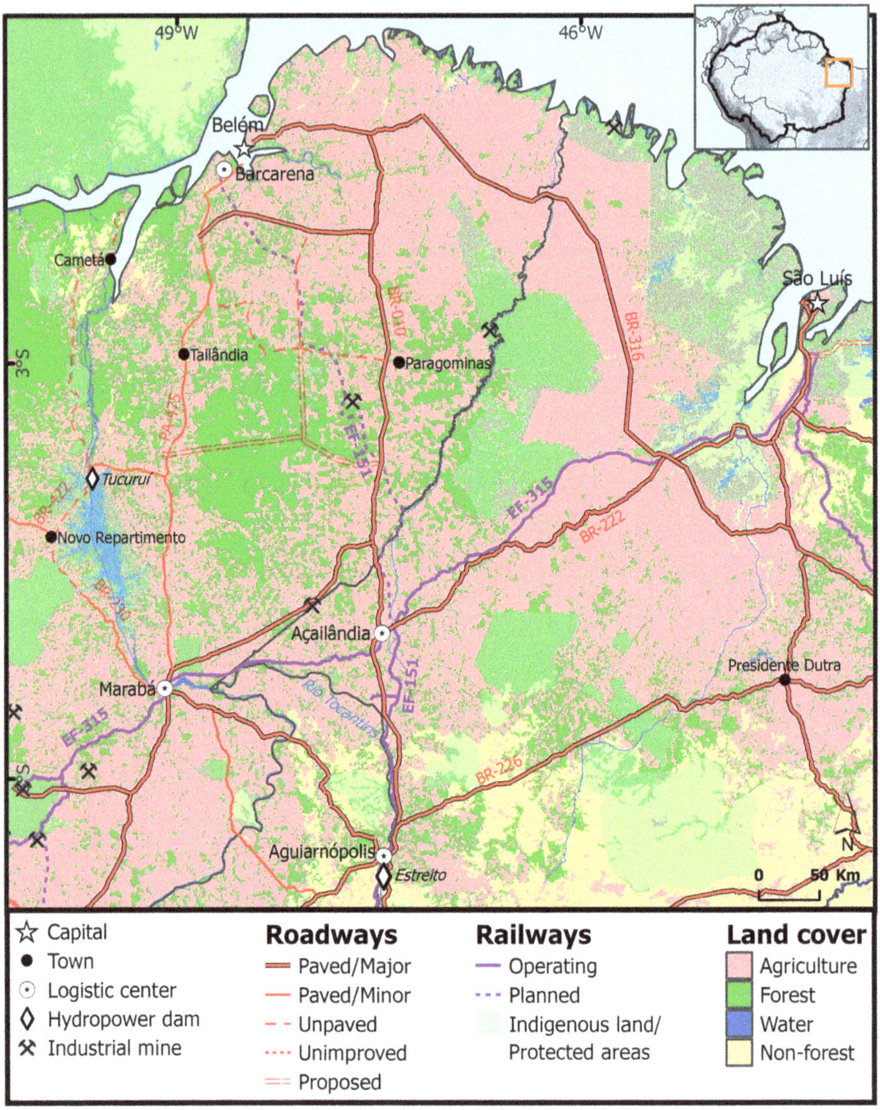

Figure 2.4: The forest remnants between the major highway corridors in eastern Pará and Maranhão are increasingly fragmented; most (but not all) are associated with indigenous territories. The most important infrastructure asset in the region is the rail line constructed between 1980 and 1985 between the mining complex at Carajás and the port of Itaquí near São Luis do Maranhão. In the near future, the North-South rail line (EF-151) will be extended to Barcarena and the Tocantins Waterway will be extended to Marabá.

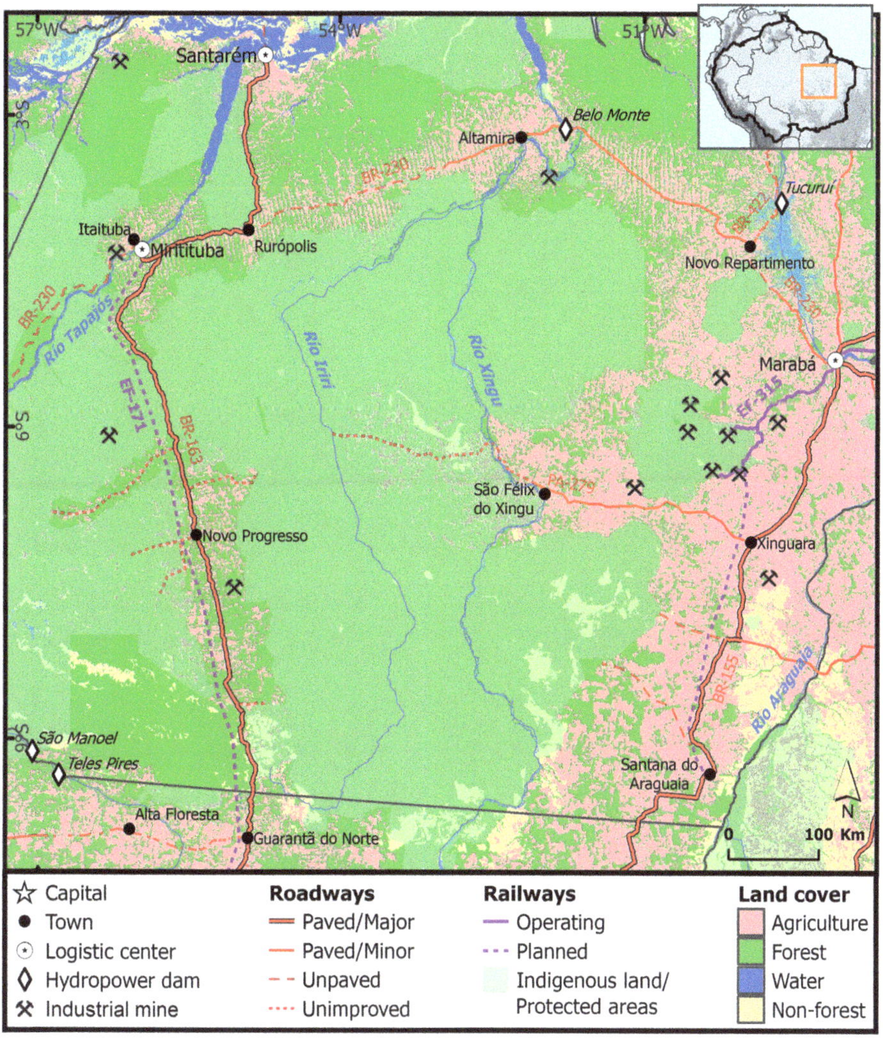

Figure 2.5: The ongoing improvement of the trunk highways in Pará (BR-163, BR-230) will accelerate deforestation on landscapes that are transitioning from forest frontiers to agricultural frontiers. Planned expansion of railroads will connect the farmland of Mato Grosso with grain terminals on the Tapajós river at Miritituba (Ferrogrão/EF-271) and foster the expansion of intensive agriculture into southeast Pará (Ferrovia Paraense).

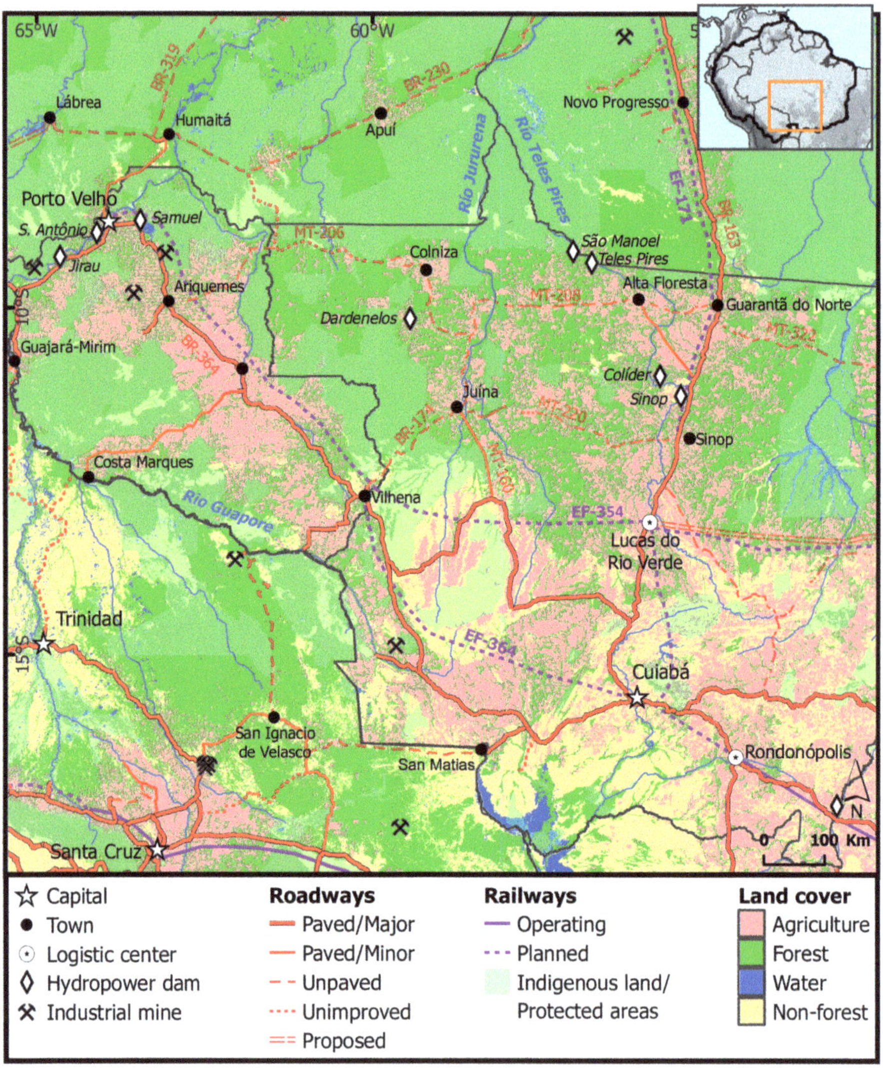

Figure 2.6: The agricultural economy of the Southern Amazon is highly dependent upon a network of paved highways and improved secondary roads. The North-South trunk highways in Brazil (BR-163 and BR-364) are major commercial arteries essential for grain exports. Regional highways in Mato Grosso (MT-208, MT-220, MT-322) are essential to that state's rural economy, as is the dense network of local roads in Rondônia. Rail expansion is planned to connect the region with southern ports via Rondonópolis and to Amazon ports via the BR-163 and BR-364 corridors. The proposed Tapajós waterway will require the construction of multiple dams on the Juruena or Teles Pires rivers.

Migrants from southern Brazil settled the southern sector (see Chapter 6), which was rapidly integrated into the national economy. Industrialised farmers occupied the landscapes with flat topography and well-drained soils ideal for the cultivation of soy (HML #15). Cattle ranchers occupied less fertile land in the hill country of central Mato Grosso (HML #13) and along the border with Pará (HML #16). The agricultural frontier expanded across Mato Gross via a gradually expanding network of state highways (Figure 2.6). Land was deeded to corporations that resold it to families organised into cooperatives or developed industrial-scale operations dedicated to farming or beef production. This is the heartland of Mato Grosso's agro-industrial complex, and it enjoys a well-maintained and extensive network of state and local highways; these support industrial infrastructure built by the private sector, including grain silos, crushing mills and animal production facilities (see Chapter 3). The secondary and tertiary road network has increased the value of rural real estate, while fostering the diversification of the rural economy. Not surprisingly, its inhabitants represent a powerful vested interest that lobbies for the improvement of road infrastructure within their state, but also for the federal highway system, which they view as essential for the growth of their production system. Like all investors, they seek to grow their economic system; as patriots, they view their production systems as a public good and a strategic national asset (see Chapter 6).

The transfer of public lands to the private sector also transferred about fifty million hectares of land that once contained more than 33 million hectares of forest, of which about half has been converted to agricultural production with the remainder distributed across tens of thousands of isolated forest fragments. Continuous forest is restricted largely to indigenous territories arranged as two north-south corridors: one along the Río Xingu and the other along the border with Rondônia. The last bit of public forest in the state is located in the northwest corner of the state (HML #21) where logging companies and land speculators are active along an unimproved road (MT-206/RO-205) between the INCRA* settlement of Colniza (Mato Grosso) and the city of Arequimes (Rondônia).

In the early 2000s, the federal government created an export corridor for the rapidly expanding soy industry of central Mato Grosso by improving the roadbed and bridges of the northern section of BR-163 (Figure 2.5).† The highway project, which would connect the croplands of central Mato

* *Instituto Nacional de Colonização e Reforma Agrária* – INCRAis the federal agency that oversees land reform initiatives and regulates land tenure for all rural properties in Brazil; see Ch. 4.

† The programme was originally conceived as a component within the *Programa de Acceleracião Económico* (PAC) and later included as a component of IIRSA: Amazon Hub, Group 5, Conexión entre la Cuenca Amazónica y el Nordeste Septentrional de Brasil: AMA33 Carretera Cuiabá–Santarém (BR-163/MT/PA)

Grosso with the grain terminals at Santarem, provoked an intense reaction from environmental advocates at a time when the country was experiencing a vigorous debate about the wisdom of Amazonian development.[12] The government responded by organising an ambitious environmental and social review (see Chapter 7), which led to the creation of several new protected areas and the recognition of indigenous land claims (see Chapters 11 and 12). Land speculators had already moved into the region, however, and created secondary roads penetrating landscapes on both sides of the highway, including one that facilitated access to the gold rush frontier in the upper reaches of the Crepori watershed (HML #18). Unlike the regional highways in Mato Grosso, these secondary roads do not appear on official maps, indicating that they were established without the participation of state planning agencies and appropriate environmental review.[13]

In 2016, the administration of Michel Temer* attempted to change the status of about half a million hectares in the Jamanxim National Forest,[14] a measure that would have granted *de facto* amnesty to the illegal appropriation of public lands. This led to a backlash from civil society organisations and the environmental prosecutor's office, who questioned the constitutionality of the executive order that authorised the modification of a protected area.[15] The government was forced to withdraw the measure in 2017 by a ruling from the Supreme Court, but neither the Bolsonaro administration nor the state authorities have interceded to combat illegal land grabbing on the landscapes surrounding BR-163 (see Chapter 4).[16]

In spite of R$ 1.5 billion expended on highway improvements between 2005 and 2015, a hundred-kilometre stretch of BR-163 remained impassable during the peak rainy season.[17] Poor road conditions caused massive traffic jams among the 3,000 trucks using BR-163 during the soy harvest. This untenable situation was exacerbated by road blockades organized by settlers seeking legal recognition for their landholdings.[18] In 2018, the federal government allocated an additional R$ 175 million in emergency funding to the Brazilian Army, which finished the paving in 2019.[19] From start to finish, it took twenty years to pave an 800-kilometre stretch of highway considered to be a vitally important strategic asset by the agro-industrial sector.

Commodity traders have responded to the improved road by building five grain terminals at Miritituba (Pará) on the east bank of the Tapajós River at what is essentially its highest navigable port. In 2020, the *Departamento*

($US 6.5 billion) and AMA34 Programa de manejo ambiental ($US 12 million). Source: http://www.iirsa.org/proyectos/Principal.aspx

* Michel Temer acceded to the Presidency following the impeachment of Dilma Rouseff, who was removed from office by a coalition of forces led by the Ruralist block in Congress who are advocates of agricultural expansion into the landscapes adjacent to BR-163 (see Ch. 6) (da Cunha et al. 2017).

The Cuiabá–Santarem corridor (BR-163) is a strategically important export corridor for grains cultivated in central Mato Grosso. It was upgraded from a poorly maintained road to a modern highway. Its construction was preceded by a strategic environmental evaluation and accompanied by a sustainable development programme. Nonetheless, the adjacent landscapes suffer from land grabbing and illegal deforestation, including within newly created protected areas.

Nacional de Infraestrutura de Transportes (DNIT)* initiated a tender process for a concession to administer BR-163 between Sinop (Mato Grosso) and Mirituba. The contract envisions an investment of an additional ~$US 600 million in highway improvements that will be financed by tolls levied on the approximately 6,000 trucks that are projected to use the highway.[20]

The most infamous highway in the Brazilian Amazon is Rodovia BR-364, which was constructed in Rondônia in the 1970s as part of a state-sponsored resettlement project supported by the World Bank (Figure 2.6).† The project triggered a wave of deforestation that was documented by newly available satellite imagery.[21] An independent review revealed that the environmental damage was compounded by social impacts that threatened indigenous communities and destined most settlers to a life of rural poverty. The resultant controversy catalysed the first serious debate about the conservation of the Amazon and the social impacts of conventional development paradigms (see Chapter 6).

In spite of the rough start, tens of thousands of smallholders eventually mastered the technological challenges of agricultural production in the Amazon. Although Rondônia is widely portrayed as a case study for misguided development policies, it also provides an example of a successful rural economy based on small family farms (HML #23). Key to that success was the creation of an extensive secondary road network that has been improved over several decades. The combination of a dense road grid and small property sizes led to the evolution of a rural landscape with an extremely low proportion of remnant forest (Figure 2.7). In central Rondônia, fifteen adjacent municipalities retain less than twenty per cent of their original forest cover and thirty have less than fifty per cent,[22] which is the approximate minimum amount allowed under the Forest Code of 2012 (see Chapter 7).

The other major highways carved out of the forest in the 1970s and 1980s are even more problematic. These include the eastern section of the Rodovia Transamazônica (BR-230), which starts at Marabá (Pará) on the Tocantins River and extends west for approximately 1,000 kilometres to Miritituba on the Tapajós River (HML #10). From there, the western section continues for an additional 1,000 kilometres through southern Amazonas state to the town of Humaitá on the Rio Madeira (HML #19). This trunk road was originally intended to integrate the three previously described transportation corridors (BR-155/158, BR-163, BR-364), but it was never

* The DNIT is a semi-autonomous agency within the Ministério da Infraestrutura, formerly the Ministério dos Transportes.

† The route largely followed the pre-existing track established by Candido Rondon in the 1900s when he built the telegraph line between Cuiabá and Porto Velho (Rohter 2019).

Source: Google Earth

Figure 2.7: A temporal comparison of deforestation associated with BR-364 in Rondônia (Left) and BR-230 in Pará (Right). Both landscapes were open to colonisation at approximately the same time, but BR-364 was paved, its producers were closer to urban markets, and they enjoyed greater extension support and better government services. When BR-230 is eventually fully paved, the forest remnants will most likely be reduced or lost, as they have been in Rondônia.

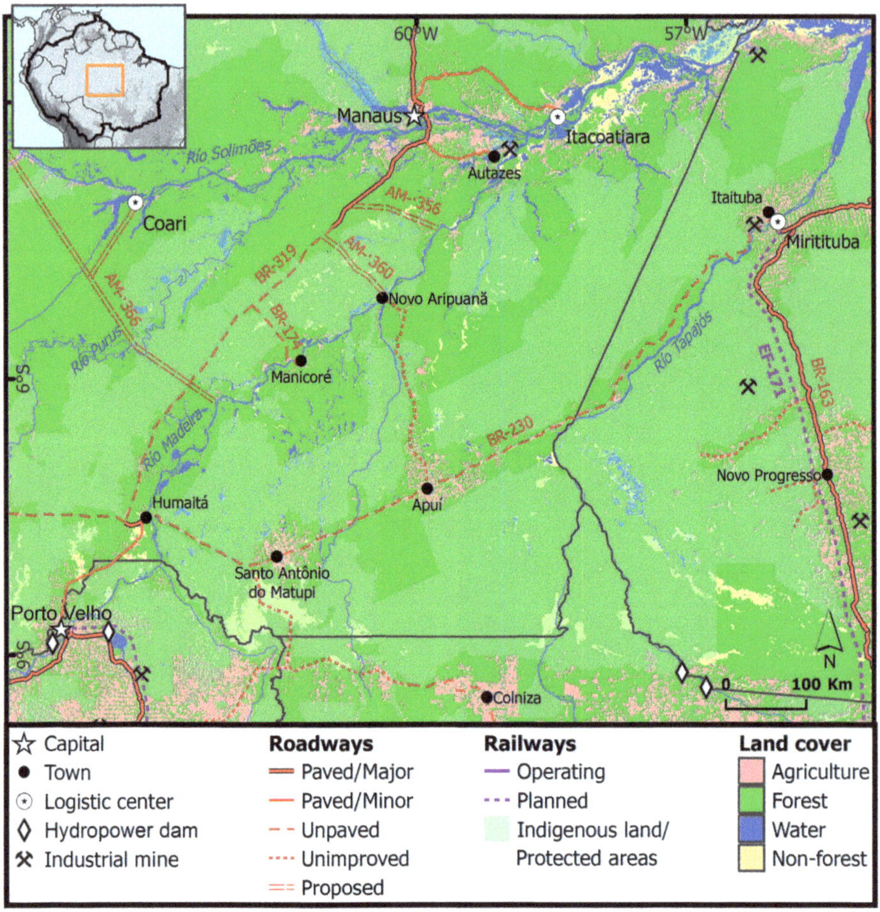

Figure 2.8: The frontier highways (BR-230 and BR-319) roughly parallel the Madeira Waterway. The improvement of existing and proposed secondary roads will accelerate deforestation, expand the footprint of agriculture and accentuate the isolation of a large block of primary forest in southeast Amazonas state.

paved, and its rapid deterioration soon left its settlers isolated and struggling to make a living (Figure 2.5 and Figure 2.8).

Land-use on the landscapes surrounding the Transamazônica in both Pará and Amazonas states is much less intensive when compared to BR-364 in Rondônia, even though all were colonised at approximately the same time and largely dedicated to beef cattle production. The difference, however, is likely to be transitory. Successive state governments have all made commitments to upgrade the highway, which is now paved between Miritituba and Rurópolis, where it overlaps with BR-163, and for another

350 kilometres between Altamira and Marabá (Figure 2.5).[23] Ongoing paving of the Transamazônica is included within the IIRSA portfolio of priority investments;* once the entire road is paved, the landscapes adjacent to the Transamazônica will almost certainly come to resemble the smallholder landscapes of Rondônia (Figure 2.7).

Other regions with a trunk highway but relatively low levels of deforestation include BR-174 between Manaus and Boa Vista (HML #4), where the presence of the Waimiri Atoari indigenous community has acted as an effective barrier to land grabbers (See Chapter 11).[24] The landscapes north of those indigenous territories in Roraima have been parcelled out to private landowners (HML #55) but have not transitioned into an agricultural frontier due, presumably, to their inherent isolation. Roraima has large extensions of natural savanna, which could transition into an agricultural frontier if political leaders succeed in their quest to replicate the agro-industrial development model exemplified by Mato Grosso.[25] An important component of their business model is the advantages conferred by the 750-kilometre paved highway (BR-174) between Boa Vista and the port of Manaus, which reduces the transportation cost of exporting soybeans and other grains.†

The historical deforestation rate in Acre has been relatively low, particularly along the western section of BR-364 between Rio Branco and Cruzeiro do Sul (HML #28). This 700-kilometre stretch of highway is currently unpaved for about 450 kilometres, but its completion has been a political priority for every state government for the last thirty years. During most of that period, successive administrations have promoted the sustainable use of forest resources, as exemplified by the agro-extractive reserves that both the state colonisation institution (INCRA) and the national protected area system (ICMBio) have sponsored.‡ Nonetheless, extensive landholdings have been distributed to small and medium-sized producers dedicated to cattle ranching, which contributes almost eight times more to Acre's GDP than the forest sector (see Chapter 1). Eventually, BR-364 will be paved in its entirety, and this will lead to increased deforestation along its margins and on the secondary roads that radiate out from a half-dozen small towns (Figure 2.9).

---

* IIRSA, Amazon Hub, Group 5; Conexión entre la cuenca amazónica y el nordeste septentrional de Brasil; AMA85, Marabá – Itaituba ($US 1.0 billion): http://www.iirsa.org/proyectos/detalle_proyecto.aspx?h=1388

† IIRSA, Guiana Hub; Group 1, Interconexión Venezuela – Brasil: GUY01, Rehabilitación de la Carretera Caracas – Manaos ($US 405): http://www.iirsa.org/proyectos/detalle_proyecto.aspx?h=175

‡ INCRA: *Instituto Nacional de Colonização e Reforma Agrária* distributes land to landless families via several land-use models, including the sustinabel use of forest resources. ICMBio: *Instituto Chico Mendes de Conservação da Biodiversidade* recognises the territorial rights of traditional people who depend on forest livelihoods (see Chs 4 and 12)

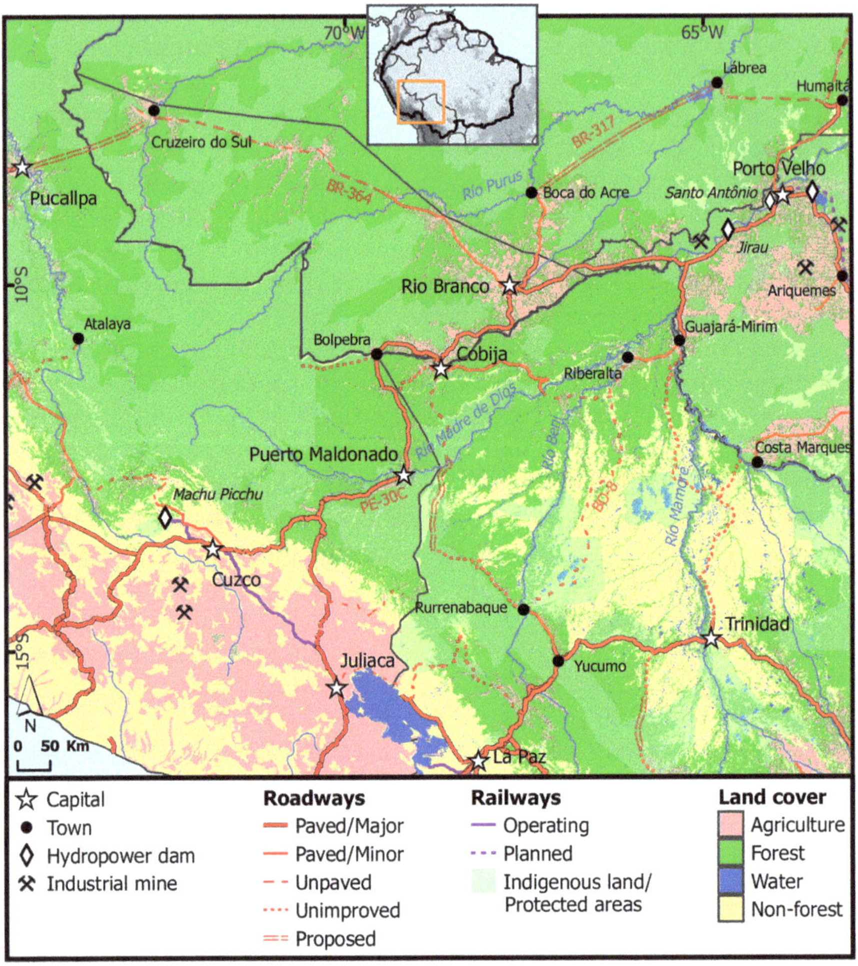

Figure 2.9: The Corridor Interoceanico is the quintessential IIRSA-sponsored initiative, with connections between the Amazonian river port at Porto Velho (Rondônia, Brazil), Cobija (Pando, Bolivia) and Puerto Maldonado (Madre de Dios), as well as to the Pacific coast of Peru and Chile. The proposed connection between Pucallpa and Cruzeiro do Sul would effectively isolate the forest ecosystems of the southwest Amazon from the central and northern Amazon.

Even more problematic is the proposal to extend BR-364 to the Peruvian border, one of two proposals recently incorporated into the IIRSA portfolio. On the Brazilian side, this includes the ongoing effort to complete paving between Rio Branco and Cruzeiro do Sul, a project that was recently included within the subgroup entitled Improving Access to the Ucayali Waterway, revealing the intention to link the Brazilian highway system with Peru via BR-364.* Even more explicit was the designation a 'terrestrial connection' between Cruzeiro do Sul (Acre) and Pucallpa (Ucayali, Peru).† The use of the term 'terrestrial' is purposefully indistinct, because it can refer to a either a highway or a railroad, which has been proposed by advocates of a transcontinental railroad (see below). The construction of the road has the support of Jair Bolsonaro and the governor of Acre, as well as civic leaders in the Peruvian city of Pucallpa.[26]

Acre figures prominently in another high-profile IIRSA initiative, referred to by media outlets as the Corridor Interoceanico, a flagship proposal that includes highway improvements in Brazil, Bolivia, and Peru (Figure 2.9). Efforts to manage the environmental and social impacts of those investments led to the organisation of the MAP‡ initiative, a novel planning process that coordinated actions among sub-national jurisdictions (see Chapter 7). Initiated in the early 2000s, it was at first viewed as a strategic environmental planning process that could identify a pathway to a sustainable forest economy. Efforts to transform the regional economy have had limited success, however, and the MAP region suffers from moderate to high levels of deforestation, a change that is particularly notable in Madre de Dios (Peru) and Pando (Bolivia), which were relatively isolated until the completion of these IIRSA-sponsored highway corridors.

One of the most controversial highway projects in the Brazilian Amazon is the ongoing programme to pave BR-319, the federal highway that links Manaus (Amazonas) with Porto Velho (Rondônia). This 1,000-kilometre corridor (HML #20) has the lowest level of deforestation of all of the trunk highways created in the 1970s (Figure 2.8). Unlike most of the other trunk highways of the epoch, however, it was completely paved in the original construction contract. The work was poorly done, and the roadbed rapidly fell into a state of disrepair. Two stretches have been 'reconstructed' and paved over the last decade: 200 kilometres on the northern sector near Manaus and 165 kilometres near Humaitá. The southern sector is at risk of being the next deforestation hotspot, because of the confluence of three trunk highways (BR-319, BR-230, and BR-364), which will attract settlers and land speculators, particularly from Rondônia, where land is no longer easily accessible.

---

* IIRSA: Amazon Hub, Group 4, Acceso a la hidrovía del Ucayali, AMA55 Conexión vial Rio Branco - Cruzeiro do Sul BR-364/AC ($US 573): http://www.iirsa.org/proyectos/detalle_proyecto.aspx?h=920

† IIRSA: Amazon Hub, Group 4, Acceso a la hidrovía del Ucayali, AMA28 Interconexión Terretsre Pucallpa - Cruzeiro do Sul (budget unknown): http://www.iirsa.org/proyectos/detalle_proyecto.aspx?h=29

‡ MAP Refers to the three jurisdictional entities: Madre de Dios (Peru) Acre (Brazil) and Pando (Bolivia).

The road between Manaus and Porto Velho (BR-319) is the most remote national highway in the Brazilian Amazon. It was first paved in the 1970s, but soon became impassable (top). The highway has been rebuilt across about two thirds of its length and is scheduled to be completely upgraded by 2025 (bottom), pending approval of an ongoing environmental review financed from the President's office via the Programa de Parcerias de Investimentos.

Over the last decade, the 'reconstruction' of BR-319 has been promoted by a regional civic organisation, as well as by elected officials and functionaries in the regional government.[27] Proponents of repaving the road contend that the manufacturing sector in Manaus is handicapped by the logistically complex transportation services required to ship consumer goods to southern Brazil. The least costly alternative is the ocean-going route, but it involves the use of trucks, docks and warehouses at both ends of the supply chain. The highway option, though twenty per cent more expensive, would reduce transport time by at least fifty per cent and, perhaps more importantly, provide door-to-door service between the manufacturer and the wholesale distributor.[28]

The planned improvements to BR-319 will require the approval of the federal environmental protection agency (IBAMA), which initiated an environmental impact analysis (EIA) in 2017 and published in June 2020.[29] Among its findings was the predictable forecast that an improved road would increase deforestation along the highway corridor, but the study also identified the road's potential to catalyse societal demand for additional highway development, including pre-existing roads (AM-174, AM-254, AM-354, AM-364) and those planned for the future (AM-360, AM-366, BR-174), as well as illegal roads built by private actors.[30] Particularly problematic would be the construction of AM-366, which would impact at least two indigenous territories and, potentially, open up wilderness sections of the western Amazon to settlement and oil exploration (Figure 2.8).[31]

Not mentioned within the EIA is the long-term impact from the mega-fragmentation of the forests of the Central Amazon. Even a limited amount of deforestation along the highway corridor would create a barrier to wildlife that would isolate approximately 200,000 square kilometres of intact forest located between BR-319 and BR-230 (Figure 2.2). President Bolsonaro has advocated 'repaving' BR-319, and unless judicial action derails the project, its completion seems increasingly likely.[32]

Another controversial highway project the Bolsonaro administration is considering is the extension of BR-163 across the Amazon River to the border with Suriname. The proposed route has been shown on maps since the 1970s but was not one of the projects in the first wave of highway development. The ambitious proposal would require a three-kilometre span over the main channel of the Amazon River at Óbidos (Pará) and more than fifty kilometres of viaduct to cross the floodplain. This would be a completely new highway and open an enormous area to development (Figure 2.10).

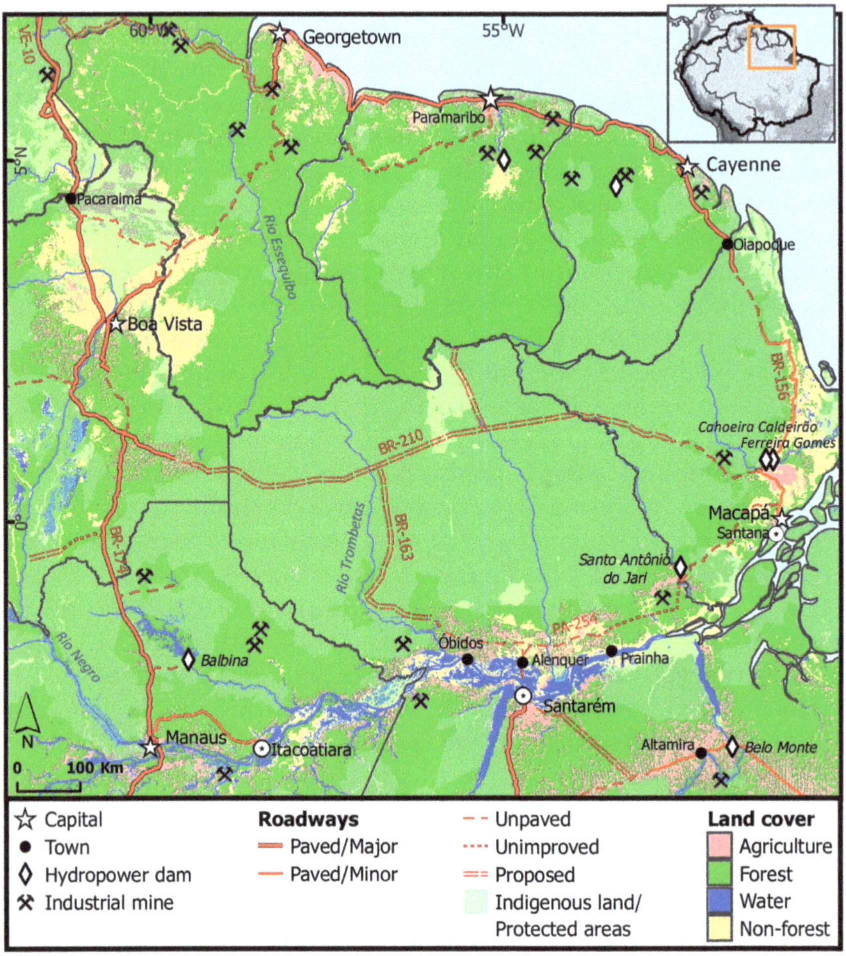

Figure 2.10: The northeast quadrant of the Pan-Amazon is relatively roadless except for BR-174, which extends from Manaus to Boa Vista, from there to the Venezuelan border, and via BR-401 to Guyana. On the coast, BR-156 will soon integrate Macapá with the coastal highway of the Guianas and, eventually, with PA-254 on the north bank of the Amazon river. A recent proposal to extend BR-163 to the border of Suriname would open vast areas to settlement and mineral exploitation.

© Dirk M. de Boer/shutterstock.com

© Carlos Grillo/shutterstock.com

(a) Landholders universally support improvements to secondary roads, particularly bridges, because they are essential for moving their production to market. (b) The Ponte Río Negro, built between 2007 and 2012 at a cost of approximately $US 350 million, is one of the longest bridges in South America. It would be surpassed in length and cost by the proposed bridge that would span the Amazon River at Obidos, a key component of President Bolsonaro's project to open the Northern Amazon to development.

The proposal will be fiercely opposed by environmental activists and indigenous groups because it would disrupt a conservation strategy assembled over three decades of planning and coordination. Nonetheless, the highway conforms to a long-held *Calha Norte* strategy espoused by the national security community, based on the objective of 'occupying' the country's northern border (Chapter 6). The concept originated with the military government of the 1970s, but some variant of it has been embraced by all of the democratically elected governments of Brazil, including the Cardoso administration in the 1990s, which included the Arco Norte development pole as part of its *Eixos Nacionais de Integração e Desenvolvimento* (ENID).

One motivation for building the highway is to create momentum to change the status of the RENCA mineral reserve, a globally significant deposit of copper and other industrial minerals.* The proposed northern leg of BR-163 would connect with PA-254, the regional highway that provides access to the settlement zones located between Óbidos and Prainha (Pará). This would almost certainly increase land values and could facilitate the pursuit of industrial agriculture on the arable soils on the upland landscapes located between the Amazon River and the hill country of the Guiana Shield (HML #1). Although it is largely unimproved, the regional road network of northern Pará is already linked to a similarly rustic network of roads in western Amapá (Figure 2.10). The improvement of these existing roads would create an uninterrupted highway from Óbidos (Pará) to Macapá (Amapá) and the coastal provinces of French Guiana, Suriname, and Guyana.† Although this chain of events might seem unlikely, history demonstrates that existing roads attract settlers who lobby for improvements from local and regional governments that can lead to their eventual development into a transportation corridor.

The Andean republics

Bolivia, Peru, Ecuador, and Colombia all invested in major highway building initiatives in the last half of the twentieth century, motivated in part to project sovereignty over their Amazonian provinces. These areas had poorly defined borders and societies remembered the trauma of the rubber boom, when Brazilian agents encroached upon their territories or when they quarrelled among themselves over the disposition of their frontiers.

* In 1984, the military government created the *Reserva Nacional de Cobre e Associados* (RENCA) as a strategic reserve for future exploitation; President Michel Temer tried to open it up to mining in 2019, but withdrew the initiative due to opposition from environmental and indigenous groups (see Ch. 5).

† IIRSA, G04: G4 - Interconexión Guyana - Suriname - Guaiana Francesa – Brasil, GOY26, Mejoramiento de la Carretera Georgetown - Albina; Carretera de Macapá a Oyapock ($350 million): http://www.iirsa.org/proyectos/Principal.aspx?Basica=1

Unlike the integrated highway network of Brazil, however, these nations built widely separated roads that connected discreet regions of the highlands with adjacent lowland landscapes (Figure 2.2).

Bolivia

Successive governments sought to connect the sparsely populated lowland provinces with the densely populated rural communities of the Altiplano; this policy started in earnest in the 1960s with the construction of all-weather roads to Santa Cruz (HML #31), the Chapare lowlands of Cochabamba (HML #32), and the Yungas of La Paz (HML #33).* Subsequently, the military regimes of the 1970s went on a spending spree that led to a default of Bolivia's sovereign debt in the 1980s, an outcome that limited the country's ability to build infrastructure throughout the 1990s. The most prescient investments occurred on the alluvial plain of Santa Cruz, where public and private resources were leveraged with loans and grants from multilateral agencies to create a secondary road network and industrial infrastructure that triggered a geometric increase in deforestation rates between 1990 and 2010.[33] This landscape now supports the most diversified agricultural production in the Pan Amazon and is a pillar of the Bolivian economy (see Chapter 3); its organic growth is driving the expansion of regional highways north toward the Beni (HML #30) and eastward into Chiquitania (HML #29) (Figure 2.6).

The commodity boom of the 2000s provided the Bolivian state with unprecedented revenues, which the government of Evo Morales used to invest in highway construction across the country. One of the most ambitious projects targeted the northern part of the country, with the objective of linking its administrative capital (La Paz) with communities and landscapes on the border with Brazil and Peru (Figure 2.9). These highways follow transportation routes that have existed for decades, and there is ample support across the region from both settler and indigenous communities. These trunk highways are the Bolivian components of the IIRSA-sponsored Corredor Interoceanico (HML #24, #25, #26, #27), which connects Porto Velho (Rondônia) and Rio Branco (Acre) with the Pacific coast.†

---

* Although Bolivia assigns alpha-numerical identities to its roads, nobody uses them; roads are identified by geographical descriptors, such as La Antigua Carretera a La Paz (BO-7), La Carretera al Norte (BO-4), La Carretera a las Yungas (BO-3).

† IIRSA, Peru-Brazil-Bolivia Hub, Group 2, Corredor Rio Branco - Cobija - Riberalta - Yucumo - La Paz; PBB05, Carretera Guayaramerín - Riberalta / Yucumo - La Paz ($US 594 million); PBB06, Carretera Cobija – Riberalta ($US 696 million); PBB07: Yucumo – Trinidad ($US5.5 million); PBB08 Cobija – Extrema ($US 29 million); PBB60 Puente Mamore ($US75 million): http://www.iirsa.org/proyectos/Principal.aspx

Almost all elements of Bolivian society are energetic supporters of highway construction, and national, regional and local governments place road construction near the top of their budget priorities. The stated goal is to link agriculture production with both domestic and export markets, but multiple social actors also seek to open remote landscapes for agricultural development and land speculation. There is one conspicuous exception, however. The Moxeño people have steadfastly opposed the construction of a highway that would dissect their territory: Tierra Indigina y Parque Nacional Isiboro – Securé (TIPNIS). The proposed road was a priority investment of the government of Evo Morales, who sought to open the area to settlement for his constituents in the coca-cocaine frontier of the Chapare (HML #32). The Moxeños have resisted by using non-violent tactics of civil disobedience and, although the government has never formally abandoned the project, it has been removed from the priority list of highway projects.*

Other highway projects were specifically designed to open wilderness landscapes to agricultural development. This includes those on the piedmont of the Andes in Irurralde Province of La Paz, whose proponents hope to develop into an industrial sugarcane complex.† Even more ambitious are the regional highways being built across the Llanos de Moxos, which will facilitate the conversion of approximately ten million hectares of savanna and forest landscape into soy and rice farms, as detailed in the recently released *Plan de Uso de Suelos del Beni* (see Chapter 4).

Peru

The earliest highway projects penetrated tropical valleys situated between the high Cordillera and the tropical valleys and foothills of the Andean Amazon. In the first half of the twentieth century, roads were built into the cloud forest regions east of Lima in an area known as the Selva Central (HML #37), the lower Huallaga Valley (HML #43), and the Marañon Canyon (HML #45). More rapid change came in the 1970s with the construction of two trunk highways named after a distinguished Peruvian historian (Carretera Federico Basadre [P-18]), and the dominant political figure of the era

* This project was not included within the IIRSA portfolio but was financed by a $332 million loan from BNDES to the Bolivian government; construction was to be executed by a Brazilian construction company (OAS). The conflict eventually motivated the company to abandon the project and BNDES to withdraw its financing in 2012 (see Ch. 11). https://www.americasquarterly.org/blog/brazil-displeased-at-bolivian-decision-to-revoke-highway-contract/

† This highway (BO-16) is also not included within the IIRSA portfolio but was financed by the IDB and the World Bank; a Chinese company (SINPOEC) built a key bridge over the Río Beni, apparently with funds from the national treasury (Molina 2014).

(Carretera Fernando Belaunde [PE-5N]).* Both connect the central highlands with ports on the Amazon River and opened lowlands to settlement and deforestation in the 1970s and 1980s (Figure 2.11).

The Carretera Federico Basadre is part of an IIRSA investment cluster that links the port of Callao (Lima) with Pucallpa on the Rio Ucayali.† The lowland landscapes adjacent to the highway have attracted tens of thousands of settlers over five decades and continue to be among the most active agricultural frontiers in the Peruvian Amazon (HML #40 and HML #41). The port of Pucallpa provides access to the Amazon waterway via the Rio Ucayali and an associated expanding forest frontier that is the source of most of Peru's timber (HML #39).

The Carretera Fernando Belaunde is sometimes referred to as the Carretera Marginal de la Selva (PE-5) because its namesake was the statesman who originally proposed the construction of an international highway integrating the Amazonian regions of the Andes. In Peru, the Carretera Marginal de la Selva is a sinuous route that weaves in and out of the Andean foothills. This road was first constructed in the Upper Huallaga Valley (HML #42) to connect with the previously established agricultural settlements of the Lower Huallaga Valley (HML #43) and was eventually extended eastward to connect with the Port of Yurimaguas on the Rio Huallaga (HML #44).

At the regional city of Tarapoto (San Martín), it merges with an IIRSA investment group collectively referred to as the Corridor Interoceanico del Norte.‡ This has two Amazonian spurs: one originates at Yurimaguas on the Rio Huallaga (PE-5NB). The other starts at Saramiriza on the Rio Marañon (PE-5NC).§ The two segments converge in the Marañon valley (HML #45) before passing over the Cordillera Occidental at the Huancabamba Depression, a geological feature with the lowest elevational point on the continental divide (2,145 metres above sea level). Known as the Paso de Porcullo, this route has been used for centuries as a gateway into the Amazon (see Chapter 6) and provides a significant logistical advantage when

* The Peruvian highway notation system is confusing and named highways often change numerical designation in different regions or sections.

† IIRSA, Amazon Hub, Group 4: Accesso a la Hydrovia de Ucayali; $US 3.6 billion; AMA26 (Tingo María - Pucallpa Road): AMA31 (El Callao Port); AMA55 (Rio Branco Cruzeiro do Sul); AMA63 (La Oroya / Cerro De Pasco / Huancayo); AMA64 (Pasco - Tingo María) and other non-highway projects: http://www.cosiplan.org/proyectos/Principal.aspx

‡ This highway is managed as a concession by *Odebrecht Perú Operaciones y Servicios*; the 25-year contract is based on investment of $US 630 million: https://www.ositran.gob.pe/anterior/carreteras/iirsa-norte/

§ IRSA, Amazon Hub, Group 3: Acceso a la Hidrovía del Huallaga – Marañón; $US 1.3 billion; AMA16 (Tarapoto - Yurimaguas Road); AMA19 (Reposo Saramariza); AMA25 (Paita - Tarapoto Road) and other non-road projects: http://www.cosiplan.org/proyectos/Principal.aspx

Figure 2.11: The highways of the Peruvian Amazon connect different sectors of the high Andes with disjunct landscapes in the Amazon piedmont. They are partially integrated via the Carretera Marginal de la Selva, which threads its way through the Andean foothills. The Amazon Waterway connects the city of Iquitos with the rest of Peru via the port cities of Pucallpa, Yurimaguas and Saramarisa.

compared to other Andean mountain passes that typically occur between 4,000 and 5,000 metres above sea level (Figure 2.11)

The highway to Saramiriza was originally built in the 1960s during the construction of the Oleoducto del Norte and is a major access point to the northern Peruvian Amazon (HML #46). This relatively remote village plays a prominent role in the Regional Government of Loreto's plan to connect Iquitos with the national road network. The proposed highway includes a 200-kilometre section from Saramiriza that would follow the existing pipeline right-of-way to the oil fields near the Ecuadorian border; here it would connect with another proposed road along the border with Ecuador, as well as a 220-kilometre spur to Nauta, a village on the Rio Marañon with an existing paved road to Iquitos (Figure 2.11).[34] At first glance, the proposed route would seem circuitous, but a more direct one would cross the massive peat swamp of the Pastaza Delta, increasing construction costs and undermining the economic viability of the project.[35]

The construction of any of these roads would open vast areas of primary forest to logging and, almost certainly, settlement by subsistence farmers and land speculators (see Chapter 4). The proposed roads would traverse land deeded to dozens of indigenous communities, while bordering both national (Zona Reservada Santiago – Conaima, Reserva Natural Oucacuro) and regional (Area de Conservación Regional Alto Cona – Pintuyaco Chambira) protected areas. The initiative has been vigorously opposed by environmental advocates and indigenous organisations; nonetheless, elected officials in Iquitos have successfully lobbied the Peruvian Congress to declare the construction of the Saramiriza - Iquitos highway a national priority.[36]

The central section of the Carretera Marginal de la Selva (PE-5) extends south from the Carretera Federico Basadre through the rapidly expanding agricultural frontier of Huanuco and Pasco (HML #40), before ascending the foothills to the coffee-producing landscapes near Oxapampa (HML #37, #38). The southern section (PE-5S) is the main trunk highway of the Selva Central and eventually crosses over the foothills again to connect to the Rio Ucayali at Atalaya (MHL #38), a major logistical centre for the timber industry (Figure 2.11).

The other major trunk highway in the Peruvian Amazon is a component of the IIRSA-sponsored Corridor Interoceanico (Figure 2.9),* which connects the Peruvian coast with the Puerto Maldonado on the Madre de Dios River and the frontier landscapes of Pando (Bolivia), Acre, and Rondônia (Brazil). In Peru, this group of highways is referred to as the

* This should not be confused with the *IIRSA Eje Interoceanico* (Interoceanic Axis), which extends from São Paulo across Mato Grosso do Sul to Bolivia, Chile and Southern Peru, which incorporates a highway between Santa Cruz and Campo Grande that is sometimes referred to as the *Corridor Bioceanico.*

Figure 2.12: The Ecuadorian Amazon has the densest and most improved road network in the Andean Amazon; however, it is more static when compared with Peru and Colombia, where new roads are being carved into wilderness areas. The proposed highway between Saramariza and Iquitos would follow a pipeline right-of-way for about 50% of its route. The Amazon Waterway is connected to two river ports in Ecuador (Puerto Morona and Puerto Providencia) and one in Colombia (Puerto Asis).

Corridor Interoceanico del Sur,* which, like the similarly named highway corridor in northern Peru, is managed by a consortium of Brazilian and Peruvian construction companies that obtained 25-year concessions in exchange for building the project.†

Like most such projects, it spawned a land rush that predated the completion of the highway. The looming impact from deforestation and social displacement motivated regional leaders in Madre de Dios to participate in the MAP initiative with like-minded individuals in Acre (Brazil) and Pando (Bolivia).[37] As in Brazil and Bolivia, the MAP initiative enjoyed success in improving protected-area management and recognising the rights of forest communities but was not successful in changing non-sustainable patterns of land use and natural resource management.[38,39] The new highway contributed to the gold rush then underway on the piedmont of Madre de Dios and, although the gold rush would have occurred regardless, the multilateral financers of the highway failed to take into account the potential for the highway to accelerate illegal mining (HML #36).[40]

The justification for developing the Corridor Interoceanico is an example of 'infrastructure hype', whose proponents exaggerate the economic benefits of a development project. In this case, they overstated the potential for exporting commodities from Rondônia and Mato Grosso via Pacific ports to Asian markets, which ignored (a) the high cost of truck transport and (b) the energy cost of moving bulk commodities over a 5,000-metre pass in the High Andes. The light traffic that has characterised the highway since its completion in 2010 demonstrates that this was never a viable option.[41]

Ecuador

The first road from the highlands to the Amazon lowlands was built by Royal Dutch Shell in 1947, a precursor to a formal policy articulated in the 1960s that linked oil exploration with road construction and colonisation (Chapter 6). Throughout the 1970s and 1980s, the government consolidated the road network in Amazonian Ecuador, which can be divided into two sectors: (1) the Sucumbíos – Orellana polygon, which sits above the major petroleum-producing formation of the country (HML #49 and HML #50) and (2) the Ecuadorian Piedmont (HML #48), which starts in the foothills near

---

* IIRSA: Peru-Brazil-Bolivia Hub; Group 1 - Corredor Porto Velho - Rio Branco - Assis - Puerto Maldonado - Cusco / Juliaca - Puertos Del Pacífico; $US 2.9 billion: PBB01 (Carretera Iñapari - Puerto Maldonado - Cuzco ($1.9 billion); PBB03, Puente Rio Acre ($US 12 million): http://www.cosiplan.org/proyectos/Principal.aspx

† The corridor is divided into four sections (*tramos*), each managed by a different consortium; the lowland tramo is operated by IIRSA Sur S/A, a subsidiary of Odebrecht Perú Operaciones y Servicios S.A.C valued at approximately $US 640 million: https://www.ositran.gob.pe/anterior/carreteras/iirsa-sur-t3/

Peru and extends northward to the Colombian border.* This north- south paved highway, known as the Troncal Amazónica (E-45), is the Ecuadorian component of the Carretera Marginal de la Selva (Figure 2.12).

In the south, two roads extend east from the Troncal Amazónica into the lowland plains situated north of the Peruvian border. The most important (E-40) connects to a port on the Rio Morona that was originally built to supply military outposts along a highly contested border (see Chapter 6). Puerto Morona is now the terminus of an IIRSA-sponsored initiative to link the ocean port of Guayaquil with the Amazon waterway† and a large-scale copper mine under development in the Cordillera del Condor (see Chapter 5). The military also built an alternative supply line along the northern stretch of the Rio Morona that passes through the heartland of the Shuar indigenous people (HML #47). The construction of the highway motivated some Shuar families to clear forest as a defensive strategy, legalise land claims and limit incursions by settlers from highland communities (see Chapter 11). The regional government is in the process of paving this road, using credit provided by the *Banco de Desarrollo del Ecuador*.[42]

Two major east-west highways (E-10 and E-20) were built to connect the highlands with the Sucumbíos – Orellana polygon. This region also has the most fertile soils in the Ecuadorian Amazon, which spurred settlement and the development of a secondary road network that parallels the collector pipeline system (Figure 2.12). Annual deforestation continues at relatively constant rates due to the conversion of remnant forest within smallholdings created in the 1980s.[43] The agricultural frontier continues to expand into the landscapes surrounding Yasuní National Park (MHL #50), although the government is imposing stricter controls over settlement along new access roads, and indigenous communities are aggressively seeking to limit the expansion of the petroleum sector (see Chapter 11).

Colombia

Deforestation is less strongly associated with the construction of major highways in the Colombian Amazon because successive governments have embraced a policy that precludes large-scale road-building. Instead, the country has decided that its most remote regional capitals – Leticia (Amazonas), Mitu (Vaupes) and Inírida (Guiania) – can more effectively be integrated into the national economy via air transportation systems. There

---

* IIRSA: Andean Hub, Group,6, Conexión Colombia - Ecuador II, AND35, Rehabilitación del Tramo Bella Unión – Gualaquiza, ($US 23 million): http://www.cosiplan.org/proyectos/detalle_proyecto.aspx?h=80

† IIRSA, Amazonas Hub, Group 7 - Acceso a la Hidrovía Amazonas; $US 420 million: AMA 45 (Puerto Morona); (AMA45); AMA 48: (Road Prto Bolívar -Prto Morona); AMA 47 (Road to Prto Bolívar – Méndez); AMA 46 (Guayaquil - Puerto Morona): http://www.cosiplan.org/proyectos/Principal.aspx

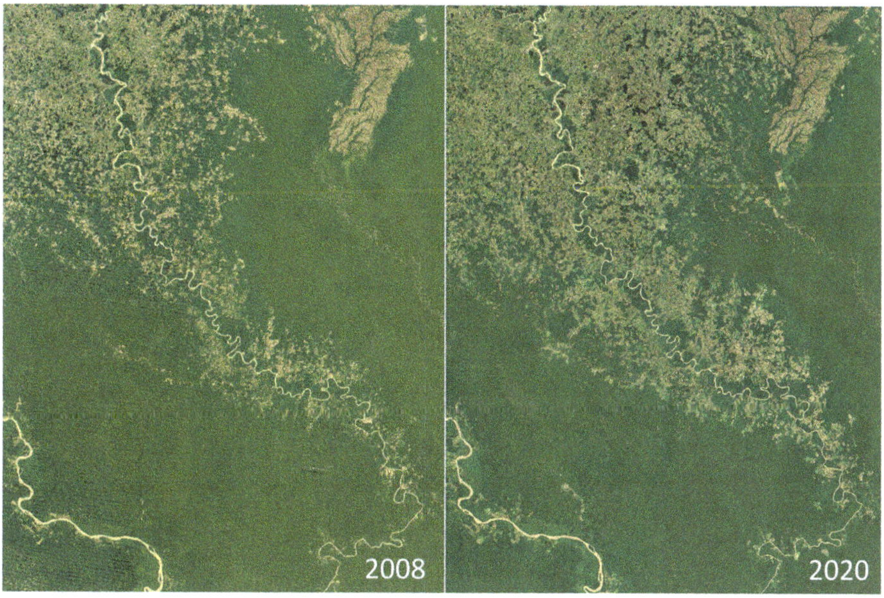

Source: Google Earth

Figure 2.13: Rivers also act as deforestation vectors, particularly in Colombia where the lack of roads and the cultivation of coca leaf coincide to create a forest frontier associated with the Rio Caquetá.

is still considerable loss of forest, but it tends to occur along rivers or in roadless landscapes where the state is only marginally present (Figure 2.13).

Although most of the lowlands are roadless, several key highways connect towns on the Andean piedmont with the rest of the country. These include CR-45 between Neiva (Huila) and Puerto Asis (Putumayo) (HML #51), which is being upgraded by the *Agência Nacional de Infraestructura* (ANI), a public-private partnership programme that accelerates investment in infrastructure assets.* Only a few select highways are commercially attractive, and improving access via highway construction to conflict areas is considered to be an essential component of the peace and reconciliation process (see Chapter 6). The *Instituto Nacional de Vias* (INVIAS) oversees the construction and operations of all other national highways, and in Caquetá, this includes two highways (CR-30 and CR-20) that connect Neiva to Caquetá (Figure 2.12), where the potential to expand industrial agriculture is attracting significant new investments (HML #52 & #53).

* A Chinese construction company recently obtained a concession valued at $US 1.0 billion for this section of R45.

Caquetá is otherwise isolated from the national highway network, but its major towns are linked by a trunk highway that runs along the base of the Andes, referred to as the Carretera Marginal de la Selva (CR-65). As the name implies, this is a component of the international highway envisioned in the 1960s. As of 2020, it had been paved for 250 kilometres in Caquetá and 165 kilometres in the adjacent department of Putumayo,* but the two segments remain separated by about twenty kilometres of back roads and the 1,000-metre width of the Rio Caquetá. Once a bridge is built, the Carretera Marginal de la Selva will connect all the major towns of Caquetá and Putumayo, as well as link to its counterpart highway in Ecuador.[44] Near the border, it intersects with CR-10, part of an IRSA-sponsored initiative to link Pacific ports with the Amazon waterway.†

The northern section of the Carretera Marginal de la Selva extends from San José de Guaviare (Guaviare) to Villavicencio (Meta) and from there along the base of the Andes to the Venezuelan border. There is no connection – yet – to Caquetá. There are, however, two road-building processes underway that will make that link, both of which will isolate Parque Nacional Natural Serranía de Macarena and, in the process, disrupt Colombia's only intact biological corridor connecting the forest ecosystems of the Andes and the Amazon. Along the northern border, INVIAS is financing the construction of the Transversal de la Macarena (R-65A), a regional highway that will facilitate the export of agricultural commodities from the Department of Meta via the Pacific ports of Buenaventura (Valle de Cauca) and Tumaco (Nariño). South of the park, approximately 200 kilometres separate the two sectors of R-65 at San José (Guaviare) and San Vicente de Caguán (Caquetá). INVIAS has no plans to close this gap, but unplanned road-building by local landholders has narrowed it to a mere fifty kilometres (see Figure 2.12).[45]

The eastern terminus of the Transversal de la Macarena will connect with the northern segment of CR-65 about 100 kilometres north of San José de Guaviare. All these landscapes are populated by coca-growing peasants and cattle ranchers, who have created a vast informal network of small roads that are slowly encircling Colombia's oldest national park.‡ South of San José de Guaviare, a regional highway (CR-75) extends to the

---

* IIRSA: Andean Hub, Group 06; Conexión Colombia - Ecuador II (Bogotá – Loja); AMA 090, Tramo San Vicente del Caguán - El Porvenir ($US 240 million) and AND79, Mocoa - Santa Ana - San Miguel Road Section ($US 210 million): http://www.iirsa.org/proyectos/Principal.aspx?Basica=1

† IIRSA: Amazon Hub, Group 01; Acceso a la Hidrovía del Putumayo; AMA01: Corredor Vial Tumaco - Puerto Asís ($US 291 million): http://www.iirsa.org/proyectos/Principal.aspx?Basica=1

‡ PNN Serranía de la Macarena was established as a biological reserve by legislative action in 1948 and was constituted as a national park in 1989. See http://www.parquesnacionales.gov.co/portal/es/

town of Calamar, the gateway to a forest frontier with the highest rate of deforestation in the Colombian Amazon (HML #54).

Carretera Marginal de la Selva

Although never organised as a specific project, the Carretera Marginal de la Selva has emerged from multiple projects that have been established in the Andean foothills from Colombia to Bolivia (Figure 2.14). In addition to the gap across the Serranía de Macarena, there is a small gap between Ecuador and Peru that conserves a biological corridor between Parque Nacional Podocarpus in the Andes to the Reserva biológica Cerro Plateado in the Cordillera del Condor and several Awajún indigenous territories in Peru.*

Within Peru, whose president proposed the idea, the Carretera Marginal de la Selva traces a sinuous route through the foothills of the Andes, the Marañon Valley (HML #45), the Huallaga Valley (HML #42 and #43), the piedmont landscapes of Huanuco and Pasco (HML #40) and the Selva Central (#37). Disjunct from this continuous highway are segments of the Peruvian Yungas (HML #35) and the piedmont of the Madre de Dios (HML #36). Virtually all Peru's lowland tropical agriculture is located within fifty kilometres of this road, which – despite its idiosyncratic route – makes it a strategic asset supporting domestic food security (see Chapter 3).

Official highway maps show a future potential route for the Carretera Marginal de la Selva, (PE-5S), which would extend south from the Selva Central ~800 kilometres toward the Camisea gas fields to the Corridor Interoceanico del Sur near Puerto Maldonado (Madre de Dios).[46]

This 'gap' in the highway network is, perhaps, the world's most important biological corridor, because it facilitates biological interchange between the mega-diverse rainforests of the Southeast Amazon and the montane forests of the Central Andes. The corridor experiences some of the highest annual rainfall on the planet and is considered to be resilient to climate change due to inherently stable, continental-scale patterns of wind flow (See Chapter 10).[47] A highway, either on the piedmont or in the foothills, would interrupt the ability of species to adapt to changes in temperature by migrating up a topographic gradient across the forest-covered slopes of three massive cordilleras (Vilcabamba, Urubamba, Vilcanota).[48]

The importance of the region for biodiversity conservation has been known for decades, and most of the area has been set aside either as a protected area or indigenous reserve. Any attempt to compromise the integrity of these reserves will be met with fierce resistance from academics, civil society and indigenous people. Nonetheless, the Peruvian government, or at least the functionaries within its highway planning agency, continue

---

* Comunidad Nativas Naranjos, Supayaco, Alto Tuntus and Tuyankuwas. Source: *Red Amazónica de Información Socioambiental Georreferenciada* (RAISG).

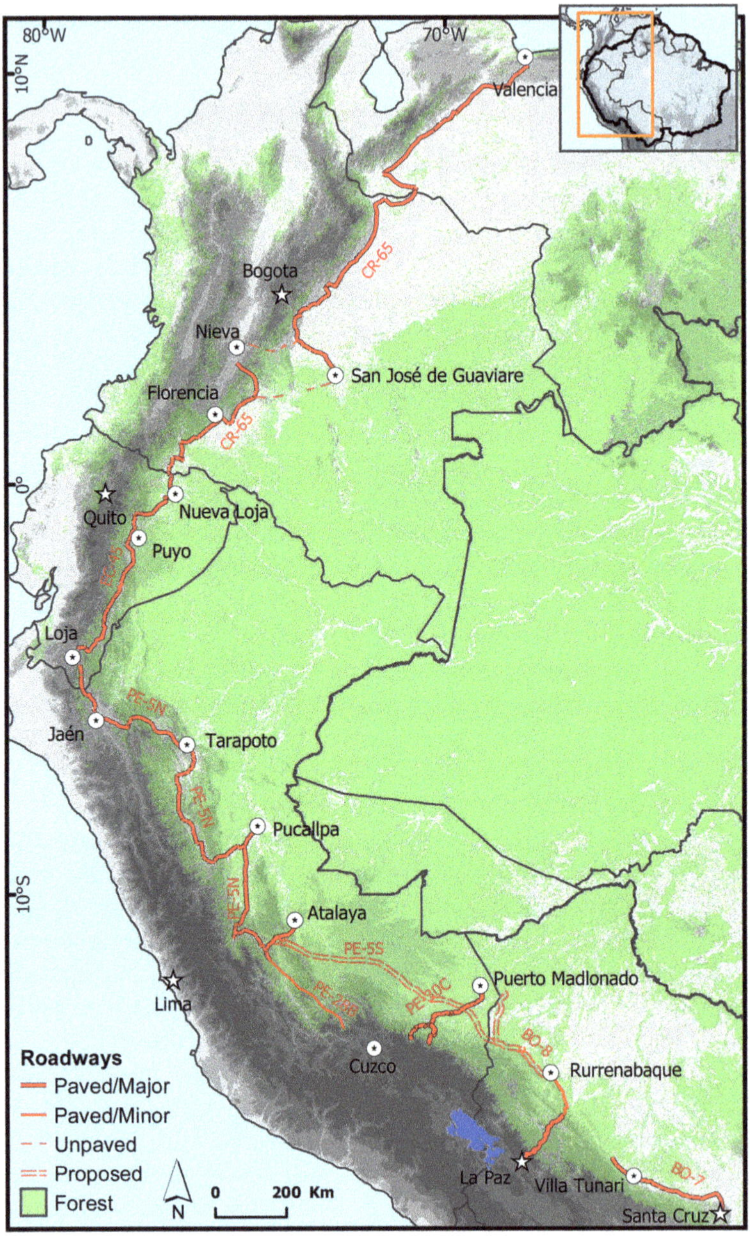

Figure 2.14: The Carretera Marginal de la Selva was first proposed in 1969 by Fernando Belaunde in order to promote the integration of the economies of the Andean Amazon. The concept has influenced the design of national highway systems, but there remain significant gaps among its different components.

to place the proposed highway on official maps and, presumably, in the investment portfolios of its future infrastructure agenda.[49]

The proposed extension of PE-5S to the Bolivian border would likewise intrude on protected areas but would link up with a planned extension of the road currently under construction between Yucumo and Ixiamas (BO-16). Unlike the other Andean nations, Bolivia does not have an explicit plan to construct a highway corridor named 'Carretera Marginal de la Selva'; nonetheless, trunk highways in La Paz (HML #34) and the Chapare (HML #32) essentially conform to the original concept. The 200-kilometre gap between these two sections is, as in Peru, an important biological corridor that has been incorporated into the protected area network and/or indigenous territory.

The Guiana Shield and the Coastal Plain

Roads are scarce in the Northern Amazon, and surprisingly, the few that exist have not triggered widespread deforestation (Figure 2.10). This apparent anomaly is largely the consequence of a development dynamic that has kept these countries from seriously pursuing agricultural development in their Amazonian provinces. Venezuela chose to build an economy based on mineral extraction, mainly oil, and its leaders have viewed its landscapes of the Guiana highlands as a giant national protected area. The colonial history of Guyana, Suriname and French Guiana has caused their residents to look to Europe or North America for economic opportunity, which has suppressed the demand for terrestrial connections to Venezuela and Brazil.

The first truly modern highway in the entire region, the Ruta a la Gran Sabana (VR-10), serves as a transitway for commerce between the Venezuelan Coast and Brazil as well as an entry point for tourists visiting the Gran Sabana and *tepuis*.[*] Most of Venezuela's tropical timber is extracted using this road, and it is a key infrastructural asset for the mining industry (see Chapter 5). This highway connects with BR-174 in Roraima and is now almost forty years old; its renovation and maintenance are included within the IIRSA portfolio of investments.[†]

The other major highway corridor is the route between Boa Vista (Roraima) and Georgetown (Guyana). The section in Brazil is paved, but the Guyana component is a gravel road between the mining centre of Linden and the border town of Letham. Although it lacks the attributes of a modern

---

* A *tepui* is a table-top mountain of the Guiana Highlands of South America. The word is derived from the Pemon indigenous people of the Gran Sabana and translates as the 'house of the gods': http://www.bbc.com/travel/story/20121020-venezuelas-lost-world

† IIRSA; Guianese Shield Hub, Group 1, Interconexión Venezuela – Brasil, GUY01 Rehabilitation of the Caracas - Manaus Road ($US 407 million): http://www.cosiplan.org/proyectos/Principal.aspx

highway, it does include several heavy-load bridges that make industrial transport between Roraima and the Atlantic Coast a viable option. Its completion in 2009 opened up remote landscapes to logging and motivated a limited number of farmers to cultivate rice on the flooded savannas near the Brazilian border. The landscapes surrounding the highway have not yet experienced significant deforestation. This road is considered to be a high priority 'anchor project' in the 2017 IIRSA investment portfolio and will be paved in the near future.* The economic logic for the modernisation of the highway rests, in part, on the assumption that Roraima will become an agricultural exporting region similar to Mato Grosso. Truck transport from the farm landscapes around Boa Vista to the Atlantic Coast (700 kilometres terrestrial) would be more cost-effective than the truck and fluvial transport options via Manaus (850 kilometres terrestrial plus 1,540 kilometres waterway).

Most of the population of Guyana, Suriname and French Guyana live near the coast, where they communicate via a road that has existed for decades (HML #57). Two IIRSA-sponsored highway initiatives seek to improve this terrestrial connection. The Ciudad Guyana – Georgetown – Paramaribo Corridor would create a new highway from Venezuela through the gold-mining landscapes of northeastern Guyana (HML #58).† It is not a priority project, however, because Venezuela does not recognise Guyana's sovereignty over the disputed area. The second initiative is to upgrade the road between Georgetown and Macapá, which would include bridges across the Corentyne River between Guyana and Suriname and the Marwijne River between Suriname and French Guiana; it would connect with BR-156 in Amapá, Brazil.‡

Currently, there are very few settlements along BR-156 (HML #60), although it is used by miners exploiting gold in the greenstone belt near the border with French Guiana. In the near future, settlement could be spurred by the development of offshore oil and gas fields, which presumably will require logistical facilities on the coast. Brazil is in the process of upgrading BR-156 as part of an IIRSA-sponsored corridor on the coasts of the Guianas. Construction began in 2010 and is expected to conclude by 2020 at a cost of

---

* IIRSA; Guianese Shield Hub, Group 2, Interconexión Brasil - Guyana; GUY09, Lethem - Linden Road ($US 250 million): http://www.cosiplan.org/proyectos/Principal.aspx

† IIRSA; Guianese Shield Hub, Group 03; GUY18: Routes Interconnecting Venezuela (Ciudad Guayana) - Guyana (Georgetown) - Suriname (Apura - Zanderij - Paramaribo): http://www.iirsa.org/proyectos/detalle_proyecto.aspx?h=200

‡ IIRSA; Guianese Shield Hub, Group: 04 – GUY 28, Routes interconnecting Guyana - Suriname – French Guiana - Brazil, $US 350 million: http://www.iirsa.org/proyectos/detalle_proyecto.aspx?h=200

approximately R$ 1 billion.[50] Once a modern highway exists between Macapá and French Guiana, land speculation and agriculture will likely follow.

Hydropower: A Shift toward Reduced Impact Facilities, But the Controversy Continues

After highways, investments in large-scale hydropower facilities are the most controversial infrastructure investments in the Pan Amazon. Governments pursue hydropower as a sovereign source of renewable energy and driver of economic growth; opponents object due to the environmental and social impacts associated with large-scale projects. There are elements of truth in both of these affirmations, and the debate surrounding hydropower usually focuses on the trade-offs in the costs and benefits that have caused some projects to be approved, others to be modified and a few to be cancelled.

Historically, Brazil has been overly reliant on hydropower. In the late 1990s, it represented a remarkable ninety per cent of installed generating capacity, a situation that provoked the electricity crisis of 2001/2002 when water levels in reservoirs were reduced by a prolonged nationwide drought.[51] The government responded by diversifying its electrical generation capacity in natural gas, biomass, wind and solar (Figure 2.15), as well as by increasing hydropower capacity in Amazonian rivers deemed to be less susceptible to the risk of periodic drought.[52] Brazil is still overly reliant on hydropower, as evidenced by the weather-induced power rationing that rocked the national economy in 2015.[53]

In 2005, the civil engineers within the federal energy agency of Brazil estimated the potential hydropower capacity of Brazil at approximately

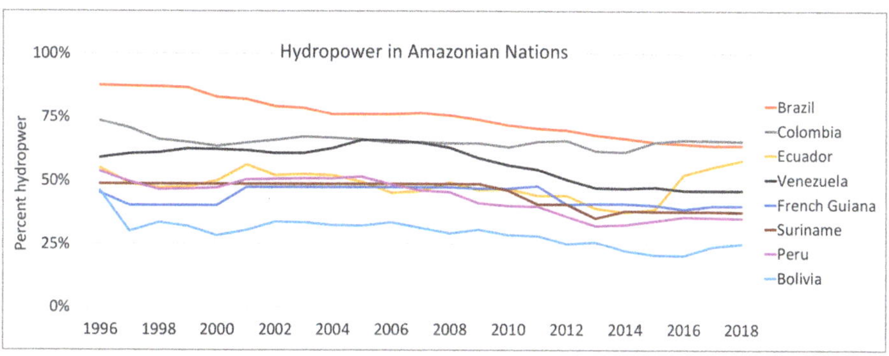

Figure 2.15: The declining dependence on hydropower in Amazonian countries. Data source: The United Nations – Energy Statistics Database (http://data.un.org/Data.aspx?d=EDATA&f=cmID%3AEC).

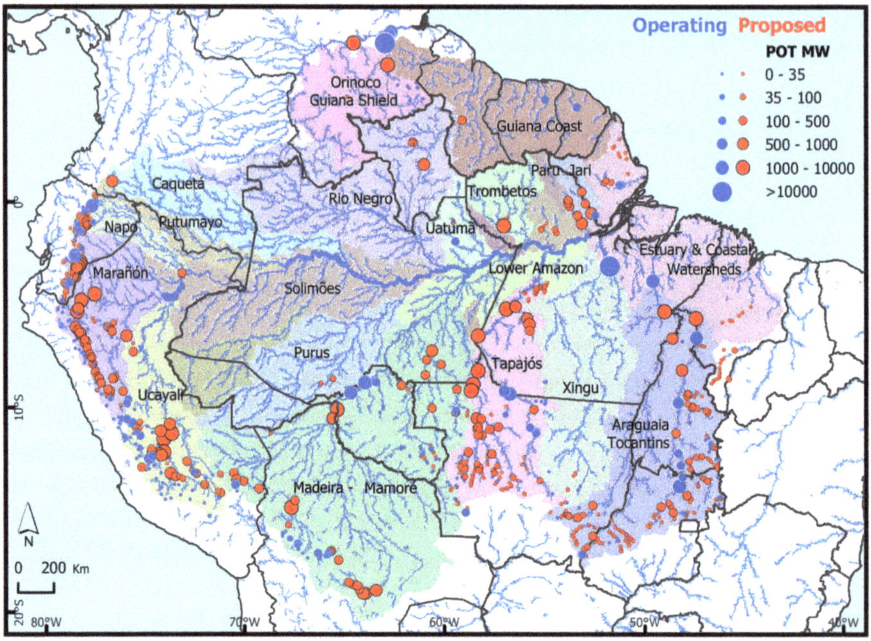

Figure 2.16: The distribution of existing and planned hydropower plants in the Pan Amazon. With the exception of about ten large-scale facilities, most existing power plants are located relatively high in individual watersheds.
Data source: RAISG 2021.

251 GW,* of which about half was located within the Amazon basin.[54] By 2020, however, they had reduced that estimate to 176 GW, while reporting that installed capacity had increased from more than 78 to 108 GW; most of that expansion had occurred within the Amazon, where installed capacity increased from 7 to 43 GW (Figure 2.16).[55] The decrease in estimated potential capacity was due not to a decline in Brazil's hydraulic resources but to a recalculation, after planners eliminated projects that were no longer deemed feasible based on regulatory criteria (See Annex 2.2).

This determination followed a 2018 decision by the government to call a halt to future large-scale development of dams in the Amazon, citing the need to reconcile social and environmental impacts with economic criteria and energy demand.[56] This policy was reversed following the election of

* The capacity of a power system represents the maximum power the system can generate when operating at any point in time; a gigawatt (GW) is a billion watts; for reference, that signifies about 3.1 million state-of-the-art photovoltaic (PV) panels in 2020. Source: Dept. of Energy USA, https://www.energy.gov/eere/articles/how-much-power-1-gigawatt

Jair Bolsonaro, who embraced many of the projects sidelined in 2018 and proposed additional projects that had not been incorporated in the national energy plans.[57] Current plans are in flux, but the database of proposed hydropower facilities under consideration in the Brazilian Amazon totals 112, with a potential installed capacity of 44 GW.[58] In 2021, Brazil once again runs the risk of experiencing an electricity crisis and, once again, the culprit is lower-than-average water levels in reservoirs.[59]

The nations of the Andes are also highly dependent on hydropower and seek to increase that commitment over the next decade. A recent study documented 142 dams in operation or under construction and an additional 160 in various stages of planning (Figure 2.16).[60] This is twice the number reported in 2012[61] and would represent a 500 per cent increase in installed generating capacity.

Peru's National Energy Plan 2014–2025 projects that 54 per cent of its electricity supply will be generated from hydropower; most will come from dams built in the Ucayali and Marañon basins (Figure 2.16).[62] Ecuador hopes to increase the proportion of hydroelectric power from fifty per cent in 2015 to approximately ninety per cent by 2025, at least seventy per cent will come from Amazonian basins.[63] Colombia obtains about sixty five per cent from hydropower, although none of that is obtained from an Amazonian tributary.[64] Bolivia has progessively reduced its reliance on hydropower over the last twenty years as it exploited the natural gas fields discovered in the 1990s; however, future plans rely almost exclusively on hydropower. In 2019, the government announced plans for quadrupling the country's installed capacity, from 1.2 to 5.1 GW, which would increase its reliance on hydropower from thirty to eighty per cent.[65]

In spite of its economic advantages, the physical attributes of the Amazon and its tributaries make hydropower problematic for the conventional dam-and-reservoir (D&R) facilities favoured by civil engineers and energy managers. In the lowlands, broad floodplains limit the potential to create reservoirs in confined areas that store large volumes of water; this impedes operators' ability to regulate reservoir levels for power management. In contrast, the valleys in the Andean foothills provide almost ideal conditions for creating massive reservoirs; however, high sediment loads cause them to lose storage capacity over time, which limits their lifespan as economic assets.*

The retention of sediments also impacts ecosystem function in downstream riparian habitats. This is particularly problematic for dams on 'white-water rivers' that are ecologically defined by high sediment

* The reduction in lifespan does not occur within the temporal framework of financial analysis (50 years) and thus does not impact feasibility studies (Ho et al. 2017).

loads.* These rivers, which originate in the Andes, drain only twelve per cent of the basin's surface, but contribute more than eighty per cent of the sediments that enter the Amazonian floodplain ecosystem (Figure 2.17).[66] The proposed construction of multiple dams within the Marañon, Ucayali, Madre de Dios and Madeira basins would have long-term consequences on biogeochemical processes in floodplain habitats along the entire course of the river and eventually would impact the intertidal zones of the delta and the marine ecosystems located above the continental shelf at the mouth of the Amazon.[67]

There are other social and environmental impacts associated with large-scale D&R facilities. Large reservoirs displace rural families, forcing them to abandon villages they have inhabited for decades or even centuries. Many dams are built just below a topographic discontinuity to maximise energy production, but these localities are often inhabited by indigenous communities that exploit a natural concentration of fish or occupy an essential portage around non-navigable rapids. Reservoirs not only force these families to move but alter the ecosystem function that sustains the local economy above and below the dam.

The day to day operation of a D&R power plant alters the natural habitats located below the dam because managers manipulate water flows to balance the demand for electricity. These are always substantially different from natural flood regimes that regulate the life cycles of species in floodplain habitats.[68] Amazonian rivers are renowned for the movement of fish and other species between the river channel and the backwater habitats, which are defined by the length and depth of seasonal inundation.[69] Power management disrupts the natural cycles that support wildlife and, consequently, affect the human communities that depend on them. The most obvious impacts occur locally, but alterations to flooding regimes can extend far downstream, while upstream communities suffer impacts when dams block the migration of economically important fish species (Figure 2.17).

Some impacts are global in scale. Amazonian rainforests are characterised by massive quantities of biomass, and if the standing vegetation is not cleared prior to flooding, the reservoir will generate substantial methane emissions via anaerobic decomposition at the bottom of the reservoir. These emissions can last for decades and nullify any potential savings of greenhouse gases (GHG) associated with hydropower as a renewable energy.[70]

One alternative for managing environmental and social impacts is to build dams that employ a run-of-the-river (R-o-R) design that minimises both the size of the reservoir and sediment removal.[71] These configurations still

---

* There are three broad classes of rivers in the Amazon basin: white water rivers have high sediment loads and neutral pH; black water rivers have low sediment loads and an acidic pH; clear water rivers have low sediment loads and neutral pH; see Chs 8 and 9.

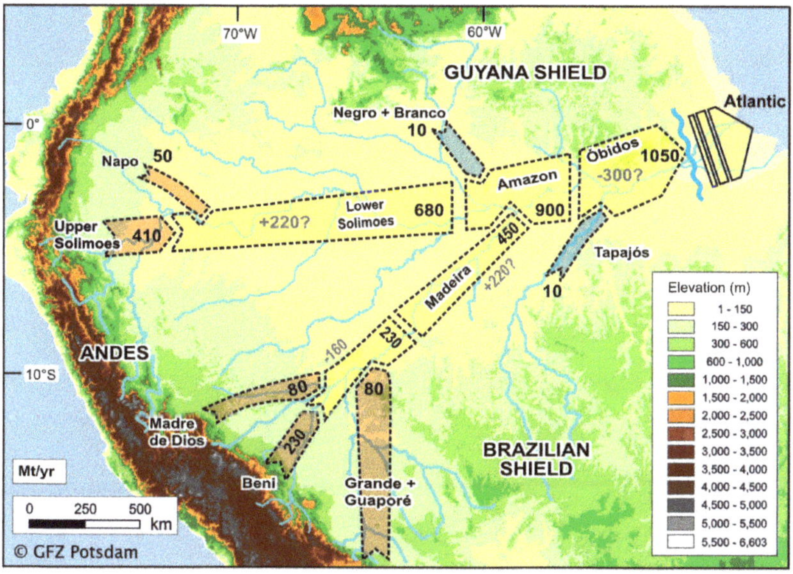

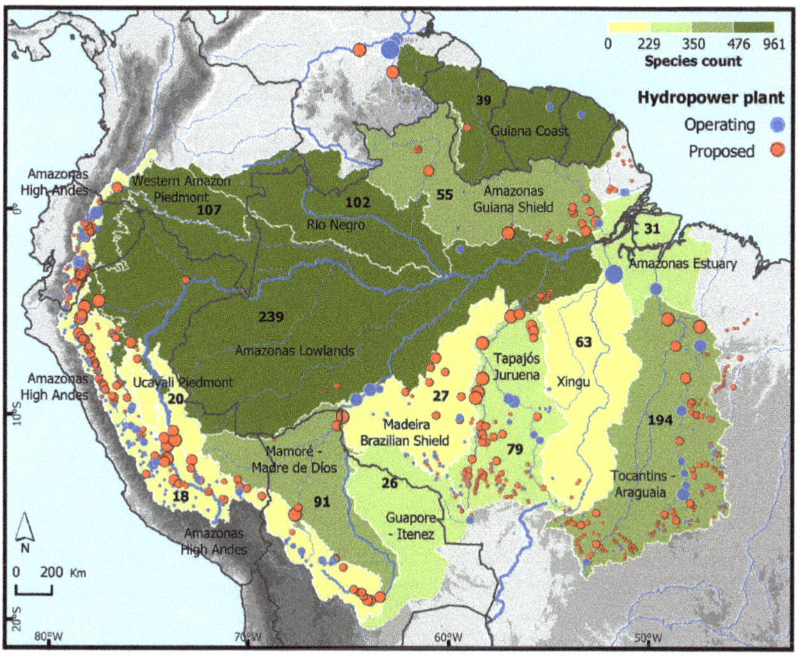

Figure 2.17: Key biophysical attributes of the Amazon Watershed. Top: Source of sediment (see Wittmann et al. 2010). Bottom: Potential fragmentation within freshwater ecoregions; species count refers to species diversity (shade of green), while numbers within polygons are endemic species unique to that ecoregion.

Data sources: Winemiller et al. 2016; RAISG 2021.

Hydropower facilities that employ dam-and-tube designs avoid most of the impacts associated with dam-and-reservoir and run-of-river designs. They are used in Bolivia at Corani/Santa Isabela (top) and Zongo/Harpa (middle). The concept has been deployed at a few locations in Brazil (Dardanelos, Mato Grosso).

cause impacts linked to biodiversity loss and the disruption of commercial fisheries.[72] From an engineering perspective, R-o-R hydropower facilities are inefficient because they do not store energy in a large reservoir, which limits operators' ability to compensate for seasonal variability in water flows. Moreover, the lack of storage capacity exposes the R-o-R facilities and their linked electrical grids to episodic crises caused by drought, a risk that will be significantly greater in future decades due to climate change.

Dam and tunnel (D&T) designs avoid the pitfalls of both D&R and R-o-R configurations by delivering water to a power plant located several hundred metres below the dam. These configurations are popular in mountainous areas because they generate large amounts of energy per cubic metre of water. Their superior efficiency reduces the need for a large reservoir, particularly in geographies characterised by abundant rainfall. Sediment removal is often close to zero because many D&T combinations are located high in the watershed, where sediment loads are naturally low or because storage times in reservoirs are short. Similarly, their impact on fish populations is minimal because these types of rivers are characterised by waterfalls and rapids that act as natural barriers.

Civil engineers favour large-scale projects because they resolve supply and demand issues over many years and create an infrastructure legacy that appeals to their professional pride. Utility companies prefer them because they conform to their preferred business model of producing energy and commercialising it to urban and industrial centres. Politicians like them because their construction generates tens of thousands of low-skilled jobs. Financial analysts at multilateral institutions approve them because they can allocate significant capital to an industry with guaranteed cash flow that obviates investment risk.

Experience has shown, however, that some projects in the Amazon are just too large or the climatic assumptions that underpin the energy model are inaccurate – or out of date. Unfortunately, corrupt practices have tainted the objectivity of feasibility and environmental studies that are used to evaluate their economic, social and environmental sustainability. Once hydropower was seen as a sign of progress and embraced by a broad sector of society, but that view has changed in recent decades as environmental and human rights advocates have questioned the sustainability of conventional business models.[73] In advanced economies, there is an emerging consensus that some facilities must be dismantled to restore ecosystem function.[74]

The Guri complex and the Caroni Cascade

The largest hydropower complex in Venezuela is the oldest and least sustainable facility in the Pan Amazon. The complex of dams on the Rio Caroni is operated by *Electrificación del Caroni C.A.* (EDELCA), a subsidiary of the state-owned *Corporación Venezolana de Guayana* (CVG) that broke ground on the first D&R power plant at Macagua in 1956 and initiated the construction of the high dam at Guri in 1963.* Energy production began in 1961 at Macagua and at Guri in 1968; both expanded capacity by adding additional turbines over the next two decades to meet demand for electricity during a period of sustained economic growth. Subsequently, EDELCO added an R-o-R facility below Guri at Caruachi that began operations in 2006, and started construction on a fourth power plant at Tocoma in 2009.

When finished, the four facilities combined will have an installed capacity of 18 GW, making the Caroni Cascade the second-largest hydropower complex in the World.† The 162-metre main dam at Guri is almost twice as high as any other dam in the Pan Amazon; it has flooded 425,000 hectares to create the largest man-made lake in South America. The reservoir has a volumetric capacity of 135 cubic kilometres and flow rates that average about 4,850 $m^3 s^{-1}$, which should be sufficient, theoretically, to maintain maximum water flow for about 320 days per year.

The Caroni hydropower cascade provides almost half of Venezuela's electrical energy, and when rainfall within the watershed falls below normal, waterflow shortages reverberate through the system, immediately plunging the nation into a power crisis. A prolonged drought caused by the *El Niño* phenomenon forced the national power company to impose energy rationing between 2009 and 2013 and again in 2015 and 2016.[75]

The over-reliance on the mega-scale hydropower complex provides three basic lessons: (1) hydropower carries an explicit risk linked to climate variability; (2) diversification of energy sources is essential; and (3) large-scale hydropower suppresses diversification because it outcompetes alternative energy sources during 'normal' years. One positive environmental outcome of the Venezuelan reliance on the Guri complex has been the decision to protect the Caroni watershed with a hundred per cent of the upstream area set aside as a national park or national monument, or zoned as a special watershed reserve with restricted land-use.

The failure to complete the Tocoma dam and power plant is an example of the problems associated with mega-scale hydropower: cost overruns and corruption. The original budget of $US 3.1 billion had increased to $US 9 billion by 2018; an estimated $US 1.5 billion is believed to have been embezzled by governmental functionaries in cahoots with the construction

---

* This roughly coincides with the construction of Glen Canyon Dam on the Colorado River in Arizona, when building large-scale hydropower was considered to represent 'progress'.

† Itaipú (14 GW) on the border between Brazil and Paraguay is larger than Guri (10.2 GW), but is a stand-alone D&R asset, while the Caroni cascade includes four dams over about 80 km that exploit the same resource.

consortium led by Oderbrecht, the Brazilian firm found guilty of fraud and bribery in the *Lava Jato* scandal.[76]

Tucuruí and the Tocantins Cascade

The oldest hydropower facility in the Brazilian Amazon is the Tucuruí D&R complex (8.4 GW) on the lower Tocantins River, about 200 kilometres south of its confluence with the Amazon River delta.* The dam and power plant were built between 1976 and 1984, and its capacity was doubled in 2007; current plans call for capacity to be expanded by another 2.5 GW over the next few years.[77] Tucuruí is owned by Electronorte,† a subsidiary of Electrobras,‡ which supplies most of the electrical energy consumed in the Brazilian Amazon.

Tucuruí was built before the environmental laws that required the completion of an environmental impact study, which allowed its proponents to discount the impacts of a reservoir covering 280,000 hectares and the relocation of an estimated 30,000 citizens, including several indigenous communities.[78] The massive reservoir flooded intact tropical forest and the subsequent methane emissions from rotting vegetation have been estimated at 2.5 million tonnes of carbon annually – a GHG footprint approximately equivalent to a gas-fired power plant.§ The dam has radically altered the ecology of the river and caused massive disruptions to fish populations. Species richness has fallen by 25 per cent below the dam and by fifty per cent within the reservoir, changes that reflect the composition of fish communities and the decline of migratory species. Total fish catch in the reservoir increased in the years immediately following its impoundment

* Some geographers do not include this as a sub-basin of the Amazon watershed, but it is similar in biophysical attributes to major Amazonian tributaries, such as the Tapajós or Xingu rivers.

† Electronorte (*Centrais Elétricas do Norte do Brasil S.A*) is composed of 10 subsidiaries and multiple joint ventures that generate or distribute electricity in the different states of the Legal Amazon: http://www.eletronorte.gov.br

‡ Eletrobras (*Centrais Elétricas Brasileiras S.A.*) is the largest electrical utility in Latin America; via its subsidiaries, it owns about 30% of Brazil's generation capacities and controls 43% of the national transmission grid; the Brazilian state owns 42% of the voting shares, which traded on several stock markets; as of January 2021, the administration of Jair Bolsonaro intended to completely privatise the company by the end of 2021. Source: Reuters, July 2020; https://www.reuters.com/article/eletrobras-privatization/brazil-minister-calls-eletrobras-privatization-a-priority-idUSL2N2ET28H

§ Values are for total carbon based on methane emissions expressed as CO_2-equivalents. By comparison, a coal-fired power plant would release about 4 million tons of carbon to produce the same amount of electrical energy, giving Tucuruí a GHG efficiency 2X compared to coal; in contrast, a more efficient hydropower system, such as Itaipú, has a GHG efficiency ratio of about 150X (dos Santos et al. 2006).

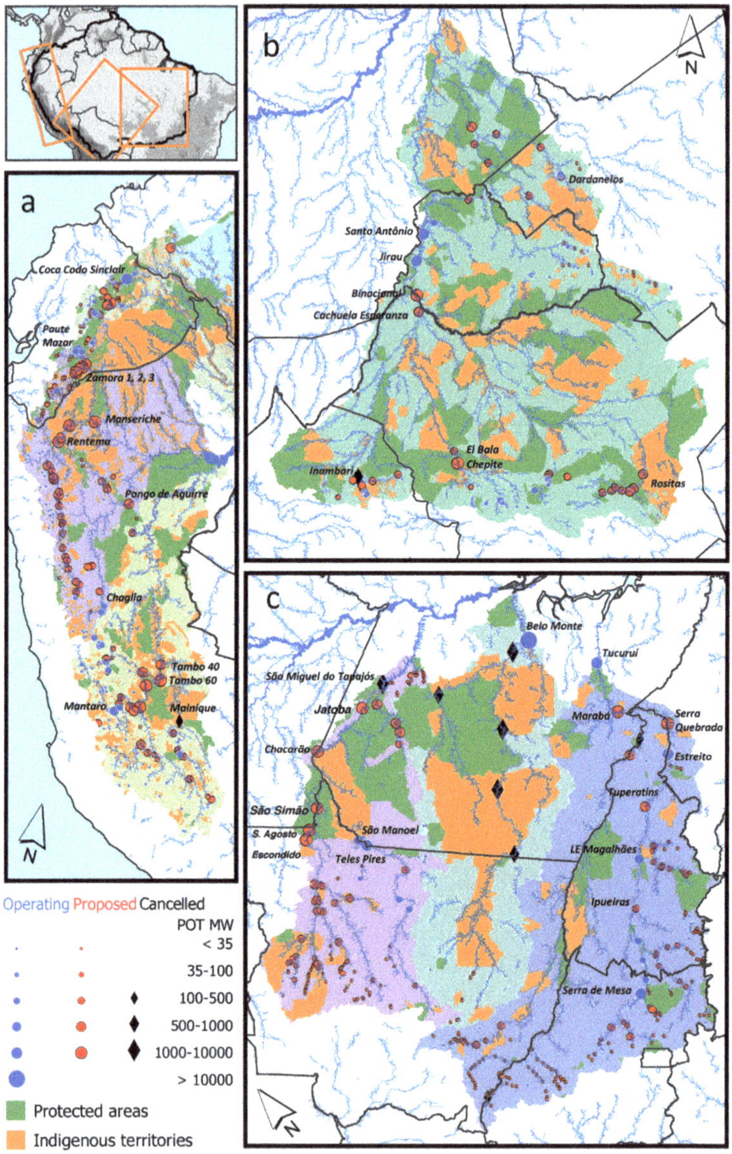

Figure 2.18: The proximity of existing, proposed and cancelled hydropower plants with indigenous lands and protected areas in: (a) the Ucayali, Marañon, Napo, Putumayo and Caquetá watersheds; (b) the Madeira–Mamoré watershed; and (c) Tapajós, Xingu and Tocantins–Araguaia watersheds.

but have declined over time, having stabilised at about eighty per cent of the original value. [79]

The Tocantins is the most heavily exploited watershed in the Brazilian Amazon (Figure 2.18) The Serra da Mesa, a D&R unit, was built in the headwaters near Brasília simultaneously with Tucuruí. These investments were followed by four large-scale D&R projects inaugurated between 2000 and 2010 (Lajeada / Luis Eduardo Magalhães, Cana Brava, Peixe Angical and Sao Salvador)[80] and an R-o-R facility at Esterito in 2014.[81] There are four additional sites on the central sector of the river that are candidates for large-scale dams: Marabá, just below the confluence of the Araguaía and Tocantins, followed by Serra Quebrada, Tuparatins, and Ipueiras (see Annex 2.2; Figure 2.12). The construction of these four dams is required for the development of the Tocantins waterway (see below); none overlap with an indigenous territory or conservation unit. The national energy agency (ANEEL)* has identified 24 additional sites as candidates for medium-scale facilities (< 150 MW), all of which are located relatively high in the watershed. If all these proposed dams were completed, the total installed capacity of the Tocantins would increase from about 13.2 GW to 20 GW.[82]

The Araguaía River, the western branch of the Tocantins basin, is free of dams between its mouth at Marabá and central Mato Grosso, although there are several projects planned for the upper watershed. Two controversial projects have been cancelled by the environmental regulatory agency (*Instituto Brasileiro do Meio Ambiente e dos Recursos Naturais* – IBAMA) based on observations made during the environmental impact studies: Santa Isabela, on the lower river near Marabá and at Cachoeira Couto Magalhães, the site of a scenic waterfall on the border between Goiás and Mato Grosso. There are no plans to establish any dams over the mid-section of the river, a broad flat floodplain that includes the Ilha do Bananal, a massive wetland complex that has been set aside as a protected area or indigenous reserve.†

The Madeira hydropower complex

The Rio Madeira was the next Amazonian tributary to attract the attention of Brazil's hydropower developers. The river is free of rapids as it flows along the western edge of the Brazilian Shield for about 1,300 kilometres between Porto Velho (Rondônia) to its junction with the Amazon River near Itacoatiara (Amazonas). Upstream, the watershed is drained by four massive tributaries (Itenez / Guaporé,‡ Mamoré, Beni and Madre de Dios);

---

* *Agência Nacional de Energia Elétrica* (ANEEL).

† The area is thought to be inhabited by a group of *Avá-Canoeiro* indigenous people who are in voluntary isolation (see Ch. 11).

‡ The river forms the international boundary between Bolivia and Brazil; the Río Itenez is its name in Bolivia and Guaporé is its name in Brazil.

the upper and lower basin are separated by approximately 250 kilometres where four groups of rapids provide a unique opportunity to generate energy from an enormous volume of water collected from a watershed of approximately one million square kilometres.*

Between 2005 and 2015, during the administrations of President Lula da Silva and his successor, Dilma Rousseff, the Brazilian state built two mega-scale dams: (1) Santo Antônio, which is located just above Porto Velho; and (2) Jirau, which is located 110 kilometres further upstream near the border with Bolivia. Eventually, the Madeira hydropower complex may include two additional dams: (3) Binacional, which would be located 150 kilometres upstream from Jirau on the border with Bolivia; and (4) Cachuela Esperanza, which would be located another fifty kilometres upstream on the Madre de Dios within the national borders of Bolivia (Figure 2.18). All of these facilities are, or will be, R-o-R facilities because the sites are not well suited to high dams and large reservoirs. Each will be located just below the rock rapids where a low dam will drive a power plant with small reservoirs between 20,000 and 25,000 hectares.[83]

The Brazilian government fast-tracked the design, environmental review and construction of both Santo Antônio and Jirau. Environmental licences were approved in 2008, and the first turbines were operating by 2012; the inauguration of the last fleet of turbines was finalised in 2017. In order to manage the financial and operational risks inherent in two massive projects constructed simultaneously, the government created parallel bidding processes. Each dam was built and is now operated by different consortia: *Energia San Antonio*† and *Energia Sustentable do Brasil*.‡ Financing was spread among multiple domestic banks with BNDES§ acting as the main source of investment capital; both were included within the IIRSA portfolio due to their potential to facilitate the development of the Madeira waterway (see below).¶ The original combined cost was projected at R$ 25 billion, but technical challenges caused the budget to balloon to an estimated R$ 43 billion by the end of the project.[84]

* This stretch of the river became infamous during the first rubber boom as a strategically important bottleneck that provoked a war between Bolivia and Brazil and the construction of misbegotten railroad between Porto Velho and Guayamirim (see Ch. 6).

† Energia San Antonio: Odebrecht, SAAG Investimentos, Furnas Centrais Elétricas, CEMIG and Caixa- FIP,

‡ Energia Sustentable do Brasil; Engie (formerly Suez), Mitsui, Eletrosul Centrais Elétricas and Companhia Hidro Elétrica do São Francisco

§ BNDES: *Banco Nacional de Desenvolvimento Econômico e Social*

¶ IIRSA, - IIRSA, Peru, Bolivia Brazil Hub, Group 3, Corredor Fluvial Madeira - Madre de Dios – Beni. PBB16, Complejo Hidroeléctrico Del Río Madeira (Hidroeléctrica Santo Antonio E Hidroeléctrica Jirau, $US 18.2 billion: http://www.iirsa.org/proyectos/Principal.aspx

The two facilities are considered to be the most expensive hydropower energy on the continent. The cost of energy is exacerbated by the distance from consumer markets, which required an additional investment of $US 3.8 billion for a 2,400-kilometre high-voltage direct-current (HVDC) (600 kV) * transmission line between Porto Velho and Araraquara (São Paulo).[85] Construction of the HVDC† line was completed in 2013; however, a technical audit in 2017 revealed a design flaw that reduced transmission capacity by 25 per cent. Ironically, the power plants had never operated at full capacity and, consequently, the system avoided damage that could have been caused by an imbalance between generation and transmission capacity.[86]

Both the bidding and the construction oversight processes have been called into question by allegations – and eventually proof – of corruption in the *Lava Jato* scandal. Evidence provided in court indicates that (at least) two per cent of the original contracts were paid in bribes by the construction companies to individual politicians and their parties (see Chapter 5). However, this amount does not capture the inflated cost of the actual construction that is reflected in the eighty per cent cost overruns. Theoretically, these losses should be assumed by the concessionaires operating the power stations, but, as regulated utilities, they will probably be allowed to pass on the total (non-penal) cost to the consumer through inflated electricity bills.

Similarly, the costs associated with environmental impacts are not likely to be assumed by the operating companies. For example, the height of the dams was increased after the conclusion of the formal environmental review, which led to a miscalculation in the capacity of the spillway and the size of the water body located between the two dams. This design flaw led to unanticipated flooding in wet years with extreme levels of waterflow. Although funds were allocated to assist families to rebuild or relocate, the companies have managed to avoid legal liability for the irregularities in the environmental review process. [87]

During the planning and construction phase, the Brazilian government essentially ignored the legal issues related to the potential environmental impacts on an international river. The Bolivian government declined to protest or request an international environmental impact study, which would have been amply justified considering the well-known potential impact on migratory fish (Text Box 2.1). Four consecutive Bolivian governments

---

* IIRSA, Peru, Bolivia Brazil Hub, Group 3 - Corredor Fluvial Madeira - Madre de Dios – Beni, PBB18: Linea de transmisión entre las dos centrales hidroeléctricas del río madeira y el sistema central, $US 3.8 billion: http://www.iirsa.org/proyectos/detalle_proyecto.aspx?h=336

† HVDC: High Voltage Direct Current is well suited for the bulk transmission of electrical power over long distances, because energy losses average about 3% per 1,000 km, compared to 30 to 40% from AC lines at the same voltage level; source: http://edisontechcenter.org/HVDC.html

Text Box 2.1: 'The dams are blocking the fish' – by Michael Goulding

Dams create impassable obstacles to many, if not most, migratory fish species. Unfortunately, this was recently demonstrated for an iconic Amazonian species: the goliath catfish or dourado (*Brachyplatystoma rousseauxii*), which is remarkable for its size (~2 metres long) and long-distance migration. Adult fish swim upstream from the Amazon Delta to the base of the Andes to spawn while the larvae swim downstream to feed in the rich estuarine waters near the mouth of the river. The Madeira population has the longest migratory route of any freshwater fish.

Tragically, the dams at Jirau and Santo Antônio have ended this ancient evolutionary behavior. The Santo Antônio dam has a fish ladder, which operators hoped would allow individuals to bypass the dam; unfortunately, this strategy has failed. In a recent study, only eleven of 471 tagged individuals were recaptured in the fish passage and only two made it into the reservoir above the dam. Their inability to climb the ladder is probably related to their instinctual navigational talent to track natural features of river morphology. Unless biologists and engineers discover how to trick the fish into climbing the ladder, this species or at least this population, is doomed to extinction.

Michael Goulding, Ph.D. is one of the world's leading experts on Amazonian rivers and their biodiversity and was lead author of the *Smithsonian Atlas of the Amazon*.

Sources: Barthem et al. 2017 and Cella-Ribeiro et al. 2017.

remained silent during the planning stages, presumably because they all hoped for the eventual completion of the two upstream dams that would include Bolivia as a partner. The collaboration of the Brazilian and Bolivian governments to complete the two remaining hydrodams is an explicit component of the IIRSA portfolio of investments.* The two projects on the Bolivian border remain on hold, in part because the supply of energy in Brazil currently satisfies regional and national consumption, but this will inevitably change as the Brazilian economy grows. The existing HVDC transmission line can be expanded to provide additional capacity, which enhances the feasibility of the Bolivian projects.

Belo Monte and the Río Xingu

The most controversial hydropower project in the Pan Amazon is the complex on the Rio Xingu near the city of Altamira (Pará). The proposal to build a dam on the Rio Xingu dates from 1979 and, as originally conceived,

* IIRSA: Perú-Brasil-Bolivia Hub; G3 - Corredor Fluvial Madeira - Madre de Dios – Beni; PBB12 - Hidroeléctrica Cachuela Esperanza ($US 1.2); PBB17 - Hidroeléctrica Binacional Bolivia – Brasil ($US 5 billion): http://www.iirsa.org/proyectos/Principal.aspx

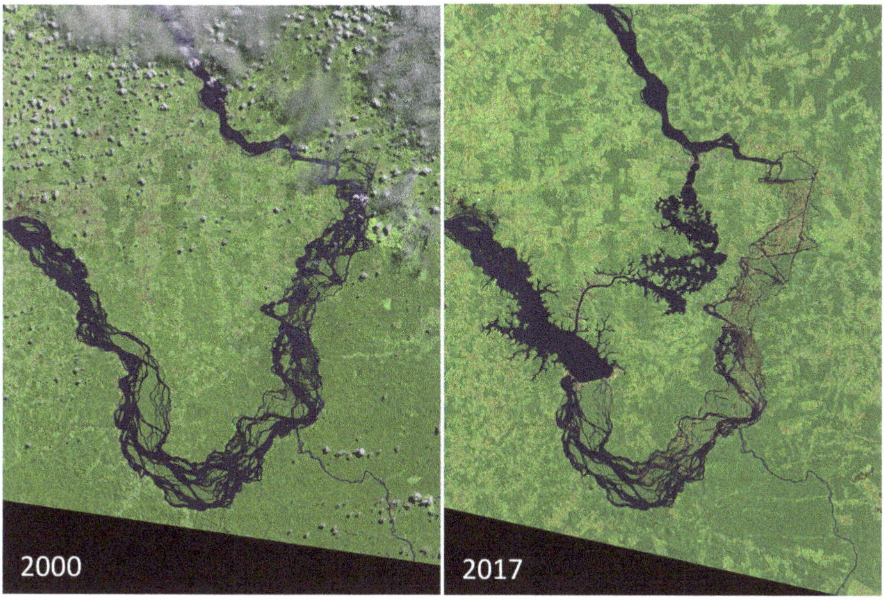

NASA Earth Observatory

Figure 2.19: The Belo Monte hydropower facility generated enormous civil and legal conflicts. It was redesigned to avoid some of the impacts linked to water impoundment.
Source: NASA Earth Observatory, https://earthobservatory.nasa.gov/images/91083/reshaping-the-xingu-river

consisted of a cascade of multiple dams located at different points along the 2,500-kilometre river. Opposition by indigenous groups, academics and civil society held up the project for decades and caused the government to develop, propose and modify multiple iterations of a constellation of dams, reservoirs and power plants.[88]

In 1987, the government proposed a configuration composed of five D&R units that would have flooded more than 1.7 million hectares of forest and the installation of 24 GW of generating capacity.* This proposal was doomed because it would have flooded hundreds of thousands of hectares of forest within indigenous territories, which were being formally constituted following the constitutional reforms of 1988.† The international

* On the Río Xingu: Kararaõ (11.3 GW / 125,000); Babaquara (6.3 GW / 627,400 ha); Ipixuana (2.3 GW 327,000 ha), Kokraimoro (1.9 Gw / 177,000 ha), Jarina 559 MW / 190,000 hectares) and on the Río Iritri (910 MW / 4060,000 hectares) (Fainguelernt 2016).

† Eventually, the watershed impacted by the dams would encompass 16 separate indigenous territories; in 1991 the list included TI Kayapo, TI Menkragnoti, TI Arara, TI Capoto / Jarina, TI Paquimba: https://terrasindigenas.org.br

attention caused by the *Encontro das Nações Indígenas do Xingu* in 1989 forced Electronorte to abandon four of the five proposed sites; nonetheless, the company insisted on building the lowest dam on the river, which had the greatest energy potential due to the volume of water and an elevational drop of more than 100 metres (Figure 2.18). Coincidentally, it was located at a spectacular set of rapids, where the Xingu flows off the Brazilian Shield onto the flood plain of the Lower Amazon, known as the rapids of the Volte Grande (Figure 2.19).*

The revised project still provoked fierce opposition from domestic and international groups, as well as technical observations by the environmental protection agency (IBAMA), which were translated into legal petitions filed by the environmental division of the public prosecutor's office (see Chapter 6). The resistance to the dam was organised by the Kayapó, an indigenous nation led by a particularly astute set of tribal leaders. The government planners changed the name of the facility from Kararaõ, the name of a Kayapó tribe, to Belo Monte, the name of the village at the bottom of the Volta Grande. The final version of the Belo Monte hydropower facility is an unusual two-stage D&R facility: an upper dam (Pimental)† that diverts water via a canal to the lower dam and power plants (Belo Monte). The two dams jointly flood 51,600 hectares, of which 38,000 are located on the floodplain above the Pimental dam.[89] An important aspect of the final design was the determination to maintain (reduced) waterflow in the Volte Grande, a measure intended to mitigate the impact on the livelihoods of the indigenous communities residing between the upper and lower dams.

The final scaled-down version of the project was proposed during the administration of Fernando Cardoso and was enthusiastically embraced in 2002 by President Lula da Silva, who, as leader of the workers' party, was attracted by the opportunity to create 50,000 direct and indirect jobs.[90] Construction started in 2011 and the first of eighteen turbines was inaugurated in 2016 during the administration of Dilma Rousseff. Jair Bolsonaro celebrated the completion of the project in 2019.[91]

The Usina Hidrelétrica de Belo Monte became a reality because it had the support of a popular president and his allies in Congress, and the backing of powerful economic interests. It was built by a consortium of construction companies, all of whom would become ensnared in the *Lavo Jato* scandal.‡ The dam and associated infrastructure are operated by *Norte*

---

* Volte Grande is Portuguese for Big Bend

† At one stage of the redesign process, this was known as the Altamira dam and before that as *Babaquara* (a demeaning term for a rural inhabitant); the name change occurred when that dam was converted from an energy producing D&R facility to a storage impoundment to support the power plant at Belo Monte.

‡ The shareholders of the Consorcio Construtor Belo Monte are: Andrade Gutierrez (18%), Odebrecht (16%), Camargo Corrêa (16%), Queiroz Galvão (11.5%),

The Belo Monte hydropower complex houses 18 turbines with an installed capacity of 11.2 GW (top). The dam was carved out of primary forest between 2008 and 2012 (bottom).

Energia, a consortium of energy utilities, mining companies and pension funds* Most of the energy generated from the unit will be consumed in Southeastern Brazil via two HVDC transmission lines built by an energy conglomerate from China (see below).

The original estimate for the construction of the entire facility was estimated at R$ 16 billion in 2010, but the final bill is estimated to be about R$ 40 billion in 2020.[92] The total cost is difficult to know due to the complex nature of the contracts and cost overruns that characterise hydropower projects.[93] Converting those numbers into dollars is problematic because the Brazilian Real (R$) lost ~70% of its nominal value when compared to the US dollar over the same period.

In spite of its design to minimise upstream impacts, the dam has drastically curtailed the extent, duration and timing of annual floods in the seasonally inundated forests below the dam, which has changed the ecological functionality of key habitat that supports the river's commercial fisheries.[94] Biologists are particularly concerned about how modified water flows will impact the Tabuleiro do Embauba, a beach located just below the dam where tens of thousands of endangered Giant Amazon River Turtles (*Podocnemis expansa*) congregate annually during the breeding season.[95]

A dramatic example of the biological impact of the dam occurred when operators diverted water to the hydropower facility in 2012. Water flow through the Volte Grande was reduced by eighty per cent and provoked the death of sixteen million tonnes of fish from oxygen starvation when they were stranded in isolated pools in the main channel.[96] A $US 1 billion fund was supposed to compensate the local communities for these impacts, but legal disputes and administrative inefficiencies impeded its disbursement, and the communities were forced to manage what presumably was a foreseen impact without the assistance they had been promised in the environmental action plan.[97]

Unfortunately, the hydraulic models used in the design of the complex two-stage facility failed to take into account the impact of periodic droughts and, in 2019, reduced water flows forced operators to shut down all but one of its eighteen turbines.[98] In late 2020, the environmental protection agency (IBAMA) ordered the company to increase water flows through the Volte Grande, which placed it in danger of defaulting on its energy

OAS (11,5%), Contern (10%), Galvão (10%), Serveng (3%), J. Malucelli (2%) e Cetenco (2%). Source: Bloomberg, https://www.bloomberg.com/research/stocks/private/snapshot.asp?privcapid=48811102

* The shareholders of *Norte Energia* are: Eletrobras (49.98%) with the remainder distributed among Companhia Energética de Minas Gerais, Light S.A., Vale, Siderúrgica Norte Brasil, J.Malucelli Energia S.A., Fundação Petrobras de Seguridade Social, Fundação dos Economiários Federais (total 50.02%). Source: https://www.norteenergiasa.com.br/pt-br/ri/composicao-acionaria

supply obligations. If this situation recurs or becomes a chronic event, the financial viability of the entire enterprise will be compromised.[99] The loss of electricity from Belo Monte would cause an energy deficit in the national grid that would be particularly serious in the Southeast and Centre West regions of the country.[100]

The Tapajós Basin and the prevalence of indigenous rights

The Tapajós is a clear-water river and the fifth largest tributary of the Amazon; it drops in elevation from about 800 metres above sea level in the highlands of central Mato Grosso* to less than seven metres at its confluence with the Amazon River near the city of Santarem. More than 45 sites have been identified as suitable for the installation of hydropower facilities within the basin, which includes the lower river and its two major tributary-branches, Teles Pires and Juruena.[101] As with the Caroni, Xingu, and Tocantins, the largest dam with the greatest potential energy is at the bottom of the watershed, where large volumes of water flow off the Brazilian Shield. Ten sites were selected for development, of which four were under construction by 2015 (São Manoel, Sinop, Colíder, Teles Pires); however, the three largest units (São Luis do Tapajós, Jatoba and Chacarão) were sidelined because they are located within or adjacent to indigenous territories (see Figure 2.18).[102]

The proposed dam at *São Luiz do Tapajós* is located at a set of rapids about 350 kilometres upstream from the mouth of the river; it would have consisted of two dams with a total installed capacity of 8.1 GW.[103] In 2016, the environmental protection agency (IBAMA) rejected the request for a construction permit on the advice of the public prosecutor (MPF) and the federal agency overseeing indigenous affairs (FUNAI).[104] The denial was based on the impacts to indigenous communities and their explicit protection by the Constitution of 1988. The proposed R-o-R dam would have permanently flooded about 72,000 hectares, including 12,500 hectares claimed by a Munduruku community. The decision was noteworthy because the community had yet to gain formal recognition for its territorial claim; the ruling thus extended the concept of protection to include indigenous lands outside of indigenous territories (see Chapter 11).

In 2018, the agency that regulates hydropower facilities (ANEEL) announced that all large-scale projects in the Amazon were being placed on hold because of the challenges of obtaining environmental licences for projects that impacted indigenous communities. The decision to forgo

---

* Mato Grosso contains the watershed divide between the Amazon and Paraná-Paraguay basins; its most notable geomorphological features are flat-topped plateaus (*planaltos*) and escarpments (*chapadas*) such as the Planalto de Parecis (Northwest), the Chapada de Guimaraes (Central) and the Planalto de Alto Araguaia (Southeast).

development of large-scale projects in the Amazon was influenced by the poor economic returns and the corruption scandals that plagued the Belo Monte complex and two dams on the Rio Madeira.

It is not clear whether the decision to halt the development of São Luis do Tapajós, Jatoba and Chacorão will be reviewed following the election of Jair Bolsonaro, but the construction of all three dams is essential for the proposed Tapajós/Teles Pires waterway (see below).[105] Less important for waterway development, but with greater potential for large-scale hydropower, are sites on the Rio Juruena (São Simão Alto, Salto Augusto Baixo, Escondido), all of which are located within or adjacent to an indigenous territory. Also under consideration are four medium-scale D&R facilities (Jamanxim, Cachoeira do Caí, Cachoeira dos Patos, Jardim de Ouro) located within the confines of Parque Nacional Jamanxim; these sites were presumed to be included within the category of cancelled projects, but a review by the Bolsonaro administration has revived the possibility they will be put back on the list for future development.*

The Río Trombetas and the Calha Norte

In 2010, Electronorte initiated the construction of a high-tension (500 kW) transmission line to connect the power plants at Tucuruí and Belo Monte; this line was extended north to the Amazon River with connections to Macapá (Amapá) and Manaus (Amazonas).† This ambitious undertaking required the construction of extraordinarily tall tower paillons (300 metres) in order to cross the 2.5-kilometre width of the Lower Amazon.‡ The line roughly parallels the right-of-way of PA-254, an unimproved regional highway, and supplies electricity to dozens of towns and villages that relied previously on expensive energy from small-scale diesel generators (HML #1).

The transmission line will eventually be extended to Boa Vista (Roraima), following the right-of-way of BR-174; however, its completion has been stalled because it traverses the territories of the Waimiri-Atroari indigenous people who have questioned aspects of the environmental impact

---

* These are under development by a consortium of companies that include Eletronorte (Eletrobras), ENGIE, Camargo Corrêa, Cemig, Copel, Électricité de France (EDF) and Neoenergia. Source: Universo Online, https://economia.uol.com.br/noticias/reuters/2020/05/25/estudos-sobre-hidreletricas-no-tapajos-tem-prazo-prorrogado-ate-2021-pela-aneel.htm

† Prior to its connection to the grid, Manaus received about 20% of its energy from the Balbina hydropower dam and 80% from thermoelectric plants powered by natural gas from the Urucú fields located in the Western Amazon (see Ch. 5).

‡ IIRSA, Amazon Hub, Group 05, AMA87, Línea de Transmisión de 500 kV de Tucuruí a Manaos ($US 1.3 billion): http://www.iirsa.org/proyectos/detalle_proyecto.aspx?h=1390

study (See Chapter 11). Ironically, the need for electricity in Roraima has been exacerbated by the termination of a long-standing supply of electricity from the Guri complex in Venezuela.*

Hydropower development has relatively strong support among the inhabitants of both Amapá and Roraima.[106] In Amapá, the D&R unit at Ferreira Gomes (250 MW) and the R o R dam at Cachoeira Caldeirão (219 MW) were built between 2008 and 2016 with minimal opposition.† In Roraima, the proposed development of the Bem Querer (650 MW) on the Rio Branco has strong support from the business sector in Boa Vista.[107] The proposed D&R unit would flood approximately 56,000 hectares but is located on a landscape long impacted by smallholder settlements (HML #55) and would not infringe on any existing protected areas or indigenous territories.‡ Environmental advocates oppose the dam because studies show that it will impede the annual migration of the goliath catfish (see Text Box 2.1) and impact the geochemistry of the Rio Negro.[108]

Even more controversial is the recently announced plan to resurrect the development of the Cachoeira Porteira (3 GW) D&R complex on the Rio Trombetas. First proposed in the 1980s, it was included in the national inventory of potential dam sites in 2006 by the Empresa de Pesquisa Energética (EPE), the agency within the *Ministerio de Minas and Energia* responsible for developing the nation's long-term energy strategies.[109] In that report, however, the EPE observed that two of these potential sites were located within a conservation unit, two in indigenous territory, and two in an INCRA-sponsored settlement deeded to Quilombola§ communities. The Cachoeira Porteira project was subject to a preliminary evaluation in 2014, and following consultation with the local communities, the public prosecutor's office recommended that all of the proposed sites in the Trombetas watershed be removed from consideration for future development.[110] The development of Cachoeira Porteira was resurrected in January of 2019 as part of Jair Bolsonaro's plan to extend the BR-163 highway to the border with Suriname and open the Calha Norte to development.[111]

---

* In 2020, the Bolsonaro administration was considering declaring an emergency to force the construction of the transmission line. Source: Reuters, https://www.reuters.com/article/energia-linhao-indios-idLTAKBN1ZK2LN

† Both these facilities are 100% privately owned and include investors from Europe and China; financing was provided by BNDES.

‡ The environmental review is being paid for by the *Programa de Parcerias de Investimentos* (PPI), a federal programme that supports the private sector investment. Source: PPI, https://www.ppi.gov.br/uhe-bem-querer-rr

§ The quilombolas are descended from escaped slaves who settled in remote communities across the Brazilian Amazon; they are considered to be a 'traditional community' with special use rights. See Ch. 4.

Bolivia seeks an energy export model

Bolivia's hydropower is based on medium-scale facilities located in a geographical region optimally suited for D&T systems. The oldest of these is in the Zongo Valley, which starts at 4,700 metres above sea level with a small reservoir (~20 hectares) that feeds water into one of eleven power plants, with a total installed capacity of 188 MW. Several similar D&T systems have been built in a region known to geographers as the Elbow of the Andes, where annual rainfall exceeds 6,000 millimetres across an altitudinal drop of 4,000 metres and a horizontal distance of less than forty kilometres (Figure 2.18). Within this area, the state-owned electrical company, *Empresa Nacional de Electricidad Bolivia* (ENDE), has recently undertaken a series of investments that will double the nation's hydropower capacity over the next few years by expanding capacity at Corani (275 MW) and Miguillas (250 MW),* as well as adding a new unit at the Ivirizo cascade (290 MW).†

Bolivia privatised its electrical energy sector in the 1990s, but Evo Morales renationalised the industry in 2006 as part of a policy to use public investments in energy and infrastructure to drive economic growth and, more importantly, generate revenues for the national treasury. In the first decade of its reincarnation, ENDE focused on building out the national grid while relying on subsidised natural gas to generate power. Eventually, however, ENDE began to focus its investments on hydropower with the explicit goal of creating a surplus of electricity for export to neighbouring countries.[112] Most of these investments were financed from the national treasury and are leveraged with loans from multilateral institutions; however, ENDE has engaged Chinese companies and hopes to entice Brazilian institutions to finance the mega-scale projects on its northern border (see below).

In addition to the D&T facilities under construction in the highlands of the Elbow of the Andes, ENDE plans to build a 600 MW D&R facility on the Rio Grande where it emerges from the Andes. Originally conceived in the 1970s, the dam will have serious environmental impacts, including the displacement of 500 Guaraní families who inhabit the valley, which will be flooded by a 40,000-hectare reservoir. Like all such dams located in the Andean foothills, its reservoir will capture massive amounts of sediment (Figure 2.17) and block the migration of important commercial fish species.[113] Rositas will be a dual-purpose dam and divert water for irrigation that will

* Corani Cascade: Santa Isabel (148 MW) San José (124 MW), Banda Azul (133 MW), Villa Jorka / Santa rita (44 MW), Ambrosia (85MW), Santa Barbara (85MW); Miguilleas Cascade (440 + 200 MW); Misicuini (120 MW), Oquitos (125 MW) and Molineros (100MW): see http://www.ende.bo/#

† In 2017, Synohydro was awarded a $US 553 million contract to build the Ivirizu D&T facility (290 MW); this is not a loan and is financed with funds from the Bolivian treasury: https://fundacionsolon.org/2020/02/18/china-bolivia-deuda-comercio-inversiones

catalyse the expansion of industrial agriculture across 500,000 hectares of dry forest. Ironically, the diversion of water for irrigation will reduce the replenishment of the aquifers that underlie the alluvial plain of Santa Cruz and limit the irrigation potential on the country's most important agricultural landscape.[114]

The Rositas project enjoys the support of all major political parties and, more importantly, the enthusiastic backing of the business community in Santa Cruz. Potential future hydropower investments on the Rio Grande include five more upstream dams that would add another 2.5 GW to the system.* The EIA for the project was executed internally by ENDE with support from the IDB, while the construction contract has been awarded to a consortium led by Synohydro with financing from the ExIm Bank of China.[115] A small but determined group of social and environmental activists has organised a campaign to stop its development by deploying a seldom-used class-action civil procedure that theoretically can halt initiatives that do not comply with health, safety, environmental or social regulations (see Chapter 6). The five upstream D&R units will prolong the economic life of Rositas because they will capture an enormous amount of sediment but, if they are not built, the lifespan of Rositas will be among the shortest in the Amazon – 135 years.[116]

Other investments that are part of Bolivia's strategy to export electrical energy include the development of the two R-o-R dams on the Madeira River† (see above) and two large-scale D&R projects on the Rio Beni. The original concept for the Rio Beni, first proposed in 1952, envisioned a 200-metre dam at the Angosto de El Bala, which would have created a massive reservoir covering almost 400,000 hectares. Different iterations of the project were rejected as uneconomic in 1958, 1976 and 1998; meanwhile, the project became even more controversial due to the creation of two high-profile protected areas: Parque Nacional y Tierra Indígena Pilón Lajas (1992) and Parque Nacional Madidi (1995).

The latest configuration is based on a feasibility study contracted by the government in 2015, which calls for a two-stage D&R design with a 168-metre dam and 68,000-hectare reservoir at the Angosto de Chepite, a gorge located about fifty kilometres upstream from an R-o-R facility at Angosto de El Bala. This scaled-down version would limit the total area flooded to 78,000 hectares overall and to 10,000 hectares within the two

---

* Rositas (600 MW); La Pesca (740 MW); Peña Blanca(520 MW); Ocampo (320 MW); Las Juntas (172 MW); Cañahuecal (500 MW) y Seripona (420 MW). See, Fundación Solon: https://fundacionsolon.org/rositas/

† IIRSA: Perú-Brasil-Bolivia Hub; G3 - Corredor Fluvial Madeira - Madre de Dios – Beni; PBB12 - Hidroeléctrica Cachuela Esperanza ($US 1.2);; PBB17 - Hidroeléctrica Binacional Bolivia – Brasil ($US 5 billion): http://www.iirsa.org/proyectos/Principal.aspx

protected areas; an estimated 4,000 individuals would need to be relocated.[117] Opposition among indigenous communities, the tourist industry and environmental advocates is strong, but the poor economics of the facility are the largest obstacle to its development.[118]

As recently as 2018, the Bolivian government hoped to invest about $US 25 billion by 2025 to quintuple installed capacity from about 1.2 GW to more than 10 GW, approximately five times greater than the estimated domestic demand in 2025.[119] Electricity exports would require significant investment in regional transmission systems, such as those proposed by the IDB in 2017: Bolivia – Brazil (500 kV), Peru – Bolivia (250 kV) and Bolivia – Chile (250 kV).[120] As of 2020, however, Bolivia remains isolated from potential markets, in spite of being a signatory to the *Sistema Andino de Interconexión Eléctrica* (SINEA), an IIRSA-like initiative to integrate the regional energy grids. The ability to pursue these capital-intensive investments is limited by Bolivia's deteriorating financial status,* and it is unlikely that financial and technical assistance from China or Brazil will allow the country to implement its ambitious schemes in the short to medium term.[121]

Peru embraces the private energy sector

Peru has enjoyed historic levels of economic growth for more than two decades, mostly due to the expansion of the minerals sector, which is a large consumer of electrical energy. Peru has also made strides in providing affordable electrical energy to its citizens, including to small towns and rural areas via a national grid that now integrates most coastal and highland regions as well as the major colonisation zones in the tropical lowlands.† Iquitos remains the only large urban area unconnected to the national grid, and there are plans to make that link over the short-term.‡

The consolidation of the Peruvian electrical sector has been accompanied by robust growth in generation, which has increased by about 150 per cent since 2005. Approximately 35% of the country's electricity comes from hydropower, which is sourced approximately equally from the Pacific and Amazon watersheds. Projects on the western slope of the Cordillera Occidental often enjoy subsidies because of their irrigation potential, but

---

* The balance of payments has fallen from a high of $US 2 billion annually in 2012 to a negative $US 2 billion in 2019; the reversal has caused the country's foreign reserves to fall from of $US 14 billion to less than $US 2 billion in 2020. Source: The World Bank Group, https://data.worldbank.org/country/bolivia

† IIRSA COSPILAN Perú-Brasil-Bolivia Hub, G01: G1 - Corredor Porto Velho - Puertos del Pacífico; PBB59 Línea de Transmisión San Gabán - Puerto Maldonado ($US 23 million).

‡ That connection would require a high tension line through a roadless wilderness area characterised by swamp forest and has generated opposition from environmental advocates.

those resources are inherently limited due to the rain shadow of the Andes. Consequently, most future expansion will take place on the eastern slope of the Andes (Figure 2.18).

In the early 1990s, the state-owned, vertically integrated electricity monopoly was broken into multiple units dedicated to generation, transmission and distribution. The role of the state was limited to developing policy, enforcing regulations and granting concessions.* By 2017, 54 different corporate entities were supplying energy to the Peruvian grid; the largest of them supplies about twenty per cent of the country's electrical energy.[122] The build-out of Peru's electricity sector occurred after the discovery of the Camisea gas field (see Chapter 5), which favoured investment in thermal generation because it is less capital-intensive when compared to hydropower. Moving forward, however, the state intends to decrease its reliance on thermal energy and promote investment in renewable energy. The official plan includes small-scale hydropower but does not consider medium- and large-scale hydropower as a 'renewable' energy.[123]

The exclusion of Peru's traditional hydropower sector is ironic, because it can objectively be characterised as 'sustainable'. Most existing power plants are D&T systems that exploit topographic drops located high in watersheds, which limits their environmental and social impact. Even projects with relatively large reservoirs have a much smaller spatial footprint when compared to D&R facilities located at lower elevations. As in Bolivia, they consist of two to three medium-scale units organised as a cascade, with water recycled from one power plant to another – for example, Mantoro I & II (1,000 MW),† Santa Teresa I & II (500 MW) and San Gabín II & III (313 MW).[124]

The opposition to conventional hydropower is the consequence of failed attempts to develop mega-scale D&R facilities in the foothills of the Andes, where large rivers pass through a narrow gorge (see Annex 2.2).‡ In 2008, the governments of Alan García (Peru) and Lula da Silva (Brazil)

---

* The regulatory agency OSINERGMIN (*Organismo Supervisor de la Inversión en Energía y Minería*) oversees the technical, administrative and financial activities of three sectors: mining, hydrocarbons and electrical energy; concessions are managed via the *Dirección de Concesiones Eléctricas*, a separate entity within the ministry.

† Peru's largest hydropower complex, Mantaro I & II, collects water at the 100-ha reservoir behind the Tablachaca dam and diverts it though a 19-km tunnel to the Santiago Antúnez de Mayolo power plant (798 MW), which then recycles that water via the river channel to the Restitución power plant (210 MW) located 10 km downstream.

‡ Madre de Dios Basin: Inambari Gorge (2.2 GW). Ucayali Basin: Urubamba: Pongo de Manique (942 MW); Tambo: Tambo-40 (1.2 GW), Tambo-60 (579 MW), Tambo-P Prado (620 MW); Ene: Sumabeni (1.1 GW), Paquitzapango (1.4 GW); Mantaro: Viscatani (750 MW), Cuquipampa (800 MW). Huallaga Basin: Pongo

signed a memorandum of understanding that sought to integrate the two countries' electricity markets by developing Peruvian hydropower resources using Brazilian technology and capital.* The agreement was signed during the surge in investments in dams on the Tocantins, Xingu, and Madeira and involved the same corporate entities that were designing, financing, and building the hydrodams in Brazil. The initiative focused on localities in southern Peru, where the Corridor Interoceanico offered a right-of-way so transmission lines could connect with the HVDC line that services the dams on the Madeira River.†

The first mega-project to be pursued was a D&R complex on the gorge where the Rio Inambari exits the Andes (Figure 2.18).‡ The project was immediately embroiled in controversy as opponents coordinated actions at local, national and international forums. They successfully demonstrated that the project was seriously flawed and the national environmental agency (OSINERGMIN) declined to approve its EIA in 2011. The impact of that outcome reverberated through the business community, and none of the other proposed mega-scale projects was seriously considered for development. The Peruvian Congress shelved the agreement in 2014 and all of the concessions have subsequently expired.

Although mega-scale projects have proven unviable, several medium- and large-scale investments have been completed successfully. Prior to 2005, the Amazonian watersheds housed eleven power plants with about 1.7 GW of capacity; this was increased by six units and 1.3 GW by 2018 and is projected to grow by another nine plants and 2.7 GW by 2023 (see Annex 2.2). The sector's largest operators are domestic companies, many of which have formed joint ventures with investors from Norway, Italy, Spain, France, Israel, Chile, and the United States. Most of the development has taken place on the Marañon watershed, which has the greatest potential among the three major Amazonian tributaries (Figure 2.20). The large-scale facility at Chaglla on the Rio Huallaga (456 MW), which was inaugurated by Odebrecht of Brazil in 2016, was sold to the China Three Gorges Corporation in 2017 following the *Lava Jato* corruption scandal (see Chapter 6).

---

de Aguirre (750 mW). Marañon Basin: Pongo de Manseriche (7.5 GW), Pongo de Escuptebraga (1.8 GW), Pongo de Retama (1.5 GW).

* The proposed scheme required ElectroPeru and Electrobras to sign long-term purchase agreements, while Brazilian construction companies would build and operate the facilities (Odebrecht, OAS, Andrade Gutiérrez, Camargo Correa); key financing would be provided by BNDES.

† The proposed dams included: Inambari (2 GW), Sumabeni (1.1 GW), Paquitzapango (2 GW), Mainique (940 MW), Tambo 40 (1.3 MW) and Tambo 60 (600 MW) (Orcotorio-Figueroa 2020).

‡ The dam would have been located just a few kilometres upstream from the Huepetuhe gold field (see Ch. 4) and the reservoir would have flooded part of the Interoceanic Corridor highway (see above).

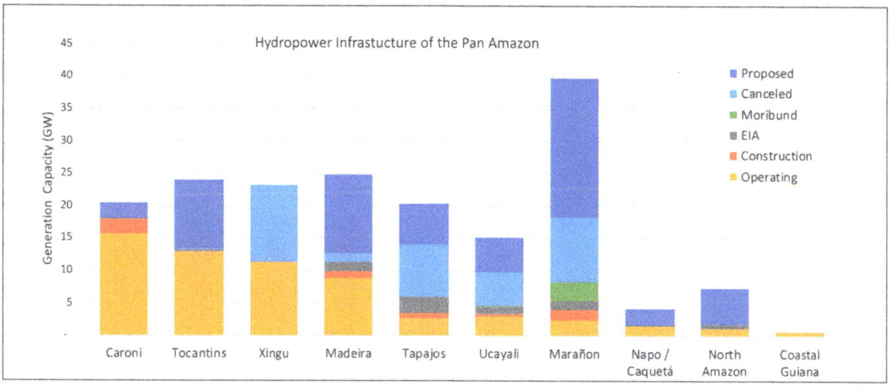

Figure 2.20: The existing, planned and potential hydropower capacity of the major Amazonian river basins.

Data source: modified from RAISG 2021.

Although impacts from hydropower development have been limited to date, that may change if the industry starts exploiting the hydrological resources of the Rio Marañon. Following the demise of the project on the Rio Inambari, President Alan García issued an executive order declaring the development of hydropower and irrigation infrastructure on the Rio Marañon to be in the national interest, a designation that accelerates development by easing environmental reviews and facilitating access to public moneys.[125] The configuration of that river basin makes it particularly attractive for conventional D&R facilities; unfortunately, its geochemistry and biodiversity make it particularly susceptible to environmental impacts (Figure 2.17).

As in southern Peru, the most problematic projects are located in a gorge where the Marañon transects the Andean foothills; this 200-kilometre-long section of the river is the site of three putative mega-scale projects.* It is highly unlikely that any of the three D&R projects would ever pass an environmental review, because they would inundate land deeded to Awajún communities who are renowned for their opposition to projects that infringe upon their territorial rights.† More likely is the development of the 25 D&R projects on the 500-kilometre section of the river above the regional capital

* The Quechua word for narrow gorge is *pongo* and these sites are known as: Pongo de Manseriche (7 GW), Pongo de Escuprebraga (1.8 GW), Pongo de Rentama (1.5 GW)

† Opposition to hydropower during this period was galvanised by the Baguazo, a peaceful protest that turned violent when the Awajún objected to policies that promote private investment in the Peruvian Amazon (see Ch. 11).

of Jaen into the highlands of Central Peru. This section is attractive to civil engineers because the river collects run-off from the mountains situated to the east and west, while the V-shaped valley provides multiple opportunities for deep reservoirs with significant storage capacity. [126] In 2014, Odebrecht initiated feasibility studies on four potential dam sites;* however, none had advanced beyond the EIA stage when the *Lava Jato* corruption scandal effectively ended that company's ability to execute projects in Peru.†

Only a single hydropower dam is being promoted for the non-Andean sections of the Peruvian Amazon: The Mazán R-o-R project would be located on a narrow isthmus separating the Napo and Amazon rivers 25 kilometres downstream from Iquitos and forty kilometres upstream from the mouth of the Río Napo. It would include an eleven-metre dam across the Napo that would divert a fraction of that river's current through a canal across the three-kilometre isthmus to the main channel of the Amazon River. The power would satisfy energy demand from Iquitos and, potentially, supply a proposed transmission line between Yurimaguas and Iquitos.‡ There are numerous reasons to doubt the technical and financial viability of this project,[127] but the governor of Loreto continues to insist it remains viable and important for the development of the region.[128]

Ecuador chooses hydropower with assistance from China

The recent history of hydropower in Ecuador is similar to that described for Bolivia and Peru, particularly with respect to its recent expansion and the predominance of D&T systems that exploit the geographic advantages of the Andean Cordillera. Like Bolivia, the state has assumed a near monopoly on generation and has turned to China for technological assistance and financial capital. In 2013, Ecuador sourced about forty per cent of its electricity from hydropower but increased its contribution to 58 per cent by 2019, while national consumption grew by forty per cent. Approximately, eighty per cent of Ecuador's installed hydropower is generated by power plants on the Napo, Pastaza and Santiago watersheds.[129]

---

* Cumba-4 (825 MW), Lorena (634 MEW), Chadin-2 (600 MW); Rio Grande I (600 MW); Río Grande II (150 MW) (Orcotorio-Figueroa 2020).

† Odebrecht maintains only one hydropower asset in Peru: The Olmos Project on the Huancabamba River, a tributary to the Marañon that diverts water through a 19-km tunnel under the Andes to a pair of power plants, prior to supplying water to approximately 40,000 hectares of irrigated farmlands near the Pacific Coast.

‡ Critics observe that the building of both a tension line and a dam is redundant, while the city recently inaugurated a 100 MW thermal power plant powered with fuel produced by the Iquitos Refinery.

Much attention has focused on the recently inaugurated large-scale project with the unusual name of Coca Codo Sinclair.* The design of this new power plant takes advantage of a 650-metre vertical drop that includes a natural waterfall and multiple rapids across about 75 kilometres of river channel. The D&T design diverts water through a 25-kilometre tunnel to turbines at the bottom of the bend with installed capacity of 1.5 GW. It is the largest power plant in Ecuador and supplies about 25 per cent of the national electricity demand. Like most mega-scale projects, it was conceived in the 1950s and underwent multiple iterations before being commissioned by the Ecuadorian government in 2009. It is operated by a subsidiary of the national electrical monopoly *Corporación Eléctrica del Ecuador* (CELEC), but was built by a Chinese construction company, Sinohydro and financed by the Export-Import Bank of China.[130]

The direct environmental impacts appear to be moderate. The reservoir is only 300 hectares and sediments are flushed back into the Coca River periodically; the barrier to fish migration was pre-existing because of the presence of the San Rafael Falls, which are located between the dam and the powerhouse. The facility is located between two national parks, PN Cayambe Coca and the PN Sumaco Galeras, but all of the infrastructure is located on a river valley previously impacted by the right-of-way for an oil pipeline and deforestation along the Troncal Amazónica (see above). The area is sparsely populated, and there is little evidence that local populations have objected to its development.

An unexpected event now threatens the physical integrity of the hydropower complex. In February 2020, the river eroded a channel underneath the lava dyke that created the waterfall and left the once magnificent cascade a mere trickle of water. Once the solid rock of the lava dyke was isolated from the erosional forces of the river, however, the river started to erode through the loosely consolidated sediments of the Coca river floodplain.[131] By December 2020, the top of the waterfall, now a series of rapids, had migrated approximately three kilometres upstream towards the dam and intake tunnel.†

At the current rate of erosion, the top of the cascade could reach the dam in approximately two years, which could force the operators to close

---

* The facility bears the name for its location at a large river bend (codo = elbow) on the Río Coca, which was first surveyed in the 1920s by the geologist Joseph H. Sinclair (Sinclair and Wasson 1923).

† The erosion caused the failure of an oil pipeline that crossed the river Coca, causing significant oil spill that impacted dozens of indigenous communities downriver and eventually reached the Amazon River. Source: Reuters, 8 Apr. 2020. https://www.reuters.com/article/us-ecuador-oil-spill/ecuador-scrambles-to-contain-oil-spill-in-amazon-region-idUSKCN21R2JU

the tunnel leading to the power plant.* The vertical drop between the dam and the bottom of the falls is similar to equivalent stretches of the river below the falls and above the dam; consequently, civil engineers will probably be able to safeguard the installation. Nonetheless, the incident calls into question the competence of the original feasibility study and the EIA, as well as the wisdom of building a strategically important infrastructure asset at the base of an active volcano (El Revantador).[132]

In spite of its size, the Coca Codo Sinclair is not the largest hydropower complex in Ecuador; that distinction belongs to a cascade of dams and powerplants on the Rio Paute, a tributary of the Santiago River. The Centro Hidroeléctrico Hidropaute began with the inauguration of the Molina D&R unit in 1983 (500 MW) and its expansion in 1991 (600 MW), which was followed by the construction of the D&T units at Mazar in 2010 (170 MW) and Sopladora in 2018 (385 MW). A fourth unit, Cardanillo (596 MW), is under development, and when it is completed, the total combined capacity from the four facilities will exceed 2.1 GW.† The CH Hidropaute is another subsidiary of CELEC, and its newest addition was built by the China-Gezhouba Group with loans from the ExIm Bank of China.

Further expansion is being planned for the Rio Zamora, the southern branch of the Santiago River, with a projected combined capacity of between 5 and 7 GW. The most likely design foresees three D&R units in a cascade though a narrow valley that transects the Cordillera del Condor. Such a design would not be unlike the proposed 'Pongo' dams on the Rio Marañon and the El Bala/Chepite project on the Rio Beni.[133] Ecuador has many options for renewable energy, particularly wind and geothermal, but hydropower is the largest component of its future development plans.[134]

The future of hydropower in the Pan Amazon

The past two decades saw a massive increase in hydropower across the Pan Amazon. The Brazilian government has scaled back investment in mega-scale hydropower projects but continues to pursue development of medium and large-scale projects.[135] Cost overruns at hydropower dams across the world have exposed them as poor investments (Figure 2.21),

---

* The dam is located approximately 19 km and 205 vertical metres above the riverbed below the falls, which translates into a vertical drop of 0.0103 m per metre of riverbed, compared to 0.0112 m/m over a similar stretch of the river below the falls and 0.0091 m/m above the dam.

† Mazar is a dam-and-reservoir facility at the top of the cascade (2100 m.a.s.l with a 159-m drop), from which water is fed into dam-and-tunnel systems located at Molina (1,310 m.a.s.l and 6-km tunnels), Sopladora (930 m.a.s.l. and 4.7-km tunnel) and Cardanillo (925 m.s.a.sl. and 4-km tunnel). Source: CELEC, Hidropaute, https://www.celec.gob.ec/hidropaute/

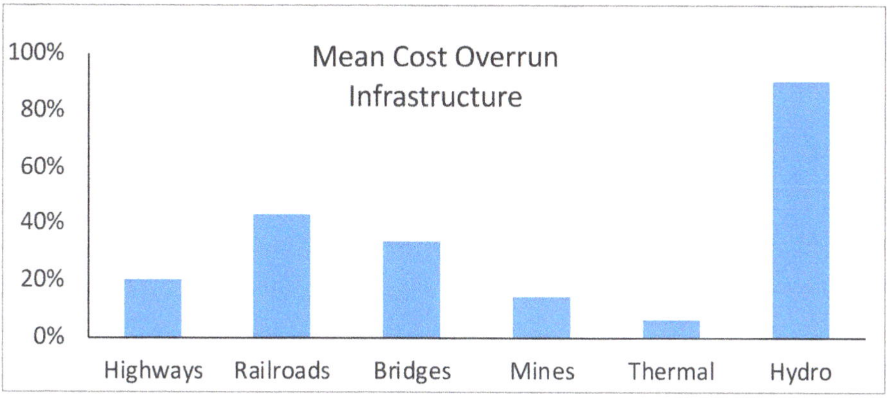

Figure 2.21: Global averages for cost overruns on infrastructure.
Data source: Ansar et al. 2014.

and the national development bank (BNDES) has curtailed investments because it does not have the financial resources for mega-scale projects.

Another factor that may impact future development is the pending privatisation of Electrobras, the state-owned corporation that was the instigator of most hydropower projects. A corporation managed for the profit of its shareholders may be less willing to allocate private capital for large-scale infrastructure investments with dubious economic returns, particularly if the global trend toward solar and wind lessen the need for large-scale hydropower within national development plans. [136]

A policy contingent on economic and political considerations can be reversed using similar criteria, and resistance to the policy to end mega-scale hydropower in the Amazon appeared within days of its announcement.[137] The administration of Jair Bolsonaro has sent clear signals that large-scale hydropower is part of its development agenda, including a prolongation of the window to complete environmental reviews for dams on the Rio Tapajós and reinitiating feasibility studies for a large-scale dam on the Rio Trombetas that had been abandoned in 2014.

Bolivia's strategy to export electricity will face tough competition from corporations investing in state-of-the-art industrial solar parks on the Pacific coast.* The government is motivated by the belief that a macro-economic

* Chile's largest solar powered facility (220 MW) is located in the Atacama near the Bolivian border (http://www.acciona-energia.com/areas-of-activity/photovoltaic/major-projects/el-romero-solar-pv-plant/); Peru's largest solar power plant (144 MW) is near Moquegua in Southern Peru: https://www.enelgreenpower.com/where-we-are

policy based on infrastructure investments will drive economic growth, but the proposed model is dependent on foreign investment capital. It is not clear, however, whether Chinese investors will risk their capital on Bolivia's highly uncertain business model. Geopolitical criteria might convince Brazil to purchase the energy from the mega-scale dams on the Rio Madeira where Brazil has compatible energy assets.

In Peru, the metallurgical sector will continue to drive investment in energy infrastructure, a phenomenon that will accelerate as demand for copper from advanced economies increases with their transition to electric vehicles.* The growth of solar and wind energy will limit the demand for high-impact hydropower, but their intermittent nature may motivate investments in hydropower to ensure grid stability. This energy model would favour D&R facilities because they are most adept at managing variations in supply. Ecuador will continue its expansion of hydropower and could experience a step-change in demand if the government follows through on the proposed electric railroad system (see below).

Colombia will continue to develop its hydropower resources, but most of these will focus on non-Amazonian watersheds; Venezuela is unlikely to invest in hydropower over the medium-term due to its current economic crisis. Neither Surinam nor Guyana will need to invest in hydropower because they can generate electricity at very low cost using natural gas that will soon be abundant from offshore platforms (see Chapter 5). Nonetheless, in 2020, the newly elected government of Guyana expressed a desire to resurrect the abandoned Amalia Falls hydropower project that died a natural death in 2013 due to financial considerations.[138]

The role of China adds an element of uncertainty to the trajectory of the hydropower sector across the Pan Amazon. Its engineering companies have a record of building facilities on time and on budget, which could change the economic calculus that currently makes mega-scale hydropower unattractive. Similarly, its state-owned financial institutions have the capacity to mobilise the capital required by large-scale projects and can leverage their status with the Chinese state to mitigate the risk of default. Finally, they have shown the flexibility to operate in multiple regulatory environments: they have acted both as investors purchasing distressed assets in Brazil and Peru and contractors providing turn-key solutions for the design and construction of hydropower assets in Ecuador and Bolivia.

* The current generation of electrical vehicles use approximately three times as much copper as a standard gasoline powered vehicle

Global Competition Drives Bulk Transport Systems

The modification of the rivers in Brazil has been driven by energy development, but investment in dams has the potential of creating an economically attractive option for shipping Brazil's farm exports to overseas markets. In the 2019/2020 crop year, Mato Grosso produced 35 million tonnes of soybeans, an increase of about forty per cent over 2015/2016.[139] In 2017, nine per cent was consumed within Mato Grosso, six per cent was shipped to other Brazilian states and the remainder was exported directly to overseas markets, mainly China (66 per cent) and Europe (twelve per cent). Soybeans are transported to export terminals by a combination of truck, rail and barge. Producers in Mato Grosso are overly reliant on truck transportation because they operate on frontier landscapes without access to modern rail systems. Until recently, most exported their production via ports in São Paulo and Paraná due to logistical constraints that limited transit to ports on the Amazon River. The need for bulk transport systems has increased over the last decade, not only due to an increase in soy production, but because of a parallel growth in exports of maize, which is increasingly grown in rotation with soybeans (see Chapter 3).

Truck transportation is inherently inefficient, and the cost of moving grain by truck 2,000 kilometres to the Port of Santos in São Paulo has fluctuated between $US 80 and 120 per ton over the last decade.[140] Farmers in Mato Grosso compete in global markets with producers from other counties with significantly lower transportation costs. In Argentina, the distance is typically less than 400 kilometres, and its legacy railroads provide producers with an efficient transport option; for example, the cost between Cordoba and the port of Rosario is only $US 30 per ton.[141] The distance between producing landscapes and export terminals in the United States is longer, but the US has highly efficient bulk transportation systems, including an extensive rail network and the Mississippi waterway, where the shipping costs from Iowa to New Orleans (2,300 kilometres) is only $US 20 per ton.[142]

Opportunities to lower transportation costs are the most obvious intervention point for improving the competitiveness of Brazil's soybean exports. Lower transportation costs will enhance profitability for Mato Grosso's farmers because commodity traders pay producers the international price for their harvest minus the cost of transport and logistics.* High transportation costs are essentially a rent that benefits truckers and

---

* This includes storage fees at regional grain silos, transport, storage and other fees at the export terminal, as well as the cost of marine transport. Depending upon volatile global commodity markets, logistics and transport can represent as much as 40% or as little as 10% of the FOB price paid at the export terminal. The cost of transportation is a powerful incentive for adding value to grain production within Mato Grosso (see Ch. 3).

Investment in bulk transport systems is a priority as inefficient trucking systems are replaced by waterways and railroads: Parking lot of a truck service centre on BR-163 in northern Mato Grosso.

impedes farmers' ability to invest in their productive capacity, which explains why agribusiness has made investments in bulk transport systems a national priority.

As the cultivation of soy exploded across the Southern Amazon, the global commodity traders* and logistical companies† began to invest in transportation systems via the Amazon River. By 2013, approximately thirty per cent of the soy cultivated in central Mato Grosso was exported via Amazonian ports, and by 2017 this proportion had increased to seventy per cent.[143] The shift in transportation patterns has not been due to a reduction in grain shipped via southern ports – that amount has remained relatively constant – but to increased production within the Amazon.

Amazonian options are organised into three logistical corridors with different multimodal combinations of truck, barge and rail (Figure 2.22; Table 2.1). Simultaneously, the cost competitiveness of the southern option has been improved by the extension of a rail line into southern Mato Grosso that has reduced the truck transport component by 1,400 kilometres. Producers

* Often referred to as the ABCD traders, because of the dominance of **A**DM, **B**unge **C**argill, and Louis **D**reyfus; in Brazil the group also includes Amaggi (Brazil), Glencore (Swiss) and COFCO (China).

† Each commodity trader operates their own logistical supply chain, but they share the market with Brazilian shipping companies: *Companhia Norte de Navegação e Portos* (Cianport); *Hidrovias do Brasil; Transportes Bertolin and Chibatão.* Source: ANTAQ - *Agencia Nacional de Transportes Aquaviáriuas.*

Investment in bulk transport systems is a priority as inefficient trucking systems are replaced by waterways and railroads: A barge convoy on the Río Madeira (top); rail cars being loaded at the grain terminal at Rondonópolis, Mato Grosso (bottom).

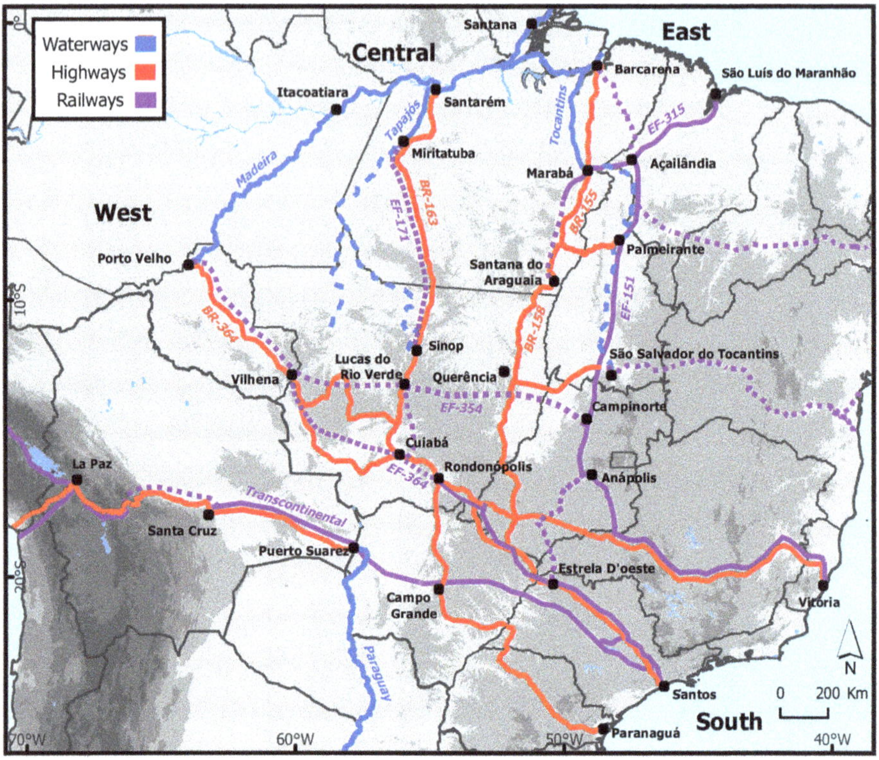

Figure 2.22: Bulk transport systems under evaluation for moving grain production from Mato Grosso to export terminals. These can be organised into four different corridors: West, via BR-364, Porto Velo and the Madeira River to Itacoatiara; Central, via BR-316 to Miritituba on the lower Tapajós or Santarem; East, via BR-158/EF-151/EF-315 to São Luis do Maranhão; and South, via Rondonópolis to Santos via EF-364. Rail and waterways are replacing truck-based systems along all three routes (see text). Data source: Instituto Mato Grossense de Economia Agropecuária (IMEA).

across central Mato Grosso now have four options for commercialising their production:

1). West: Farmers truck their grain via BR-364 to Port Velho, where it is loaded on barges for trans-shipment to Itacoatiara (Amazonas), a port located 75 kilometres east of Manaus, just across the channel from the mouth of Rio Madeira.

2). Central: Farmers truck their grain north via BR-163 to Santarem or Miritituba (Pará) on the Rio Tapajós for loading onto barges for trans-shipment to ocean-going ports on the Amazon River.

3). East: Farmers can use truck transport (BR-158 / 155) to Marabá (Pará) or the recently completed rail line, Ferrovia Norte – Sur (EF-151), both of which connect with the pre-existing rail line, Estrado Ferro Carajás (EF-315), between Marabá and São Luis do Maranhão.

4). South: Farmers across the region have the option of trucking their grain to a large logistical complex at Rondonópolis (Mato Grosso) where they can transfer their grain to the Ferrovia Norte (EF-364)[*] for trans-shipment to the grain terminals at Santos (São Paulo). [144]

The expansion of Amazonian ports has improved the profitability of Mato Grosso farmers, but those benefits remain limited by a combination of factors, including a lack of silos and port facilities, poorly maintained roads and, in the case of Porto Velho, an additional 1,000 kilometres of barge transport.[145] The long-awaited completion of BR-163 in 2019 eased some of these constraints and motivated commodity traders and logistic companies to invest in silos and barge-loading facilities at Miritituba,[†] fleets of high-capacity barges and the expansion of the grain terminals at Barcarena and Santana.[‡]

Between 2013 and 2017, exports via the three northern corridors represented an annual saving of about $US 100 million when compared to the previous option of trucking production to Santos or Paranaguá (Paraná). The transition from truck to rail for the 1,400 kilometres between Rondonópolis and Santos, however, represented an even greater savings of approximately $US 200 million.[§] The transition to rail highlights the potential savings that will accrue as the logistical systems continue a transition away from truck transport and toward rail and barge systems. The estimated savings from future investments in bulk transport systems, perhaps as large as $US 1 billion annually, will lower the logistical cost of producers from central Mato Gross to levels that are cost competitive with their global competitors (Table 2.2). Not surprisingly, farmers are strong advocates for the construction of railways and the modification of rivers so they can function as commercial waterways.

---

* Sometimes referred to as the Malha Norte (Northern Network)

† Estação de Transbordo de Cargas (ETC) at Miritituba in 2020: ETC-Cargill, ETC-Cianport; ETC-Hidrovias do Brasil; ETC-Unitapajós (Amaggi & Bunge); ETC-LDC (Louis Dreyfus Company); ETC-Bertolini (Transportes Bertolini); ETC-Chibatão. Source: ANTAQ.

‡ Barcarena: Terminais de Uso Privado (TUP) Ponta da Montanha (ADM & Glencore); TUP Unitapajos (Amaggi & Bunge), TUP Villa do Conde (Hidrovias do Brasil,); Santana: Cia Norte de Navegação e Portos. There are also gran terminals at Itacoatiara (TUP Hermasa / Amaagi) and Santarem (TUP Cargill).

§ The potential savings are even greater, because rail transports only about 50% of the grain shipped from Mato Grosso to Santos. Source: *Globorural*, 10 May 2018, Safra de Mato Grosso usa mais ferrovia para chegar ao Porto de Santos: https://revistagloborural.globo.com/

Table 2.1: Estimated transportation options available to producers between 2010 and 2016 for grain shipped from three municipalities in Mato Grosso; cost estimates are derived from the USDA.[218]

Point of Origin *Corridor*	**Export terminals (Route options)**	**Truck (km)**	**Barge (km)**	**Rail (km)**	**Sea (km)**	**Truck ($/t)**	**Barge ($/t)**	**Rail ($/t)**	**Sea ($/t)**	**Total ($/t)**
Sapazal										
West	Itacoatiara via Porto Velho (BR-364 + R. Madeira)	1,450	1,100		21,760	66.14	19.00		26.67	**112**
South	Santos (various highways)	2,260			20,400	100.86			25.00	**126**
	Santos via Rondonópolis (EF-364)	725		1,400	20,400	35.07		24.00	25.00	**84**
Sinop										
Central	Santarem (BR-163)	1,300			21,160	59.71			25.93	**86**
	Santarem via Miritituba (BR-163 + R. Tapajós)	1,000	300		21,160	46.86	8.09		25.93	**81**
South	Santos (various highways)	2,250			20,400	100.43			25.00	**125**
	Santos via Rondonópolis (EF-364)	700		1,400	20,400	34.00		24.00	25.00	**83**
Querencia										
East	Itaqui via Marabá (BR-226)	1,989			16,602	89.24			20.35	**110**
	Itaqui via Palmas (EF-151 + EF-315)	718		1,150	21,203	34.77		20.43	25.98	**81**
	Barcarena via Marabá (BR-155)	1,875			20,818	84.36			25.51	**110**
South	Santos (various highway)	1,789			20,400	80.67			25.00	**106**
	Santos via Rondonópolis (EF-364)	853		1,400	20,400	40.56		24.00	25.00	**90**

Table 2.2: Estimated cost of transportation options for grain shipped from three municipalities in Mato Grosso based on probable future development; cost estimates are derived from the USDA.[219]

Point of Origin / *Corridor*	Export terminals (Route options)	Truck (km)	Barge (km)	Rail (km)	Sea (km)	Truck ($/t)	Barge ($/t)	Rail ($/t)	Sea ($/t)	Total ($/t)
Sapazal										
West	Itacoatiara via Porto Velho (EF-364 + R. Madeira)	50	1,100	1,400	21,760	6.14	19.00	23.09	26.67	**75**
	Itacoatiara via Pimenteiras (R. Guaporé + R. Madeira)	250	2,500		21,760	14.71	38.09		26.67	**79**
South	Santos (EF-354 + EF-364)	50		2,400	20,400	6.14		36.73	25.00	**68**
Sinop										
Central	Santarem (EF-171)	50		1,300	21,160	6.14		21.73	25.93	**54**
	Santarem via Miritituba (EF-171)	50	300	1,000	21,160	6.14	8.09	17.64	25.93	**58**
	Santarem (R. Tapajós/ Tele Pires)	50	1,600		21,160	6.14	25.82		25.93	**58**
South	Santos (EF-364)	50		2,250	20,400	6.14		34.68	25.00	**66**
Querencia										
North	Itaqui (EF-360 + EF-151+ EF-315)	100		1,780	21,203	8.29		28.27	25.98	**63**
	Barcarena via Marabá (BR-155 + R. Tocantins)	1,314	550		21,203	60.31	11.50		25.98	**98**
	Barcarena via Palmerirantre (R. Araguaía + R. Tocantins)	872	1,250		20,818	41.37	21.05		25.51	**88**
South	Santos (EF-354 + EF-151 + EF-364)	100		1,800	20,400	8.29		28.85	25.00	**62**

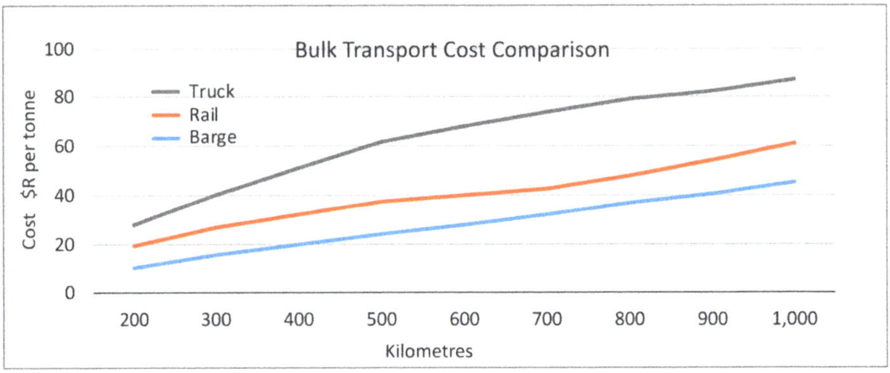

Figure 2.23: The comparative cost of three different bulk transport system as a function of distance for farmers in central Mato Grosso, Brazil.
Data source: ANTAQ 2008.

The Institute of Economic Analysis (IPEA) of Brazil assessed the comparative costs and benefits of waterways when compared to railroads.[146] Based on a standard unit of 1,000 kilometres, a projected lifespan of 25 years, and an average transport of 10 million tonnes of goods per year, fluvial transportation was estimated to be about 35 per cent less expensive when compared to railways (Figure 2.23). This estimate excluded capital investment in dams, however, which are essential components on all Brazilian waterways; for this reason, the construction of hydropower assets constitutes a massive subsidy. The IPEA included neither the cost of the environmental damage that this type of infrastructure would inflict on the Amazon ecosystem, including modifications to the hydrology of its river systems, nor the increased deforestation that might be stimulated by the expansion of agriculture.

Waterway options

The main stem of the Amazon River has provided access to ocean-going cargo ships for centuries, including modern container ships that service the manufacturing sector in Manaus and ore-carriers that haul bauxite from near Oriximiná (Pará) and iron ore and manganese from Santana (Amapá). The first modern grain terminal, built at Itacoatiara in 1998 across from the mouth of the Madeira River (Amazonas), was followed in 2003 by one at Santarem (Pará) at the mouth of the Tapajós, in 2014 at Barcarena near Belem (Pará) at the mouth of the Tocantins, and in 2016 at Santana (Amapá) on the north side of the Amazon delta. These terminals are currently receiving grain from barge loading facilities located at the top of the three major

© Caio Pederneiras / shutterstock.com

The Amazon Waterway is a transportation corridor with a service sector based in Manaus that caters to dozens of communities.

transportation corridors, each of which has developed or hopes to expand an industrial waterway, known as a *hidrovía*.*

Above Manaus, on the section of the river known as the Solimões, river traffic declines by several orders of magnitude because there are no industrial mines or agricultural landscapes producing commodities at scales required to support a bulk transport system. A few ocean-going cargo ships hauling timber are known to operate sporadically from Iquitos (Peru)[147] or deliver heavy machinery required by the oil and gas industry at Coari (Amazonas) and Iquitos. River traffic consists largely of riverboats providing fuel and consumer goods to riparian communities (HML #3), cruise liners catering to tourists on the Rio Negro and timber for the manufacturing sector in Manaus or for export to overseas markets. There is an uptick in activity on the tri-border area around Tabatinga (Brazil), Leticia (Colombia) and Santa Rosa de Yavarí (Peru).

The aspiration of creating an industrial waterway between Brazil and the Andean republics is a major component of the IIRSA investment portfolio (Figure 2.1), which includes eighteen projects organised in four groups with a total budget of $US 530 million.† This basket of proposed

---

* The term *hidrovía* is used in both Spanish and Portuguese; Brazil also employs the term *aquivía* as a synonym.

† IIRSA. Amazon Hub: G1 - Acceso a la hidrovía del Putumayo ($Us 3 million); G2 - Acceso a la hidrovía del Napo ($25 million); G3 - Acceso a la hidrovía del

The extractive industries use the Amazon waterway as a bulk transport system and for logistical support for the gas plant at Coari (top); the grain terminal at Santarem is one of several ports that exports soy and maize cultivated in the Southern Amazon (bottom).

and completed projects represents a laudable effort to provide sustainable transportation options that minimise the need for roads. Unfortunately, most river port investments on the Andean piedmont are akin to the *Field of Dreams* approach to infrastructure planning: *If you build it, they will come.*[*] There is no commercially relevant bulk cargo in either direction between Brazil and the Andean nations, while the manufactured goods produced in Manaus are not likely to be competitive with similar products from East Asia.

That does not mean, however, that these are not 'good' investments. They provide essential services to many isolated communities of the region that are being denied road access in the name of forest conservation. As such, their development should not be viewed as an investment that will pay for itself but as a subsidy to support nature-based livelihoods. Ironically, the one example where a waterway might be able to support a self-sufficient commercial transportation system is opposed by environmental and social advocates: the Hidrovía Amazónica between Iquitos, Yurimaguas, Pucallpa and Saramariza in Peru.[148] Resistance is based, in part, on the impacts caused by dredging operations on problematic sections of the river but also from current operators and inhabitants, including indigenous communities, who fear they will be exploited by the concessionaire that has been awarded the contract to upgrade and operate the waterway.[†]

Río Madeira

The largest tributary of the Amazon[‡] has functioned as a fluvial corridor for millennia and was a major commercial artery during the rubber booms of the nineteenth and twentieth centuries. Since 1998, it has experienced a commercial revival due to the transport of soy and maize, and Porto Velho is now one of the busiest ports in the Amazon (Figure 2.22). Historically, navigation between the Madeira and its tributaries of the upper watershed

---

Huallaga – Marañón ($53 million); G4 - Acceso a la hidrovía del Ucayali ($Us 55 Million); G6 - Red de Hidrovías Amazónicas ($us 305 million): http://www.iirsa.org/proyectos/Principal.aspx?Basica=1

* The film *Field of Dreams* is a 1989 American sports fantasy drama starring Kevin Costner, whose character builds a baseball diamond in a cornfield. The term 'if you build it, they will come' refers to a common error by would-be entrepreneurs who invest time in money based on a passionate belief in a vison rather than rigorous due diligence and feasibility studies.

† The waterway will be operated by *Sociedad Concesionaria Hidrovía Amazónica S.A.* (COHIDRO), a joint venture between the Hidalgo Group of Peru and Sinohydro of China; the consortium signed a 20-year contract in 2017 valued at approximately $100 million. Source: https://www.ositran.gob.pe/anterior/hidrovias/hidrovia-amazonica/

‡ It is the largest by volume of water at 31 m^2/sec, which represents 14% of the total at the mouth of the Amazon (Ziesler and Ardizzone 1979).

(Itenez/Guaporé,* Mamoré, Beni and Madre de Dios) were blocked by multiple rock outcrops distributed along approximately 200 kilometres between Porto Velho and Guajará-Mirim.† About half of these rapids have been flooded by the dams built at Santo Antonia y Jirau, while the rest would be flooded by the two dams that have been proposed for future construction (see above).‡

In 2013, the Brazilian Transportation Ministry evaluated the feasibility of extending the Madeira waterway beyond Porto Velho and confirmed that the construction of the two additional dams would resolve the physical blockages that impede navigation.[149] If all four dams were enhanced by the construction of locks, navigation via the Madeira waterway would extend fluvial transportation to central Mato Grosso (1,200 kilometres via the Guaporé), the agricultural frontiers of the Chapare and the Guayaros (1,000 kilometres via the Mamoré), the Bolivian Yungas (500 kilometres via the Rio Beni), and southern Peru (1,000 kilometres via the Madre de Dios).§

The agency that manages Brazil's commercial waterways, *Agência Nacional de Transportes Aquaviários* (ANTAQ), considers the Guaporé to be an economically important waterway and has, on occasion, supported its extension.[150] However, the 2013 feasibility study identified physical attributes, such as seasonally shallow water and excessive sinuosity, that limit its utility as an industrial waterway while highlighting the presence of indigenous territories in Bolivia and Rondônia that would complicate its development. The agribusiness sector shows no interest in developing the Guaporé as a bulk transport waterway, presumably because it would not be cost competitive with rail (see below).

The most enthusiastic supporters of a greater Madeira–Mamoré waterway have always been Bolivian politicians who dream of converting their lowland provinces into agricultural breadbaskets.[151] This aspiration may be another example of the if-you-build-it-they-will-come syndrome; however, the regional government of Beni has approved a new land-use

* The river forms the international boundary between Bolivia and Brazil; Río Itenez is its name in Bolivia and Guaporé is its name in Brazil.

† This stretch of the river became infamous during the first rubber boom as a strategically important bottleneck that provoked a war between Bolivia and Brazil and the construction of a misbegotten railroad between Porto Velho and Guayamirim (see Ch. 6).

‡ IIRSA, G3 - Corredor Fluvial Madeira - Madre De Dios – Beni, PBB12: Hidroeléctrica Cachuela Esperanza ($US 1.2 billion; PBB16, Complejo Hidroeléctrico del Río Madeira (Hidroeléctrica Santo Antonio e Hidroeléctrica Jirau), PBB17, Hidrolectica Binacional Bolivia p- Brasil: http://www.iirsa.org/proyectos/

§ IIRSA, G3 - Corredor Fluvial Madeira - Madre De Dios – Beni: PBB13, Hidrovía Ichilo Mamoré; PBB14, Navegabilidad del Río Ben; PBB 15, Hidrovía Madre de Dios y Puerto Fluvial; all are in the 'planning stage' with no projected budgets: http://www.iirsa.org/proyectos/

planning framework that legalises the conversion of one million hectares of savanna habitat for the cultivation of soy, corn and rice, which has the support of both the central government and the agribusiness sector of Santa Cruz (Chapter 4).

Río Tapajós

The Brazilian Transportation Ministry and the agribusiness sector view the Tapajós River as a strategically important waterway located between the most productive farmlands of Mato Grosso and grain terminals on the Amazon River. It is, however, a technically challenging river with multiple rapids that limit its navigability during twelve months of the year. The Tapajós can be used as an industrial waterway only if a series of dams (between three and six) are constructed to create lakes that flood the most problematic rapids and regulate water flows necessary to ensure transit by barge (Figure 2.18).[152]

The waterway was selected for priority development in the late 2010s, and construction was initiated on four dams on the upper the basin: Teles-Pires, São Manual, Sinop and Colíder.[153] The environmental licensing process was initiated for São Miguel do Tapajós, the lowest and largest dam, while development of the two dams located on the mid-section of the river (Chacarão and Jatoba) were placed on the docket for future evaluation. Following the political backlash that accompanied the constriction of the Belo Monte hydropower complex and the corruption scandals that marred the dams on the Rio Madeira, political support for the São Miguel project declined, and the environmental authorities successfully denied approval of its environmental licence (see Chapter 11). This determination has effectively killed the Tapajós–Teles Pires waterway project because the three lower dams are all keystone elements, and if any one of them is eliminated, the waterway is rendered nonfunctional.

Currently, the Tapajós is navigable for 300 kilometres between its mouth and the twin towns of Itaituba and Miritituba, located across from each other at the top of a naturally inundated valley situated at the base of the northern border of the Brazilian Shield.* Itaituba is the larger of the two towns and capital of the municipality, but Miritituba is located near the intersection of BR-163 and BR-230, making it the preferred site for building logistical facilities for loading barges for trans-shipment to grain terminals on the main stem of the Amazon River. Unless the Brazilian Congress acts

* Sea levels were about 120 m lower at the last glacial maximum during the Pleistocene (~ 20,000 years before present); this caused the Amazon to carve a deep valley that was subsequently refilled with sediments delivered by white-water rivers originating in the Andes. Clear-water rivers, such as the Tapajós and Xingu, lack high sediment loads; consequently, their paleo channels and floodplains became 'drowned valleys' that geomorphologists refer to as *ria* lakes (Fricke et al. 2017).

to create a legal mechanism that overrides the 1988 Constitution or the Murunduku indigenous people modify their opposition to dams (Chapter 11), the Tapajós waterway will be limited to the section between Miritituba and the Amazon River.

Tocantins and Araguaia

These two parallel rivers drain the landscapes between the highlands of central Mato Grosso and western Bahía; consequently, they are strategically located to provide a transportation option for two of Brazil's most important agricultural landscapes. Both are candidates for waterway development but differ in their physical characteristics and the complexity of the social and environmental challenges that accompany the development of an industrial waterway (Figure 2.18). The two rivers come together at Marabá, then flow north toward Tucuruí Dam, which was outfitted with locks in 2010, creating the first essential asset on the Araguaía–Tocantins waterway system.*

The Araguaía requires less investment in dams and locks (Santa Isabela and Araguanã) but suffers from seasonally low water levels due to its broad flat floodplain (Figure 2.18). In contrast, the Tocantins has a relatively confined river valley but with numerous rock outcrops that require the construction of several dams and their associated locks. In 2013, the transportation ministry decided to limit its waterway investments to the Tocantins, in part, because of the existence of two dams (Estreito and Lajeado) and the planned construction of three additional hydropower units (Marabá, Serra Quebrada and Tuperintins).[154] The decision not to pursue the Araguaía waterway avoided the inevitable confrontation over the operation of an industrial waterway on the border of a large indigenous territory (TI Araguaía) and two protected areas (PN Araguaía and PE Cantão).†

The inevitable confrontation with environmental advocates began with the difficulty in obtaining an environmental licence to modify the channel of the Rio Tocantins at the Pedral do Lourenço, a massive rock outcrop situated between the Tucuruí Reservoir and the city of Marabá. This stretch of the lower Tocantins is navigable during highwater seasons, but an industrial waterway must be open twelve months of the year to be economically viable. Proposals to dynamite a channel through the Pedral would impact the nesting habitat of two species of aquatic turtles, an endemic species of dolphin and migratory catfish that have already suffered population declines caused by the construction of the Tucuruí Dam.[155] Delay in the approval of the licence has frozen the use of the waterway for

---

* The locks were built 30 years after the inauguration of the dam at a cost of R$ 1.6 billion (~$US 300 million; source: G1 Globo.com, 1 July 2017, Demora nas obras do Pedral do Lourenço afeta a economia no Pará: https://g1.globo.com/

† TI: Terra Indigena (indigenous reserve); PN: Parque Nacional; PE: Parque Estadual

The lock at the Tucuruí dam, built between 2007 and 2010, has opened the Tocantins Waterway for future development (top). If the government's plans come to fruition, the run-of-river dam at Estreito will be modified to incorporate a lock that would extend the waterway several hundred kilometres upstream.

almost ten years, in spite of strong support from both national and state authorities * and the demands of private investors who have committed significant financial capital in developing port facilities in Barcarena, Marabá and Imperatriz (Tocantins).

Development of the Tocantins waterway above Marabá is less than certain. To reach the border with the state of Goiás – the goal of the transportation ministry – would require the construction of three more large-scale dams at the cost of between $US 5 and $10 billion, and the installation of locks at all five hydropower facilities at approximately $US 100 million each.[156] Other impediments include the need to build access roads between the farms in Mato Grosso and the waterway, which would entail building a highway across an enormous wetland complex (Ilha de Bananal) that is part of the Araguaía indigenous territory.

Paraná – Paraguay

This non-Amazonian waterway is navigable between Corumbá (Mato Grosso do Sul) and ports in Argentina and Uruguay; it is used by mining companies operating near Corumbá but is not an export corridor for agricultural commodities from Brazil. It is, however, an essential transportation asset for Bolivia's agroindustry, whose producers in Santa Cruz (HML #31) are 2,000 kilometres from the nearest Atlantic port. Pacific ports are closer at 1,500 kilometres but they are also situated on the other side of a 5,000-metre pass via roads that are not designed for high-capacity grain trucks. Bolivia's soybean producers are absolutely dependent upon the Paraguay River and a legacy railroad system (see below), without which they would be unable to compete in global markets.

Railroad Development

The challenges to developing waterways have focused investor's attention on railroads. In 2020, the Amazon Hub of the IIRSA portfolio included eight rail projects, which were either completed (2), under construction (1) or on the drawing boards (5). The estimated total budget ranges between $US 20 and $US 30 billion, but even the larger number is an underestimate because it excludes several of Brazil's most ambitious initiatives.† The Brazilian rail-

* The EIA was commissioned in 2014 and is financed by the Program de Parcerias de Investimentos, a federal programme that fast-tracks infrastructure investments; the R$ 650 million contract was approved in 2016. The project has the support of Jader Barlbaho, an influential senator from Pará. Source: *Associaçao Brasileira de Operadores Logisticos*, https://abolbrasil.org.br/posts/derrocamento-do-pedral-do-lourenco-deve-ter-inicio-em-2021/

† IIRSA, Amazon Hub, Group 5 G08: G8 - Conexión Ferroviaria Porto Velho - Nordeste Meridional de Brasil (5 Projects @ $US 9 Billion); G05: G5 – Conexión Entre La Cuenca Amazónica y El Nordeste Septentrional De Brasil (5 projects

road sector is an unusual mixture of private and public corporations, and a concessionaire system where public assets are leased to private companies that commit to large capital investments.* Starting in 2008, the federal government launched an initiative to expand the rail network, particularly new lines that would penetrate the agricultural landscapes of the Southern Amazon and the Amazon-adjacent landscapes of Northeast Brazil, which are collectively referred to as MATOPIBA.†

Following is a description of the major rail investments underway in the Brazilian Amazon (Figure 2.22).

Ferrovia Norte (EF-364).

This is probably the most lucrative railroad in Brazil. Built between 1998 and 2012, it has dramatically lowered the cost of commodity transport from the farms of central and southern Mato Grosso to the port of Santos (São Pualo). Operated by Brazil's largest private railroad company (*Rumo Logístico*),‡ the line currently reaches Terminal Ferroviário de Rondonópolis, a massive logistical facility with the capacity to transship twelve million tonnes per year. The ninety-year concession for EF-340 stipulates the rail line will be extended to Cuiabá and, potentially, to Porto Velho and Santarem.[157] Over the near term, Rumo plans to extend Ferrovia Norte to the town of Lucas do Rio Verde (Mato Grosso), where it will intersect with an East – West railroad under development (see below).[158]

@ $US 10 billion); cost estimates are based on an assumption of $2 million / km: http://www.iirsa.org/proyectos/

* The system is a legacy of state-sponsored corporations that have undergone cycles of nationalization (1950s) and privatization (1990s) and a hybrid model during the administrations of Lula da Silva and Rousseff who used a state-owned corporation (VALEC Engenharia Construcoes e Ferrovias S/A) to undertake the construction of strategic rail lines, including several within the Legal Amazon. VALEC was dissolved in 2019 and its regulatory functions were passed to the *Agencia Nacional de Transporte Terrestres* (ATTN); railroads directly administered by VALEC were leased to private operators via public auction: https://portal.antt.gov.br/en/ferrovias

† The acronym is derived from their postal abbreviations: MA (Maranhão), Tocantins (TO), Piauí (PI), Bahía (BA); agroindustrial farming is expanding on landscapes that are geologically and edaphically similar to the Planalto de Mato Grosso. Source: EMBRAPA, https://www.embrapa.br/en/tema-matopiba/perguntas-e-respostas

‡ Rumo Logistico is a subsidiary of Cosan S/A, a conglomerate with interests in sugar cane, bioenergy and natural gas; it was created in 2008 with the acquisition of América Latina Logística. Rumo operates an integrated rail network consisting of four concessions totaling ~ 12,000 kilometres of rail, 12 transshipment platforms and three terminals in the port of Santos: https://www.cosan.com.br/en/about-cosan/cosan-group/

The extension of the rail network has enhanced the capacity of agroindustry to export soy and maize. The Ferrovia Norte links the port of Santos (São Paulo) with the logistical centre at Rondonópolis in Mato Grosso (top). The recently completed Ferrovia Norte Sur spans Tocantins to connect with the Estrada de Ferro Carajás in Pará (bottom).

Ferrogrão (EF-171).

This is a new initiative that was not included in the strategic transportation plans formulated in 2011, nor is it included within the IIRSA portfolio.* It is a direct response to farmers' demands for an economically attractive export option from central Mato Grosso. The Ferrogrão† will run parallel to BR-163 for 935 kilometres between Sinop (Mato Grosso) and Miritituba (Pará) (Figure 2.5 and Figure 2.6). Its projected capacity of sixty million tonnes per year approximates the combined soy and maize produced in Mato Grosso in 2019.[159] The $US 1.5 billion investment is being coordinated by a Brazilian engineering firm, *Estação da Luz Participações* (EDLP) with support from the ABCD‡ commodity traders. The federal government is seeking to fast-track its construction by supporting the environmental review process via the *Programa de Parcerias de Investimentos* (PPI), which is managed from the President's office to facilitate private sector investment in public infrastucture assets.[160] A formal tender process for building and operating a 65-year concession is expected to be convened in 2021. The proposal is unusual in that it would award a monopoly to the concessionaire to operate trains over the railroad, a privilege that would be revoked if the rail line were ever linked to the national rail network.[161]

Not surprisingly, the construction of the Ferrogrão is opposed by environmental advocates and indigenous groups, who maintain that the railroad will promote settlement on the narrow corridor along BR-163 (HML #17). The region suffers from an epidemic of illegal activities, particularly land grabbing and unregulated deforestation, phenomena they contend would be supercharged by the influx of thousands of migrant workers for the railroad's construction.[162] Critics also contend that a reduction in transport costs will increase deforestation across the farm landscapes of northern Mato Grosso, which among other impacts would degrade the water resources of the indigenous territories along the Xingu River.[163]

The most conflictive zone is a 75-kilometre stretch through the heart of Parque Nacional Jamanxim, where the BR-163 right-of-way has a width of only 200 metres. Congress approved a measure that would widen the right-of-way of BR-163 through Jamanxim National Park, a precondition for obtaining an environmental licence from IBAMA.[164] Construction is planned to start in 2021 and be completed by 2025, but like most infrastructure projects in Brazil, it is not proceeding according to the projected timeline.

---

* *Sistema Nacional de Viação*, Lei nº 12.379, de 6 de Janeiro de 2011.

† The name *Ferrogrão* is a neologism that translates as the 'grain train'.

‡ ABCD: ADM, Cargill, Bunge, Luis Dreyfus and Amaggi Group.

The Ferrogrão will parallel BR-163 between Sinop (Mato Grosso) and Miritituba (Pará); the most controversial stretch will cross Jamanxim National Park where the current right-of-way is only 200 metres wide. Construction is contingent on the approval of an environmental evaluation being financed from the President's office by the Programa de Parcerias de Investimentos.

Ferrovia Norte Sul (EF-151).

As the name implies, this rail line will span the country and integrate railroads in the North, Central and, eventually, South of Brazil.* In its current configuration, the line has been split into three sections: The southern component is operated by *Rumo Logístico* for 1,500 km between Porto Nacional (Tocantins) and Estrella de Oriente (São Paulo); at its southern terminus, the line connects with the Rumo network that terminates at their massive port facilities at Santos.

* Amazonas, G 5: AMA73 Ferrovia Nova Transnordestina Fase I ($US 3 b); AMA76; Ferrovia Nova Transnordestina Fase II ($US 1.2 billion); AMA77, Ferrovia Norte-Sur Fase I, Belem - Açailândia ($ 1 billion); AMA78, Ferrovia Norte-Sur Fase II, Açailândia – Palmas, ($US 2 billion): http://www.iirsa.org/proyectos/

The central component between Porto Nacional and Açailândia (Maranhão) is operated by *Valor da Logística Integrada* (VLI),* which also owns the concession for the Estrado Ferro Carajás (EF-315) between the mining complex at the Serra de Carajás and the Port of Itaqui at São Luis de Maranhão. The combination of EF-315 (660 kilometres) and EF-151 (750 kilometres) provides the first fully integrated bulk transport option for farmers from Eastern Mato Grosso, Tocantins, western Bahía, and southern Maranhão. Its capacity was enhanced by the simultaneous construction of fifteen grain-loading platforms located between Anápolis (Goiás) and São Luis de Maranhão. Because it parallels BR-153 through a consolidated frontier (HML #6, #7, #14) inhabited by farmers and ranchers, its construction has been relatively free of social conflict. Its completion has relieved traffic bottlenecks on the regional highway network, while providing a practical alternative to the long-delayed and conflictive Tocantins waterway.

Eventually, a third section of the Ferrovia Norte Sul will be built between Açailândia and Barcarena (Figure 2.4). Its construction was postponed because of the expediency of exporting farm commodities via São Luis do Maranhão, but the state government of Pará has embraced its completion as a regional priority.

Ferrovia Paraense

In 2017, the governor of Pará presented an ambitious plan to expand the nascent rail network to more fully integrate the agricultural landscapes and mineral assets of eastern Pará with an industries park and port facilities at Barcarena. The proposed railway would complete the link between Açailândia and Belem by passing through the oil palm plantations near Tailândia (See Chapter 2) and include spurs to bauxite mines under development at Paragominas and Rondon do Pará (see Chapter 5). At Marabá, the rail line would cross the Tocantins River, proceed south to El Dorado do Carajás and then up the Araguaía valley to the border with Mato Grosso (Figure 2.4 and Figure 2.5).†

* *Valor da Logística Integrada* (VLI) is a subsidiary of the mining conglomerate Vale S/A, Brazil's fourth largest publicly traded company. Vale built the Estrado Ferro Carajás in the 1980s when it was a state owned corporation. The decision to expand its railroad operation was accompanied by a large investment in oil palm plantations with the intention of converting all of their heavy machinery to biodiesel, an initiative that was abandoned in 2018 (see Chapter 3).

† Financing for the $1.5 billion would be provided by China Communications Construction Company (CCCCSA), a state-owned construction company, a quasi-state enterprise that plays a prominent role in China's Belt and Road Initiative. The practices of CCCCSA have been questioned by the World Bank and the US Department of State has placed the company on a list known to participate in human rights violations in Xinjiang province. *Bloomberg Businessweek*, 18 Sept. 2018, A Chinese Company Reshaping the World Leaves a Troubled Trail:

Advocates of conventional development support the construction of the railroad because it would generate about 25,000 jobs over the short-term and facilitate the development of a proposed steel mill in Marabá. Agribusiness supports the initiative because it would sway the choice of production models along the railway corridor. Intensive cropping of soy and maize is already the preferred land-use in Northeast Mato Grosso, and the extension of a low-cost grain transport system would accelerate the expansion of industrial agriculture into the municipalities of southeast Pará.[165] Environmental advocates and indigenous groups oppose the initiative because they contend that the Ferrovia Paraense will catalyse another wave of deforestation in the last block of remnant forest between Marabá and Belem (HML #5), while spurring land grabbing in the indigenous territories on the headwaters of the Rio Xingu. (HML #12).[166]

Ferrovia de Integração Centro Oeste (EF-354)

This is, perhaps, Brazil's most ambitious rail project and, if completed, would extend from Port of Vitoria (Espírito Santos) west to near the western border with Peru.[167] The transportation ministry has stratified its development into three phases: the first phase will connect Capinonorte (Goaías) with Lucas do Rio Verde (Mato Grosso) across 750 kilometres through central Mato Grosso (HML #15). When completed, this railine will reduce trucking costs by linking to both the Ferrovia Norte Sul in the east and [to be extended] Ferrovia Norte in the west (Figure 2.6). The second phase will extend westward to Porto Velho following the approximate route of highway BR-364, which will ensure the Madeira waterway remains a cost-effective option by replacing 1,000 kilometres of truck transport by rail.[*] It will also accelerate the expansion of intensive agriculture into Rondônia (HML #23), where smallholders are restoring degraded soils by rotating pastures with the cultivation of soy and maize (Chapter 3).[168] The third phase will cross the Rio Madeira and extend to Cruzeiro do Sul (Acre) near the border with Peru (HML #28).[†] There is no obvious economic justification for this last segment – except as a link in a transcontinental rail line between the Atlantic and Pacific oceans.

---

https://www.bloomberg.com/news/features/2018-09-19/a-chinese-company-reshaping-the-world-leaves-a-troubled-trail

* IIRSA, Amazon Hub, Group 8, AMA90, Ferrovia de Integración Centro-Oeste Fase I, Campinorte - Lucas do Rio Verde ($US 2 billion); AMA91, Ferrovia De Integración Centro-Oeste Fase II, Lucas do Rio Verde - Porto Velho (~$US 2 billion); AMA68: Ferrovía de Integración Centro - Oeste Fase III, Porto Velho - Rio Branco - Cruzeiro Do Sul ($US 2 billion): http://www.iirsa.org/proyectos/

† There is a fourth phase that will connect Capinorte with Vittoria (Espirtu Santos) or via the port of Ilhéus (Bahía), but a more practical route to the Atlantic Ocean is via the Rumo network to the port of Santos.

Ferrovia Transcontinental

In November 2014, the governments of Brazil, China, and Peru signed an agreement to evaluate the feasibility of a transcontinental railroad.[169] The route of EF-364 via Acre was one of several projects under evaluation; its proponents contend that it is the most cost-efficient because it would transit the Andes at the Huancabamba Depression,* where the maximum elevation is only 2,150 metres above sea level, approximately half the elevational incline that exists in competing proposals.[170] A railroad between Cruzeiro do Sul (HML #28) and Pucallpa (HML #41) would cross two national parks and infringe upon indigenous lands; consequently, there is zero possibility that a multilateral agency would finance the project, which is why the participation of entities from China was viewed with alarm by environmental advocates.[171] The rationale for a transcontinental railroad is based on the assumption that the savings in marine transport would offset the increased cost of rail transport. An independent evaluation by the International Union of Railways showed that the energetic cost of crossing the Andes and the capital cost of a new rail line would make the Ferrovia Transcontinental between fifty and a hundred per cent more expensive when compared to routes through southern Brazil or via the Amazon River. †

In 2018, the Brazilian government announced it would support an alternative proposal via Bolivia; known as the Ferroviário Bioceánico Central, this route is both shorter and takes advantage of pre-existing rail lines. Regardless, functionaries within the infrastructure agencies in both Brazil and Peru inserted a nebulous infrastructure component within the IIRSA portfolio, referred to as an 'Interconexión Terrestre', a term that leaves open the option of building either by road or rail line.

Ferroviaria Oriental SA

The pre-existing rail line that made Bolivia's transcontinental proposal 'more attractive' is a legacy rail system built in the 1950s by Brazil in compensation for the [perceived] loss of territories in the first decades of the twentieth century.‡ This rail line may – or may not – be part of a transcontinental railway, but it has played an essential role in the development of

* The Marañon Depression is also known as the Huancabamba Gap (see Ch. 5 for geological discussion).

† The feasibility study commissioned by a Chinese consulting company used an estimate of $30 per ton in their economic model (1), while the model used by the International Union of Railways revealed a coast of about $160 per ton (2). Sources: (1) ejatlas.com, https://ejatlas.org/print/opposition-against-controversial-amazon-route-of-transatlantic-railway-brazil-peru and (2) https://ecoa.org.br/ferrocarril-bioceanico-entre-peru-y-brasil-seria-inviable-segun-estudio/

‡ IIRSA Amazon Hub; Group 04 Acceso A La Hidrovía Del Ucayali; AMA 28, Interconexión Terrestre Pucallpa - Cruzeiro Do Sul: http://www.iirsa.org/proyectos/detalle_proyecto.aspx?h=29

Bolivia's agroindustry (Chapter 3). Bolivia is a land-locked country and the fertile farmland in Santa Cruz (HML #31) is located 2,000 kilometres from the nearest Atlantic port, while Pacific ports are located on the other side of the Andean Cordillera. Although refined products are exported to Peru by truck, it is not economically viable to move bulk grains. Fortunately for Bolivia's farmers, the Ferroviaria Oriental connects with ports on the Paraguay – Paraná Waterway, which allows them to compete in global markets. Without the pre-existing railroad, which was built for political rather than economic reasons, the agricultural sector in Bolivia would have grown to only a fraction of its current size.*

Tren Electrico de Carga de Ecuador

In 2013, the *Instituto Ecuatoriano de Preinversión* commissioned a pre-feasibility study to evaluate the viability of an electric-powered rail network. According to descriptions in the general press, most of the track would be built on the Pacific coast with the goal of connecting the country's banana and oil palm plantations to port facilities. The idea originated during a time when ambitious plans to reduce carbon emissions were popular among government planners, and the concept caught the attention of then-President Rafael Correa. The scheme included a spur that would cross the Andes to service the copper mines under development at the Cordillera del Condor (See Chapter 5) and, presumably, would have been built with the financial and technical support of China.[172] The rail line's demand for electricity would be very large and, apparently, influenced plans to increase the construction of hydropower dams in the Amazon. The feasibility study was completed in 2017 and in the same year incorporated into the IIRSA portfolio of investments;† however, there is no other evidence the Moreno administration pursued this investment as a priority.

Finance: What is New and What is Not

In spite of the ongoing build-out of infrastructure in Latin America, investment remains well below what most economists think the region needs to spur economic growth and reduce poverty. This includes assets not only within the Pan Amazon, but more importantly, in regions with larger populations and greater economic activity. A brief visit to any major city in these

* The Ferrocarril del Oriente (FCA) was built by Brazil in compensation for the cessation of Acre during the rubber boom at the dawn of the 20th Century. This rail line operated at a loss for decades, but is now an essential component in Bolivia's agricultural economy.

† IIRSA Adean Hub; Group 05. Conexión Colombia (Puerto Tumaco) - Ecuador (Puerto Esmeraldas - Guayaquil) - Perú (Carretera Panamericana); AND95 Tren Eléctrico de Carga del Ecuador; @US17.8 billion: http://www.iirsa.org/proyectos/detalle_proyecto.aspx?h=1437

eight countries will reveal the inadequate infrastructure that plagues the region; not surprisingly, the situation is worse in the countryside.

A well-established tenet of macro-economic theory maintains that poor infrastructure constrains growth because it imposes inefficiencies on domestic production.[173] Investment in infrastructure stimulates growth by creating a short-term demand for labour, machinery and basic materials; more importantly, it increases productivity over the long-term.[174] The absence of basic infrastructure imposes opportunity costs due to lost economic growth that compound over time; in contrast, investment in well-designed projects pays dividends in greater economic growth that likewise compounds over time. It is easy to understand why the region's leaders are focused on making investments in basic infrastructure; it is less easy to understand why there has not been more sustained investment over the past half century.*

The infrastructure deficit in Latin America is due to a shortage of investment capital that is a historical legacy of failed economic models, erratic fiscal management, rampant corruption and endemic political instability. The shortage of investment capital exists in both the public and private sector. National economies are characterised by a large informal sector and a culture of tax evasion (see Chapter 6) that obligates governments to operate within budgets that preclude large-scale investment in new infrastructure. Periodic experiments with populist economic philosophies have stunted the development of domestic bond markets, an important source in advanced economies of long-term finance for both national and local governments. Economic mismanagement has created a deeply ingrained fear of inflation, which motivates central banks to pursue monetary policies based on high interest rates.† Political risk inhibits the participation of overseas private investors, who are wary of sovereign debt defaults, which also contributes to the persistence of high interest rates. The high cost of capital is the most important constraint on infrastructure investment in Latin America.

Governments have traditionally relied for investment capital to build basic infrastructure on multilateral financial institutions, such as the World Bank Group, the Interamerican Development Bank (IDB), the Development

* Brazil invests about 2.5% of GDP in infrastructure, well below other emerging economies such as China (8.5%), India (5.2%), South Africa (4.7%), Russia (4.5%) and Turkey (3.6%), and only slightly above developed nations such as the USA (2.4%), the UK (2.2%) and Germany (2%), which enjoy huge legacies from past investments: https://www.statista.com/statistics/566787/average-yearly-expenditure-on-economic-infrastructure-as-percent-of-gdp-worldwide-by-country/

† Base-line interest rates were at historical lows in 2020 (Brazil: 2%; Peru: 0.5%), but they averaged from 20% to 10% between 2005 and 2017 and were as high as 40% in Brazil in 2000; Peru's rates have been less volatile, but still fluctuated between 8% and 2% between 2005 and 2018. Source: https://tradingeconomics.com/

Bank for Latin America (CAF) and bilateral development agencies. The needs of Latin America greatly exceed the lending capacity of these multilateral institutions, however, so loans are usually leveraged with resources from private banks, domestic bond markets and national development funds.

An example of this type of combined operation was an investment by the International Finance Corporation (IFC) to support Brazil's on-again-off-again programme to privatise its electric energy sector. In 2012, the IFC approved a loan to restructure the electrical distribution utility of Pará (CELPA); as part of that agreement, the bankrupt utility was sold to a holding company (*Equatorial Energia S / A*), which was acquiring similarly distressed utilities across Brazil. The IFC's involvement represented a stamp of approval for the new holding company, which bolstered investor confidence for a public offering of *Equatorial Energia* shares in the Brazilian stock market.*

The investment was subject to the standard environmental and social evaluation that accompanies all IFC investments and was ranked category A – the highest level of risk – an unsurprising qualification considering the loan coincided with the construction of the Belo Monte dam. Although CELPA was not a participant, it was an obvious candidate for distributing at least some of the electricity generated by the controversial hydropower project. The loan was approved contingent upon the reform of CELPA's internal procedures to comply with IFC's Social and Environmental Performance Standards.[175] In 2020, the company inaugurated the expansion of its transmission lines across the frontier landscapes of the Transamazon highway (BR-230) between the Xingu substation near the Belo Monte dam and one near Rurópolis (Pará) that supplies energy to both Miritituba and Santarem (Figure 2.5).

The IIRSA initiative relies heavily on this investment strategy, particularly in Colombia, Peru and Brazil, where the state grants concessions to private companies to build and operate infrastructure assets.

Public-Private Partnerships

The concessionaire system in Peru is managed by the *Organismo Supervisor de la Inversión en Infraestructura de Transporte de Uso Público* (OSITRAN), which oversees investments in transportation infrastructure, while the electrical system is administered by the *Organismo Supervisor de la Inversión en Energia y Minería* (OSINERGMIN). Both of these regulatory agencies supervise joint ventures that have been organised by the *Agência de Promoción de la Inversión Privada – Peru* (ProInversión), a government agency

* The IFC invested $US 99 million in a complex operation that included $US 74 million in new equity and matching investments from private funds totaling $US 640 million. Source: IFC 2013.

that conceives projects, attracts investment capital and administers the auctions that launch public-private partnerships (PPP). There have been several high-profile PPP projects in the Peruvian Amazon, most notably the two IIRSA interoceanic highways that are operated by construction companies that allegedly co-financed their construction in partnership with the Peruvian state, the IDB and CAF.* The term 'allegedly' is used because the construction companies didn't actually invest a significant share of the capital upfront but assumed (partial and limited) liability for repaying debt raised in public capital markets.

The northern project (IIRSA Norte) shows the potential to leverage public and private capital for a development project.† It was co-financed by the Peruvian state and the IDB, who together paid for feasibility studies, environmental review and the engineering design required to tender the project via a concessionary system. The construction was financed by infrastructure bonds traded in public markets that are to be repaid (in part) by revenues from highway tolls collected by the concessionaire.[176] The concept has been hailed as an innovative financial product because it expands the pool of financial resources by legally obligating highway revenues to debt repayment.‡ Nonetheless, these bonds are still a form of sovereign debt and, if the business model fails, the state remains responsible for debt repayment. The IIRSA Norte project was completed on time and in relatively good fashion but at more than double the original coast estimate ($US 575 million vs. $US $250). The original budget was unrealistically low and there are multiple reasons to question the quality of the environmental review; nonetheless, the highway is an important transportation asset that

---

* IIRSA Norte was built and has been operated by Odebrecht Latinvest Peru SA since its inception in 2005; Oderbrecht also controls IIRSA Sur - Tramos 2 & 3, which was constituted originally with the participation of Peruvian companies (Graña y Montero SA, JJC Contratistas Generales SA, and Ingenieros Civiles y Contratistas Generales SA (ICCGSA)) and IIRSA Sur - Tramo 4 with Andrade Gutiérrez, Camargo Correa y Queiroz Galvão; over time Odebrecht Latinvest Peru S.A has consolidated ownership of IIRSA Sur Tramos 2,3 & 4. Source: OSITRAN, https://www.ositran.gob.pe/

† The Corridor Interoceanico del Norte was financed by a combination of debt (infrastructure bonds) of $US 367 million, a $US 60 million loan from the IDB and expenditures of $US 81 million by the Peruvian state: see http://www.iirsa.org/proyectos/Index.aspx; a similar operation was organised by the CAF for IIRSA Sur: https://www.iirsasur.com.pe/etiqueta/caf/.

‡ Formally referred to as *Certificados de Reconocimiento de Derechos del Pago Anual por Obras* (CRPAOs), they refer to the commitment of future income from the infrastructure asset. The concept was pioneered in Peru and has been used in other countries, including Mexico, Panama and Argentina. See Crédit Agricole Corporate and Investment Bank, 15 Project Bonds To Change Your Preconceptions https://www.ca-cib.com/sites/default/files/2020-03/Project-Bond-Focus-2019-15-Milestone-Transactions.pdf

links economically active areas in the northern part of the country. As of June 2020, the highway was generating significant revenues and OSITRAN listed it as income-generating asset.[177]

In stark contrast, the southern enterprise (IIRSA Sur) is emblematic of the risks associated with a highway through a wilderness area conceived via an inadequate feasibility study with mediocre environmental review, deficiencies that were compounded and expanded by self-dealing and political corruption. This highway is at the heart of the *Lava Jato* scandal in Peru, and its principle contractor, Oderbrecht, has pleaded guilty to various forms of malfeasance (see Chapter 6). The losses incurred by the project are much greater than the bribes paid to officials or the excess billing of avaricious companies* because the final cost of construction ($US 2.5 billion) was vastly underestimated ($US 879 million). Unlike the northern project, the highway has not generated any revenue, which has forced the state to assume full responsibility for debt repayment.[178]

The IIRSA Sur imbroglio has stained the reputation of the concessionaire system in Peru. Nonetheless, it is used to operate sixteen highways, 22 airports (four in the Amazon), three rail systems, and ten ports (four in the Amazon) and the Amazon waterway. The total investments channelled via the OSITRAN concession system add up to more than $US 9 billion, approximately double the amount lent to Peru for infrastructure by the IDB over the same period.[179,180]

Peru also has a national development bank that finances private sector projects, the *Corporación Financiera de Desarrollo SA* (COFIDES) has multiple strategies for raising capital, including by offering sovereign debt in international bond markets.[181] Some of these are part of the IIRSA portfolio, including IIRSA Sur -Tramo 1&5 (the non-Oderbrecht components) and others, such as the Gasoducto Sur Peruano, that fall outside the remit of the regional infrastructure integration portfolio (see Chapter 5). COFIDES is active in the electrical energy sector and coordinates its investments with OSGERMIN and ProInversión. Within the Amazon watershed, hydropower investments are valued at approximately $US 300 million.† This amount, however, is only a fraction of the OSGERMIN's projected investments, which in 2020 showed the construction of sixteen hydropower plants with a total capacity of 2.7 GW at an estimated cost of $US 5.7 billion.[182]

---

* The construction consortium is accused of over-billing the state by $US 182 million, which included a $31 million dollar bribe to then-President Alejando Toledo. Source: *Carreteras Pan Americanas*, 6 June 2019, Ex socias Odebrecht incluidas en caso de sobornos en IIRSA Sur: https://www.carreteras-pa.com/noticias/exsocias-odebrecht-incluidas-en-caso-de-sobornos-en-iirsa-sur/

† Cerro del Aguila ($US 75 million), Chaglla $US 100 million), La Virgin ($US 100 million), H-1 ($33 million), El Angel ($US 47 million); see http://www.minem.gob.pe/

In Brazil, highway and rail concessions are managed by the *Agência Nacional dos Transportes Terrestres* (ANTT), the electrical energy sector by the *Agência Nacional de Energia Elétrica* (ANEEL), and waterways by the *Agência Nacional de Transportes Aquaviários* (ANTAQ). All three agencies are key to the administrative procedures tied to the current administration's policies to privatise strategic components of the national economy.[183]

The expansion of the rail network in Brazil, outlined previously, is dominated by large corporations (Rumo / Cosan and VLI / Vale). The development of railroads is based on robust business models that have access to multiple sources of investment capital, including the Brazilian development bank (see below). Depending upon the outcome of their environmental review, two additional rail concessions should soon move into the construction phase: Ferrogrão (EF-171) and Ferrovia de Integração Centro-Oeste (EF-354). A third concession (Ferrovia Paraenese) is under development and state authorities in Pará have signed an agreement with a Chinese construction company to build the first phase of the project. Because the project does not cross state borders, its developers may escape oversight by federal regulators; consequently, the project may proceed quickly in comparison to the more problematic Ferrogrão.

The concessionaire system for highways is less lucrative and, consequently, resembles the system described for Peru, where the concessionaire manages the transportation asset for a fee that is partially funded by tolls collected by the concessionaire.

In the Legal Amazon, privately operated highway concessions are both new and uncommon. As of 2020, there was only one: Roto do Oeste (BR-163), the strategically important corridor between the farm landscapes of central Mato Grosso and the logistical hub at Rondonópolis (Figure 2.6). The thirty-year, 600-kilometre concession was awarded in 2014 to *Odebrecht Rodovias*,* which is contractually obligated to expand its capacity from two to four lanes over approximately half its length. These and other improvements are projected to cost R$ 5.5 billion,† a reasonable sum to collect from the ~70,000 trucks that operate on the transportation corridor that moves agricultural commodities south and manufactured goods north.

Four additional highway concessions in the Legal Amazon were being prepared for public auction in 2020 and, like the Roto do Oeste, all are important components in the truck-based bulk transport system that services the soy-maize complex of the Southern Amazon (Table 2.3). All four roads will be upgraded to accommodate heavy traffic, but only BR-364 in Rondônia is slated to be expanded from two to four lanes. Significantly,

---

* Another subsidiary of Odebrecht S.A., which emerged from a court-supervised bankruptcy in 2020 as NOVONOR: see https://www.novonor.com.br/

† Exchange rate fluctuations make $US dollar estimates imprecise, but this equated to about $1 billion in Feb. 2021.

Table 2.3: Infrastructure projects supported by the PPI; values are projected cost of capital investments as reported by the PPI or based on methodology used by the Ministerio de Infraestrutura to calculate costs;[220] values expressed in $US are based on exchange rates at the end of 2020.

Support for Concession Process			**Support for Environmental Review**		
	R$	$US		R$	$US
Highways - operations			*Hydropower & Grid*		
BR-163/230 (MT/PA)	1,584	317	Bem Querer (RR)	5,000	1,000
BR-158/155 (MT/PA)	1,808	362	Castanheira (MT)	1,300	260
BR-153/080/414 (GO/TO)	8,661	1,732	Trabajara (RO)	1,000	200
BR 364 (RO/MT)*	8,212	1,642	LT Manaus - Boa Vista (AM, RR)	170	34
Rail Lines – construction			*Waterway Improvement*		
FICO (EF-354)	2,730	546	Pedral do Lourenço (PA)	650	130
Ferrogrão (EF-171)	8,400	1,680	*Highway construction*		
Ports - operations			BR-080 (MT/ GO)	1,742	348
Itaqui (MA)	594	119	BR-158 (MT)	2,912	582
Santana (AP)	42	8	BR-242 (MT)	2,090	418
Barcarena (PA)	571	114	BR-174 (RO/MT)	3,483	697
Airports (Legal Amazon)	380	76	BR-319 (AM/RO)	2,786	557
	32,982	**6,596**		**21,133**	**4,227**

ANTT has stipulated that preparations for the terms of reference for BR-163/230 (MT/PA) should include measures that facilitate the eventual construction of the Ferrogrão.[184] All five projects are being supported by the PPI, a high-profile programme established in 2016 to fast-track private sector investment in strategic infrastructure.

Evidence of the current administration's priorities was the inclusion within the PPI portfolio of financial support for the environmental reviews of projects delayed by social protests. The inclusion of BR-319 suggests that efforts to complete the highway between Manaus and Porto Velho will be successful; less notable, but of greater economic significance, are four highways in Mato Grosso that will improve farmers' capacity to move their harvests to logistical platforms on the rail transport systems under development (see Figure 2.6). Presumably, their inclusion within the PPI portfolio reflects a belief that they will pay for themselves and finance their construction via the concessionaire system.

PPPs are common in the electricity sector of Brazil and were used extensively during the administrations of Lula de Silva and Dilma Rousseff.

Typically, the state is represented by a subsidiary of Electrobras[*] as a minority shareholder in a joint venture with one or more private companies. The companies can be grouped into four categories: (1) construction companies who are contractors to the project;[†] (2) pension or investments funds domiciled in Brazil;[‡] (3) metallurgical and mining companies seeking to ensure power for their processing mills;[§] and (4) energy holding companies that acquire generation or transmission assets as part of their business model.[¶]

Energy holding companies, which attract investment capital from a variety of sources, are often characterised by complex corporate structures. For example, Neoenergia S/A owns ten per cent of Norte Energia (Belo Monte) and 51 per cent of the Teles Pires and Dardanelos facilities in Mato Grosso. Neoenergia is controlled by Iberdrola SA (51 per cent) a Spanish partnership with a large pension fund affiliated with Banco do Brasil (thirty per cent) however, approximately nineteen per cent of Neoenergias equity is a 'free float' that is held by investment funds from the United States and Europe.[**]

Most of these energy companies also operate generation and transmission concessions without the participation of Electrobras. As of 2020, there were nineteen privately operated hydropower plants in the Brazilian (Legal) Amazon, ranging from the 1.1 GW mega-dam at Estreito ($US 2.4 billion) to the 20 MW power plant ($US 40 million) operated at the Pitinga mine near Manaus. Altogether, private companies operate about 3.6 GW of installed capacity that required approximately $US 5 billion in investment capital, compared to the 23 GW of capacity operated by PPP ventures that required between $US 30 and $US 40 billion.[††] Legacy public-owned and operated power plants, such as Tucuruí, Balbina and Samuel, have a total of about 11 GW.

---

* Electrobras: Electronorte, Chesf, Furnas, Electrosul

† Construction companies: Oderbrecht Energia, SAAG/Andrade Gutierrez, Camargo Correa

‡ Pension and investment funds: Petros, Fundação dos Economiários Federais (Funcef), Caixa FIP

§ Metallurgical companies SINOBRAS, Alcoa, BHP, Vale, Intercement.

¶ Energy companies: Neoenergia (Iberdarola, Spain), ENGIE (formerly Tractobel/Suez, France & Belgium), EDP- Brasil (Energia do Portugal), EDF - Flumenense (France), Alupar SA (Guarupart Participacoes Ltda), Amazonia Energia (Light/CEMIG),

** These include: America Funds/Capital Research & Management (USA), The Pictet Asset Management (Europe), BlackRock Investment Management (USA) and The Vanguard Group (USA), among others. Source: https://finance.yahoo.com/quote/NEOE3.SA?ltr=1

†† The exact amount is difficult to calculate because of cost overruns that have not been reported transparently and foreign exchange rates that ranged from near parity with the $US dollar (2000) to 0.25 (US/R) in Jan. 2021.

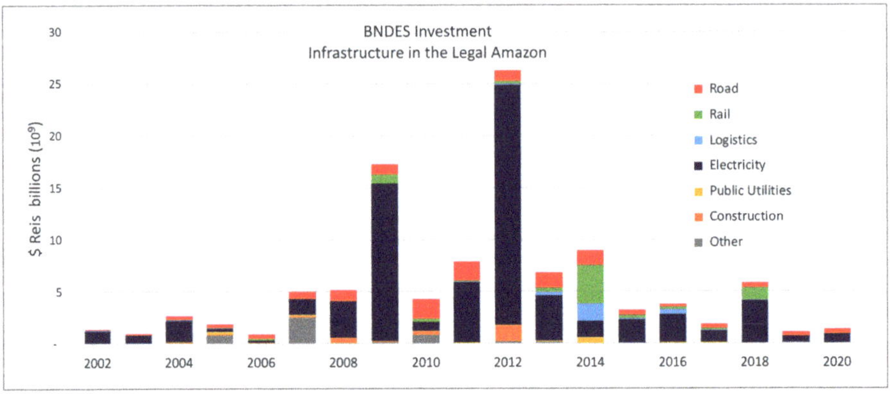

Figure 2.24: The distribution of disbursements made to infrastructure projects located in the Legal Amazon by the Brazilian Economic and Social Development Bank (BNDES). Data source: BNDES transparency portal.

Between 2000 and 2020, the most important source of infrastructure finance in Brazil was the *Banco Nacional de Desenvolvimento Econômico e Social* (BNDES), a quasi-autonomous fund that extends loans and makes grants to both public and private entities. BNDES has a broad portfolio that supports infrastructure, agriculture, technology, finance and manufacturing, as well as initiatives to reduce inequality and protect the environment (see Chapter 12). Total disbursements to projects in the Legal Amazon between 2002 and 2020 totalled R$ 171 billion (~ $US 78 billion); of this amount, approximately 65 per cent was allocated to infrastructure (Figure 2.24).[185]

Although BNDES is autonomous, its authorities respond to policy priorities established by elected officials. The administrations of Lula da Silva and Dilma Rousseff embraced policies that stressed hydropower and the extractive industries, priorities reflected in disbursements made to finance a gas pipeline between Coari and Manaus (2007) and the dams at Estreito* (2008), on the Rio Maderia (2009), and Belo Monte (2011, 2012). Investment in road construction and the electrical grid were present throughout the decade, including several regional highways in Mato Grosso (2012) and Pará (2005, 2006); investment in rail became a priority only after 2014.[186]

BNDES also functions as an export-import bank by providing credit to facilitate the sale of manufactured goods; in addition, it has a specialised division that supports the export of engineering services. Between 2009 and 2014, the bank loaned $US 2.9 billion to entities that contracted Brazilian

* The power plant at Estreito on the Tocantins is owned by a consortium of corporations that includes several mining companies – BHP, Vale and Alcoa – as well as the construction company and the operator Engie (formerly Suez).

construction companies building hydropower plants in Ecuador, Venezuela and Peru.[187]

The most noticable aspect of BNDES' recent history is the decline in its lending activities after 2014, following the collapse of commodity prices that triggered an economic crisis: National GDP fell by 3.5 per cent in 2015 and by 3.3 per cent in 2016. The economic recession was exacerbated by the inability of BNDES to act as a counter-cyclical source of fiscal stimulus, due in part to a lack of revenue from non-performing loans embroiled in the *Lava Jato* scandal. More important, however, was the macro-economic environment that caused foreign investors to abandon the country's capital markets, which limited BNDES' ability to raise fresh capital. More recently, the Bolsonaro adminstration has imposed budgetary restraints and obligated the bank to return hundred of billions of *reais*[*] to the federal government. Ironically, the government's privitasation policies may assist BNDES to renew its investment capital by liquidating its equity shares in Brazil's flagship companies.[188]

In Colombia, the concessionaire system is managed by the *Agência Nacional de Infraestructura,* which has overseen approximately $US 12 billion in highway investment in the country since 2010. Most of that activity occurred in extra-Amazonian regions and only two concessions have been awarded within the Colombian Amazon. The Malla vial del Meta, which is operated by a consortium of Colombian construction companies, is an important transportation asset that provides access to the oil palm landscapes of Meta. Portions of this highway network constitute the northern component of the Carretera Marginal de la Selva (see above) and functions as a gateway to the coca frontiers that surround La Macarena National Park (Figure 2.12).

The other concession consists of the Santana – Mocoa – Neiva (R-45) corridor that connects Bogotá with the oil fields of Putumayo. This highway was originally adjudicated to a Colombian consortium, but was recently acquired by China Construction America, a subsidiary of the world's largest construction company: China State Construction Engineering Corporation Ltd (CSCEC). The contract obligates the concessionaire to invest $US 21 million in its first year of operations (2021) and eventually reach a total target of ~ $US 440 million.[189]

Like many Chinese corporations, CSCEC is a publicly traded company; nonetheless, it is closely associated with the government of the People's Republic of China. Coincidentally, one of China's state owned oil companies, Sinochem, has acquired a half dozen oil concessions in the Putumayo region.[†] Both companies are considered to function as a branch

* The Brazilian currency is the *Real,* which is expressed in its plural form as *Reai*s.

† Sinochem acquired Emerald Energy PLC, a British company for $US 898 million in 2009.

of China's military-industrial complex and to coordinate actions to advance the strategic interest of the Chinese state.[190]

Investment from China

The increasing presence of Chinese companies in South America has become an issue of concern among social and political analysts. Security specialists argue they are a geopolitical threat to the interests of the United States and other Western nations, while supporting authoritarian governments that undermine democratic traditions.[191] Environmental advocates attest that companies from China are uninterested in sustainability and will maximise profits at the expense of biodiversity and ecosystem services.[192] Nationalists view them with suspicion because they represent a new type of imperialism that will lead to a loss of strategic industries while creating dependency via so-called debt traps.[193] Perhaps. Chinese banks and businesses have brought much-needed investment to the region, and their presence has advanced investments in infrastructure that would not have occurred or would have taken another decade or longer to finalise.

Investment from China falls into two major categories: loans to finance infrastructure and direct foreign investment (DFI) by corporations that seek to own or operate a business. Loans are the predominant form of investment in Venezuela, Ecuador and Bolivia because their 'socialist' governments are averse to foreign entities owning the means of production within their countries. There are exceptions, particularly in the mineral sector of Ecuador (see Chapter 5). DFI is the preferred model in Brazil and Peru, where Chinese companies have initiated new 'greenfield' investments or acquired existing business and infrastructure assets (Figure 2.25).*

Some, but not all, of this investment activity is occurring in the context of the 'Belt and Road Initiative' (BRI), a foreign policy and infrastructure development programme launched by the People's Republic of China in 2013. The BRI is best known for suites of coordinated projects that extend across the Eurasian landmass, Southeast Asia, and into Africa. The Latin American component, which is more modest in scope, was announced at the Asian Pacific Economic Cooperation (APEC) summit in Lima in 2016. Subsequently, Bolivia, Ecuador, Guyana, Suriname, Peru and Venezuela all signed agreements formally entering the BRI initiative.[194] Only Colombia, which has the smallest amount of Chinese investment, and Brazil, which has the largest, have declined to formally join the BRI. Nonetheless, both have bilateral agreements that facilitate trade and investment between their countries. China is Brazil's largest trading partner, and the flow of

---

* The distinction between what is a loan and what is an equity investment is not always straightforward, since many loans are secured with equity, or the entity receiving the loan is partially owned by Chinese stockholders.

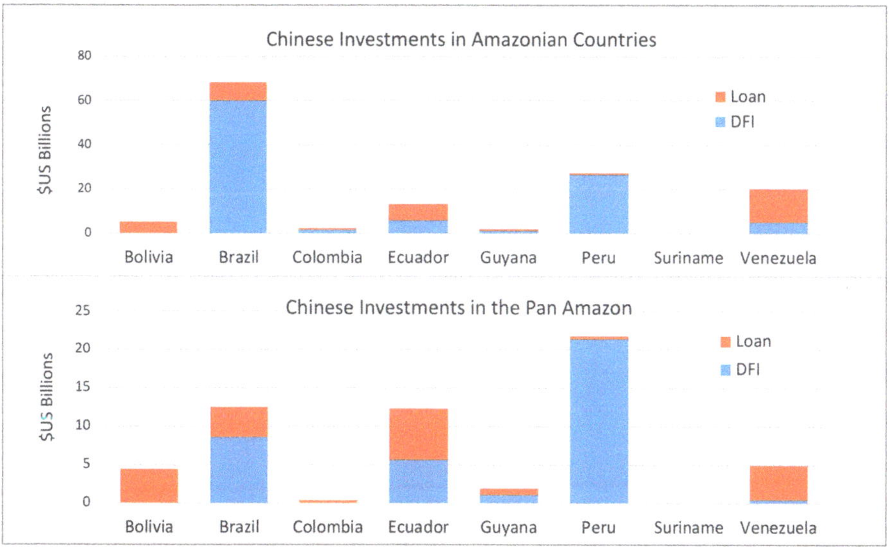

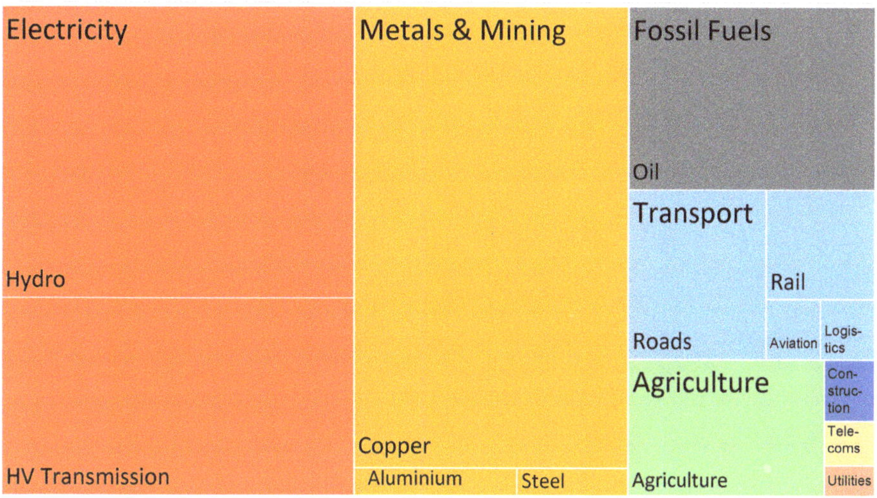

Figure 2.25: Distribution of investments by Chinese entities in Amazonian countries and within their Amazonian jurisdictions (ton) and by sector; DFI refers to direct foreign investment. Investment from China stratified by sector at the national level.

Data source: American Enterprise Institute.

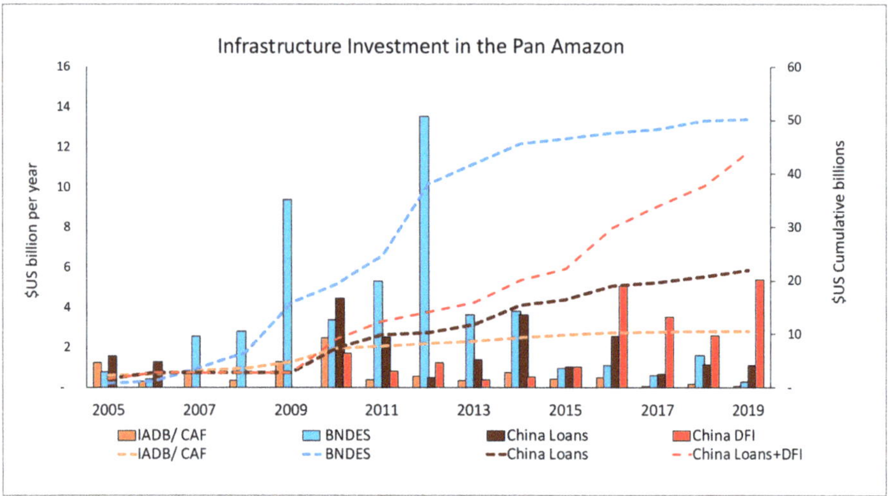

Figure 2.26: Financial flows for infrastructure investments in the Pan Amazonian countries (left axis – bar graph; right axis – line graph).

Investment from all sources includes only Amazonian projects. BNDES includes automatic and non-automatic disbursements for states within the Legal Amazon and loans to Andean countries for engineering services. Loans and direct foreign investment (DFI) from China are segregated annually, but DFI is added to the loan total in the cumulative line graph.

Data sources: IDB project database; CAF project database; BNDES transparency portal; China Global Investment Tracker, The American Enterprise Institute.

investment capital from China to Brazil is an important for maintaining the economic health of both nations.

Most of the financial assistance to Latin America has been channelled through public 'policy banks' that are superficially similar to the multilateral institutions based in Washington, Caracas and Brussels: Their primary mission is to assist client countries in developing their infrastructure and economies, but as self-financed banks they must also earn a satisfactory return on their investment. In the case of the China Development Bank and the Ex-Im Bank of China, these fundamental principles were modified between 2000 and 2015 as part of a deliberate strategy to secure supply chains for strategic commodities. After 2014, at least in Latin America, they became more pragmatic because of financial constraints in China, a surplus of commodities, and the weakened financial status of their clients.

The lending policies of Chinese institutions incorporate social and environmental standards that mimic those championed by the World Bank but, unlike the multilaterals, they are not subject to strict oversight by civil society. Chinese policy banks have a due diligence system, but their precautions against corruption and the rigour of their environmental reviews are significantly less thorough and, in some instances, may amount to window dressing. Consequently, Chinese banks and partner companies move quickly and in close coordination, which allows them to design, build and deliver a project in a fraction of the time required by a project financed by the multilateral agencies.

The total value of infrastructure investments by Chinese entities is greater in value than those deployed by the multilateral banks when considered over the same period of time (Figure 2.26). That is a false comparison, however, because Chinese policy banks tend to narrowly focus on the extractive industries, energy and infrastructure, while the multilaterals also invest in health, education, governance and water and sewage systems. Moreover, this comparison also ignores the leveraged contributions by the private sector and national treasuries, which are usually three to ten times larger than the contribution of the multilaterals. Increasingly, finance from China, the multilaterals, and the private sector are joining together to finance projects that make good 'business sense'.

Debt-trap countries

Most infrastructure lending is organised via the Chinese Development Bank (CDB) or the Export-Import Bank of China;[195] Jointly, they invested approximately $US 5 billion in South America between 2000 and 2009, a number that doubled in 2010 and by 2020 totalled more than $US 53 billion.[196] Approximately forty per cent of these loans were made to finance investments in the Pan Amazon, mainly in hydropower, electricity transmission and roads (Figure 2.25)[197] The volume of loan transactions has trickled to a stop since 2016, while DFI continues, although at reduced values.[198]

Ecuador was an early recipient of Chinese investment, in part because it had difficulty attracting investment capital due to political instability and economic mismanagement, a predicament that become particularly acute in 2008 when the country defaulted on its international debt obligations. President Rafael Correa, who was elected in 2007, aggravated the crisis by rejecting standard (neoliberal) advice to privatise state-owned infrastructure and energy assets; instead, he sought other sources of finance. The Export-Import Bank of China and the China Development Bank responded with loans that allowed the government to pursue the construction of several highways, modernise an oil refinery and expand hydropower. The loans were backed by future oil production and, eventually, revenues from a copper concession granted to a Chinese mining company.*

Although Ecuador largely used credit to finance its economic expansion during the commodity boom, it did allow a limited amount of DFI in the extractive industries. In 2006, the China National Petroleum Company (CNPC) and Sinopec created a joint venture, incorporated as Andes Petroleum, which they used to purchase one of Ecuador's most valuable infrastructure assets, the Oleoducto de Crudos Pesados (OCP), the pipeline between the Amazon and the Pacific coast, for $US 1.3 billion.[199]

Outside the extractive industries, the largest loans were used for the construction of two hydropower facilities in the Andean foothills: Coca Coda Sinclair ($US 1.5 billion) and Paute / Sopladora ($US 487 million), both of which will be operated by *Corporación Eléctrica del Ecuador* (CELEC) an electrical utility company owned and operated by the Ecuadorian state. A third complex is planned on the Santiago River that, presumably, will also be built by a Chinese company and financed by the Ex-Im Bank of China. The country has been unable to service its new debt obligations following the decline in the price of oil after 2015. Ecuador was forced to restructure its foreign debt in August 2020, which they undertook with the assistance of the International Monetary Fund (IMF) in coordination with their Chinese creditors.† As usual, oil revenues are a guarantee of future payment.

In Bolivia, Chinese credit has financed approximately ten per cent of the highways built over the last two decades. Between 2007 and 2017, fifteen separate projects were awarded to contractors from China at a combined value $1.78 billion;[200] however, the debt incurred for highway construction from Chinese banks over the same period totalled only $US 1.1 billion. Chinese construction companies were awarded all the contracts financed by the Ex-Im Bank, but several were successful in competing for contracts financed entirely by the Bolivian state (see Chapter 6).‡ Almost all of these roads are components of the IIRSA master plan and about two thirds are

---

* Chinese National Petroleum Company (CNPC) was guaranteed 90% of Petroecuador's production in return for a loan in excess of $US 9 billion; the China Railway Construction and China Nonferrous Company (CRCC) is the concessionaire of new copper mine in the Cordillera del Condor ($US 2.9 billion). Source: American Enterprise Institute 2017.

† The IMF will provide $US 6.5 billion, Chinese Banks will provide $US 2 billion; approximately ~$16 billion of bond debt will be restructured. Source: *The Financial Times*, 5 Sept. 2020, Ecuador basks in glow of debt-restructuring success, https://www.ft.com/content/1dd975c9-e3a1-4fcc-b049-f29dbd59f6fa

‡ Bolivia self-financed about 50% of the highway projects during the administration of Evo Morales, which grew from less than $125 million in 2005 to more than $1.0 billion in 2016. Source: ABC – *Administradora Boliviana de Carreteras,* Plan Estratégico Institucional 2016– 2020: http://www.abc.gob.bo/wp-content/uploads/2018/03/pei_abc_2016_-_2020.pdf

located within the Bolivian Amazon, including the Rurrenabaque–Riberalta corridor ($US 579 million) that connects the Andean highlands with the Department of the Pando and the Corridor Interoceanico (see Figure 2.8).

In the electricity sector, Sinohydro is building the dam and power plant at Ivirizo ($US 632 million), and the China Three Gorges (CTG) corporation was awarded the contract to build the Rositas Dam near Santa Cruz ($1.3 billion). Both power plants will be operated by a state-owned domestic utility company (ENDE), which has a near-monopoly on generation and transmission. The Rositas project was paralysed in 2019 because ENDE failed to conduct an environmental review that complied with the principles of Free Prior and Informed Consent (FPIC).* Indigenous communities have filed a legal challenge to halt the project, at least temporarily, but the project has strong support across multiple social and economic groups.[201]

Bolivia and China have engaged in an unusual infrastructure initiative in the form of a communications satellite, which was designed, built and launched by the Chinese space agency. Christened Tupac Katari -1, the satellite cost $US 300 million, of which 85 per cent was financed by the China Development Bank. The enterprise has been criticised as a white elephant because it has generated only $70 million in revenues since its launch in 2013.[202] Proponents argue the investment should not be evaluated solely on financial criteria, however, because it integrates remote indigenous communities into the national community by providing them with celular phone service and access to the internet.

Overall, Bolivia's foreign debt increased from $US 2.2 billion in 2007 to $US 11.3 billion in 2019, of which only about ten per cent is held by Chinese banks, an amount considerably less than either the IDB ($US 3.5 billion) and CAF ($US 2.9 billion); moreover, interest on the loans from China is concessionary, with interest rates of about 2.6 per cent compared to 3.3 to 3.6 per cent for the multilaterals.[203] Bolivia continues to service its debt, but financial analysts consider Bolivia to have reached a plateau in its ability to service any additional non-concessional debt. Chinese banks have made no new loans to Bolivian entities since 2018, and the two largest loans announced in 2016 (Rositas, $US 990 million) and 2017 (the El Mutun mine, $470 million) have not been executed due to bureaucratic delays.

The most heavily indebted country in Latin America is Venezuela. Between 2000 and 2014, the country reportedly borrowed $US 50 billion from Chinese entities to finance multiple components of the national economy, including the hydropower facilities on the Caroni river and industrial mines in Bolivar state. Most of the loans were to be repaid via direct oil shipments, but declining production and low prices caused Venezuela to

* The database curated by the American Enterprise Institute shows the loan commitment, even though no funds have actually been disbursed to initiate construction.

default in 2014, an outcome that followed several years of cosmetic debt roll-overs. Even investment in iron ore and bauxite mines of the Guyana highlands has failed to provide any type of cash flow or produced the mineral commodities that guaranteed debt payment (see Chapter 5). The debt was restructured in 2020 with a 'grace period' provided for $US 19 billion.

Both Venezuela and Ecuador could be considered to be caught in a 'debt trap'. The inability to repay loans has led to a loss of sovereign control over a strategic asset. In both countries, the state-owned oil companies now export almost all of their production to China to service past debt. Consequently, both countries are forfeiting potential revenues in an improving global market. Nonetheless, China has not benefited from these failed business ventures and, rather than acquire assets via bankruptcy proceedings, the policy banks of China have written off bad loans.* The debt trapped everybody.

The domain of direct foreign investment

The flow of capital into the private sector and PPPs was similar to the loan portfolios: Investment started in the early 2000s, with a surge in 2010 followed by more moderate flows thereafter.[204] Unlike loans, however, the flow has not stopped, as Chinese corporations continue to acquire companies with good cash flow and growth potential. Several sectors stand out. The most obvious examples are companies operating in the agricultural and the extractive industries, whose business models align with China's need for basic commodities (see Chapters 3 and 6). Investment in infrastructure is different, because it isn't built on a supply chain that extends back to China; instead, companies export engineering expertise to build and operate highways, railroads, airports, dams and electrical power systems.

Peru has welcomed significant quantities of DFI and completely avoided the debt trap that afflicts its neighbours (Figure 2.25). Approximately $US 15 billion has been invested in the polymetallic mines of the High Andes and another $US 3.5 billion in oil and gas.[205] The China Three Gorge (GTC) company owns the largest portfolio of infrastructure assets. It made its first investments in Peru via a joint venture in 2016 with a Portuguese energy firm at the San Gaban-III hydropower facility on the Rio Inambari ($US 185 million).[206] This was followed by the acquisition of the Chaglla dam and power station on the Rio Huallaga in 2018, as part of a deal forced on the beleaguered Peruvian subsidiary of Oderbrecht ($US 1.4 billion).† The largest acquisition (to date) was GTC's purchase of Luz Del

* Bolivia has avoided this outcome with its Chinese creditors, who have no lien on that country's natural resources, but Bolivia is facing a looming finanical predicament due to an over-reliance on mineral exports and a failure to invest in more sustainable enterprises.

† The Peruvian state has confiscated $US 312 million from the transaction to pay back taxes ($135 million) and a judicial fine ($US 200 million) as part of the

Sur ($US 3.6 billion), which it acquired from a US-based energy company in 2020. Among Luz Del Sur's assets are the electrical distribution system of Lima and the Santa Teresa dams under construction on the Urubamba River near Cuzco ($US 600 million). The Chinese company is the largest diversified electrical utility in Peru.

A more controversial investment is Sinohydro's participation in the consortium (*Sociedad Concesionaria Hidrovía Amazónica SA* – COHIDRO), which was awarded the contract to administer Peru's Hidrovía Amazónica in 2017. Presumably, Sinohydro is providing technical expertise and financial capital ($100 million), while their Peruvian partners manage the administrative and legal process regulated by OSITRAN. The controversy surrounding the project stems from accusations the consortium manipulated the bidding process[207] and claims by indigenous federations that they were not adequately or legally consulted during the environmental review.[208]

In Brazil, Chinese companies have invested more than $US 60 billion since 2005; this is the country's largest and most diversified investment portfolio in South America. It includes businesses that specialise in commercial real estate, finance, transportation, telecommunications, agriculture and minerals. As in Peru, a large portion of investment capital has been allocated toward minerals, including loans to Brazil's very large mining sector (> $US 5 billion) and a combination of loans and joint ventures with Petrobras (> $US 16 billion). Brazil's importance to China's food supply is reflected in investments made by COFCO, China's state-owned commodity trader (> $US 3.2 billion).

As large as these investments are, they are smaller than the combined acquisitions of CTG and State Grid Corporation of China (SGCC) in the generation, transmission and distribution of electrical energy ($US 23 billion). The vast majority of these investments are located in extra Amazonian Brazil, except for minority interest by GTC in two hydropower facilities in Amapá: Santo Antônio do Jari ($US 250 million) and Cachoeira Caldeirão ($US 130 million). The only large investments in the Brazilian Amazon is SGCC's controlling interest (51 per cent) in ultra-high voltage transmission lines between Belo Monte and Southeast Brazil: BMTE-I to Rio de Janeiro (> $US 2 billion) and BMTE-II to Estreito, Minas Gerais (> $US 2.8 billion) (Figure 2.3 and Figure 2.5).[209]

State Grid's partners in the BMTE projects are Electronorte (24.5 per cent) and Furnas (24.5 per cent), both of which are subsidiaries of Electrobras, which is being [partially] privatised by the Brazilian government.[210] At the corporate level, the government intends to reduce state ownership

---

ongoing resolution of the *Lava Jato* scandal. Source: *Latin Finance*, 16 July 201, Brazil's Odebrecht asks Peru for proceeds from Chaglla sale, Latin American Financial Publications: https://www.latinfinance.com/daily-briefs/2019/7/16/brazil-s-odebrecht-asks-peru-for-proceeds-from-chaglla-sale

The construction of mega-scale hydropower facilities on the Xingu and Madeira rivers was accompanied by investments in grid infrastructure, such as the ultra-high voltage transformers at the Altamira (Pará) sub-station (top) and high voltage direct current (HVDC) 750 kV transmission lines that move electricity from remote power plants to urban centres in Southeast Brazil (bottom).

from 62 to 45 per cent by offering shares (common stock) in public equity markets. Internally, management is also downsizing the company by selling subsidiaries and joint ventures.[211] The semi-privatisation via the stock market is not likely to attract Chinese investors, but they are most certainly interested in acquiring individual assets in distribution, transmission and generating. The presence of CTG and SGCC in the Brazilian Amazon is not extensive, but that could change if they acquired assets from Electrobras or from a private sector investor, such as those that operate the hydropower plants on the Madeira and Tocantins rivers. Conceivably, they could lead a greenfield investment if regulators (or the Brazilian Congress) open up hydropower development within the Legal Amazon.

There is less uncertainty about the role of Chinese companies in the development of railroads. In 2019, the governor of Pará signed an agreement with the China Communication Construction Company (CCCCSA)* that committed both parties to initiating construction on the first stage of Ferrovia Paraense by the end of 2021 (Figure 2.4 and Figure 2.6). Although the COVID crisis has undoubtedly delayed its implementation, the world's largest construction company will soon bring its hyper-efficient engineering systems to the construction of that railroad. Finance for the R$ 15 billion ($US 2.7 billion) project is available via the China – Brazil Fund, an investment vehicle created in 2017 that has been capitalised by the China-Latin American Production Fund ($15 billion) and BNDES ($US 5 billion). Chinese construction companies with expertise in rail can be expected to participate in the public auctions for the Ferrogrão and FICO railroads (Figure 2.6).

The activities of Chinese investors in Guyana and Suriname have historically reflected the size of their economy and, in general, could be characterised as overseas development assistance. Between 2000 and 2014, China's policy banks provided $US 350 million to Suriname and $US 309 million to Guyana. Most of these resources were allocated to road 'rehabilitation' and electricity power systems.

The discovery of the offshore oil field has led to an upsurge in direct investment. In Guyana, that includes a $950-million contribution to a joint venture between the Chinese National Oil Company (CNOC) and Exxon, a luxury hotel, improved port facilities, and, reportedly, paving the IIRSA-sponsored Letham–Georgetown highway. Less likely is the renewal of the Amalia Falls hydropower project, which was the object of a potential $US 850 million loan in 2009, co-financed by the IDB and China Development Bank. The Amelia Falls project was removed from the development pipeline in 2015 due to concerns over its financial viability.[212]

---

* The practices of CCCCSA have been questioned by the World Bank, and the US Department of State has placed the company on a list of entities known to participate in human rights violations in Xinjiang province. See Prasso 2018.

In Suriname, oil discoveries lag those in Guyana, but geologists are confident of future production, and CNOC will participate. In the meantime, the country has contracted a Chinese engineering company to modernise its international airport.[213]

Sustainable Infrastructure: In Search of an Oxymoron

The macroeconomic hypothesis that infrastructure investments stimulate economic growth assumes that these physical assets overcome a logistical or systemic constraint on production. In practice, this requires that individual projects are subject to an objective feasibility analysis, are priced fairly, and have been approved after a full evaluation of their social and environmental impacts. Unfortunately, many infrastructure investments in the Pan Amazon have not met these three fundamental criteria. Some are poorly conceived or simply unnecessary, and many are foisted on society by vested interests or corrupt politicians.

Poorly designed or unnecessary infrastructure can create a negative economic impact if scarce financial capital is funneled into projects with limited benefit to society, thus diverting investment from more deserving initiatives. A poorly conceived project may provide a short-term economic boost but fail to provide the long-term benefits of a well-designed asset. In the Pan Amazon, the lack of an economic return from poorly conceived infrastructure initiatives is compounded by their large negative environmental and social impacts.

This chapter has focused on large-scale infrastructure built in the Pan Amazon over the last several decades. Some projects have been poor investments when evaluated using only economic criteria, but many more have been economically successful and politically popular because they benefit the actual (albeit recent) inhabitants of the Amazon. The most controversial have benefited stakeholders who are not residents of the Amazon, particularly companies that cater to extra-Amazonian consumers of energy and commodities. These assets and systems may be profitable, but they are not 'sustainable' when they fail to comply with the criteria of sustainable development.* Persistent high rates of deforestation, environmental degradation and social inequality are the most obvious manifestations of the failure to incorporate the concepts of sustainability into infrastructure development.

The concept of 'sustainable infrastructure' has been around since academic discussions first defined the concept of sustainable development; however, guidelines defining specific criteria for built infrastructure appeared

* The goal of sustainable development is the long-term stability of the economy, which is only achievable through the integration of economic, environmental, and social concerns throughout the decision-making process. Source: the Brundtland Commission, http://www.un-documents.net/wced-ocf.htm

The need for basic infrastructure extends beyond transportation and energy systems, and includes water treatment plants (top) and sewage systems (bottom).

only about a decade ago. The first efforts were little more than a laundry list of high-level standards for loan officers and corporate executives to consider when evaluating a potential investment.[214] More detailed systems that address the multiple and complex challenges required to transform the global economy in the face of climate crisis are now available.[215] These systems are not, however, particularly applicable to the unique situation of the Amazon, which also has an acute deficit in basic infrastructure.

What might define sustainable infrastructure in the Pan Amazon? First and foremost, it should benefit the inhabitants of the Amazon. Equally important, investments should yield economic benefits over the short, medium and long term, particularly those that contribute to GDP but also others that improve people's quality of life without generating an immediate increase in income. In some cases, an investment may be uneconomic in conventional terms but justified on humanitarian or ecological grounds and thus worthy of a subsidy.

The infrastructure investments that most obviously meet these criteria are the physical assets that are the foundation of the region's health care systems. The Covid pandemic of 2020/2021 revealed glaring deficiencies in the health care systems of all eight nations. The impact of the disease was particularly severe in their Amazonian jurisdictions, where it fell disproportionately on marginalised populations. Shortfalls in basic health infrastructure included not just a lack of hospital beds but also of the specialised equipment needed for acute medical conditions.

The Covid-19 pandemic also revealed that a significant portion of Amazonian populations were vulnerable to the coronavirus due to chronic conditions caused by infectious diseases associated with unsafe drinking water and poor sanitation. Public utilities provide water in most large and medium sized cities, but coverage is far from universal, particularly in the peri-urban neighbourhoods that house recent migrants. Investment in basic sanitation in urban areas is abysmal, and even the largest cities have woefully inadequate sewage systems (Table 2.4). Rural inhabitants are left to their own devices.

The situation is worse in the Andean Amazon, where administrative decentralisation and revenue sharing mechanisms have just begun to address decades of under-investment by central governments. Cities show radically different outcomes. Pucallpa (~370,000 inhabitants) provides only about 48 per cent of households with potable water, while Iquitos (477,000) approached ninety per cent coverage; neither city has any system to collect and treat sewage. In contrast, Santa Cruz de la Sierra (1.7 million) provides

Table 2.4: Comparison of potable water and sewage provision in major cities.

City	Population in 2015	% potable water	% sewage system	% sewage treatment	Investment ($R/5 years)
Manaus	2,130,264	88%	12%	48%	311
Belem	1,425,275	71%	13%	0.8%	266
São Luis	1,082,935	87%	50%	11%	346
Cuiabá	590,118	100%	54%	52%	379
Porto Velho	519,436	32%	5%	3%	62
Macapá	465,495	39%	9%	17%	38
Boa Vista	332,020	100%	64%	75%	214
Rio Branco	383,443	60%	24%	34%	72
Santarem	294,447	71%	6%	1%	
Palmas	286,787	98%	86%	60%	276
Varzea Grande	271,339	97%	35%	35%	

Data source: *Instituto Trata Brasil*, http://www.tratabrasil.org.br/estudos/estudos-itb/itb/novo-ranking-do-saneamento-2021

universal coverage for potable water and has extended basic sanitation to approximately half its population.*

Amazonian cities have invested in drinking water systems, typically with the assistance of the multilateral banks, but are still dumping nearly all their wastewater into the Amazon and its tributaries, operating under the assumption that the volume of water will provide a level of protection against contamination.† Unfortunately, that practice does not protect downstream populations from diseases caused by poor sanitation.

A similar infrastructure deficit exists in the region's primary and secondary schools, particularly in the public systems that serve rural communities and impoverished urban neighbourhoods. This deficiency is exacerbated by middle-class and affluent families who send their children

* Santa Cruz has the highest population growth in the Western hemisphere, averaging 5–7% annually between 1970 and 2000 and 2–3% after 2000. Water and sewage are managed by an independent cooperative (SAGUAPAC) owned by the city's inhabitants, but the phenomenal growth motivated real estate investors to develop neighbourhoods with full services while SAGUAPAC and the municipal government coordinated investments to provide service to '*barrios populares*'. For example, streets were not paved until sewage systems were installed to avoid wasted expenditures on street repair.

† The concept 'dilution is the solution to pollution' dominated waste-water treatment for decades. It is now questioned by environmental scientists but continues to be taught in engineering schools and employed widely across the planet.

The desire for improved infrastructure is exemplified by this family taking a newly purchased freezer back home to their riverside community.

to private schools, a common practice across South America that encourages under-investment in public school systems.[216]

Investments in education and health care might not meet the classic definition of built infrastructure (roads and bridges), but few would deny that sustained economic growth is wholly dependent upon a healthy and well-educated populace. Moreover, investments in schools, clinics and sanitation systems are labour-intensive and require significant amounts of concrete, lumber, and hardware, which ensures they meet the short-term objectives of infrastructure investment programmes – job creation and economic stimulus. Since schools and clinics are relatively simple structures, their construction can be contracted to local businesses, creating a positive feedback loop since the proprietors and employees of these spend the proceeds close to home.

Both health and educational systems would benefit from improvements in digital infrastructure, another non-traditional class of investment essential for economic growth and development in the twenty-first century.* Technology is an obvious antidote to the long distances and antiquated telecommunication systems that isolate rural communities. Exposure to computer technology would allow rural students to acquire basic information management skills essential for success in modern society, while access to

* In advanced economies the consumption of electrical energy has forced the tech sector to purchase renewable energy to operate their power-hungry server farms so as to avoid high GHG emission from conventional power grids.

the internet would democratise learning opportunities for tens of thousands of students and teachers. High-speed internet is the gold standard. It would not only allow students to participate in seminars and virtual events via the rapidly evolving 'zoom' technology, but would also permit health-care professionals to diagnose and treat many more patients remotely.

Providing high-speed internet to remote areas constitutes a significant technological challenge, and only a limited number of Amazonian cities have internet connections faster than 10 Mbs.* Speed is inherently limited because most connections are mediated by geostationary satellites with a response time limited by the speed of light. The cost of extending a fibre-optic network to the far-flung communities across the Pan Amazon would be exorbitant and precludes any attempt to expand high-speed internet service. This is about to change.

In 2020, the Brazilian government launched the *Projeto Amazônia Conectada*, an ambitious initiative to lay 8,000 kilometres of fibre-optic cable on the bottom of the Amazon river and its major tributaries (Figure 2.27). The initiative is being coordinated by the Brazilian army;[217] more importantly, it is being implemented as a public-private partnership with the backing of several large domestic and international telecom companies.† This initiative has the potential to provide a digital backbone that extends far beyond the region's major cities and could conceivably be expanded upstream to include isolated urban areas in the Andean Amazon. It will not, however, provide a solution for thousands of communities that are not located on a major river or connected to a cell phone network that could provide cost-effective internet services to the populace. Fortunately, there is a Plan B.

Two of the most innovative entrepreneurs on the planet, Jeff Bezos and Elon Musk, are competing to launch a network of low-orbit communication satellites with the express purpose of providing affordable high-speed internet services to remote regions.‡ If successful, these will provide an alternative to the fibre-optic backbones and cell-phone towers that constrain the expansion of high-speed internet. Low-orbit satellites will communicate with a receiver that consists of an antenna and router, which will interface with a local area wifi network. The expected cost of a terminal is estimated at about $US 500, with a monthly service fee of $US 99. The business model

* Until 2020, Manaus, Belem, Cuiabá, Porto Velho and Rio Branco were connected through fibre-optic cables mounted on the electrical transmission grid. Source: Electronorte 2017.

† *Oi*, a Brazilian telecom company emerging from bankruptcy, has adopted a business model that will focus on fibre- optic networks; Huawei, the Chinese telecom giant is seeking to introduce its 5G technology into Brazil. Source: https://www.datacenterdynamics.com/br/not%C3%ADcias/huawei-fornece-infraestrutura-para-conectar-a-amaz%C3%B4nia/

‡ Starlink is a subsidiary of SpaceX Corporation (Elon Musk is CEO); Project Kuiper is a subsidiary of Amazon (Jeff Bezos is the chairman of the board).

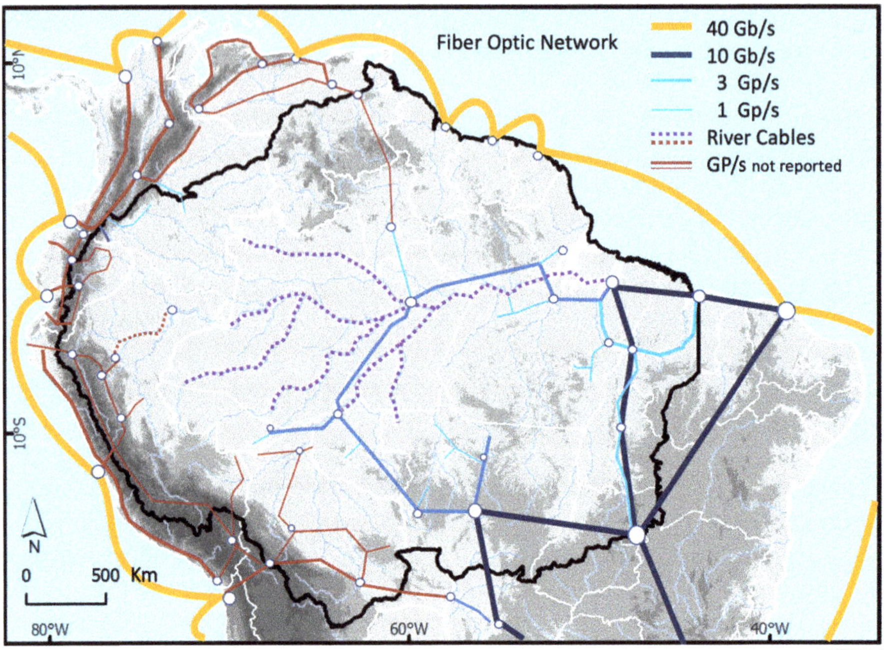

Figure 2.27: The major components of the fibre optic network in the Pan Amazon. The Projeto Amazônia Conectada, coordinated by the Brazilian Army in collaboration with private companies, is laying cable on the floor of the Amazon River and its major tributaries in the Central Amazon (Solimões, Madeira, Aripuanã, Purus, Juruá and Negro). Peru will lay a similar cable between Yurimaguas and Iquitos.

Data sources: Ministério da Defesa 2021; Electronorte 2018; Technoblog 2018; BNAmericas 2021.

is based on providing service to marine vessels, remote mine sites, industrial farms, vacation homes and rural communities isolated from fibre-optic networks. The technology represents a step-change in cost and quality; nonetheless, it will still be too expensive for most of the rural schools and clinics across the Amazon.* Presumably, the cost is negotiable, and citizens of the Amazon may be able to obtain a discount from the wealthiest men

* In 2018, there were 33,000 public schools in the Brazilian Amazon (source https://cidades.ibge.gov.br), and there are similar numbers of basic health clinics (*Unidad Básica de Saúde*). The Andean Amazon has about a quarter of the population and, presumably, health and educational infrastructure. A 'back-of-the-envelope' calculation for the cost of outfitting each school and clinic with a Starlink system would sum to only $US 25 million, and the monthly fees would sum to about $US 55 million annually. However, outfitting schools and clinics with computers to make the system functional would cost between $US 500 million and $1 billion.

on the planet.* Perhaps Mr Bezos could be convinced to donate systems to the communities of the region in compensation for appropriating their regional identity to brand his commercial enterprise.

Roads and bridges can also be sustainable. They are anathema to environmental advocates because of their association with logging, deforestation and settlement. However, these classic examples of built infrastructure contribute to sustainable economic growth when they improve secondary road networks within long-established agricultural frontiers. Almost all farmers lose a portion of their harvest due to spoilage, a problem that is particularly severe in humid tropical climates that accelerate decay and disease infestation. Supply and demand govern commodity markets and, according to macroeconomic theory, post-harvest losses will be replaced by production elsewhere; consequently, spoilage drives crop production onto frontier landscapes. Just as energy conservation should be part of a green energy strategy, so should investments in traditional infrastructure be part of a holistic development strategy when it enhances long-term productivity and profitability on existing farms.

The agricultural and consolidated frontiers across the Southern and Andean Amazon all suffer from poor secondary road networks. Improved secondary road networks will do more than reduce waste because bad roads delay planting, cause wear-and-tear on farm equipment and add to the cost of moving commodities to market. Small farmers suffer the most because they are more likely to depend on truckers who pro-rate their services based on the quality of the roads. Most small farmers would welcome the option to transport coffee or cacao via truck rather than mule, and the shift might motivate many of them to expand production.

Another underappreciated infrastructure class consists of the airports that support the aviation transportation system. Regional carriers have benefited from investments in airports in larger cities,† but the development of smaller airstrips is largely managed *ad hoc* by the military, loggers and (illegal) miners. The region once had a flourishing air taxi system organised by Evangelical and Catholic missionaries (see Chapter 11), but this has been replaced by commercially motivated operators, many of whom are complicit in the illicit drug trade. Examples in Alaska and Canada show that

---

* Starlink is providing free internet to a limited number of Native American communities in Canada and Alaska; see https://nativenewsonline.net/business/rural-alaska-natives-hope-elon-musk-s-starlink-internet-service-can-level-playing-field

† The IIRSA programme has supported investments in Ecuador (2) and Peru (2); the *Programa de Aceleração do Crescimento* supported improvements in Amazonas (9), Pará (3), Rondônia (3), Acre (1), Mato Grosso (3), Amapá (2), Tocantins (1); the Ex-Im Bank of China financed work in Bolivia (1), Guyana (1), and Suriname (1).

air taxis can provide a cost-effective transportation solution for a roadless wilderness populated by indigenous villages, settlers and vacation resorts. Key to that system is an airstrip in every village and operating subsidies to ensure the services are affordable for the region's indigenous inhabitants. The lack of affordable air taxi services is the largest single constraint to the expansion of ecotourism, which is currently clustered around a handful of Amazonian cities with large airports (Manaus, Iquitos, Puerto Maldonado, Leticia, and Coca; see Chapter 7).

Energy systems will continue to figure prominently in future investments in the Pan Amazon. Historically, economic growth has been tightly correlated with energy consumption. As families become more affluent, they buy electrical appliances and consume more energy. Refrigeration is the first type of appliance purchased by families emerging from rural poverty; once they are solidly middle class, they buy air conditioning. The relationship between GDP and energy consumption may change in the next fifty years at the global scale as societies transition from fossil fuels to a low-carbon economy. It is not, however, the path that will characterise development in the Pan Amazon. Energy consumption will grow as the economy expands, because access to affordable and reliable sources of electrical energy improves the quality of the lives of its inhabitants.

The rapid decline in the cost of solar power – and the advantages of distributed solar – provide an interesting option for communities and households that are not connected to the electrical grid. However, the capital cost of solar will hinder its adoption by most Amazonian families, who prefer to be linked into a grid system and pay a small monthly fee. Some electrical utilities have technologies and consumer packages designed to build and expand distributed systems, but the incentive system inherent in their business models continues to favour the expansion of the grid. Utility-scale solar will become more important and the construction of large-scale hydropower plants may become less attractive as an investment due to opposition by environmental advocates and indigenous groups. If so, utility companies may increase their investments in medium- and small-scale hydropower, which will contribute resilience to a diversified portfolio of electricity generation assets.

The pandemic of 2021 highlighted the social and economic inequalities of Amazonian society, but no well-informed observer was surprised by the suffering that Covid-19 wrought on remote indigenous villages, smallholder landscapes and the marginalised neighbourhoods in rapidly growing urban centres. The asymmetric impact was not unlike the scenario that played out in the advanced economies, where other disadvantaged populations suffered disproportionately from a legacy of under-investment. The (promised) response, particularly in the United States, is to implement stimulus and recovery policies that benefit under-served populations. In addition, the Biden administration seeks to channel resources to the long-delayed campaign to invest in renewable energy and other technologies required to avoid a climatic catastrophe.

The same logic of favouring marginalised populations and promoting sustainable production models can – and should – be applied to the pandemic recovery in the Pan Amazon. There is a palpable need to compensate for longstanding inequalities and the need to reform the conventional economy is similarly urgent.

In July of 2020, a progressive think tank launched a petition signed by seven former presidents* calling on the International Monetary Fund (IMF) and other multilateral organisations to cancel the external debt of Latin American countries and for bondholders to accept a restructuring of sovereign debt that included a two-year interest-payment holiday. The petition argues that such actions are 'fair and necessary' given the extraordinary challenge posed by the pandemic. Multilateral institutions have recognised the need to respond to the crisis but, unfortunately, none has the financial capacity to provide (real) debt forgiveness, much less to allocate the capital resources required to transform the economies of the Pan Amazonian nations. Institutions from China are not capable or inclined to alleviate the debt burden. More likely, governments and their private sector partners will invest in mining ventures and industrial agriculture that can generate export revenues needed to shore up their economies. Consequently, the 2020/2021 pandemic will foster investments that promote another cycle of conventional economic development.

---

* All are associated with socialist or progressive political parties: Dilma Rousseff (Brazil), Evo Morales (Bolivia), Rafael Correa (Ecuador), Ernesto Samper (Colombia), Jose Luis Rodriguez Zapatero (Spain), Fernando Lugo (Paraguay), Luis Guillermo Solis (Cost Rica)

Annex 2.1: IIRSA investments

A summary of investments organised via the The Initiative for the Integration of the Regional Infrastructure of South America (IIRSA), also known as Consejo Suramericano de Infraestructura y Planeamiento (COSIPLAN); it is organised into Hubs, Groups and Projects (see Figure 2.1).

Group		# projects	US$ million
	Amazon Hub		
Am-1	Access to the Putumayo Waterway	5	378
Am-2	Access to the Napo Waterway	5	105
Am-3	Access to the Huallaga–Marañón Waterway	10	1,299
Am-4	Access to the Ucayali Waterway	14	3,634
Am-5	Connection Between the Amazon Basin and Northeastern Brazil	12	15,197
Am-6	Amazon Waterway Network	12	321
Am-7	Access to the Morona–Marañón–Amazon Waterway	5	415
Am-8	Porto Velho–Southern Northeastern Brazil Rail Connection	7	6,150
	Andean Hub		
Ad-6	Colombia–Ecuador (Bogotá–Mocoa–Tena–Zamora–Palanda–Loja) Connection	5	496
Ad-7	Peru–Ecuador (Loja–Puente de Integración–Yurimaguas) Connection	2	147
	Guiana Shield Hub		
Gu-1	Venezuela–Brazil Interconnection	5	407
Gu-2	Brazil–Guyana Interconnection	6	277
Gu-3	Venezuela (Ciudad Guayana)–Guyana (Georgetown)–Suriname	3	302
Gu-4	Guyana–Suriname–French Guiana–Brazil Interconnection	6	3,596
	Central Interoceanic Hub		
CI-2	Optimisation of the Corumbá–São Paulo–Santos–Rio De Janeiro Corridor	8	6,307
CI-3	Santa Cruz–Puerto Suárez–Corumbá Connection	4	443
CI-4	Santa Cruz–Cuiabá Connection	5	141
CI-5	Connections of the Hub to the Pacific: Ilo/Matarani–Desaguadero–La Paz + Arica–La Paz + Iquique–Oruro–Cochabamba–Santa Cruz	29	11,164
	Peru - Brazil - Bolivia Hub		
PBB-1	Corridor Porto Velho–Rio Branco–Assis–Puerto Maldonado–Cusco/Juliaca–Ports In the Pacific	8	2,934
PBB-2	Rio Branco–Cobija–Riberalta–Yucumo–La Paz Corridor	9	1,482
PBB-3	Madeira–Madre De Dios–Beni River Corridor	7	28,232
	Total	**167**	**83,426**

Annex 2.2: Selected attributes of the Human Modified Landscapes recognised in the Pan Amazon

This table shows the categories of information contained in Online Supplement 2.1 at https://www.whpress.co.uk/Books/Killeen.html. *See Chapter 1 for definition, criteria and spatial distribution (Figure 1.1) and Figure 2.1 for location with respect to major highways.*

Column	Title	Information
1	HML Code	1 to 60 (key to maps)
2	HML name	Geographic descriptor
3	Total Area within HML	Hectares
4	Original forest area within HML	Hectares
5	Total historical deforestation within HML	Precent of original forest cover
6	Stage of Development of HML	Forest Frontier
	See Chapter 1	Agricultural Frontier
		Consolidated Frontier
		Coca Frontier
		Gold Rush Frontier
7	Strategic Development Initiatives	POLOAMANZOIA, PAC, PIN, ENID, IIRSA, etc.)
	See Chapter 6	
8	Flagship Infrastructure Assets	Names of key highways, hydrodams, railways, etc.
9	Planned Development Initiatives	includes proposed and planned roads, railways, dams, etc.
10	Major Highways	Including those that are existing and actively beimng considered (planned but not proposed)
11	Paved (referring to highways)	Yes/no as of 2021
12	Secondary Road network	No/limited/moderate/extensive
13	Agricultural Production Models See Chapter 3	Beef, soy-maize, oil palm, coffee, cacao, food crops, plantation forestry, timber, non-timber forest products
14	Key Mineral Resources See Chapter 5	Iron ore, copper, bauxite, cassiterite, gold, diamonds polymetallic

Bibliography

Agostinho, C.S., A. Akama and P.H.F. Lucinda. 2009. 'Capítulo 1. Inserção da UHE Peixe Angical na bacia Araguaia-Tocantins e metodologia de amostragem'. In C.S. Agostinho, F.M. Pelicice and E.E. Marques. *Reservatório de peixe angical: bases ecológicas para o manejo da ictiofauna*. São Carlos: RIMA - Relatório de Impacto Ambiental. pp. 5–14.

Alves, F., G. Elacqua, M. Koslinki, M. Martinez, H. Santos and D. Urbina. 2015. 'Winners and losers of school choice: Evidence from Rio de Janeiro, Brazil and Santiago, Chile'. *International Journal of Educational Development* **41**: 25–34.

The American Enterprise Institute & Aid, Data Research and Evaluation Unit. 2017. Version 2.0. Williamsburg, VA: AidData at William & Mary.

The American Enterprise Institute, China Global Investment Tracker, https://www.aei.org/china-global-investment-tracker/

Anderson, E.P., C.N. Jenkins, S. Heilpern, J.A. Maldonado-Ocampo, F.M. Carvajal-Vallejos, A.C. Encalada ... and N. Salcedo. 2018. 'Fragmentation of Andes-to-Amazon connectivity by hydropower dams'. *Science Advances* 4 (1): eaao1642.

Ansar, A., B. Flyvbjerg, A. Budzier and D. Lunn. 2014. 'Should we build more large dams? The actual costs of hydropower megaproject development'. *Energy Policy* **69**: 43–56.

Arias, M.E., T.A. Cochrane, M. Kummu, H. Lauri, G.W. Holtgrieve, J. Koponen and T. Piman. 2014. 'Impacts of hydropower and climate change on drivers of ecological productivity of Southeast Asia's most important wetland'. *Ecological Modelling* **272**: 252–263.

Assahiraa, C. et al. 2017. 'Tree mortality of a flood-adapted species in response of hydrographic changes caused by an Amazonian river dam'. *Forest Ecology and Management* **396**: 113–123.

ANTAQ. 2008. *Escoamento da produção agrícola da região centro-oeste. Vantagens comparativas do transporte hidroviário*. Brasília.

Bandura, R. and S. McKeown. 2020. *Sustainable Infrastructure in the Amazon*. Washington, DC: Center for Strategic and International Studies.

Barber, C.P., M.A. Cochrane, C.M. Souza and W.F. Laurance. 2014. 'Roads, deforestation, and the mitigating effect of protected areas in the Amazon'. *Biological Conservation* **177**: 203–209.

Barthem, R.B., M. Goulding, R.G. Leite, C. Cañas, B. Forsberg, E. Venticinque, P.Petry, M.L.D.B. Ribeiro, J. Chuctaya and A. Mercado. 2017. 'Goliath catfish spawning in the far western Amazon confirmed by the distribution of mature adults, drifting larvae and migrating juveniles'. *Scientific Reports* **7**: 41784.

Bermúdez-Liévano. B. 2019. China's Belt and Road advances in Latin America's Andean region, Dialogo Chino: https://dialogochino.net/en/infrastructure/27815-chinas-belt-and-road-advances-in-latin-americas-andean-region/

Berti, L. and A. Gatto Boueri. 6 Dec. 2020. Low reservoirs force Brazil to buy energy from Argentina. The Brazilian Report: https://brazilian.report/latin-america/2020/12/06/low-reservoirs-force-brazil-to-buy-energy-from-argentina/?mc_cid=05deb41048&mc_eid=e4c13bc18e

BNAmericas. 2021. Peru embarks on new rural connectivity push: https://www.bnamericas.com/en/news/peru-embarks-on-new-rural-connectivity-push

BNDES. https://www.bndes.gov.br/wps/portal/site/home/transparencia/centraldedownloads

BNDES. 2011. *Informe Anual, Programa de Aceleración del Crecimiento*, https://www.bndes.gov.br/SiteBNDES/bndes/bndes_es/Hotsites/Informe_Anual_2011/Capitulos/actuacion_institucional/o_bndes_politicas_publicas/pac.html

Bôas, A. 28 Aug. 2020. Ferrogrão, a Path of Illusion, Pulitzer Center: https://pulitzercenter.org/blog/ferrograo-path-illusion

BNDES transparency portal, https://www.bndes.gov.br/wps/portal/site/home/transparencia/centraldedownloads

Borges, T. and S. Branford. 26 Aug. 2019. Bolsonaro expresses 'love' for Amazon as it burns, offers no policy shift. Mongabay.com: https://news.mongabay.com/2019/08/bolsonaro-expresses-love-for-amazon-as-it-burns-offers-no-policy-shift/

Branford S. 2018. Brazil announces end to Amazon mega-dam building policy, Mongabay Series: Amazon Infrastructure, Global Forests: https://news.mongabay.com/2018/01/brazil-announces-end-to-amazon-mega-dam-building-policy/

CAF. Project database, https://www.caf.com/en/projects/

Carneiro da Cunha, M., R. Caixeta, J.M. Campbell, C. Fausto, J.A. Kelly, C. Lomnitz, C.D. Londoño Sulkin, C. Pompeia and A. Vilaça. 2017. 'Indigenous peoples boxed in by Brazil's political crisis'. *HAU: Journal of Ethnographic Theory* **7** (2): 403–426.

Cayetano, S., C. Alvarado and PO. Tuesta. 6 Dec. 2019. Personajes vinculados al 'Club de la Construcción' detrás de licitación de la Hidrovía Amazónica. *CONVOCCA - Periodismo de Investigación y Análisis De Datos*: https://convoca.pe/investigacion/personajes-vinculados-al-club-de-la-construccion-detras-de-licitacion-de-la-hidrovia

Cella-Ribeiro, A., C.R. da Costa Doria, J. Dutka-Gianelli, H. Alves and G. Torrente-Vilara. 2017. 'Temporal fish community responses to two cascade run-of-river dams in the Madeira River, Amazon basin'. *Ecohydrology* **10** (8): e1889.

Cochrane, T.A., T.J. Killeen and O. Rosales. 2006. *Agua, Gas y Agroindustria, la Gestion Sostenible de la Riego Agrícola en Santa Cruz, Bolivia*. La Paz, Bolivia: Conservation International.

Corpolec. 2019. Iluminando la fibra social, Cobertura Nacional (Venezuela): http://edcnetwork.co.ve/servicios/

Cruz Cossetti, M. 2019. 'O que é um backbone', Tecnoblog: https://tecnoblog.net/277282/o-que-e-um-backbone/

Cruz, D. 2019. Cooperación de China en América Latina y el caribe en Infraestructura: discursos, modalidades y conflictos socioambientales, Asociación Ambiente y Sociedad. Bogotá, Colombia: https://www.chinaenamericalatina.info/es/inicio/

Cui, Shoujun and Zheng Zhang. 2019. 'China and Brazil development finance cooperation. A case study of the Belo Monte Transmission Line project'. In E. Duseel Peters (ed.) *China's Financing in Latin America And The Caribbean*. Universidad Nacional Autónoma de México: https://dusselpeters.com/146.pdf

Dammert, J.L. 2013, Financing Infrastructure Projects in the Southern Amazon of Peru: its relation with environmental and social safeguards, Global Development Policy Center, Working paper 017: https://www.bu.edu/gdp/files/2018/10/GEGI_GDP-Peru-WP.pdf

Dos Santos, M.A., L.P. Rosa, B. Sikar, E. Sikar, E. and E.O. Dos Santos. 2006. 'Gross greenhouse gas fluxes from hydro-power reservoir compared to thermo-power plants'. *Energy Policy* **34** (4): 481–488.

Dourojeanni, M. 2005. MAP: Madre de Dios, Acre, Pando, Associação O Eco: http://www.oeco.org.br/colunas/marc-dourojeanni/16374-oeco-14292/

Dourojeanni, M.J. 2006. Estudio de caso sobre la carretera Interoceánica en la Amazonía sur del Perú. Conservación Internacional, Perú: https://www.academia.edu/5674644/Estudio_de_caso_sobre_la_Carretera_Interoceanica_en_la_Amazonia_Sur_del_peru

Dourojeanini, M. 1 June 2015. El ferrocarril interoceánico chino y nuestra desordenada visión de Desarrollo, SPDA Actualoidad Ambiental: https://www.actualidad-ambiental.pe/el-ferrocarril-interoceanico-chino/

Draper, F.C., K.H. Roucoux, I.T. Lawson, E.T. Mitchard, E.N.H. Coronado,). Lähteenoja ... and T.R. Baker. 2014). 'The distribution and amount of carbon in the largest peatland complex in Amazonia'. *Environmental Research Letters* **9** (12): 124017.

Duchelle, A., A. Almeyda, N. Hoyos, M. Marsik, E. Broadbent and K.A. Kainer. 2010. 'Conservation in an Amazonian tri-national frontier: patterns and drivers of land cover change in community-managed forests'. In proceedings of the conference 'Taking stock of smallholder and community forestry: where do we go from here'.

Electrobras/Electronorte 2017. As Oportunidades de compartlhar servços com a Electronorte: https://www.eventos.momentoeditorial.com.br/wp-content/uploads/2016/12/Flavio-Roberto-Antonio.pdf

Elgueta, M.E. 2020. La irrupción de China en América Latina: Atentos a la 'letra chica' o neocolonización. Infogate: https://www.infogate.cl/2020/11/18/la-irrupcion-de-china-en-america-latina-atentos-a-la-letra-chica/

Estache, A. and G. Garsous. 2012. 'The impact of infrastructure on growth in developing countries'. International Finance Corporation, IFC Economics Notes, Note 1, Washington DC.

Fainguelernt, M.B. 2016. 'The historical trajectory of the Belo Monte hydroelectric plant's environmental licensing process'. *Ambiente & Sociedade* **19** (2): 245–264.

Farinosi, F. 2017. 'Energy transitions and climate security in Brazil'. In Robert A. Looney (ed.) *Handbook of Transitions to Energy and Climate Security*. Routledge, 526 pages,

Fearnside, P.M. 1999. 'Social impacts of Brazil's Tucuruí dam'. *Environmental Management* **24** (4): 483–495.

Fearnside, P.M. 2001. 'Environmental impacts of Brazil's Tucuruí Dam: Unlearned lessons for hydroelectric development in Amazonia'. *Environmental Management* **27** (3): 377–396.

Fearnside, P.M. 2005. 'Deforestation in Brazilian Amazonia: history, rates, and consequences'. *Conservation Biology* **19** (3): 680–688.

Fearnside, P.M., 2006. 'Dams in the Amazon: Belo Monte and Brazil's hydroelectric development of the Xingu River Basin'. *Environmental Management* **38** (1): 16–27.

Fearnside, P.M. 2007. 'Brazil's Cuiabá-Santarém (BR-163) Highway: The environmental cost of paving a soybean corridor through the Amazon'. *Environmental Management* **39** (5): 601–614.

Fearnside, P. 3 Nov. 2020. BR-319: The beginning of the end for Brazil's Amazon forest (commentary), Amazon Infrastructure: https://news.mongabay.com/2020/11/br-319-the-beginning-of-the-end-for-brazils-amazon-forest-commentary/

Fearnside, P.M., 2015. 'Amazon dams and waterways: Brazil's Tapajós Basin plans'. *Ambio* **44** (5): 426–439.

Fearnside, P.M. 2016. 'Greenhouse gas emissions from Brazil's Amazonian hydroelectric dams'. *Environmental Research Letters* **11**: 011002.

Fearnside, P.M. and A.M.R. Figueiredo. 2015. 'China's influence on deforestation in Brazilian Amazonia: A growing force in the state of Mato Grosso'. In R. Ray, K. Gallagher, A. López and C. Sanborn (eds). *China and Sustainable Development in Latin America: The Social and Environmental Dimension*, pp. 229–265: http://www.bu.edu/gdp-cn/files/2017/09/Brazil.pdf

Finer, M. and C.N. Jenkins. 2012. 'Proliferation of hydroelectric dams in the Andean Amazon and implications for Andes-Amazon connectivity'. *PLOS ONE* **7** (4): e35126.

Fleck, L. 2009. Eficiência econômica, riscos e custos ambientais da reconstrução da rodovia BR-319. Conservation Strategy Fund: https://www.conservation-strategy.org/sites/default/files/field-file/CSF_Eficiencia_economica_BR319.pdf

Fonseca, V. 2013. Hidrelétrica do Bem Querer: grande lago, pouca energia. ((o))eco – Associação O Eco: https://www.oeco.org.br/reportagens/27446-hidreletrica-do-bem-querer-grande-lago-pouca-energia/

Forsberg, B.R., J.M. Melack, T. Dunne, R.B. Barthem, M. Goulding, R.C. Paiva … and S. Weisser. 2017. 'The potential impact of new Andean dams on Amazon fluvial ecosystems'. *PLOS ONE* **12** (8): e0182254.

Fraser, B.J. 20 Dec. 2007. Troubled Amazon corner, Grassroots forum attempts to alleviate negative impacts of Interoceanic Highway. Latinamerica Press: http://www.lapress.org/articles.asp?item=1&art=5441

Fricke, A.T., C.A. Nittrouer, A.S. Ogston, D.J. Nowacki, N.E. Asp, P.W. Souza Filh ... and A.M. Jalowska. 2017. 'River tributaries as sediment sinks: Processes operating where the Tapajós and Xingu rivers meet the Amazon tidal river'. *Sedimentology* **64** (6): 1731–1753.

Gallagher, Kevin P. and Margaret Myers., 2021. *China-Latin America Finance Database.* Washington, DC: Inter-American Dialogue.

García Delgado, F. 29 Jan. 2021. El futuro incierto de la Hidrovía Amazónica peruana. Mongabay.com: https://es.mongabay.com/2021/01/hidrovia-amazonia-peru-consulta-previa-pueblos-indigenas/

Gómez, A . 2018. Inversión pública y financiamiento chino en América Latina, Asociación Ambiente y Sociedad, Ecuador: https://www.chinaenamericalatina.info/inversion-publica-y-financiamiento-chino-en-america-latina/

Gonzales, J. 28 June 2019. Amazon infrastructure puts 68% of indigenous lands/protected areas at risk. Mongabay Series: Amazon Agribusiness, Amazon Illegal Deforestation, Amazon Infrastructure, Global Forests, https://news.mongabay.com/2019/06/amazon-infrastructure-puts-68-of-indigenous-lands-protected-areas-at-risk-report/

Guyot, J.L., N. Fillzola, J. Quintanilla and J. Cortez. 1996. 'Dissolved solids and suspended sediment yields in the Rio Madeira basin, from the Bolivian Andes to the Amazon'. International Association of Hydrological Sciences, Publication 236: 55–64.

Harris, M. 13 Aug. 2015. Guyana nixes plans for 165-MW Amaila Falls hydropower project. Hydro Review: https://www.hydroreview.com/world-regions/guyana-nixes-plans-for-165-mw-amaila-falls-hydropower-project/#gref

Harris, M. 12 Jan 2017. Peru's San Gaban project receives Chinese cash infusion'. Hydro Review: https://www.hydroreview.com/business-finance/peru-s-san-gaban-project-receives-chinese-cash-infusion/#gref

Hemming, J. 2003. *Die If You Must: The Brazilian Indians in the Twentieth Century*. (New York: Macmillan).

Higgins, T. 2020. Para ampliar hidrovia na Amazônia, governo planeja dinamitar habitat de botos e tartarugas, por Infraestrutura na Amazônia: https://brasil.mongabay.com/series/infraestrutura-na-amazonia/

Higgins, T. 17 Jan. 2020. Belo Monte boondoggle: Brazil's biggest, costliest dam may be unviable. Mongabay.com, Amazon Infrastructure series: https://news.mongabay.com/2020/01/belo-monte-boondoggle-brazils-biggest-costliest-dam-may-be-unviable/

Ho, M., U. Lall, M. Allaire, N. Devineni, H.H. Kwon, I. Pal, D. Raff and D. Wegner. 2017. 'The future role of dams in the United States of America'. *Water Resources Research* **53** (2): 982–998.

Horna, H. 1976. 'South America's marginal highway'. *The Journal of Developing Areas* **10** (4): 409–424.

IDB – Interamerican Development Bank. Project database, https://www.iadb.org/en/projects

IDB – Interamerican Development Bank. 2017. La Red del Futuro: Desarrollo de una red eléctrica limpia y sostenible para América Latina: https://publications.iadb.org/handle/11319/8682#sthash.lX4X5BiS.dpuf

IFC – International Finance Corporation. 2012. Sustainability policies and standards, E&S performance standards: https://www.ifc.org/wps/wcm/connect/topics_ext_content/ifc_external_corporate_site/sustainability-at-ifc/policies-standards

IFC – International Finance Corporation. 2013. Summary of investment information, Equatorial Energia SA: https://disclosures.ifc.org/project-detail/SII/32170/equatorial-energia-s-a
Inversor LATAM. 2020. Bolivia, la red troncal de fibra se ha ampliado a 26,587 km: https://inversorlatam.com/bolivia-la-red-troncal-de-fibra-se-ha-ampliado-a-26-587km-ganando-mas-de-2-000km/

Junk, W.J., F. Wittmann, J. Schöngart and M.T. Piedade. 2015. 'A classification of the major habitats of Amazonian black-water river floodplains and a comparison with their white-water counterparts'. *Wetlands Ecology and Management* **23** (4): 677–693.

Killeen T.J. 2007. *A Perfect Storm in the Amazon Wilderness, Development and Conservation in the Context of the Initiative for Integration of the Regional Infrastructure of South America (IIRSA)*. Washington, DC: Conservation International.

Killeen, T.J., M. Douglas, T. Consiglio, P. M. Jørgensen and J. Mejia. 2007. 'Dry spots and wet spots in the Andean hotspot'. *Journal of Biogeography* **34** (8): 1357–1373.

Killeen, T.J., A. Guerra, V. Calderon, M. Calzada, L. Correa, L. Soria, B. Quezada and M. Stieninger. 2008. 'Total historical land-use change in Bolivia, who, where, why and how much'. *Ecology & Society* 13 (36).

Killeen, T.J. and L.A. Solórzano. 2008. 'Conservation strategies to mitigate impacts from climate change in Amazonia'. *Philosophical Transactions of the Royal Society B: Biological Sciences* **363** (1498): 1881–1888.

Mérona, B., A.A. Juras, G.M. Santos and I.H.O. Cintra. 2010. Os peixes e a pesca no baixo Rio Tocantins: vinte anos depois da UHE Tucuruí. *Centrais Elétricas do Norte do Brasil S.A – Eletrobras Eletronorte*: https://horizon.documentation.ird.fr/exl-doc/pleins_textes/2021-05/010051893.pdf

Ministério da Defesa. 2020. Projeto Amazônia Conectada: avanços em fibra ótica beneficiam comunidades afastadas: https://www.gov.br/defesa/pt-br/centrais-de-conteudo/noticias/projeto-amazonia-conectada-avancos-em-fibra-otica-beneficiam-comunidades-afastadas

MinTEC – Ministerio de Tecnologías de la Información y las Comunicaciones. 2021. Proyecto Nacional de Fibra Óptica: https://www.colombiatic.mintic.gov.co/679/articles-63393_metodologia.pdf

Molina, S. 2014. El Puente Rurrenabaque-San Buenaventura. La responsabilidad del BID. FOBOMADE: http://fobomade.org.bo/2014/06/01/elpuente-rurrenabaque/

Myers, M. and R. Ray. 2021. 'Shifting gears: Chinese finance in LACV 2020'. The Dialogue, China-Latin America report Global Deevelopment Center: https://www.thedialogue.org/wp-content/uploads/2021/02/Chinese_Finance_LAC_2020.pdf

Oliveira, R. 10 Dec. 2019. No Xingu, finalização da última turbina de Belo Monte pode significar 'o fim do rio'. Publica, Agência de Jornalismo Investigativo: https://apublica.org/2019/12/no-xingu-finalizacao-da-ultima-turbina-de-belo-monte-pode-significar-o-fim-do-rio/

Orcotorio-Figueroa, R. 2020. Hidroeléctricas en Suramérica y en Perú: Situación de la cuenca del río Marañ*ón*. DAR - Derecho, Ambiente y Recursos Naturales, https://dar.org.pe/daropina-hidroelectricas/

Ortiz, F. 2014. Belo Monte ameaça maior sítio de desova de tartarugas da América do Sul. ((o))eco – Associação O Eco: https://www.oeco.org.br/reportagens/27941-belo-monte-ameaca-maior-sitio-de-desova-de-tartarugas-da-america-do-sul/

Paz Cardona, A.J. 14 July 2020. Ecuador: gobierno reconoce que la erosión de la cascada San Rafael podría afectar la hidroeléctrica Coca Codo Sinclair. Mongabay.com: https://es.mongabay.com/2020/07/erosion-rio-coca-ecuador-hidroelectrica-coca-codo-sinclair/

Pérez, Gabriel and Sarah Nunes. 2016. Natural resources logistics in landlocked countries in Latin America and the Caribbean. Facilitation of Transport and Trade in Latin America and the Caribbean – *FAL Bulletin* 348. CEPAL: http://repository.eclac.org/handle/11362/60

Piccone, T.J. 2016. *The Geopolitics of China's Rise in Latin America.* Washington DC, Brookings Institution.

Pompermayer, F.M., C.A. da Silva Campos Neto and J.M. Pepino de Paula. 2014. 'Hidrovias no Brasil: Perspectiva Histórica, Custos e Institucionalidade, Instituto de Pesquisa Econômica Aplicada (ipea), Secretaria de Assuntos Estratégicos da Presidência da República', http://repositorio.ipea.gov.br/bitstream/11058/2714/1/TD_1931.pdf

Ponce-Jara, M.A., M. Castro, M.R. Pelaez-Samaniego, J.L. Espinoza-Abad and E. Ruiz. 2018. 'Electricity sector in Ecuador: An overview of the 2007–2017 decade'. *Energy Policy* **113**: 513–522.

RAISG – Rede Amazônica de Informação Socioambiental Georreferenciada. 2021. Dados cartográficos: https://www.amazoniasocioambiental.org/pt-br/mapas/

Rioja, F. 2013. 'What is the value of infrastructure maintenance? A survey'. *Infrastructure and Land Policies* **13**: 347–365. The Lincoln Institute of Land Policy: https://www.lincolninst.edu/sites/default/files/pubfiles/what-is-the-value-of-infrastructure-maintenance_0.pdf

Rivera-González, L., D. Bolonio, L.F. Mazadiego and R. Valencia-Chapi. 2019. 'Long-term electricity supply and demand forecast (2018–2040): a LEAP model application towards a sustainable power generation system in Ecuador'. *Sustainability* **11** (19): 5316.

Rohter, L. (2019). *Rondon: Uma Biogtrafia.* Rio de Janeiro: Editora Schwarts.

Rubio, J., A. Escobedo, C. Vallejos, E. Oyague, A. Vera and M. Estrada. 2017. Marañón: Costo social de los impactos acumulativos de cinco proyectos hidroeléctricos. Serie Técnica Conservation Strategy Fund No. 50: https://www.conservation-strategy.org/sites/default/files/field-file/Maranon_Costo_Social_0.pdf

Sinclair, J.H. and T. Wasson. 1923. 'Explorations in eastern Ecuador'. *Geographical Review* **13** (2): 190–210.

Straub, S. 2008. *Infrastructure and Growth in Developing Countries* Vol. 4460. World Bank Publications.

Volckhausen, T. 21 June 2017. New highway brings deforestation near two Colombian national parks. Mongabay Series: Global Forest Reporting Network, Global Forests: https://news.mongabay.com/2017/06/new-highway-brings-deforestation-to-two-colombian-national-parks/

Wade, W.H. 2011. 'Boulevard of broken dreams: the inside story of the World Bank's Polonoroeste Road Project in Brazil's Amazon'. Grantham Research Institute on Climate Change and the Environment. Working Paper No. 55.

Wildlife Conservation Society. 2015. Análisis de la Evaluación de Impactos Ambientales de Hidroproyectos en Loreto, Perú: Estudio de los casos de la Central Hidroeléctrica Mazán y de la Hidrovía del Amazonas: https://www.conservation-strategy.org/sites/default/files/field-file/ES_Documento_de_Trabajo_Compensacion_Ambiental_Peru_CH3_Loreto_Sept_2017R.pdf

Winemiller, K.O., P.B. McIntyre, L. Castello, E. Fluet-Chouinard, T. Giarrizzo, S. Nam … and M.L.J. Stiassny. 2016. 'Balancing hydropower and biodiversity in the Amazon, Congo, and Mekong'. *Science* **351** (6269): 128–129.

Wittmann, H., F. von Blanckenburg, L. Maurice, J.L. Guyot, N. Filizola and P.W. Kubik. 2010. 'Sediment production and delivery in the Amazon River basin quantified by in situ-produced cosmogenic nuclides and recent river loads'. *Geological Society of America Bulletin* **123** (5–6): 934–950.

World Bank Group. 2008. World Bank Group Sustainable Infrastructure Action Plan (SIAP). Washington DC: World Bank Group.

Ziesler, R. and G.D. Ardizzone. 1979. *The Inland Waters of Latin America.* COPESCAL Tech. Paper No. 1.Rome: Food and Agriculture Organization – FAO

Notes to Chapter 2

1. BNDES. 2011. Informe Anual, Programa de Aceleración del Crecimiento: https://www.bndes.gov.br/SiteBNDES/bndes/bndes_es/Hotsites/Informe_Anual_2011/Capitulos/actuacion_institucional/o_bndes_politicas_publicas/pac.html
2. Horna, H. 1976. 'South America's marginal highway'. *The Journal of Developing Areas* **10** (4): 409–424.
3. Gonzales, J. 28 June 2019. Amazon infrastructure puts 68% of indigenous lands/protected areas at risk. *Mongabay Series*: Amazon Agribusiness, Amazon Illegal Deforestation, Amazon Infrastructure, Global Forests: https://news.mongabay.com/2019/06/amazon-infrastructure-puts-68-of-indigenous-lands-protected-areas-at-risk-report/ ; Bandura, R. and S. McKeown. 2020. *Sustainable Infrastructure in the Amazon*. Washington, DC: Center for Strategic and International Studies; Killeen T.J. 2007. *A Perfect Storm in the Amazon Wilderness, Development and Conservation in the Context of the Initiative for Integration of the Regional Infrastructure of South America (IIRSA)*. Washington, DC: Conservation International.
4. Latin Finance. 30 July 2020. Multilaterals hold the key for infrastructure investments after COVID-19. Latin American Financial Publications, Inc.: https://www.latinfinance.com/daily-briefs/2020/7/30/multilaterals-hold-the-key-for-infrastructure-investments-after-covid-19
5. Barber, C.P., M.A. Cochrane, C.M. Souza and W.F. Laurance. 2014. 'Roads, deforestation, and the mitigating effect of protected areas in the Amazon'. *Biological Conservation* 177: 203–209.
6. Pérez, Gabriel and Sarah Nunes. 2016. Natural resources logistics in landlocked countries in Latin America and the Caribbean. Facilitation of Transport and Trade in Latin America and the Caribbean – *FAL Bulletin* 348. CEPAL: http://repository.eclac.org/handle/11362/60
7. Rioja, F. 2013. 'What is the value of infrastructure maintenance? A survey'. *Infrastructure and Land Policies* 13: 347–365. The Lincoln Institute of Land Policy: https://www.lincolninst.edu/sites/default/files/pubfiles/what-is-the-value-of-infrastructure-maintenance_0.pdf
8. Fearnside, P.M. 2005. 'Deforestation in Brazilian Amazonia: history, rates, and consequences'. *Conservation Biology* **19** (3): 680–688.
9. Contas Regionais do Brasil 2013: Produto Interno Bruto, população residente e Produto Interno Bruto per capita, segundo as Grandes Regiões e as Unidades da Federação. IBGE. From Wikipedia: https://pt.wikipedia.org/wiki/Lista_de_unidades_federativas_do_Brasil_por_PIB_per_capita
10. PRODES – *Programa Monitoramento da Floresta Amazônica Brasileira por Satélite*. 8 April 2018. Desflorestamento nos Municípios da Amazônia Legal: http://www.dpi.inpe.br/prodesdigital/prodesmunicipal.php
11. SEMAS – *Secretaria de Meio Ambiente e Sustentabilidade do Pará*. 2016. Semas Embarga 16 Mil Hectares De Desmatamento Ilegal Em São Félix Do Xingu: https://www.semas.pa.gov.br/2016/07/25/semas-embarga-16-mil-hectares-de-desmatamento-ilegal-em-sao-felix-do-xingu/
12. Fearnside, P.M. 2007. 'Brazil's Cuiabá-Santarém (BR-163) Highway: The environmental cost of paving a soybean corridor through the Amazon'. *Environmental Management* **39** (5): 601–614.

13. DNIT – *Departamento Nacional de Infraestrutura de Transportes*. 2013. Mapa Rodoviário do Pará, https://www.brasil-turismo.com/para/mapa-rodoviario.htm
14. BrasilGovNews. 2 Dec. 2016. Amazon adds 230,000 hectares of protected area: http://www.brazilgovnews.gov.br/news/2016/12/amazon-adds-230-000-hectares-of-protected-area
15. ISA – *Instituto Socio Ambiental*. 8 Apr. 2018. Notícias Socioambientais Vitória do meio ambiente: STF veta redução de parques e reservas por Medida Provisória: https://www.socioambiental.org/pt-br/noticias-socioambientais/vitoria-do-meio-ambiente-stf-veta-reducao-de-parques-e-reservas-por-medida-provisoria
16. Borges, T. and S. Branford. 26 Aug. 2019. Bolsonaro expresses 'love' for Amazon as it burns, offers no policy shift. Mongabay.com: https://news.mongabay.com/2019/08/bolsonaro-expresses-love-for-amazon-as-it-burns-offers-no-policy-shift/
17. PSC – *Program Aceleracao do Crescimento*. 2010. Balanção - Anos 2007 – 2010, Pavimentação Guarantã do Norte/MT-Santarém/PA, p. 65: http://www.planejamento.gov.br/assuntos/investimento-e-pac/publicacoes-nacionais/11o-balanco-completo-do-pac-4-anos-2007-a-2010.pdf
18. G1 – Globo. 2016. Produtores rurais e garimpeiros interditam o km 313 da BR-163: https://g1.globo.com/pa/para/noticia/produtores-rurais-e-garimpeiros-interditam-o-km-313-da-br-163.ghtml
19. *Gazeta do Povo*. 9 Apr. 2018. Campeão de atoleiros, trecho da BR-163 escapa de congelamento de gastos: http://www.gazetadopovo.com.br/agronegocio/logistica/campeao-de-atoleiros-trecho-da-br-163-escapa-de-congelamento-de-gastos-d0a0vh9jrc9tc52l05cxeprjr
20. GFB – *Governo Federal do Brasil*. 9 Sept. 2020, TCU autoriza a concessão da BR-153/080/414/TO/GO, BR-163/230/PA e 22 aeroportos: https://www.gov.br/pt-br/noticias/transito-e-transportes/2020/12/tcu-autoriza-a-concessao-da-br-153-080-414-to-go-br-163-pa-e-22-aeroportos
21. Wade, W.H. 2011. 'Boulevard of broken dreams: the inside story of the World Bank's Polonoroeste Road Project in Brazil's Amazon'. Grantham Research Institute on Climate Change and the Environment. Working Paper No. 55.
22. PRODES – *Programa Monitoramento da Floresta Amazônica Brasileira por Satélite*. 8 April 2018. Desflorestamento nos Municípios da Amazônia Legal: http://www.dpi.inpe.br/prodesdigital/prodesmunicipal.php
23. GFB – *Governo Federal do Brasil*. 30 May 2020. DNIT finaliza a pavimentação de mais 32 quilômetros na BR-230/PA: https://www.gov.br/dnit/pt-br/assuntos/noticias/dnit-finaliza-a-pavimentacao-de-mais-32-quilometros-na-br-230-pa
24. *Agência Câmara de Notícias*. 11 Feb. 2013. Comissão visitará rodovia que passa por reserva indígena e é fechada durante a noite, Camra dos Deputados, https://www.camara.leg.br/noticias/400846-comissao-visitara-rodovia-que-passa-por-reserva-indigena-e-e-fechada-durante-a-noite/:
25. *Notícias Agrícolas*. 4 Sept. 2020. Governador de Roraima destaca safra recorde de soja e milho em 2020 e espera crescimento de 40% do setor para 2021: https://www.noticiasagricolas.com.br/videos/soja/267938-governador-de-roraima-destaca-safra-recorde-de-soja-e-milho-em-2020-e-espera-crescimento-de-40-do-setor.html

26. *Noticias do Acre*. 25 Sept. 2020. Governo do Acre e ministros debatem com peruanos integração rodoviária entre Cruzeiro e Pucallpa: https://agencia.ac.gov.br/governo-do-acre-e-ministros-debatem-com-peruanos-integracao-rodoviaria-entre-cruzeiro-e-pucallpa/

27. AADBR-319 – Associação 'Amigos e Defensores da BR 319'. 2 Apr. 2018. BR319 – UM DIREITO NOSSO: https://br319.wordpress.com/

28. Fleck, L. 2009. Eficiência econômica, riscos e custos ambientais da reconstrução da rodovia BR-319. Conservation Strategy Fund: https://www.conservation-strategy.org/sites/default/files/field-file/CSF_Eficiencia_economica_BR319.pdf

29. Fearnside, P. 3 Nov. 2020 BR-319: The beginning of the end for Brazil's Amazon forest (commentary). Amazon Infrastructure: https://news.mongabay.com/2020/11/br-319-the-beginning-of-the-end-for-brazils-amazon-forest-commentary/

30. DNIT – *Departamento Nacional De Infraestrutura De Transportes*. 2020. Elaboração da versão preliminar do componente indígena (CI) do processo de licenciamento ambiental da rodovia MR 319-am (trecho km 250,0 ao km 655,0). Ministério Da Infraestrutura, República Federativa Do Brasil

31. PPI –*Programa de Parcerias de Investimentos*. 14 Dec. 2020. Apoio ao licenciamento ambiental da rodovia BR-319/AM/RO (pavimentação do segmento central – KM 250 ao KM 655,70): https://www.ppi.gov.br/apoio-ao-licenciamento-ambiental-da-rodovia-br-319-am-ro-pavimentacao-do-segmento-central-km-250-ao-km-65570 ; IBAMA – *Instituto Brasileiro do Meio Ambiente e dos Recursos Naturais Renováveis*. 2015. Termo aditivo ao Termo de Acordo e Compromisso, IBAMA – DNIT: https://servicos.ibama.gov.br/licenciamento/consulta_empreendimentos.php

32. GFB – *Governo Federal do Brasil*. 30 May 2020. Ministro pede engajamento da sociedade amazonense para defender a BR-319: https://www.gov.br/infraestrutura/pt-br/assuntos/noticias/ministro-pede-engajamento-da-sociedade-amazonense-para-defender-a-br-319

33. Killeen, T.J., A. Guerra, V. Calderon, M. Calzada, L. Correa, L. Soria, B. Quezada and M. Stieninger. 2008. 'Total historical land-use change in Bolivia, who, where, why and how much'. *Ecology & Society* **13** (36).

34. GORE – *Gobierno Regional de Loreto*. 2014. Plan de Desarrollo Regional Concertado del departamento de Loreto al 2021: http://www.dar.org.pe/archivos/docs/PDRC_241115.pdf

35. Draper, F.C., K.H. Roucoux, I.T. Lawson, E.T. Mitchard, E.N.H. Coronado,). Lähteenoja ... and T.R. Baker. 2014. 'The distribution and amount of carbon in the largest peatland complex in Amazonia'. *Environmental Research Letters* **9** (12): 124017.

36. *El Peruano*. 4 Oct. 2017. Ley que declara de necesidad pública e interés nacional la construcción de la Carretera Iquitos-Saramiriza para su interconexión con la costa norte: https://leyes.congreso.gob.pe/Documentos/2016_2021/ADLP/Normas_Legales/30670-LEY.pdf

37. Dourojeanni, M. 2005. MAP: Madre de Dios, Acre, Pando, Associação O Eco: http://www.oeco.org.br/colunas/marc-dourojeanni/16374-oeco-14292/

38. Fraser, B.J. 20 Dec. 2007. Troubled Amazon corner, Grassroots forum attempts to alleviate negative impacts of Interoceanic Highway. Latinamerica Press: http://www.lapress.org/articles.asp?item=1&art=5441
39. Duchelle, A., A. Almeyda, N. Hoyos, M. Marsik, E. Broadbent and K.A. Kainer. 2010. 'Conservation in an Amazonian tri-national frontier: patterns and drivers of land cover change in community-managed forests'. In proceedings of the conference 'Taking stock of smallholder and community forestry: where do we go from here'. Montpellier, France
40. Capital Humano y Socil Alternativo. 2016. Rutas de la trata de personas en la Amazoniía Peruana: http://www.chsalternativo.org/images/publicaciones/01_trata_de_personas_en_el_peru/2016_rutas_trata_de_personas_amazonia_peruana_prtg.pdf
41. Dourojeanni, M.J. 2006. Estudio de caso sobre la carretera Interoceánica en la Amazonía sur del Perú. Conservación Internacional Perú: https://www.academia.edu/5674644/Estudio_de_caso_sobre_la_Carretera_Interoceanica_en_la_Amazonia_Sur_del_peru
42. BDE – *Banco de Desarollo del Ecuador*. 2018. Vias Rurales – Caminos Vecinales, Canton Taisha, Provincia De Morona Santiago: https://consulta.bde.fin.ec/geoportal/CreditoProyecto.aspx#
43. MAAP –Monitoring of the Andean Amazon Project. 2018. MAAP #82: Oil-related Deforestation in Yasuni National Park, Ecuadorian Amazon: http://maaproject.org/yasuni_eng/
44. INVIAS – *Instituto Nacional de Vías*. 12 Apr. 2018. Programa Estratégico de Obra Pública – Vías para la Equidad: https://www.invias.gov.co/index.php/seguimiento-inversion/proyectos-invias
45. Volckhausen, T. 21 June 2017. New highway brings deforestation near two Colombian national parks. *Mongabay Series*: Global Forest Reporting Network, Global Forests: https://news.mongabay.com/2017/06/new-highway-brings-deforestation-to-two-colombian-national-parks/
46. DGCF – *Dirección General de Caminos y Ferrocarrils*. 2015. Sistema Nacional de Carrateras (SINAC), Vice Ministero de Transporte, Ministero de Transporte y Comunivcaciones, Gobierno de Perú: https://sinia.minam.gob.pe/mapas/mapa-sistema-nacional-carreteras-peru
47. Killeen, T.J., M. Douglas, T. Consiglio, P. M. Jørgensen and J. Mejia. 2007. 'Dry spots and wet spots in the Andean hotspot'. *Journal of Biogeography* **34** (8): 1357–1373.
48. Killeen, T.J. and L.A. Solórzano. 2008. 'Conservation strategies to mitigate impacts from climate change in Amazonia'. *Philosophical Transactions of the Royal Society B: Biological Sciences* **363** (1498): 1881–1888
49. DGCF – *Dirección General de Caminos y Ferrocarrils*. 2015. Sistema Nacional de Carrateras (SINAC), Vice Ministero de Transporte, Ministero de Transporte y Comunivcaciones, Gobierno de Perú: https://sinia.minam.gob.pe/mapas/mapa-sistema-nacional-carreteras-peru
50. *Diario do Amapá*. 22 Sept. 2017. 'Asfaltar a BR 156 do Jari a Oiapoque custa R$ 1 bilhão', diz DNIT: https://www.diariodoamapa.com.br/cadernos/cidades/asfaltar-a-br-156-do-jari-a-oiapoque-custa-r-1-bilhao-diz-dnit/

51. International Hydropower Association. Country Profile, Brazil: http://www.hydropower.org/country-profiles/brazil
52. EPE – *Empresa de Pesquisa Energética*. 2006. Plano Nacional de Energia – 2030, Secretaria de Planejamento e desenvolvimento Energetico Minsterio de Minas e Energia: https://www.epe.gov.br/pt/publicacoes-dados-abertos/publicacoes/plano-nacional-de-energia-pne
53. Farinosi, F. 2017. 'Energy transitions and climate security in Brazil'. In Robert A. Looney (ed.) *Handbook of Transitions to Energy and Climate Security*. Routledge
54. EPE – *Empresa de Pesquisa Energética*. 2006. Plano Nacional de Energia – 2030, Secretaria de Planejamento e desenvolvimento Energetico Minsterio de Minas e Energia: https://www.epe.gov.br/pt/publicacoes-dados-abertos/publicacoes/plano-nacional-de-energia-pne
55. EPE – *Empresa de Pesquisa Energética*. 2020. PNE2050, Plano Nacional de Energia, Secretaria de Planejamento e desenvolvimento Energetico, Minsterio de Minas e Energia: https://www.epe.gov.br/pt/publicacoes-dados-abertos/publicacoes/plano-nacional-de-energia-pne
56. Branford S. 2018. Brazil announces end to Amazon mega-dam building policy, Mongabay Series: Amazon Infrastructure, Global Forests: https://news.mongabay.com/2018/01/brazil-announces-end-to-amazon-mega-dam-building-policy/
57. Hydro Review . 26 Aug. 2019. Brazil to support construction of four new hydropower plants: https://www.hydroreview.com/2019/08/26/brazil-to-support-construction-of-four-new-hydropower-plants/#gref
58. ANEEL – *Agência Nacional de Energia Elétrica*. 2017. Inventários Hidrelétricos, obtained from Rede Amazônica de Informação Socioambiental ... – RAISG, Maps and Publications: https://www.amazoniasocioambiental.org/en/maps/
59. Berti, L. and A. Gatto Boueri. 6 Dec. 2020. Low reservoirs force Brazil to buy energy from Argentina. The Brazilian Report: https://brazilian.report/latin-america/2020/12/06/low-reservoirs-force-brazil-to-buy-energy-from-argentina/?mc_cid=05deb41048&mc_eid=e4c13bc18e
60. Anderson, E.P., C.N. Jenkins, S. Heilpern, J.A. Maldonado-Ocampo, F.M. Carvajal-Vallejos, A.C. Encalada ... and N. Salcedo. 2018. 'Fragmentation of Andes-to-Amazon connectivity by hydropower dams'. *Science Advances* **4** (1): eaao1642.
61. Finer, M. and C.N. Jenkins. 2012. 'Proliferation of hydroelectric dams in the Andean Amazon and implications for Andes-Amazon connectivity'. *PLOS ONE* 7 (4): e35126.
62. International Hydropower Association. Country Profile: Peru: http://www.hydropower.org/country-profiles/peru
63. International Hydropower Association. http://www.hydroworld.com/articles/2009/10/hydropower-is-centerpiece.html
64. International Hydropower Association. Country Profile: Colombia: http://www.hydropower.org/country-profiles/colombia
65. Hydroworld.com. http://www.hydroworld.com/world-regions/latin-america/infographic.html

66. Wittmann, H., F. von Blanckenburg, L. Maurice, J.L. Guyot, N. Filizola and P.W. Kubik. 2010. 'Sediment production and delivery in the Amazon River basin quantified by in situ-produced cosmogenic nuclides and recent river loads'. *Geological Society of America Bulletin*: B30317-1.
67. Winemiller, K.O., P.B. McIntyre, L. Castello, E. Fluet-Chouinard, T. Giarrizzo, S. Nam … and M.L.J. Stiassny. 2016. 'Balancing hydropower and biodiversity in the Amazon, Congo, and Mekong'. *Science* 351 (6269): 128–129; Arias, M.E., T.A. Cochrane, M. Kummu, H. Lauri, G.W. Holtgrieve, J. Koponen and T. Piman. 2014. 'Impacts of hydropower and climate change on drivers of ecological productivity of Southeast Asia's most important wetland'. *Ecological Modelling* **272**: 252–263.
68. Assahiraa, C. et al. 2017. 'Tree mortality of a flood-adapted species in response of hydrographic changes caused by an Amazonian river dam'. *Forest Ecology and Management* **396**: 113–123.
69. Junk, W.J., F. Wittmann, J. Schöngart and M.T. Piedade. 2015. 'A classification of the major habitats of Amazonian black-water river floodplains and a comparison with their white-water counterparts'. *Wetlands Ecology and Management* 23 (4): 677–693.
70. Fearnside, P.M. 2016. 'Greenhouse gas emissions from Brazil's Amazonian hydroelectric dams'. *Environmental Research Letters* 11: 011002.
71. International Hydropower Association. Types of Hydropower: http://www.hydropower.org/types-of-hydropower
72. Anderson et al. 'Fragmentation of Andes-to-Amazon connectivity'
73. International Rivers network. 1 Jan 2021. Protecting rivers and defending the rights of the communities that depend on them: https://www.internationalrivers.org/
74. US Fish & Wildlife Service. 2018. Dam removal: An opportunity for our rivers: https://www.fws.gov/southeast/pdf/fact-sheet/dam-removal.pdf
75. Kurmanaev, A. and J. Otis. 3 Apr. 2016. Water shortage cripples Venezuela. *The Wall Street Journal*: https://www.wsj.com/articles/water-shortage-cripples-venezuela-1459717127
76. Siverio, R.U. 4 Apr. 2019. The rise and fall of Venezuela's hydro system. *Caracas Chronicles*: https://www.caracaschronicles.com/2019/04/04/the-rise-and-fall-of-venezuelas-hydro-system/
77. BNAmericas. 19 Apr. 2018. Tucuruí hydroelectric expansion: https://www.bnamericas.com/project-profile/en/ampliacion-hidroelectrica-tucurui-tucurui
78. Fearnside, P.M. 1999. 'Social impacts of Brazil's Tucuruí dam'. *Environmental Management*: 24 (4): 483–495; Fearnside, P.M. 2001. 'Environmental impacts of Brazil's Tucuruí Dam: Unlearned lessons for hydroelectric development in Amazonia'. *Environmental Management* 27 (3): 377–396.
79. Mérona, B., A.A. Juras, G.M. Santos and I.H.O. Cintra. 2010. Os peixes e a pesca no baixo Rio Tocantins: vinte anos depois da UHE Tucuruí. *Centrais Elétricas do Norte do Brasil* S.A – Eletrobras Eletronorte: http://www.eletronorte.gov.br/opencms/export/sites/eletronorte/publicacoes/publicacoes/Os_Peixes_e_a_Pesca_no_Baixo_Rio_Tocantins.pdf

80. Agostinho, C.S., A. Akama and P.H.F. Lucinda. 2009. 'Capítulo 1. Inserção da UHE Peixe Angical na bacia Araguaia-Tocantins e metodologia de amostragem'. In C.S. Agostinho, F.M. Pelicice and E.E. Marques. *Reservatório de peixe angical: bases ecológicas para o manejo da ictiofauna*. RIMA - Relatório de Impacto Ambiental, São Carlos, pp. 5–14.

81. ENGIE. 31 Dec 2020. Usina Hidrelétrica Estreito: https://www.engie.com.br/complexo-gerador/usinas/usina-hidreletrica-estreito/

82. ANEEL – *Agência Nacional de Energia Elétrica*. 2017. Inventários Hidrelétricos, obtained from Rede Amazônica de Informação Socioambiental ... – RAISG, Maps and Publications: https://www.amazoniasocioambiental.org/en/maps/

83. Dams in the Amazon. 3 Apr. 2018. Search Dams database: http://dams-info.org/en

84. ABIAPE – *Associação Brasileira dos Investidores em Autoprodução de Energia*. 3 Apr. 2018. Usinas do rio Madeira têm custo estimado em R$ 43 Bilhões: http://www.abiape.com.br/newsletter/54-ano-1-2007/maio/316-usinas-do-rio-madeira-tem--custo-estimado-em-43-bilhoes.html

85. ANEEL – *Agencia Nacional de Energia Electrica*. 2008. Características e Requisitos Técnicos Básicos das Instalações De Transmissão: http://www2.aneel.gov.br/aplicacoes/editais_transmissao/documentos/04_LOTE_LD_CC_Madeira.pdf

86. Estadão. 23 Oct. 2017. Error de projeto limita a geração de usinas do Madeira, que custaram R$ 40 bi: http://economia.estadao.com.br/noticias/geral,falha-limita--geracao-de-usinas-do-madeira,70002056408

87. EDF – Environmental Defense Fund. 2017. Brazil and the CDM: Defrauding Investors, Cheating the Atmosphere?: https://www.edf.org/sites/default/files/documents/brazil_cdm_report.pdf

88. Fearnside, P.M., 2006. Dams in the Amazon: Belo Monte and Brazil's hydroelectric development of the Xingu River Basin. *Environmental Management* 38 (1): 16–27.

89. Relatorio de impacto ambiental do aproveitamento hidroeletrico de Belo Monte. 2009: https://www.eletrobras.com/elb/data/Pages/LUMIS46763BB8PTBRIE.htm

90. Fainguelernt, M.B. 2016. 'The historical trajectory of the Belo Monte hydroelectric plant's environmental licensing process'. *Ambiente & Sociedade* 19 (2): 245–264.

91. BBC. 19 April 2010. Entenda a polêmica envolvendo a usina de Belo Monte: https://www.bbc.com/portuguese/noticias/2010/04/100419_belomonte_qandanovo_cq

92. Oliveira, R. 10 Dec. 2019. No Xingu, finalização da última turbina de Belo Monte pode significar 'o fim do rio'. Publica, Agência de Jornalismo Investigativo: https://apublica.org/2019/12/no-xingu-finalizacao-da-ultima-turbina-de-belo-monte-pode-significar-o-fim-do-rio/

93. Raposa, K. 7 May 2014. Was Brazil's Belo Monte dam a bad idea?. Forbes: https://www.forbes.com/sites/kenrapoza/2014/03/07/was-brazils-belo-monte-dam-a-bad-idea/?sh=48692e692a4e

94. Mongabay. 13 Mar. 2016. Fish kills at Amazon's Belo Monte dam point up builder's failures. https://news.mongabay.com/2016/07/fish-kills-at-amazons-belo-monte-dam-point-up-builders-failures/

95. Ortiz, F. 2014. *Belo Monte ameaça maior sítio de desova de tartarugas da América do Sul*, ((o))eco – Associação O Eco: https://www.oeco.org.br/reportagens/27941-belo-monte-ameaca-maior-sitio-de-desova-de-tartarugas-da-america-do-sul/
96. IBAMA – *Instituto Brasileiro do Meio Ambiente e dos Recursos Naturais Renováveis.* 2017. Ibama multa Norte Energia em R$ 35 milhões por mortandade de peixes em Belo Monte: http://www.ibama.gov.br/noticias/58-2016/171-ibama-multa-norte-energia-em-r-35-milhoes-por-mortandade-de-peixes-em-belo-monte
97. Mongabay. 12 Aug 2016. Promised US$1 billion in Belo Monte dam compensation largely unpaid?: https://news.mongabay.com/2016/08/promised-us1-billion-in-belo-monte-dam-compensation-largely-unpaid/
98. Higgins, T. 17 Jan. 2020. Belo Monte boondoggle: Brazil's biggest, costliest dam may be unviable. Mongabay.com, Amazon Infrastructure series: https://news.mongabay.com/2020/01/belo-monte-boondoggle-brazils-biggest-costliest-dam-may-be-unviable/
99. Reuters. 11 Dec. 2020. Brazil's Belo Monte dam owners challenge order to raise outflow. https://www.reuters.com/article/brazil-dam/brazils-belo-monte-dam-owners-challenge-order-to-raise-outflow-idUSL1N2IR2M8
100. Berti and Gatto, Low reservoirs force Brazil to buy energy from Argentina.
101. ANEEL – *Agência Nacional de Energia Elétrica.* 2017. Inventários Hidrelétricos, obtained from Rede Amazônica de Informação Socioambiental ... – RAISG, Maps and Publications, https://www.amazoniasocioambiental.org/en/maps/
102. Fearnside, P.M., 2015. 'Amazon dams and waterways: Brazil's Tapajós Basin plans'. *Ambio* **44** (5): 426–439.
103. *Ministerio de Minas y Energía* (MME): http://www.mme.gov.br/ and *Empresa de Pesquisa Energética* (ENE): http://www.epe.gov.br/geracao/Paginas/default.aspx
104. IBAMA – Despacho 02001.017118/2016-68 DILIC/IBAMA
105. Reuters. 25 May 2020. Estudos sobre hidrelétricas no Tapajós têm prazo prorrogado até 2021 pela Aneel: https://economia.uol.com.br/noticias/reuters/2020/05/25/estudos-sobre-hidreletricas-no-tapajos-tem-prazo-prorrogado-ate-2021-pela-aneel.htm
106. Roraima em Coco. 24 June 2020. Empresa apresenta os aspectos técnicos da usina hidrelétrica do Bem Querer: https://roraimaemfoco.com/empresa-apresenta-os-aspectos-tecnicos-da-usina-hidreletrica-do-bem-querer/
107. Reuters. 26 Mar. 2019. Governo retoma projeto de hidrelétrica de R$6 bi em Roraima e prevê leilão em 2021: https://www.reuters.com/article/energia-hidreletrica-bemquerer-idLTAKCN1R72I4
108. Fonseca, V. 2013. Hidrelétrica do Bem Querer: grande lago, pouca energia, ((o))eco – Associação O Eco: https://www.oeco.org.br/reportagens/27446-hidreletrica-do-bem-querer-grande-lago-pouca-energia/
109. EPE – *Empresa de Pesquisa Energética.* 2006. Plano Nacional de Energia – 2030, Secretaria de Planejamento e desenvolvimento Energetico Minsterio de Minas e Energia: https://www.epe.gov.br/pt/publicacoes-dados-abertos/publicacoes/plano-nacional-de-energia-pne

110. CPISP – *Comissão Pro-Indio de São Paulo.* 2014. Estudos das hidrelétricas no Rio Trombetas são retomados sem consulta previa: https://cpisp.org.br/estudos-das-hidreletricas-no-rio-trombetas-sao-retomados-sem-consulta-previa/

111. CPISP – *Comissão Pro-Indio de São Paulo.* 2019. Governo anuncia nova hidrelétrica na Amazônia que impactará Terras Indígenas e Quilombolas: https://cpisp.org.br/estudos-das-hidreletricas-no-rio-trombetas-sao-retomados-sem-consulta-previa/

112. *Viceministerio de Comunicación – Bolivia.* 16 Nov. 2016. Brasil ofrece comprar todo el excedente eléctrico de Bolivia: https://comunicacion.gob.bo/?q=20161115/22458

113. Mongabay – LATAM. 12 Nov 2017. Tres claves para entender por qué las comunidades se oponen a la hidroeléctrica Rositas en Bolivia: https://es.mongabay.com/2017/11/tres-claves-entender-las-comunidades-se-oponen-la-hidroelectrica-rositas-bolivia/

114. Cochrane, T.A., T.J. Killeen and O. Rosales. 2006. 'Agua, Gas y Agroindustria, la Gestion Sostenible de la Riego Agrícola en Santa Cruz, Bolivia'. La Paz, Bolivia: Conservation International.

115. ENDE – *Empresa Nacional de Electricidad.* 2017. Servicios de Consultoría para el diseño final y documentos de licitación del Proyecto 'Construcción Central Hidroeléctrica Rositas'. La Fundación Solon: https://fundacionsolon.org/rositas/

116. Guyot, J.L., N. Fillzola, J. Quintanilla and J. Cortez. 1996. 'Dissolved solids and suspended sediment yields in the Rio Madeira basin, from the Bolivian Andes to the Amazon, International Association of Hydrological Sciences, Publication 236: 55–64; Forsberg, B.R., J.M. Melack, T. Dunne, R.B. Barthem, M. Goulding, R.C. Paiva … and S. Weisser. 2017. 'The potential impact of new Andean dams on Amazon fluvial ecosystems'. *PLOS ONE* **12** (8): e0182254.

117. ENDE – *Empresa Nacional de Electricidad.* 2017. Ficha Ambiental completa 'Componente 1 angosto Chepete 400' and Ficha Ambiental completa 'Componente 2 angosto El Bala 220': https://fundacionsolon.org/el-bala-chepete/

118. Fundación Solon. 2019. Exportación de electricidad: Silencio sobre mega hidroeléctricas y excedentes sin mercado: https://fundacionsolon.org/2019/08/19/exportacion-de-electricidad-silencio-sobre-mega-hidroelectricas-y-excedentes-sin-mercado/

119. ENDE – *Empresa Nacional de Electricidad.* 20 Jan. 2021. Proyectos en Estudio: https://www.ende.bo/proyectos/estudio

120. IDB – Interamerican Development Bank. 2017. La Red del Futuro: Desarrollo de una red eléctrica limpia y sostenible para América Latina: https://publications.iadb.org/handle/11319/8682#sthash.lX4X5BiS.pdf

121. Fundación Solon. 19 Jan. 2021. Energía: Perdidos en el laberinto, Fundación Solon, La Paz, https://fundacionsolon.org/2020/01/24/hidroelectricas-entre-la-necesidad-y-la-pesadilla/

122. Osinergmin – *Organismo Supervisor de la Inversión en Energía y Minería.* 2016. Participación de las Empresas Generadoras en el Mercado del Sistema Eléctrico, Goberrono de Perú: https://www.osinergmin.gob.pe/seccion/centro_documental/electricidad/Documentos/Publicaciones/Participacion-Empresas-Mercado-Electrico-2016.xlsx

123. Osinergmin – *Organismo Supervisor de la Inversión en Energía y Minería*. 2016. *La industria de la energía renovable en el Perú: 10 años de contribuciones a la mitigación del cambio climático:* https://www.osinergmin.gob.pe/seccion/centro_documental/Institucional/Estudios_Economicos/Libros/Osinergmin-Energia-Renovable-Peru-10anios.pdf

124. *Luz del sur*. 26 April 2018. Proyecto Central Hidroeléctrica Santa Teresa – Collpani Grande: https://www.luzdelsur.com.pe/desarrollo-de-proyectos/proyecto-central-hidroelectrica-santa-teresa-ccollpani-grande.html

125. Orcotorio-Figueroa, R. 2020. *Hidroeléctricas en Suramérica y en Perú: Situación de la cuenca del río Marañón*. DAR – Derecho, Ambiente y Recursos Naturales: https://dar.org.pe/daropina-hidroelectricas/

126. Rubio, J., A. Escobedo, C. Vallejos, E. Oyague, A. Vera and M. Estrada. 2017. *Marañón: Costo social de los impactos acumulativos de cinco proyectos hidroeléctricos*. Serie Técnica Conservation Strategy Fund No. 50: https://www.conservation-strategy.org/sites/default/files/field-file/Maranon_Costo_Social_0.pdf

127. Wildlife Conservation Society. 2015. Análisis de la Evaluación de Impactos Ambientales de Hidroproyectos en Loreto, Perú: Estudio de los casos de la Central Hidroeléctrica Mazán y de la Hidrovía del Amazonas: https://peru.wcs.org/es-es/WCS-Perú/Publicaciones

128. *Periódico Luz Verde de Iquitos*. 15 July 2019. Elisbán Quiere levantar Proyecto Cadáver de Yvan: Hidroeléctrico de Mazán sin consultar al Pueblo: http://www.periodicoluzverdeloreto.com/

129. Ponce-Jara, M.A., M. Castro, M.R. Pelaez-Samaniego, J.L. Espinoza-Abad and E. Ruiz. 2018. 'Electricity sector in Ecuador: An overview of the 2007–2017 decade'. *Energy Policy* 113: 513–522; ARCE – *Agencia Regulación y Control de Electricidad*. 2018. Estadística Anual y Multianual del Sector Eléctrico Ecuatoriano, Agencia de Regulación y Control de Energia y Recuros Naturales No renovables 2018: https://www.regulacionelectrica.gob.ec/boletines-estadisticos/

130. Coca Codo Sinclair, *Ministerio de Electricidad y Energía Renovable*: http://www.energia.gob.ec/coca-codo-sinclair/

131. Paz Cardona, A.J. 14 July 2020. Ecuador: gobierno reconoce que la erosión de la cascada San Rafael podría afectar la hidroeléctrica Coca Codo Sinclair. Mongabay.com: https://es.mongabay.com/2020/07/erosion-rio-coca-ecuador-hidroelectrica-coca-codo-sinclair/

132. *Primacia.ec*. 31 Aug. 2020. Coca Codo Sinclair, el principal sospechoso de la erosión en el río Coca: https://www.primicias.ec/noticias/economia/erosion-rio-coca-coca-codo-sinclair/

133. Hidropaute. 1 Apr. 2018. Información Técnica Proyecto Río Zamora – Santiago: https://www.celec.gob.ec/hidropaute/proyectos/informacion-tecnica-rio.html

134. Rivera-González, L., D. Bolonio, L.F. Mazadiego and R. Valencia-Chapi. 2019. 'Long-term electricity supply and demand forecast (2018–2040): a LEAP model application towards a sustainable power generation system in Ecuador'. *Sustainability* 11(19): 5316.

135. Branford, Brazil announces end to Amazon mega-dam building policy.

136. EPE – *Empresa de Pesquisa Energética*. 2020. Plano Nacional de Energia 2050, Ministério de Minas e Energia: https://www.epe.gov.br/pt/publicacoes-dados-abertos/publicacoes/Plano-Nacional-de-Energia-2050

137. *Gazeta Online*. 5 Feb. 2018. TCU abre camnho para retomada de hgrandes hidroeléctrica na Amazonia: https://www.gazetaonline.com.br/noticias/economia/2018/02/tcu-abre-caminho-para-retomada-de-grandes-hidreletricas-na-amazonia-1014117823.html

138. Starbroek news. 17 Aug. 2020. Amaila Falls hydro back on agenda – Jagdeo: https://www.stabroeknews.com/2020/08/17/news/guyana/amaila-falls-hydro-back-on-agenda-jagdeo/

139. USDA – United States Department of Agriculture. 2020. Brazil: Oilseeds and Products Update. USDA, Foreign Agriculture Service: https://www.fas.usda.gov/data/oilseeds-world-markets-and-trade

140. Salin, Delmy. Sept. 2018. Soybean Transportation Guide: Brazil 2017. US Dept. of Agriculture, Agricultural Marketing Service: http://dx.doi.org/10.9752/TS048.09-2018

141. Télam. 22 Mar. 2017. El uso del ferrocarril mejoraría los precios de los granos entre 19% y 27%. Télam S.E. Agencia Nacional de Noticias: https://www.telam.com.ar/notas/201704/186481-agro-granos-costos-ferrocarril-julio-calzada-bolsa-rosario.html

142. Informa Economics. 2016. Farm to market: Cost comparison, a soybean's journey. United Soybean Board, US Soybean Export Council, Soy Transportation Coalition: https://www.soytransportation.org/Stats/Brazil_CostComparison.pdf

143. CONAB – *Companhia Nacional de Abastecimento*. 2017. Estimativa do escoamento das exportações do complexo soja e milho pelos portos nacionais. Compêndio de estudos Conab, V6 2017: https://www.conab.gov.br/institucional/publicacoes/compendio-de-estudos-da-conab/item/download/2516_c5823fadb57a5c99b622e40fbb8ceac2

144. *Agencia Transporta Brasil*. 15 June 2014. Terminal ferroviário de Rondonópolis transporta 6,5 toneladas em oito meses: http://www.transportabrasil.com.br/2014/06/terminal-ferroviario-de-Rondonópolis-transporta-65-toneladas-em-oito-meses/?

145. *Associação Brasileira dos Terminais Retroportuários e das Empresas Transportadoras de Contêineres*. 23 Mar. 2015. Notícias do setor: Apesar de atrasos em licenças, OTP prevê escoar grãos pela região na safra 2017/18: http://www.abttc.org.br/noticias-do-setor-integra.asp?id=8080; Midiamax. 21 Dec. 2015. Trecho da BR-163 bloqueado após chuvas só deve ser liberado no início do ano: http://www.midiamax.com.br/cotidiano/trecho-br-163-bloqueado-chuvas-deve-ser-liberado-ate-dia-5-285122

146. *Secretaria de Assuntos Estratégicos da Presidência da República – Instituto de Pesquisa Economica* (IPEA). 2014. Hidrovias no Brasil: perspectiva histórica, custos e institucionalidade: http://repositorio.ipea.gov.br/bitstream/11058/2714/1/TD_1931.pdf

147. Global Witness. 2017. 'Buyers in good faith': How timber exporters are complicit in plundering Peru's Amazon: https://www.globalwitness.org/en/campaigns/forests/buyers-good-faith/

148. Amazon Waters. 15 Nov. 2018. The Hidrovía Amazónica: Who wins and who pays?: https://amazonwaters.org/noticia/hidrovia-amazonica/
149. MdT – *Minestério dos Transportes*. 2013. Plan Hidroviario Estrategico, Inland Waterway Strategic Plan, Product 3 - Assessment and Diagnosis Report. ARCADIS logos: https://www.gov.br/infraestrutura/pt-br/assuntos/transporte-aquaviario/plano-hidroviario-estrategico
150. ANTAQ – *Agencia Nacional de Transportes Aquaviáriuas*. 2018. Vias Econicmante Navegadas: http://portal.antaq.gov.br/index.php/informacoes-geograficas/; SINICOn – *Sindicato Nacional da Industria da Construção Pesada* . 2007. ANTAQ defende construção de eclusas no rio madeira: https://www.sinicon.org.br/ANTAQ-DEFENDE-CONSTRUCAO-DE-ECLUSAS-RIO-MADEIRA.pdf
151. Asemblea Legisltiva del Beni. 15 Oct. 2018. Beni apuesta por la Hidrovía de exportación Mamoré-Madeira.
152. MdT, Plan Hidroviario Estrategico.
153. MdT, Plan Hidroviario Estrategico.
154. MdT, Plan Hidroviario Estrategico.
155. Higgins, T. 2020. Para ampliar hidrovia na Amazônia, governo planeja dinamitar habitat de botos e tartarugas, por Infraestrutura na Amazônia: https://brasil.mongabay.com/series/infraestrutura-na-amazonia/; Paschoalini, M., R.M. Almeida, F. Trujillo, G. Melo-Santos, M. Marmontel, H.J. Pavanato, F.M. Guerra, N. Ristau and A.N. Zerbini. 2020. 'On the brink of isolation: Population estimates of the Araguaian river dolphin in a human-impacted region in Brazil'. *PLOS ONE* **15** (4): p.e0231224.
156. Folha de São Paulo. 22 June 2010. Obra da eclusa começou no regime militar: https://www1.folha.uol.com.br/fsp/brasil/fc2201201005.htm
157. ANTT – *Agencia Nacional de Transporte Terrestres*. 5 Feb. 2020. Rumo Malha Norte S.A., Edital e contratos: https://portal.antt.gov.br/en/rumo-malha-norte-sa
158. Olhar Agro & Negocioa. 17 Aug. 2020. Empresa confirma investimentos em ferrovia que irá passar por Cuiabá, Lucas do Rio Verde e Nova Mutum: https://www.agroolhar.com.br/noticias/exibir.asp?id=27548&noticia=empresa-confirma-investimentos-em-ferrovia-que-ira-passar-por-cuiaba-lucas-do-rio-verde-e-nova-mutum
159. Agro Olhar. 13 Sept. 2016. Michel Temer anuncia concessão da Ferrogão ligando Mato Grosso ao Pará: http://www.olhardireto.com.br/agro/noticias/exibir.asp?noticia=temer-anuncia-concessao-da-ferrogao-entre-mato-grosso-ao-para&id=24172
160. PPI – *Programa de Parcerias de Investimentos*. 2017. Concessão da Ferrogrão trecho Sinop/MT a Itaituba/PA, Programa de Concessão de Ferrovias Federais: https://www.poder360.com.br/wp-content/uploads/2017/09/cartilha-ferrograo-final.pdf
161. IRJ – *International Railway Journal*. 16 July 2020. Brazil's Ferrogrão project moves forward ahead of public bidding: https://www.railjournal.com/infrastructure/brazils-ferrograo-project-moves-forward-ahead-of-public-bidding/
162. Xingu +. 5 Feb. 2021. Ferrogrão (Ef-170) - Trecho: Sinop/Mt - Itaituba/Pa, Impactos Socioambientas: https://xingumais.org.br/obra/ferrograo-ef-170-trecho-sinop-mt-miritituba-pa

163. Bôas, A. 28 Aug. 2020. Ferrogrão, a path of illusion. Pulitzer Center: https://pulitzercenter.org/blog/ferrograo-path-illusion

164. ICMBio – *Insituto Chico Mendes de Conservaçao da Biodiversidade.* 2016. Nota Técnica Conjunta N°O /2016/Diman/Icmbjo/Mma, Assunto: Implantação Da Ferrovia Ef 170 – Ferrogrão Ministerio do Meio Ambiente: http://www.oeco.org.br/wp-content/uploads/2017/02/Nota_Tecnica_Conjunta_03_2016.pdf

165. SEDEMA - *Sectraria de Estado de Desenvolvimento Econômico, Mineração e Energia.* 2019. Relatório de Impacto Ambiental (RIMA), Ferrovia Paraense SA, Governp do Pará: https://www.semas.pa.gov.br/wp-content/uploads/2017/06/rima/RIMA_FerroviaParaenseSA.pdf

166. *Agencia Pará.* 28 Mar. 2017. Sedeme apresenta projeto da Ferrovia Paraense a gestores e setor produtivo: https://agenciapara.com.br/noticia/2531/

167. Fearnside, P.M. and A.M.R. Figueiredo. 2015. 'China's influence on deforestation in Brazilian Amazonia: A growing force in the state of Mato Grosso'. In R. Ray, K. Gallagher, A. López and C. Sanborn (eds). *China and Sustainable Development in Latin America: The Social and Environmental Dimension*, pp. 229–265: http://www.bu.edu/gdp-cn/files/2017/09/Brazil.pdf

168. EMBRAPA – *Empresa Brasileira de Pesquisa Agropecuária.* 3 Mar. 2017. A soja se expande em Rondônia e produtores têm acesso tecnologias para melhorar a produção'. Ministério da Agricultura, Pecuária e Abastecimento: https://www.embrapa.br/en/busca-de-noticias/-/noticia/21235286/a-soja-se-expande-em-rondonia-e-produtores-tem-acesso-a-tecnologias-para-melhorar-a-producao

169. EJAtlas.org. 18 Aug. 2018. A controversial Amazon route of Transcontinental Railway Brazil-Peru: https://ejatlas.org/print/opposition-against-controversial-amazon-route-of-transatlantic-railway-brazil-peru

170. Dourojeanini, M. 1 June 2015. El ferrocarril interoceánico chino y nuestra desordenada visión de Desarrollo. SPDA, Actualidad Ambiental: https://www.actualidadambiental.pe/el-ferrocarril-interoceanico-chino/

171. Cultural Survival. 27 July 2015. Proposed twin ocean railroad through the Amazon raises concerns: https://www.culturalsurvival.org/news/proposed-twin-ocean-railroad-through-amazon-raises-concerns

172. SENPLADES – *Secretaría Nacional de Planificación y Desarrollo.* 2017. Subsecretaría de Inversión Pública Dirección de Planes de Inversión Pública, Proyecto: K025, Senplades/Inp - Estudio de prefactibilidad del tren de carga eléctrico del Ecuador: http://www.planificacion.gob.ec/wp-content/uploads/downloads/2018/01/Estudio-de-prefactibilidad-del-tren-de-carga-electrico-del-Ecuador.pdf

173. Straub, S. 2008. *Infrastructure and Growth in Developing Countries* Vol. 4460. World Bank Publications.

174. Estache, A. and G. Garsous. 2012. The impact of infrastructure on growth in developing countries. International Finance Corporation Economics IFC-Notes, Note 1, Washington DC.

175. IFC- International Finance Corporation 2012. Sustainability Policies & Standards E&S Performance Standards, https://www.ifc.org/wps/wcm/connect/topics_ext_content/ifc_external_corporate_site/sustainability-at-ifc/policies-standards

176. Dammert, J.L. 2013. Financing Infrastructure Projects in the Southern Amazon of Peru: its relation with environmental and social safeguards. Global Development Policy Center, Working paper 017: https://www.bu.edu/gdp/files/2018/10/GEGI_GDP-Peru-WP.pdf

177. OSITRAN – *Organismo Supervisor de la Inversión en Infraestructura de Transporte de Uso Público*. 2020. Informe de desempeño 2019: Eje Multimodal Amazonas Norte – IIRSA Norte: Paita – Yurimaguas: https://www.ositran.gob.pe/anterior/wp-content/uploads/2020/07/id-iirsa-norte-2019-1.pdf

178. OSITRAN – *Organismo Supervisor de la Inversión en Infraestructura de Transporte de Uso Público*. 2020. Informe de desempeño 2019: Concesión del Corredor Vial Interoceánico Sur Perú Brasil, Tramo N° 3: Inambari – Iñapari: https://www.ositran.gob.pe/wp-content/uploads/2020/07/id-iirsa-sur-t2-2019.pdf

179. OSITRAN – *Organismo Supervisor de la Inversión en Infraestructura de Transporte de Uso Público*. 2021. Inversiones Valorizadas en los Contratos De Concesión: https://www.ositran.gob.pe/anterior/wp-content/uploads/2021/02/01-inversiones-contratos-ene-2021.pdf

180. IDB – Interameric Development Bank. 2020. Access to Information, PROJECT DETAILS: https://www.iadb.org/en/projects-search?country=VE&sector=&status=&query=

181. COFIDE – *Corporación Financiera de Desarrollo S.A.* 2017. Corporate Presentation: http://www.cofide.com.pe/COFIDE/pdfs/relacion_CORPORATE%20PRESENTATION-jun2017.pdf

182. OSERGMIN – *Organismo Supervisor de la Inversión en Energía y Minería*. 2020. Unidad de Supervisión de Inversión en Electricidad División de Supervisión de Electricidad, Centrales De Generación Eléctrica - En Construcción: https://www.osinergmin.gob.pe/empresas/electricidad/proyectos/generacion

183. PPI – *Programa de Parcerias de Investimentos*. 21 Aug. 2019. Brazil to speed up privatizations with billions on the line: https://www.ppi.gov.br/brazil-to-speed-up-privatizations-with-billions-on-the-line

184. ATTN – *Agência Nacional de Transportes Terrestres*. 22 Feb. 2021. Novos projetos em rodovias, Lista de concessões: https://portal.antt.gov.br/en/novos-projetos-em-rodovias

185. BNDES – *Banco Nacional de Desenvolvimento Econômico e Social*. 25 Feb. 2021. Central de Downloads, Dados sobre Operações de Financiamentos: https://www.bndes.gov.br/wps/portal/site/home/transparencia/centraldedownloads

186. BNDES – *Banco Nacional de Desenvolvimento Econômico e Social*. 25 Feb. 2021. Operações contratadas na forma direta e indireta não automática: https://www.bndes.gov.br/wps/portal/site/home/transparencia/centraldedownloads

187. BNDES – *Banco Nacional de Desenvolvimento Econômico e Social* 25 Feb. 2021. Desembolsos BNDES Pós-embarque por destino das exportações financiadas servicios de engengeria: https://www.bndes.gov.br/wps/portal/site/home/transparencia/centraldedownloads

188. Lucchesi, C., V. Andrade and S. Valle. 2020. Business Brazil pushes its great downsizing and rakes in $5.2 billion. Bloomberg: https://www.bloomberg.com/news/articles/2020-02-05/petrobras-said-to-price-shares-at-30-reais-apiece-in-jumbo-sale

189. Carreteras Pan-Americanas. 4 Jan. 2021. Colombia: Aliadas para el Progreso cede parcialmente a CCA el proyecto Neiva – Santana – Mocoa: https://www.carreteras-pa.com/noticias/colombia-aliadas-para-el-progreso-cede-parcialmente-a-cca-el-proyecto-neiva-santana-mocoa/
190. United States Department of State. 8 Dec. 2020. FACT SHEET. US investors are funding malign PRC companies on major indices: https://www.state.gov/wp-content/uploads/2020/12/20201208-FACT-SHEET-U.S.-Investors-Are-Funding-Malign-PRC-Companies-on-Major-Indices-1.pdf
191. Piccone, T.J. 2016. *The Geopolitics of China's Rise in Latin America*. Brookings Institution. https://www.brookings.edu/wp-content/uploads/2016/11/the-geopolitics-of-chinas-rise-in-latin-america_ted-piccone.pdf
192. Gómez, A . 2018. Inversión pública y financiamiento chino en América Latina, Asociación Ambiente y Sociedad, Ecuador: https://www.chinaenamericalatina.info/inversion-publica-y-financiamiento-chino-en-america-latina/
193. Elgueta, M.E. 2020. La irrupción de China en América Latina: Atentos a la 'letra chica' o neocolonización. Infogate: https://www.infogate.cl/2020/11/18/la-irrupcion-de-china-en-america-latina-atentos-a-la-letra-chica/
194. Bermúdez-Liévano. B. 2019. China's Belt and Road advances in Latin America's Andean region. Dialogo Chino: https://dialogochino.net/en/infrastructure/27815-chinas-belt-and-road-advances-in-latin-americas-andean-region/
195. China – Latin America Report. 2017. Chinese finance to LAC in 2016. The Dialogue, Leadership for the Americas and Global Economic Governance Initiative (GEGI), Boston University: http://www.the dialgie.org
196. GCIT – China Global Investment Tracker. 2020. The American Enterprise Institute and The Heritage Foundation. Washington DC: https://www.aei.org/china-global-investment-tracker/
197. Cruz, D. 2019. Cooperación de China en América Latina y el caribe en Infraestructura: discursos, modalidades y conflictos socioambientales. Asociación Ambiente y Sociedad. Bogotá, Colombia: https://www.chinaenamericalatina.info/es/inicio/
198. Myers, M. and R. Ray. 2021. Shifting gears: Chinese finance in LACV 2020. The Dialogue, China-Latin America report Global Development Center, https://www.thedialogue.org/wp-content/uploads/2021/02/Chinese_Finance_LAC_2020.pdf
199. Dummitt, B. 14 Sept. 2015. 'Chinese firms to pay $1.42 billion For EnCana oil assets. *The Wall Street Journal*: https://www.wsj.com/articles/SB112660924522839142
200. *Pagina Siete*. 24 Apr. 2018. 15 Empresas Chinas se Adjudican Mayoría de los Proyectos de ABC: https://www.paginasiete.bo/economia/2018/4/24/15-empresas-chinas-se-adjudican-mayoria-de-los-proyectos-de-abc-177715.html
201. ANF – *Agencia Nacional Fides*. 17 May 2018. Rositas: tercer juez rechaza acción popular de guaraníes y dice 'no se vulneró consulta previa': https://www.noticiasfides.com/economia/rositas-tercer-juez-rechaza-accion-popular-de-guaranies-y-estos-anuncian-ir-a-la-cidh--388103

202. Bolivia.com. 1 Aug. 2017. Satélite Túpac Katari: ¿Cuánto ha producido desde su lanzamiento?: https://www.bolivia.com/actualidad/economia/sdi/159926/satelite-tupac-katari-cuanto-ha-producido-desde-su-lanzamiento
203. El Deber 20 Feb 2020. Créditos chinos y de otros seis organismos tienen tasas más altas que las del FMI, https://eldeber.com.bo/economia/creditos-chinos-y-de-otros-seis-organismos-tienen-tasas-mas-altas-que-las-del-fmi_221521
204. Myers and Ray, Shifting gears.
205. Gallagher, Kevin P. and Margaret Myers. 2021. *China-Latin America Finance Database.* Washington, DC: Inter-American Dialogue.
206. Harris, M. 12 Jan 2017. Peru's San Gaban project receives Chinese cash infusion, HydroReview: https://www.hydroreview.com/business-finance/peru-s-san-gaban-project-receives-chinese-cash-infusion/#gref
207. Cayetano, S., C. Alvarado and PO. Tuesta. 6 Dec. 2019. Personajes vinculados al 'Club de la Construcción' detrás de licitación de la Hidrovía Amazónica. *CONVOCCA - Periodismo de Investigación y Análisis De Datos*: https://convoca.pe/investigacion/personajes-vinculados-al-club-de-la-construccion-detras-de-licitacion-de-la-hidrovia
208. García Delgado, F. 29 Jan. 2021. El futuro incierto de la Hidrovía Amazónica peruana. Mongabay.com: https://es.mongabay.com/2021/01/hidrovia-amazonia-peru-consulta-previa-pueblos-indigenas/
209. Cui, Shoujun and Zheng Zhang. 2019. 'China and Brazil development finance cooperation. A case study of the Belo Monte Transmission Line project'. In E. Duseel Peters (ed.) *China's Financing in Latin America And The Caribbean*. Universidad Nacional Autónoma de México: https://dusselpeters.com/146.pdf
210. Costa, L. 24 Feb. 2021. Brazil to cut stake in Eletrobras to 45% from 61% in privatization – energy secretary. Reuters: https://www.reuters.com/article/eletrobras-privatisation/brazil-still-plans-to-privatize-state-utility-eletrobras-sources-say-idUSL1N2K32PN
211. Electrobras. 10 March 2021. What is the capital structure of Eletrobras? Investor Relations, IR Services, FAQ, Centrais Elétricas Brasileiras S/A: https://eletrobras.com/en/ri/Paginas/FAQ.aspx
212. Harris, M. 13 Aug. 2015. Guyana nixes plans for 165-MW Amaila Falls hydropower project. Hydro Review: https://www.hydroreview.com/world-regions/guyana-nixes-plans-for-165-mw-amaila-falls-hydropower-project/#gref
213. MENAFIN – Caribbean News Now. 28 Nov. 2018. Suriname seeks US$205 million for airport expansion and modernization: https://menafn.com/1097761405/Suriname-seeks-US205-million-for-airport-expansion-and-modernisation
214. World Bank Group. 2008. World Bank Group Sustainable Infrastructure Action Plan (SIAP). Washington DC: World Bank Group.
215. The Sustainable Asset Valuation (SAVi) developed by the The International Institute for Sustainable Development (IISD) and the ENVISION rating system developed by the Zofnass Program for Sustainable Infrastructure at Harvard University.

216. Alves, F., G. Elacqua, M. Koslinki, M. Martinez, H. Santos and D. Urbina. 2015. 'Winners and losers of school choice: Evidence from Rio de Janeiro, Brazil and Santiago, Chile'. *International Journal of Educational Development* 41: 25–34.

217. Ministerio de Defensa. 12 Dec. 2021. Projeto Amazônia Conectada: avanços em fibra ótica beneficiam comunidades afastadas: https://www.gov.br/defesa/pt-br/centrais-de-conteudo/noticias/projeto-amazonia-conectada-avancos-em-fibra-otica-beneficiam-comunidades-afastadas ; DCD – Data Center dynamics. 14 Aug 2020. Huawei fornece infraestrutura para conectar a Amazônia: https://www.datacenterdynamics.com/br/not%C3%ADcias/huawei-fornece-infraestrutura-para-conectar-a-amaz%C3%B4nia/

218. Salin, Soybean transportation guide: Brazil 2017; Informa Economics, Farm to market: Cost comparison, a soybean's journey.

219. Ibid.

220. DNIT – *Departamento Nacional de Infraestrutura de Transportes*. 2019. Manual de Custos Médios Gerenciais, Coordenação-Geral de Custos de Infraestrutura de Transportes, Ministério Da Infraestrutura: https://www.gov.br/dnit/pt-br/assuntos/planejamento-e-pesquisa/custos-e-pagamentos/custos-e-pagamentos-dnit/documentos/ManualdeCMGVol02ManualdoUsurio_.pdf

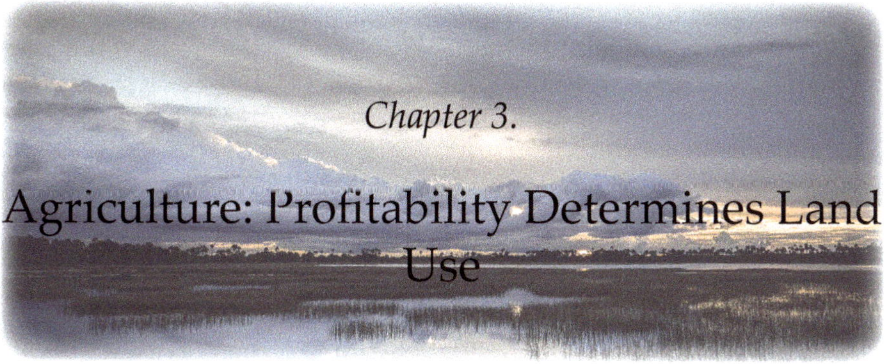

Chapter 3.

Agriculture: Profitability Determines Land Use

Highway infrastructure initiates the deforestation process, but it is almost always accompanied by some sort of agricultural activity. Depending upon circumstances, deforestation can proceed rapidly or slowly, lead to large or small forest clearings, and create forest remnants of different sizes and configurations. Agricultural production models vary from huge ranches and plantations spanning tens of thousands of hectares to small plots consisting of less than a single hectare. These differences are rooted in cultural traditions and business models as well as incentive systems and land tenure regimes imposed by the state. The Pan Amazon covers a vast geographic area with a diversity of landforms, soil types and climates that support a wide range of production systems supplying food, fibre and biomass energy to local, national and global markets.

The diversity of landholdings, production systems and business models leads to a similarly wide range of social and environmental impacts. Agriculture supports the livelihoods of millions of families across the Pan Amazon. For many migrant families, it provides a pathway out of poverty. For the middle class, it is a way to accumulate wealth and ensure prosperity for future generations. For entrepreneurs, it represents a business opportunity with a proven technology and manageable levels of risk. As long-term investments, agricultural production systems can assume a role as key elements of a sustainable economy. When practised as a speculative enterprise, however, extractive practices that yield cashflow tend to degrade productive potential over the medium term. Under-investment in smallholder communities forces the adoption of non-sustainable options that function as a poverty trap and rob the regional economy of an important driver of growth.

A great deal of emphasis is placed on the first phase of deforestation, which follows soon after the construction of a highway through a remote area. The amount of deforestation that occurs in later years depends large-

doi: 10.3197/9781912186228.ch03

ly on the ability of producers to move their production to market. Some landscapes remain remote due to distance or a physical barrier, such as a river without a bridge or mountain range subject to landslides. These forest frontiers may experience relatively low deforestation rates over many years. Those that enjoy more direct market access usually have a higher rate of deforestation, which creates a larger agricultural landscape but also one with fewer forest remnants. Over time, agricultural frontiers evolve into consolidated frontiers as they become incorporated into the global economy as producers of basic food commodities.*

The role of the private sector becomes increasingly important in landscapes with viable transportation networks and expanding agricultural production. Landholders are the primary actors in determining land use, but commercial agents play an essential role by supplying inputs, such as seeds, fertiliser and pesticides, as well as by acting as intermediaries between the producers of basic food commodities and consumer markets. Commodity traders invest in logistical facilities and bulk transport systems that are key components in global supply chains. Industrial infrastructure, such as crushing mills and slaughterhouses, adds value to primary production that increases demand for basic commodities. As the rural economy grows and becomes more diversified, incomes increase for farmers and ranchers, which acts as a catalyst to accelerate deforestation.

Agricultural production systems can be classified as a proximate cause of deforestation, but the demand for commodities is the ultimate driver that motivates producers to expand production.[1] The markets that influence agricultural production are as varied as the commodities produced by the farmers of the Pan Amazon. Global markets dominate the supply chains of soy, coffee and cacao, while maize, rice, manioc and perishable fruits are largely commercialised in national markets. Beef and palm oil are global commodities and are influenced by international markets, but most of the production in the Pan Amazon is commercialised domestically to meet the needs of national consumers. Understanding the market dynamics of each commodity is essential to devise strategies that promote sustainability and to eliminate deforestation.

The global market for food commodities has been greatly influenced by demand from China. Policies designed to improve the standard of living of its citizens combined with phenomenal economic growth created positive synergies that led China to import large quantities of basic food commodities.

---

* A commodity is a basic good used in commerce that is interchangeable with other materials of the same type; they are used as inputs in the production of other goods. Commodities are uniform and interchangeable in the production of goods and services. 'Hard' commodities refers to minerals, including petroleum and natural gas. 'Soft' commodities refers to basic foodstuffs and fibre produced by the agriculture and forest product sectors.

© Alexandre Laprise/Shutterstock.com

Forest is cleared to establish some type of agricultural activity, like this cattle ranch near Riberalta, Bolivia.

Unlike mineral commodities, which experienced a unique super-cycle linked to a one-time buildout of that nation's basic infrastructure, the demand for food commodities will continue to grow for the foreseeable future. China is transitioning from a dependence on manufacturing and infrastructure investment to a more diversified economy with an emphasis on consumer spending. Food commodities will continue to be subject to short-term volatility but, over the medium to long term, demand for food will grow. The impact of China is most noticeable in Brazil, but producers in the Andean republics all aspire to sell their production to China, be it beef (Bolivia and Colombia) or coffee and cacao (Ecuador and Peru).

Many conservation advocates focus their attention on forest communities and promote policies that will foster a more robust forest economy. This is well and good. A coherent conservation strategy, however, must also address the proverbial 800-pound gorilla that dominates the domestic policies governing land use on frontier landscapes. The commodities that flow from the region's farms, ranches and plantations are essential to the financial wellbeing of all the Pan Amazonian nations. Consequently, there is an enormously powerful group of stakeholders who are not likely to abandon the policies and practices that underpin their business models. Like most economic interest groups, they are not satisfied with the status quo; much to the contrary, they hope to expand their financial wellbeing and their economic power. This is true of the corporate sector, typically represented by groups like the chamber of commerce, as well as the smallholders' associations and syndicates who represent the rural poor, particularly the migrant populations on forest and agricultural frontiers.

Some environmental and social activists, particularly those who hold progressive (socialist) views, advocate for a regulatory approach to constrain the negative forces emanating from an expanding agricultural production. That approach, however, must overcome the real political and economic power of these vested interest groups, many of whom righteously believe their systems are in the national interest. A more realistic option is to convince these groups there are growth-positive options that do not include an expansion of the forest frontier, which will outperform the conventional options they seek to protect.

With that goal in mind, this chapter seeks to describe and understand the predominant agricultural production systems that dominate the conventional economy of the Pan Amazon in the second decade of the twenty-first century.

Beef Production Models

The single largest cause of deforestation in the Pan Amazon is the clearing of forest to establish cultivated pasture to produce beef cattle. By some estimates, as much as eighty per cent of previously deforested landscapes are covered by non-native forage grasses (Figure 3.1). In most cases, grasses have been seeded directly into the recently deforested soil in newly established cattle ranches, but smallholders also use pastures as a rotational fallow as part of a production system based on food crops.

The use of technology among livestock producers ranges from extremely rudimentary to highly sophisticated; not surprisingly, technology improves productivity and economic return but requires 'know-how' and financial capital. Brazil has the most sophisticated beef production system, which includes three overlapping phases that correspond to the life stages of a typical bull or cow:

A. Cow-calf operations span gestation (9.5 months), birth and early growth until calves are weaned from their mothers (8–12 months);*

B. Grow-out operations start when yearlings are sold or moved to separate pastures until they are full-sized in stature but not weight (~12 months); these animals are known as *gado magro* (skinny cattle).

C. Fat-cattle operations describe the finishing stage, which varies depending upon feed ration and breed (6–12 months) until they reach optimum slaughter weight (375–425 kilograms).

---

* Breed cows typically live four to six years before being sent to slaughter.

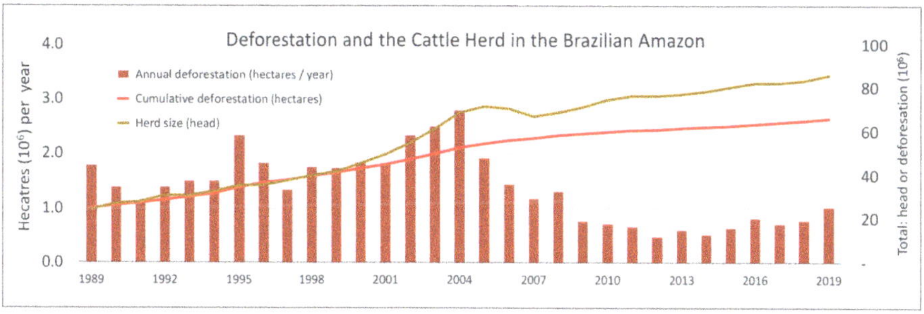

Figure 3.1: A comparison of the size of the cattle herd with the total accumulated deforestation in the Legal Amazon of Brazil.
Data sources: IBGE / SIDRA and PRODES.

Ranchers can specialise in a single phase, but more often they combine two or more production phases into a business model appropriate to their geographic location and predilection to technology. For example, ranchers in remote areas with poor infrastructure are almost uniformly dedicated to running cow-calf operations combined with grow-out operations (A+B). For them, the only realistic option is to drive cattle to a market on foot because poor roads make moving live animals by truck uneconomic and risky. Animals lose weight if they are confined in a truck over several days and can die if the truck becomes stuck on a poorly maintained road. Calves are too fragile for long drives, while fat cattle lose weight if they are forced to walk. Cattle drives are still common on forest frontiers, and many landholders maintain pastures explicitly for rental to drovers, who move their herds to market at a relaxed pace to avoid subjecting animals to undue stress. Drovers will move a herd to a town or small city, typically located on a trunk highway or improved secondary road highway, where they will be sold to a cattle trader or another producer via auction.

Producers who wish to avoid the risk of birthing calves purchase yearlings and keep them on pasture until they reach slaughter weight (B+C). Some ranchers integrate them all on the same property (A+B+C), which allows them to avoid middlemen and maximise the return on a per-animal basis. Others specialise in producing calves of known genetic background (A), which are sold at a premium for qualities linked to productivity, meat yield or disease tolerance.* Although the practice is still rare in the Amazon,

* There are various Brazilian breeds, but all are Zebu types derived from South Asian stock. The most popular breed is the *Nelore*, which was developed in Brazil in the first decades of the 19th century from *Ongole* breed stock imported from the state of Andhra Pradesh in India. Source: *Associação dos Criadores de Nelore do Brasil – ACNB*; http://www.nelore.org.br/Raca/Historico

© Hermes Justiniano

© Paralaxis/Shutterstock.com

The beef supply chain extends from remote ranches dedicated to producing calves (top) to feedlots where cattle are fattened prior to slaughter (bottom).

feedlot operators specialise in fattening cattle using balanced rations (C), a practice more common in consolidated frontiers with an ever-increasing supply of feed grains (see below).

The economics of cattle ranching is calculated on a per animal basis. The sale price for a calf ranges from $US 180–250 per animal and from $US 1,000–1,200 for a full-grown breeding bull; most steers are slaughtered at about 400 kilos, with a value of between $US 600 and 800. A smallholder in Ariquemes (Rondônia) with fifty hectares of pasture specialising in the production of calves would have gross income of between $US 8,000 and 12,000 annually. A middle-class rancher with 3,000 hectares in Alta Floresta (Mato Grosso) with a similar cow-calf production model would gross between $US 375,000 and 425,000 per year.* For an integrated ranch (A+B+C) where cattle are held for the entire 36 months, revenues should be about ten to twenty per cent greater.

The net worth of a producer would depend upon land values and capital improvements, but at $US 2,000 per hectare, the small farm would be worth about $US 200,000, while the larger ranch would bring approximately $US 3 million.† Although these numbers look plausible, the viability of smallholder production is dependent upon family labour, and if those producers had to pay market value for their labour, they would barely break even. Similarly, many medium- to large-scale cattle ranching operations enjoy the legacy of past decades, when land was obtained at a large discount and would be hard-pressed to establish a ranch if they had to purchase land at its current market value. The difference in the value of land between the forest frontier and agricultural frontiers is the primary driver of rural real estate markets and, arguably, the greatest single driver of deforestation (see Chapter 4).[2]

The cattle herd in the Legal Amazon grew from fourteen million head in 1980 to more than 85 million in 2019; along the way, its growth caused the deforestation of more than seventy million hectares (Figure 3.1). Between 1980 and 2000, approximately one hectare of forest was sacrificed for the possibility of maintaining one live animal; however, tropical grass-fed cattle require three years to reach slaughter weight. Consequently, it requires about two hectares of pasture to produce 100 kilograms of live animal per year, which actually represents only fifty kilograms of [bone-in] dressed beef.[3]

These numbers are phenomenally unimpressive in terms of productivity when considered based on a kilogram of protein per hectare. By way of comparison, soy yields about four tonnes of beans per hectare, which

* Calculations are based on exchange rates of approximately R$ 5 per $US.

† These valuations assume 50% of the land is planted to pasture and include the value of the land, cattle and on-farm infrastructure; apparently, land values are higher in densely populated Rondônia compared to Alta Floresta. (Values taken from online rural real estate markets).

© Pelizzeri / Shutterstock.com

© Cysun / Shutterstock.com

Amazonian pastures are notorious for their low productivity and frequently show signs of overgrazing and soil degradation (top). The cattle herd in the Brazilian Amazon are derived from South Asian breeds, particularly Nelore, Gir, Zebu and Brahma; those pictured are nearing slaughter weight of approximately 400 kilograms (bottom).

upon refining produces approximately one tonne of vegetable oil and three tonnes of soymeal, of which the latter is about fifty per cent protein. In other words, soy produces about fifty times the amount of protein per hectare as grass-fed beef raised in the Amazon. Aquaculture is even more productive when calculated on a per hectare basis, producing between two to three times as much as soy (see Chapter 8).

As disappointing as beef productivity numbers are, there was a noticeable improvement in overall stocking rates after 2000.[4] The surge in productivity was the result of several factors, one of which was a surplus of available pasture caused by land speculation, which allowed ranchers to rapidly expand herd size in response to booming demand. There was a short-term reduction in the herd following the economic crisis of 2008, but the ratio between the total herd size and the pasture area has been continuously improving over the last decade.*

Increased stocking rates are only one aspect of the improved productivity of the Brazilian beef industry. At the national level, between 2000 and 2019, the cattle herd increased by about 26%, while total spatial extent of its pastures declined by 12%, a gain in efficiency of more than 44%. The use of technology and management practices is most notable in South and Southeast Brazil, where stocking rates surpass three head per hectare.† There have been similar improvements by Amazonian producers, who increased stocking rates by 62%; however, they started from a much lower baseline and still lag their counterparts in Southern Brazil by ~50%. Surprisingly, the highest stocking rates in the Legal Amazon have been obtained by smallholder cattle producers in Rondônia and Acre, who have also made impressive gains in overall efficiency (Table 3.1).[5]

Improving the productivity of cattle ranchers is a major component of initiatives designed to improve the image of the Brazilian beef industry and, allegedly, reduce deforestation. The goal is to channel future growth into technological improvements that allow producers to expand production without increasing the spatial extent of pasture area. There are essentially six technologies that can be deployed to increase the efficiency of beef production systems:

- *Pasture management*: stocking rates can be improved via rotational grazing, agroforestry and rotating pasture with crops. Stocking rates of up to four head per hectare have been achieved by experiment stations.

* The use of deforestation data is a crude measure of grazing intensity because it ignores deforestation linked to other crops, such as oil palm (Pará) and soy (Mato Grosso), as well as Cerrado lands that were converted to pasture (Maranhão, Mato Grosso and the Tocantins).

† Santa Catarina, Rio Grande do Sul, Paraná, São Paulo

Table 3.1: Cattle stocking rates in the Legal Amazon.

State	Total Area (km²)	Forest Area (km²)†	Accumulated Deforestation†	Reported Pasture††	Pasture as % of Deforestation†††	Cattle Populations	Cattle Density head / ha
Acre	170,895	148,700	21,611	18,387	85%	3,509,682	1.91
Amapá	142,926	110,266	2,283	3,081	135%	54,296	0.18
Amazonas	1,589,691	1,452,267	43,718	16,099	37%	1,455,842	0.90
Maranhão	264,714	40,127	50,455	63,291	125%	8,008,643	1.27
Mato Grosso	904,865	312,691	224,736	209,192	93%	31,973,856	1.53
Pará	1,249,718	876,635	228,414	175,004	77%	20,881,204	1.19
Rondônia	240,398	125,926	98,573	78,966	80%	14,349,219	1.82
Roraima	226,224	152,469	11,474	6,610	58%	879,007	1.33
Tocantins	278,998	9,803	25,296	75,246	297%	8,480,724	1.13

† Forest cover and accumulated deforestation date from PRODES.

†† From *Laboratório de Processamento de Imagens e Geoprocessamento* (Lapig) or IBGE (AP, MA).

††† States with positive values where pasture / deforestation > 100% occur because they contain areas of Cerrado savanna that have been converted to pasture.

- *Health and reproductive success*: Cows sometimes fail to be impregnated, or suffer a miscarriage, or lose their calves to illness or predation. Reproductive success on modern commercial ranches ranges from 60% to 80% but can reach levels as high as 95% under ideal conditions.
- *Nutrition management*: Cattle are ruminants and have the capacity to metabolise cellulose, but they need vitamins to thrive; combining soil analysis with vitamin supplements will improve the daily weight gain, a key metric used to monitor productivity.
- *Supplemental feed:* Cattle will increase their daily weight gain when their ration of cellulose is complemented with starch (maize) and protein (soy). Providing cattle on pastures with feed rations will increase daily weight gain, but greater gains are obtained when they are finished in feedlots.
- *Genetics*: Cattle breeders have multiple avenues for enhancing productivity by improving resistance to disease and boosting physiological efficiency as well as by increasing meat yield per animal, which is measured by the ratio of carcass weight to live weight, currently about 50–55%.
- *Reducing time to slaughter*: This is a function of growth measured by daily weight gain, which is dependent upon genetics and

nutrition; shortening the lives of animals increases the proportion of the total herd harvested each year.

Industrial infrastructure

Packing plants are also important to producers because the construction of a modern facility will stimulate the growth and diversification of the livestock sector. Modern packing plants must be located on a good (preferably paved) road to avoid wear-and-tear on refrigeration trucks that transport the meat to urban markets. The number of modern industrial-scale packing plants is a good measure of the productivity and sophistication of the beef sector: Mato Grosso has forty packing plants that slaughter about 100,000 head per week, followed by Rondônia with 22 packing plants processing 50,000 head per week. In contrast, Pará has only thirteen modern industrial-scale plants but still manages to harvest 45,000 head per week, mainly in smaller-scale packing plants. Acre, far removed from urban markets, has only three industrial plants that slaughter about 5,000 animals per week.[6] Packing plants must be operated at or near capacity to be profitable, and their construction will change the cattle market in the surrounding landscapes. Having a slaughterhouse nearby increases the production options for ranchers. They can pursue a fully integrated production model (A+B+C), but many opt to specialise in fat-cattle operations (C), which are less risky and more profitable. Increased profitability will motivate most cattlemen to increase production, either by increasing herd productivity using technology or by expanding pasture area or both.

As of 2021, there were no industrial slaughterhouses along the entire length of the Transamazônica (HML #10 and #19) or in the municipalities of São Felix de Xingu (HML #9) or Novo Progresso (HML #17),* where cattle ranchers only have the option of pursuing the A+B production paradigm (Figure 3.2). Still, these remote communities play an essential role in the beef supply chain because they export their *gado magro* to producers near slaughterhouses specialising in the production of fat cattle. The municipality of São Felix de Xingu is home to the largest herd of cattle in Brazil, with more than two million head grazing on approximately 1.8 million hectares of pasture. Coincidentally, this municipality has suffered the highest annual deforestation rate in Amazonian Brazil since 2001.[7]

Another type of industrial infrastructure is the feedlot, known as *confinamentos* in Brazil. These industrial facilities increase daily weight gain and shorten the time-to-slaughter, two key metrics that track improvement in beef productivity. The use of feedlots has increased in Brazil from 500,000 head in 2003 to 4.5 million in 2016 and 6.2 million by 2020.[8] In 2019, Mato Grosso led the nation in feedlot development with more than 175 facilities

* HML: human modified landscape (see Chapter 1, Figure 1.2).

Source: Google Earth

Figure 3.2: Deforestation adjacent to BR-163 over three decades. There was no respite after 2000, following the implementation of the Plano BR-163 Sustentável, a high-profile project managed by the Casa Civil da Presidência da República; or the Cattle Agreement in 2010.

and an installed capacity for 800,000 head. Since feedlots shorten time-to-slaughter, the total number of animals fed in *confinamentos* exceeded 1.2 million.[9]

The impact of feedlots on land use is complex. Their growing popularity has contributed to the improvement in land-use intensity of the beef supply chain; however, the expansion of the feedlot model is dependent on the soy-maize production paradigm, which also has an expanding spatial footprint (see below). Simultaneously, feedlots increase the demand for *gado magro* supplied by ranchers from the forest frontier. Feedlots are also a source of pollution, due to the concentration of nitrogen-rich runoff from manure, which contributes to the hydrological degradation of the Tapajós, Xingu, Araguaia and Tocantins rivers.

National versus global markets

Brazil is both a massive producer and consumer of beef. Domestic consumption has expanded steadily year on year with slight variations linked to periodic recessions, but most of the recent growth is caused by Brazil's increasing dominance in global export markets. Most of that growth has occurred within the Legal Amazon (Figure 3.3). Prior to 2000, exports largely consisted of processed meat[*] and fluctuated between five and six per cent of total production; after 2000, beef packing companies started exporting fresh and frozen beef to Europe, the Middle East and East Asian markets.[†] Within five years, exports represented about twenty per cent of total national production. Exports fell by 25 per cent during the global economic crisis of 2008 and 2009. Domestic consumption buffered the market shock, but the drop in demand reverberated through the supply chain. The contraction coincided with an international boycott of Brazilian beef that motivated the three largest meat packing companies in Brazil to embrace the 'Cattle Agreement' and to eliminate deforestation from their supply chains (see Text Box 3.1).

The growth in exports resumed after 2012, growing at a mean annual rate of ~7 per cent and now represent about 25 per cent of national production. The Peoples Republic of China began to import Brazilian beef in 2016 and rapidly scaled-up its purchases to almost 900,000 tonnes in 2020.[10] China is now the largest international market for Brazilian beef and approximately half of those sales orginate in Mato Grosso. Exports to China from Amazonian Brazil will increase in 2021 as packing plants in

---

* Corned beef, salted beef and canned beef.

† This coincided with the eradication of foot and mouth disease (aftosa), which was accomplished by the mass vaccination of cattle in Mato Grosso (2001), Rondônia (2003), Acre (2005) and Pará (2007). Source: Marques et al. (2016).

Pará have been authorised to sell to the the world's largest growing market for dressed beef.

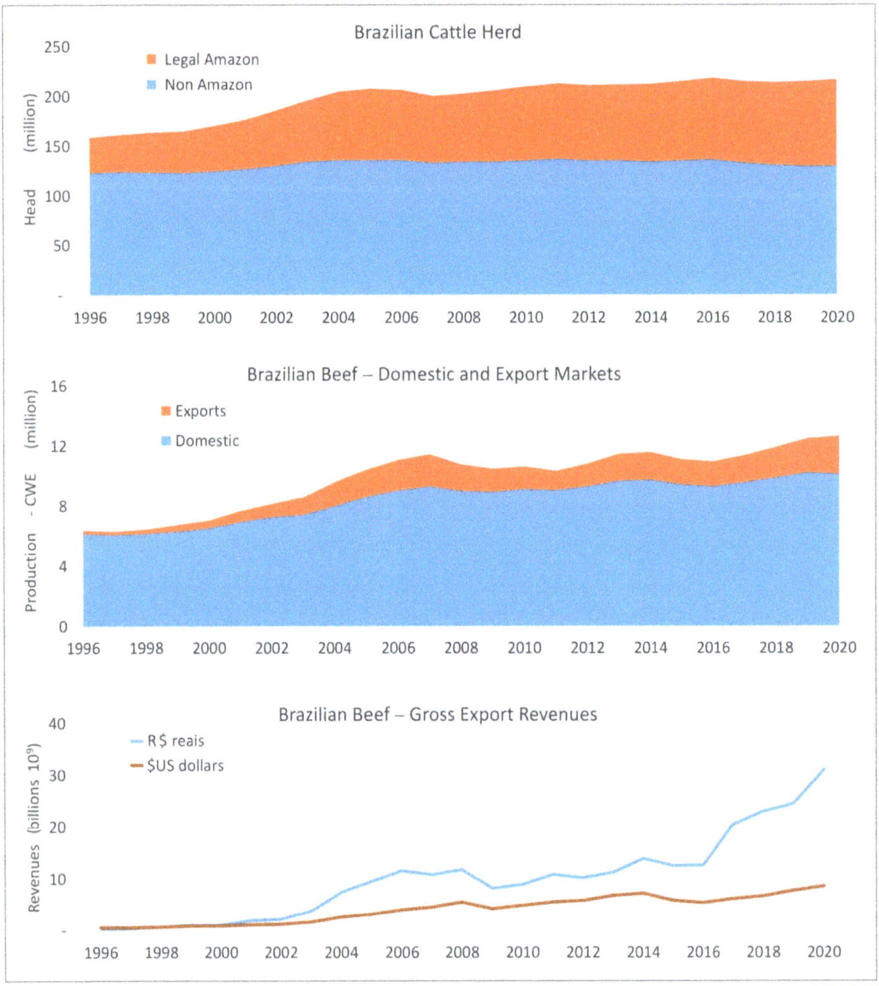

Figure 3.3: Brazilian beef production: (a) The evolution of the national herd; (b) demand from domestic and export markets at national scale; (c) Gross revenues expressed in R$ and $US at national scale.

Data sources: IBGE / SIDRA and ABIEC.

Text Box 3.1: The Cattle Agreement

Greenpeace declared a boycott of Brazilian beef in 2009 as part of its campaign to halt deforestation in the Amazon. It ended the boycott when Brazil's four largest beef-packing companies (JBS, Marfrig, Minerva and Mercurio) agreed to adopt measures to exclude from their supply chains cattle that were raised on landholdings that were not in compliance with the environmental laws of Brazil. Among its provisions were three commitments on the part of beef-packing companies:

- Obligate suppliers (direct and indirect) to provide geographically precise data identifying their rural properties.
- Accept only suppliers who can prove they are in possession of properly registered land tenure documents.
- Validate the origin of all cattle by a tracking system that demonstrates suppliers do not engage in illegal deforestation or unfair labor practices or encroach on indigenous lands and protected areas.

The agreement contributed to the dramatic eighty per cent reduction in deforestation in the Brazilian Amazon between 2006 and 2012. It did not succeed, however, in eliminating illegal deforestation from the beef supply chain. There are several sources of non-compliance, including the approximately thirty per cent of animals that are slaughtered by smaller companies who commercialise their beef entirely within the domestic market. There is also evidence that some ranchers game the certification system by the sale of immature cattle (*gado magro*) from non-compliant ranches to long-established operations that adhere to the law. Producers from the frontier landscapes presumably sell their animals at a discount, while the compliant ranches closer to packing plants mix the illegal animals with their own herds to avoid detection – and increase their bottom line.

The failure of the Cattle Agreement motivated Greenpeace to withdraw as an active participant in 2017 and to renew its campaign to boycott Brazilian beef. Meanwhile, an upsurge in deforestation in 2018 and the election of Jair Bolsonaro in 2019 increased consumer attention on the Brazilian beef industry. This motivated the three largest of the packing companies to move their self-imposed deadline for eliminating illegal deforestation forward from 2035 to 2025. They have adopted a system that will unite vaccination certificates and blockchain technology to identify properties where calves are born and, presumably, eliminate rogue cattle from their supply chains.

Sources: Zero Deforestation Working Group 2017; Klingler et al. 2018; Alves 2021; Conecta – Parcerias de Agropecuária Responável 2021.

The Andean Amazon and the Guianas

Beef production technology in all other Amazonian regions lags Brazil except for Bolivia, which has largely adopted the Brazilian approach to beef production (Figure 3.4).[11] Bolivian producers can be classified into two

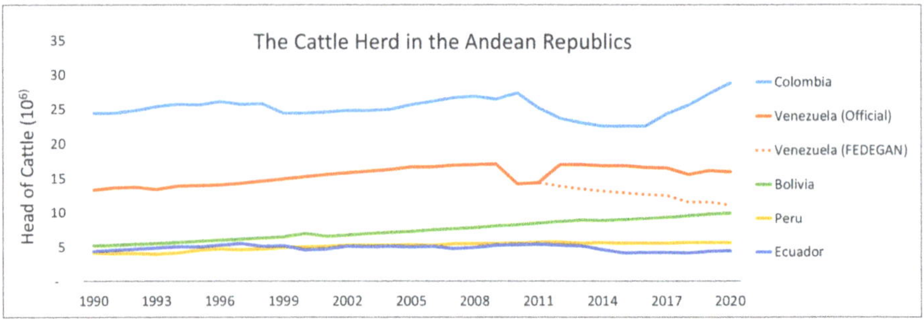

Figure 3.4: Cattle herds have declined in Venezuela, while expanding in Bolivia and in Colombia after 2016; in both nations, beef production is a major driver of deforestation. Cattle herds are small and stable in Ecuador and Peru; in neither country is beef production a major driver of deforestation.

Data sources: FAOSTAT and USDA / FAS

major types, each of which is in a different region: (1) extensive ranches on seasonally flooded natural grasslands in the Beni;* and (2) intensive ranches on cultivated pastures in Santa Cruz. Both groups include large and medium-sized producers, and both have adopted Brazilian breeds and veterinary technology. The ranches in the Beni have very low stocking rates due to the poor forage quality of the native grasses. Consequently, they largely produce calves and immature males (A+B), which are sold to *Cruceño* producers located closer to population centres and packing plants.

Cruceño producers have embraced the Brazilian production model in its entirety, including the practice of seeding cultivated pasture grasses directly into recently deforested soils. The sector is responsible for about 35 per cent of the total historical deforestation in Bolivia,[12] which amounts to about four million hectares (HML # 29, #30 and #31). In 2016, the government ended its decade-long ban on beef exports as part of a strategy to diversify the national economy.[13] China initiated imports from Bolivia in 2019 with an initial purchase of about a thousand tonnes of dressed beef, a volume that increased to 7,900 tonnes in 2020, eighty per cent of total beef exports.[14] The value of those exports in 2020 was $US 42 million, a small but significant number in the rural economy of Santa Cruz.† Looking forward,

* These seasonally inundated wetlands, also known as the Llanos de Moxos, cover about 100 km2 and are the third largest wetland complex in South America after the flood plain of the Amazon river (250,000 km^2) and the inundated sections of the Llanos del Orinoco (3,200 km^2).

† The GDP attributed to the livestock sector in 2017 was $US 250 million.

the government hopes to expand that number by a factor of five by 2025;[15] this and other policies have stimulated deforestation,[16] which reached a historical high of 240,000 hectares in 2020.[17]

In Peru and Ecuador, cattle raising is more accurately described as an artisanal activity than a modern production model. Herds are typically a genetic mixture of traditional stock, dairy cows and Brazilian breeds; reproductive rates are low, and mortality is high. Stocking rates are usually below 0.5 head per hectare; forage is of very poor quality and weed infestation is a universal problem. Most pastures are fallow, part of a production system where pastures and second-growth forest occupy a temporal stage in a rotation cycle centred on annual crops. The economic return on these low-tech cattle production practices is notoriously poor, with a net cash flow of only about $50 per hectare, compared to about $US 300 for maize and $US 850 for cacao.[18]

Small farmers in Peru and Ecuador raise cattle on land that has essentially no other economic activity. Cattle are viewed as a liquid asset that accrues value over the short-term and can be monetised easily for medical emergencies or milestone events. Essentially, a savings account with hooves. Both countries have programmes designed to improve productivity, but neither country exports beef, nor are they likely to do so in the near future.

Pastures occupy about seventy per cent of previously deforested land in the Amazon in both countries, but cattle production is not a cause of deforestation; rather, it is a by-product of land clearing by small farmers who grow food crops for national markets or cultivate a perennial cash crop for international markets (see below). There are some exceptions. In Morona-Santiago Province in Southeastern Ecuador, Schuar and Achuar families adopted cattle farming in the 1970s as a tactic to formalise land tenure (see Chapter 11). The motivation for clearing the forest was not to pursue cattle ranching as a livelihood but to protect their lands from encroachment by immigrants.[19] They are, nonetheless, an important source of high-quality beef in Ecuador.

In Colombia, the beef supply chain is more sophisticated than in Ecuador and Peru but still lags the productivity of cattle ranchers in Bolivia and Brazil. In part, this represents the diversity of its rural communities and producers, but it also reflects a lack of investment caused by its decades-long civil conflict. The cattle sector is undergoing profound changes, stimulated in part by the peace process that began in 2016, but also by the free trade agreement with United States that is obligating producers to increase efficiency or lose market share.[*]

---

* Between 2012 and 2019, beef imports from the USA increased from $US 3.5 to $US 25 million; tariff barriers are being phased out and will end in 2022. Source: USDA/FAS, https://www.fas.usda.gov/colombia-2019-export-highlights

Prior to 2010, the Colombian cattle herd comprised about 25 million head, a population that was stable for approximately thirty years. About 55 per cent of the herd was destined for beef and four per cent for dairy operations, with the remainder raised for both milk and meat, a characteristic typical of traditional artisanal systems.[20] The most sophisticated producers were the large-scale ranchers on the natural savannas of the *Llanos del Orinoco*[*] where native grasses imposed low stocking rates. In the Andes and on the Caribbean Coast, tens of thousands of small, medium and large-scale producers raised cattle on poor pasturage cultivated on degraded soils. In the Colombian Amazon, three departments have a legacy of deforestation linked to the cattle industry: Putumayo (HML #51), Caquetá (HML #52) and Guaviare (HML #54). The most important, Caquetá, was settled by migrants starting in about 1960 and is now characterised by medium to large-scale cattle ranchers on approximately 1.3 million hectares.

The government and *Federación Colombiana de Ganaderos* (FEDEGAN) have embarked on an ambitious programme to expand and modernise the cattle sector. The sector is essentially adopting the Brazilian technology and production model, which has led to a sustained increase in the national herd over the last several years (Figure 3.4).[21] This effort has revitalised the cattle industry in Caquetá, where the cattle herd grew from 1.3 million in 2016 to more than 2.3 million in 2019.[22] The initiative is notable for its intent to create a zero-deforestation production model and most of the expanded production has been obtained by improving animal husbandry and pasture management.[23] Nonetheless, the Colombian Amazon has experienced an upsurge in deforestation that reached record levels in 2017 (130,000 ha), 2018 (155,000 ha) and 2020 (145,000).[24] The spike in deforestation is often linked to the production of illicit coca, but pastures are being simultaneously established on landscapes where law enforcement is poor and land tenure is characterised by chaos (see Chapter 4).[25]

Venezuela has a relatively large cattle sector located in the Llanos del Orinoco (Figure 3.4), but the country still imports about forty per cent of its national consumption – or did prior to the economic crisis that became particularly acute after 2016. Government-generated statistics show a stable herd, but unofficial sources report a drop of about 45 per cent.[26] Regardless, the Amazonian region has never developed an economically significant cattle industry. Similarly, Guyana and Suriname have minuscule cattle sectors and no history of deforestation linked to the cattle industry. Roraima in Brazil has a moderate-sized cattle industry based on its native savanna with about 880,000 head; nonetheless, the municipality with the largest population of

* The Llanos del Orinoco is one of three savanna ecoregions located within the Greater Amazon: the others include the Cerrado (Brazil), Llanos de Moxos (Bolivia) and Gran Sabana (Venezuela). There are also scattered white-sand savannas across the Northern Amazon.

cattle is Mucajaí (100,000 head), a heavily deforested smallholder landscape located south of the state capital of Boa Vista (HML #51).[27]

© PARALAXIS / Shutterstock.com

© PARALAXIS / Shutterstock.com

In Mato Grosso, farmers can plant two crops per year, and many are choosing to rotate soybeans (top) with maize (bottom) as part of an integrated strategy to manage plant pathogens; this approach also maximises profits.

Intensive Cultivation: Soy, Maize and Other Field Crops

The most important production system in the Pan Amazon when measured by GDP is the cultivation of annual crops: particularly soy, but also maize, rice, sorghum, wheat and cotton. In Brazil and Bolivia, annual cropping is organised around soy, because export markets provide the potential for a very substantial return on investment. Industrial-scale farming is much riskier than cattle ranching because it requires a considerable capital outlay to sow and harvest a crop. A successful harvest depends upon weather, which is unpredictable, and price, which is determined by commodity markets that are notoriously volatile. A poor harvest during the bottom of the commodity price cycle can bankrupt a farmer, especially those that are overly reliant on short-term credit to finance operations. The increase in risk is offset, however, by the potential return.

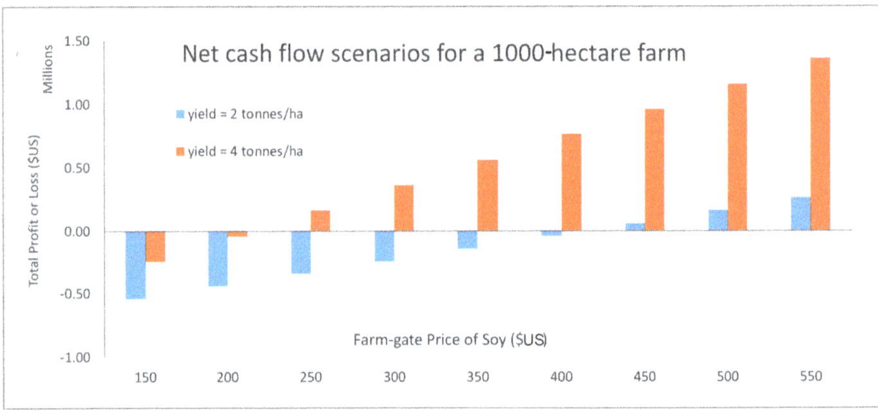

Figure 3.5: Risk profile of a 1000-hectare farm in Mato Grosso showing different potential outcomes depending on yield (tonnes/hectare) and price of soy ($US/tonne) at the farm gate.

Source of cost estimates: Instituto Mato-Grossense de Economia Agropecuária (IMEA).

The cost of production in 2020, including fuel, fertilisers, pesticides, labour and on-farm operations was approximately $US 700 per hectare in Mato Grosso.[28] Yields range between two to four tonnes per hectare, while the international price of soybean has fluctuated between $US 200 and $US 600 per tonne since 2000.* Farmers in the hinterlands of South America

* Between March 2020 and March 2021, the price surged from $US 375 to $US 575 per ton; the previous year the value of the real fell by 30% – these types of fluctuations are out of the control of producers. Source: https://www.indexmundi.com/commodities/

are paid a discounted price that reflects the cost of transport to the export terminal, where it is loaded onto ocean-going grain ships (see Chapter 2). In a good year, soybean farmers can double their money, but in a bad year, some will go bankrupt (Figure 3.5). Although these back-of-the-envelope calculations do not include capital investments in farm equipment or land, they do reveal the risk-reward potential of the industry.

Although the cultivation of soy is lucrative, it can be grown only in rotation with other crops, due to the proliferation of plant pathogens in monoculture production systems. Farmers in both Bolivia and Brazil sow two crops per year, which also allows them to spread climate risk between a summer (wet season) and winter (dry season) harvest. Many opt to plant a cover crop for one of the two cropping seasons; this allows them to improve the organic matter in the soil as well as reduce the risk from pest outbreaks. Farmers increasingly are choosing to cultivate a feed grain as a rotational crop because it can improve their bottom line and diversify their market opportunities. Maize is the most common rotational crop in Brazil, while drought-tolerant sorghum is preferred in Bolivia. The rotation of soy with feed grains has brought substantial benefits to the farm economy because it has increased the supply and affordability of feed rations for poultry and swine (see below).

The expansion of the soy / maize production model continues apace in the Brazilian Amazon (HML #4, #5, #6, #7, #11, #12, #13, #14, #15 #16 and #23) and Bolivia (HML #30 and #31). There have been dips in production, but overall, the sector has expanded its spatial footprint year after year for more than forty years. In Mato Grosso, mean yields have increased from around 3.1 in 2000 to 3.5 tonnes per hectare in 2019; producers in Bolivia tend to use less fertiliser and other inputs and average between 1.8 to 2.2 tonnes per hectare. Mato Grosso produces about 27 per cent of Brazil's total soy crop, a proportion that has remained stable over the last decade, although total production has increased by fifty per cent since 2010 (Figure 3.6).

The municipalities that produce the most are Sorriso, Nova Mutum and Nova Ubiratã, which are situated along BR-163, or Diamantino, Sapezal, and Campo Novo do Parecis, which are located further west along BR-364 (HML#15). Farmers in each of these municipalities harvested between one and two million tonnes of soy in 2019. Expansion has been most pronounced in the municipalities associated with BR-158, where soy plantings expanded by 500,000 hectares between 2016 and 2019 (HML #11 and #12).* A similar phenomenon is underway in Tocantins, northeast Pará and Maranhão to take advantage of the lower transportation costs

* Mato Grosso: São Felix deo Araguaia, São Jose do Norte; Pará: Santana de Araguaia, Conceição do Aaragaia, Santa Maria das Barreiras.

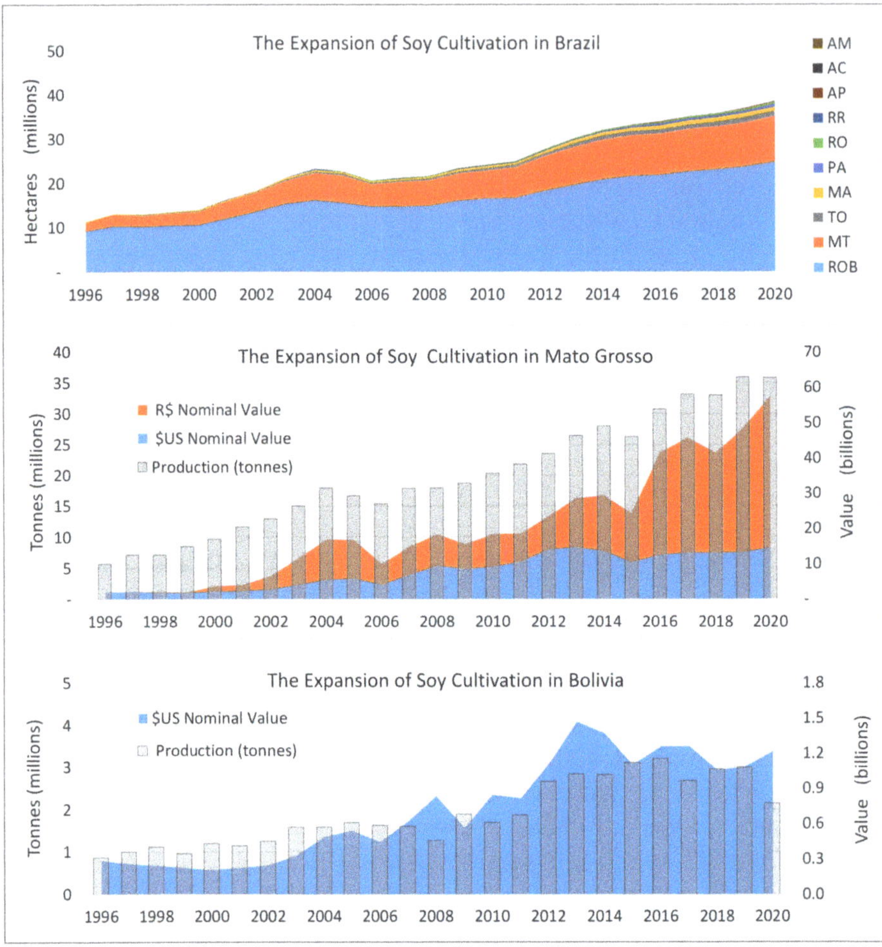

Figure 3.6: The evolution of the cultivation of soybeans in (a) Brazil, (b) Mato Grosso and (c) Bolivia. Mato Grosso is the largest producer of soy within the federal union, producing about 27% of national production. The cultivation of soy has expanded to all nine states in the Legal Amazon; abbreviations in (a) are postal codes (except ROB which stands for 'rest of Brazil'). Nominal value is based on total production and the mean annual value of soybeans in international markets; producers are paid that value minus the cost of transportation.

Data sources: CONAB and IBGE/SIDRA (Brazil); FAOSTAT and IBCE (Bolivia).

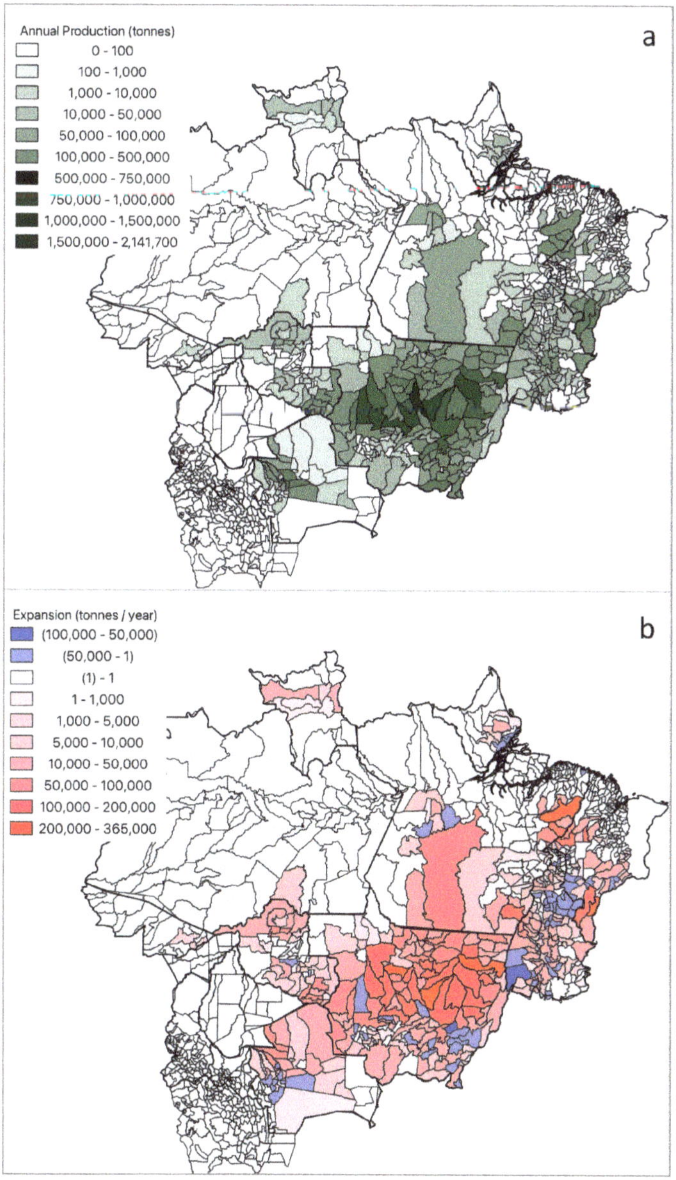

Figure 3.7: The dimensions and expansion of soybean production are reflected in municipal harvest data. The municipalities of Central Mato Grosso continue to expand production via both intensification and extensification, while bulk transport systems facilitate expansion in Northeast Pará and Western Rondônia. The relatively modest production statistics in Santa Cruz, Bolivia reflect lower yields and highlight the potential to expand via intensification.

Data sources: CONAB and IBCE.

provided by the *Ferrocarril Norte-Sul* (see Chapter 2).* Landscapes with the highest rate of conversion to soy include several located in north-central Mato Grosso, where farming became more attractive following the paving of BR-163.† Similarly, the proliferation of soy cultivation in municipalities adjacent to the river ports on the Madeira, Tocantins, Tapajós, Xingu and Amazon rivers reflects investors' desires to improve returns by lowering transportation costs;‡ more surprising, and worrisome, is the installation of soy fields in remote municipalities of the northern Amazon (Figure 3.7).

The cultivation of soybeans in the Brazilian Amazon was linked by academics and journalists to deforestation in the early 2000s when annual deforestation in the Brazilian Amazon surpassed 2.5 million hectares per year. The revelation of soy-related deforestation happened to coincide with a period when European imports of soy from Brazil reached an all-time high of fifty million tonnes per year.[29] The public linkage between soy and Amazonian deforestation led to a high-profile campaign by Greenpeace and other NGOs, which led to the Soy Moratorium (see Text Box 3.2).

The Soy Moratorium contributed to and coincided with a multi-faceted policy, referred to as the PPCDAm,§ which was organised by the administration of Lula da Silva to reduce deforestation in the Brazilian Amazon (See Chapter 7). The declines were particularly impressive in Mato Grosso, where forest clearing by farmers fell to near zero, a success story essential for the future of one of Brazil's most important export industries. Nonetheless, the actual land-use change associated with the expansion of soy/maize production model is a more nuanced story.[30] Of the approximately ten million hectares of soy planted in Mato Grosso in 2020, about thirty per cent was cultivated on land originally covered by forest vegetation, while the other seventy per cent was established on landscapes within the Cerrado savanna biome (Figure 3.8).[31] In neither case, however, was conversion always a direct operation that cleared native vegetation to establish a working soybean farm; approximately 75 per cent of the land involved was first cleared to plant pasture grasses as part of a beef production operation that was subsequently converted to the cultivation of annual crops. Farmers expand via the conversion of pasture rather than forest because it is more cost-effective. Forest properties tend to be more remote, which

* Pará: Paragominas, Dom Elise, Ulianópolis, Rondon do Pará, Dom Eliseu; Tocantins: Campos Lindos, Porto Nacional, Mateiros, Monte do Carmo, Caseara; Maranhão: Açailândia.

† Novo Horizonte do Norte (122% annually), Paranaíta (76%), Alta Floresta (43%), Peixoto de Azevedo (32%).

‡ Porto Velho (10,000 ha), Santarem (19,000 ha) and Mojuí dos Campos (64,500 ha), Macapá (8,000 ha).

§ PPCDAm: Plano de Ação para Prevenção e Controle do Desmatamento na Amazônia Legal. See http://www.mma.gov.br/component/k2/item/616?Itemid=1155

Text Box 3.2: The Soy Moratorium

In 2006, Greenpeace released a report exposing the links between deforestation in the Amazon and the expansion of the cultivation of soy. The report highlighted how soy was used as a feedstock to produce meat and other consumer goods and called for a boycott of the brands and enterprises whose supply chains extended into the Brazilian Amazon. The boycott focused the attention of major corporations (MacDonald's, Walmart, Carrefour and many others), which exerted pressure on the five major commodity traders (ADM, Bunge, Cargill, Louis Dreyfus and Amaggi Group) to devise a system that would exclude producers who were clearing forest to establish industrial farms dedicated to the cultivation of soy.

The ban was originally organised as a ten-year moratorium that committed the traders to organise a system that would eliminate any farmer guilty of clearing forest, using slave labour or encroaching on indigenous lands or other protected areas. The moratorium participants succeeded in developing a robust and transparent monitoring system that essentially eliminated the practice of clearing forest to establish industrial-scale farms. The ten-year moratorium was made permanent in 2017, and its success has been used to inform similar efforts to reform the supply chains of other commodities, notably beef and palm oil.

Source: Greenpeace: https://www.greenpeace.org/usa/victories/amazon-rainforest-deforestation-soy-moratorium-success/

increases transportation costs, and clearing them requires contracting heavy machinery. The business preference of farmers coincides with the interest of cattle ranchers who decide to monetise the capital appreciation they had obtained by being early-stage pioneers. Some ranchers relocate to forest frontiers where land is cheap.

In 2016, the amount of pasture in Mato Grosso was estimated at approximately twenty million hectares, while total cropland was reported to be ten million hectares. Most analysts project that the soy/maize production model will continue to expand and pasture area will decrease over the short term. Within Mato Grosso, that expansion is most likely to occur in the northern tier of municipalities (HML #16) where (1) landscapes are well-suited for mechanised agriculture because they have flat topography and deep, well-drained soils,* and (2) land is privately-owned and farmers can expand into the region without violating the criteria of the Soy Moratorium (Figure 3.8). The construction of the *Ferrograu* will likewise catalyse the

* With very few exceptions, tropical soils are acidic and require large inputs of agricultural lime ($CaCO_3$) to make them suitable for intensive cropping; after deforestation, this is the largest component of the GHG emission profiles of industrial agriculture in Brazil (see Ch. 4).

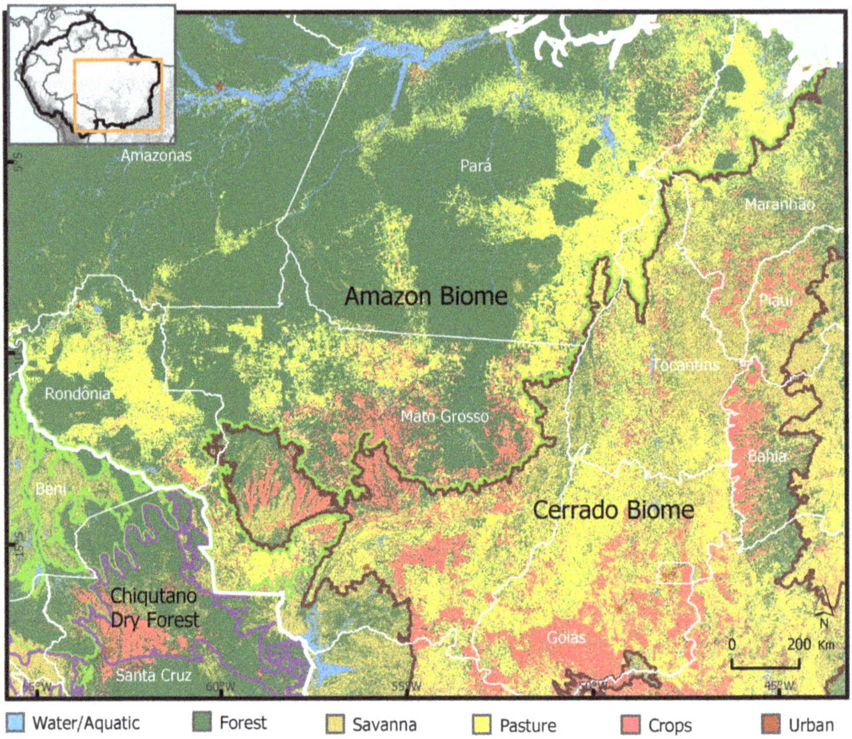

Figure 3.8: Across the Southern Amazon, intensive agriculture predominates on landscapes with flat topography and arable soils, particularly the tablelands of Central Mato Grosso, Bahia, Piaui and Maranhão. Current sustainability protocols allow for the cultivation of soybeans on pastureland within the Amazon biome, a process underway in Northern Mato Grosso, Rondônia, Eastern Pará and on the landscapes near Santarem. Within the Cerrado Biome, the Forest Code allows landholders to clear between 65% to 80% of the native vegetation; currently, about 50% of its original savanna habitat has been converted to pastures or cropland. In Bolivia, intensive agriculture has transformed the alluvial plain of Santa Cruz, largely at the expense of the Chiquitano Dry Forest, a transitional ecoregion distinct from both the Amazon and Cerrado biomes.

expansion of intensive agriculture into these previously remote landscapes that will have an advantage due to lower transportation costs. Similarly, if the *Ferrovia Paraensis* is extended south to the border between Mato Grosso and Pará, the cattle ranching landscapes along BR-158 will most likely be converted into farmland (see Chapter 2)

Land-use change also impacts the watersheds that drain these landscapes. Both deforestation and Cerrado conversion degrade the physical properties of topsoils, which makes croplands susceptible to surface erosion and increases lateral transport of nutrients to the stream network. This increases the potential for eutropication in aquatic habitats due to nitrogen enrichment from the use of fertiliser or nitrogen-fixing crops such as soy. Pesticides are present throughout the entire aquatic system, sometimes at levels that may pose serious health risks.[32] The growing adoption of centre-pivot irrigation systems threatens to seriously diminish water flows, particularly in the Tapajos basin (See Chapter 4).

In Bolivia, the expansion of soy is largely the consequence of direct deforestation, which includes not only humid forests forest near the Andean foothills but also the seasonal and dry forests of Chiquitania and the Gran Chaco (Figure 3.8). Unlike Mato Grosso, where the conversion of pasture to cropland predominates, the expansion of soy cultivation can be directly linked to new deforestation.[33] The expansion of the farm sector has been a top priority of successive governments, including that of Evo Morales, who adopted a policy to double the area under cultivation by 2025.[34] In spite of the favourable policies, expansion is constrained by the realities of farming in Bolivia. Producers face significant risk linked to weather, as well as challenges linked to poor secondary road infrastructure, inadequate storage facilities and contradictory government policies that limit the use of genetically modified organisms. Bolivian producers can compete in export markets due to fertile soils and inexpensive land, both of which are the consequence of the expanding agricultural frontier. They also benefit from a multi-modal bulk transport system (rail and waterway), which mitigates the high cost of transport that is a consequence of their geography (see Chapter 2)

Industrial infrastucture

The soy/maize supply chain is dependent on the existence of privately owned logistical facilities that are essential for receiving, drying and storing soy and feed grains post-harvest. Without silos, the entire harvest would have to be transported immediately to distant processing plants or export terminals. This would cause inefficiencies in the supply chain and lead to traffic jams on the limited number of highways that connect producing landscapes with export terminals.* Silos absorb production during harvest and then permit grains to be dispatched to markets over the following weeks and months. Soy will ferment and spoil when access to storage is restricted due to hauling distance or long lines at logistical facilities, particularly if

* Traffic jams are a common occurrence due to the shortage of bulk transport systems, particularly on BR-163 and BNR-364. Source: Reuters Commodities, 1 Mar. 2017, http://www.reuters.com/article/us-brazil-soybeans-road-idUSKBN1685AN

the beans have a high water content. Estimates of the loss due to spoilage range from five to six per cent per year, which in terms of monetary value ranges from $US 200 to 500 million annually for the state of Mato Grosso.[35]

According to *Companhia Nacional de Abastecimento* (CONAB), Mato Grosso had a total storage capacity of 38.6 million tonnes in 2016, approximately 53 per cent of the combined harvest of soy and maize in 2020. Much of this capacity is owned and operated by one of the four global trading companies: ADM, Cargill, Luis Dreyfus and Bunge. However, the largest operator is the Amaggi Group, a family-held Brazilian company, which has an integrated supply chain that spans production (280,000 hectares), silos (25 facilities), crushing mills (3) and transportation assets (barges and trucks). Other Brazilian companies that focus on the domestic market compete with these larger companies, as does a Chinese company, COFCO, with silos in Sorriso and San Lucas do Rio Verde, as well as a crushing mill in Rondonópolis.* In addition, the corporations that operate key railroads (Rumo and VLS) have silos at their logistical facilities both in the field and at the port terminals they operate in Santos (SP) and São Luis de Maranhão (MA).

There is currently an initiative organised by the Brazilian soybean association *(Aprosoja Brasil)* to promote investment in on-farm silos, which would reduce wastage and provide farmers with the option to hold their beans for later sale; this would benefit their bottom line because farm-gate prices are lowest during the harvest season.[36] In Bolivia, which is the world's tenth largest producer of soybeans and sixth largest exporter of soybeans, producers likewise depend on the presence of the four global traders, but about half of its production is processed and traded by companies based in Bolivia, Venezuela or Peru.

Crushing mills are the other major soybean industrial asset. As the name implies, these facilities process the beans by crushing them to extract vegetable oil and separate it from soybean meal (also referred to as soy cake). Soy oil is consumed as food or as an ingredient in food products or other consumer goods; in Brazil, it is used as a feedstock for the production of biodiesel (see below). Soy cake, which is high in protein, is used as an ingredient in animal feeds. Soy is exported as a seed (soybean), as vegetable oil and as soy cake. When exported as a seed, it is processed at its destination, where its two major products are consumed, but when it is processed near the farm, it provides the traders with the opportunity to export the sub-products to different markets. More importantly, crushing mills create a strategically important opportunity to add value to soy production by transforming it into a product with a higher market value. This

---

* China National Cereals, Oils and Foodstuffs Corporation (COFCO) is China's state-owned food processing holding company and is China's largest food processing, manufacturer and trader.

On-farm silos reduce wastage and allow farmers to commercialise production later in the year, rather than during the harvest period when prices tend to be low.

is the basis of the poultry and swine industries in both Mato Grosso and Bolivia (see below).

Global markets

Because soybeans are annuals, soybean prices can vary sharply over relatively short periods of time. Farmers can choose to expand soy cultivation when prices are 'good', which eventually leads to oversupply and a drop in prices; this motivates farmers to switch crops or leave land idle. For this reason, soybean farmers are more attuned to global markets when compared to beef producers (see above) and more agile when compared to palm oil producers (see below).

Brazil exports between seventy and eighty per cent of its national harvest of soybeans, while Bolivia exports as much as 85 per cent of its production. Consequently, the global market has an overriding impact on the farm economy in both countries. The global demand for soy is driven by the combined consumption of vegetable oil and vegetable protein. Soy

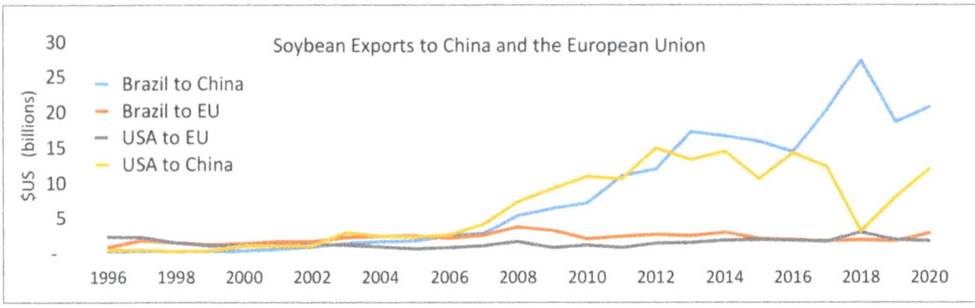

Figure 3.9: China is the world's largest importer of soybeans and Brazil is the world's largest exporter. Tensions between the United States and China have led to an increase in exports from Brazil. Approximately 35% of Brazil's production comes from the Legal Amazon, where harvests are growing at approximately double the rate as that seen for extra-Amazonian producers.

Data sources: FAOSTAT and the USDA/FAS.

oil competes with palm oil, which is more competitive based on price, but the revenue producers enjoy from soy cake ensures that it will continue to be widely cultivated for the foreseeable future.

Geopolitics also plays a role. The trade war between the United States and China initiated by the Trump administration in 2017 led to a dramatic increase in the export of soybeans from Brazil to China. Subsequent agreements led to a return of exports from the United States to China, but Brazil has consolidated its position as the world's largest exporter of soybeans (since 2013) and displaced the United States as the largest producer in 2019 (Figure 3.9).

Demand for soy is expected to double by 2050.[37] In 2020, Brazil exported approximately fifty per cent of total production to China;[*] however, increased growth in future exports is more likely to come from South Asia as this region becomes wealthier and increases its consumption of animal protein dependent upon soy cake. Similarly, population growth and increased per capita income in Sub-Saharan Africa are projected to increase consumption of chicken and pork. If future supply chains remain tied to current production landscapes, South American producers will provide most of the increase in future production. There are other options, but if the projected increase in global demand for soy is met only by production in South America, this would almost inevitably lead to new deforestation in

* Leading Chinese soy importing companies include the Jiusan Group, Hopefull Group, Sinograin, COFCO, CP Group and Chongqing Grains, but the four major trading companies (ADM, Bunge, Cargill, Dreyefus) are intermediaries for most of the transactions.

the Amazon or, more probably, large-scale conversion of natural grasslands in the Cerrado, Campos and Pampa biomes.*

Over the short term, Brazilian producers will continue to expand production because their technological know-how, abundant land resources and post-harvest infrastructure makes expansion an attractive investment. The improvement of bulk transport systems will facilitate the expansion of production in Northern Mato Grosso and increase that region's competitive position in global markets. Increases in cultivation area will most likely occur on non-forest landscapes, including both cultivated pasture within the Amazon biome , as well as by the conversion of Cerrado landscapes, especially in the northeast sector of the Pan Amazon (Figure 3.8).

Over the medium term, pressure on the Amazon from soy and feed grains will depend on the evolution of the global market. Future supply chains could be radically different than those that dominate today. For example, the increase in demand from Sub-Saharan Africa could be met by policies designed to ensure that region remains self-sufficient in food production. Similarly, the EU may choose to embrace an emerging trend to source non-GMO and organic soy from producers in the Ukraine.[38]

Swine and Poultry: Adding Value to Farm Production

The cultivation of soy and maize has brought additional benefits to the farm economy in Mato Grosso because it has increased the supply and affordability of feed rations for livestock producers. Although there is a robust international market for both commodities, the potential return for farmers in Mato Grosso is not as lucrative for maize when compared to soy. This is due mainly to the very substantial gap between maize yields obtained by farmers in the United States when compared to producers in Brazil,† but also to the steep transportation costs that limit the profitability of producers in Mato Grosso. Consequently, the agricultural industry has a strong incentive to create livestock production systems that convert crop commodities into a product with greater market value. The poultry and swine industries have expanded at about the same rate as the soy / maize complex in Mato Grosso (Figure 3.10). In contrast, poultry production in Rondônia and Pará, which do not (yet) cultivate significant areas of soy, declined over the same period in both states.

---

* In Argentina, soy is encroaching on the Pampas biome at the expense of cattle ranching, while the largest growth in soy cultivation in Brazil is occurring via the conversion of Cerrado savannas in the states of Maranhão, Piauí and Bahia, which are collectively known as MATOPIBA (Modernel et al. 2016; Spera et al. 2016).

† Yields in Iowa average about 15 tons per hectare, compared to between 9–7 tons per hectare in Mato Grosso.

© Rafael Tomazi/Shutterstock.com

© Alf Ribeiro/Shutterstock.com

Industrial-scale production of pork and chickens has expanded by about 400% over the last thirty years due, in part, to the affordability of feed grains in Mato Grosso (soy + maize) and Santa Cruz Bolivia (soy + sorghum).

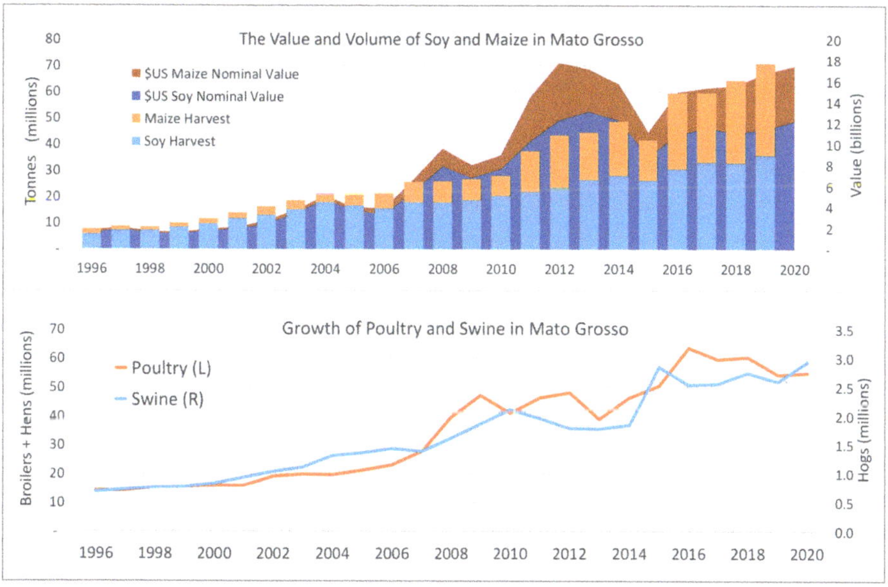

Figure 3.10: The soy boom in Mato Grosso has been accompanied by an increase in the cultivation of maize, which has catalysed the development of the livestock sector, particularly swine and poultry, both of which have grown ~7% annually for more than two decades.

Data source: IBGE/SIDRA.

Brazil is the second-largest producer of poultry in the world and the largest exporter of processed chicken meat. The largest producing states are Paraná, São Paulo, Santa Catarina and Rio Grande do Sul, which are closer to urban markets and ports. Nonetheless, Mato Grosso enjoys the fastest growth rate of poultry production in Brazil: Between 2000 and 2020, the total chicken population in Mato Gross grew an average of seven per cent per year, from ten million to sixty million birds.[39] About 75 per cent of these are broilers (chicken raised for meat) that have a life span of between six and seven weeks, which means that Mato Grosso produced about 330 million chickens for slaughter in 2020. This translates into approximately 700,000 tonnes of meat, about five per cent of Brazil's annual poultry production.[40]

The amount of maize and soy consumed by the production of broilers is dependent upon two factors: the composition of chicken feed and the feed conversion ratio (FCO), which is the metric animal scientists use to calculate the quantity of feed required to produce one kilogram of meat. Chicken feed is about sixty per cent grain and twenty per cent soy cake: the FCO for broilers is about 1.67, while hens have an FCO of 2.0 per kilogram

Source: Google Earth

Figure 3.11: Industrial livestock production facilities near Sorriso in Mato Grosso: (a) Swine barns are accompanied by wastewater treatment ponds; (b) poultry sheds are separated by eucalyptus plantations and a patch of remnant forest.

of egg. This suggests that Mato Grosso's poultry industry was consuming about 0.8 per cent of the state's production of soy and about 1.9 per cent of its maize harvest in 2015.* The situation for the swine industry is similar, except feeder pigs live between six to eight months and have an FCO of about 2.5.† Taking into account the feed intake of sows, a swine herd of about three million pigs would consume about one million tonnes of feed, constituting about one per cent of soy and 1.7 per cent of Mato Grosso's maize production annually.

---

* This calculation assumes that broilers live about 7 weeks and yield 2.2 kilograms of meat at slaughter.

† The take-off rate is the number of heads slaughtered each year compared to the total herd size. In Brazil that value is 0.98; the take-off rate is 1.68 in the U.S., because pigs have shorter lifespans and higher reproductive success.

Although local consumption of soy and maize by the livestock sector is not enormous, it is significant when viewed from an economic perspective. The swine and poultry industries generated about $US 1 billion dollars in gross sales in 2020, a value about three times greater than the unprocessed maize and soybeans fed to those animals. This is why the livestock sector can be viewed as a value-added component of the soy/maize production system. The livestock sector injects significant capital into the rural economy, most of which will be reinvested to expand production.

The growth of the livestock sector has been accompanied by the creation of the industrial infrastructure linked to these industries, specifically facilities dedicated to different stages of the livestock life cycle: breeding, brooding, grow-out, packing plants and associated logistics (Figure 3.11). The growth of these industries increases the demand for soy and maize, which will be met either by expansion through extensification or intensification (see Chapter 4). In the first case, this will lead to new deforestation or, more likely, the conversion of remnant Cerrado habitat. Intensification also comes with the risk of environmental degradation, through either the conversion of forest remnants within agricultural frontiers or the displacement of beef ranching operations to the forest frontier. The poultry and pork industries can be viewed as indirect drivers of deforestation because they contribute to the expansion of both the farm and beef sector.

Palm Oil

Oil palm is highly suited to cultivation in the ecological and climatic conditions of the tropics and its cultivation in South America has expanded steadily over the last three decades. The area under cultivation has shown a marked increase in its rate of growth during the last decade (Figure 3.12). Palm oil has been the second-largest global driver of deforestation in the tropics, and its expansion in the Amazon has been accompanied by widespread concern that land-use practices that characterise the industry globally will be replicated there. The concern is based on the predominant business model within the industry, which combines ownership of large-scale plantations with the operation of industrial processing mills. The modern palm oil corporation is a classic example of the benefits of vertical integration and the economies of scale. Some companies also invest in transportation systems, refineries and manufacturing enterprises that transform crude palm oil into consumer goods. The industry brings multiple social and economic benefits to its host nations by improving food security, balance of trade, tax revenues, job creation and economic growth in rural landscapes.

The industrial-scale business model depends upon the acquisition of large landholding – minimally 5,000 hectares but often as large as 50,000

Large-scale oil palm plantations require a work force of unskilled, semi-skilled and professional employees. Smallholders like them because the trees produce fruit and revenue throughout the year. They are replanted after about 25 years because they are too tall to harvest efficiently; smallholders tend not to replant, however, because they cannot afford to lose five years of production between planting and the initiation of fruit production.

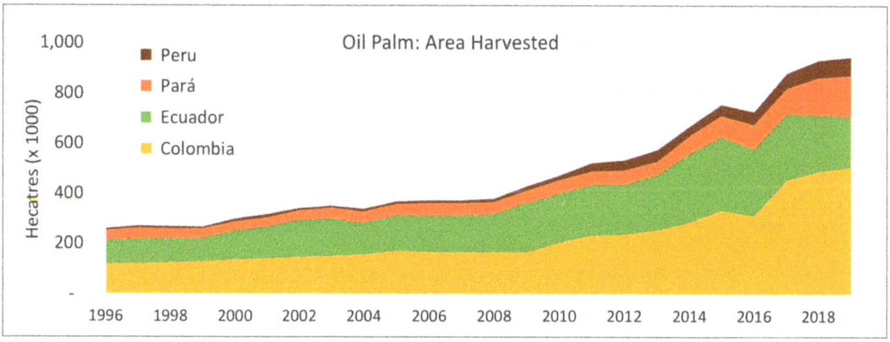

Figure 3.12: Oil palm plantations harvested each year; excludes new plantings that are not yet producing fruit, while the data for Colombia and Ecuador include non-Amazonian landscapes.
Data source: FAOSTAT.

hectares. These are difficult to acquire on landscapes that have already been deforested, especially in those developing countries where past migratory phenomena and weak administrative systems have made land tenure systems chaotic and insecure (see Chapter 4). Palm oil corporations have used their political influence to gain access to public lands in the forest estate to avoid the extra cost of acquiring land on previously deforested land.* Historically, this was the predominant business model for palm oil companies in the Pan Amazon, but the countries have diverged in their practices since about 2000 (see below).

Palm oil corporations act as a proximate (direct) cause of deforestation when they acquire forest land and establish new plantations; however, large vertically integrated corporations also function as an ultimate (indirect) driver of deforestation because they manage the supply chains that commercialise palm oil in consumer markets. The obvious linkages between the cultivation of oil palm and deforestation have motivated multiple environmental organisations to forcefully criticise the industry, which has been the focus of numerous campaigns and boycotts over the last two decades.

Although large corporations dominate most oil palm landscapes, particularly in Southeast Asia, they usually coexist with smallholders and independent producers who are also seeking to benefit from a highly profitable production system (see Text Box 3.3). In most cases, smaller producers

---

* In many cases, this would require resolving conflicting land-titling claims for dozens or hundreds of small individual properties, as well as purchasing the properties at prices that would increase once residents learned of the investment plans of the company.

rely on the corporate mill to commercialise their production. Sometimes this coexistence can evolve into a stronger partnership where the company provides technical support or commits to long-term purchase agreements.

Text Box 3.3: Oil palm growers can be classified into three groups:

Smallholders are growers with less than 20 hectares and, typically, incompletely documented legal tenure. They depend upon family labour and manage groves with limited technology and yields between 50% to 60% of industrial plantations. They are dependent upon public programmes for technical assistance, tend not to renovate plantings and have limited access to financial credit. Annual cash flow varies from $5,000 to $10,000.

Independent producers are family-owned enterprises with between 20 and 1,000 hectares; most have secure land tenure and contract both permanent and temporary employees. Fresh fruit yields tend to be between 70% and 90% of industrial plantations due to technical assistance contracted from specialised service providers. Growers are more risk tolerant when compared to smallholders and are more willing to take on credit to expand or replace senile palm trees. Annual cash flow varies from $50,000 to $500,000.

Industrial-scale plantation and mill complex are corporate entities, usually subsidiaries of family-held holding companies with landholding of 5,000 hectares or larger. Land tenure is secure in all but a few cases; plantations and mills are managed by professionals with university education who oversee hundreds of permanent employees with benefits and low turnover; day labor is contracted as needed and paid minimum wage. Companies have almost unrestricted access to credit and new capital; annual cash flow can vary between $50 and $100 million.

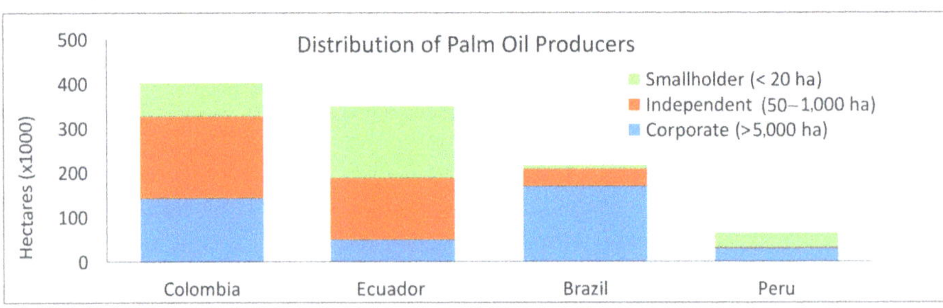

Figure 3.13: The distribution of oil palm plantations stratified by size and social group; data based on a 2016 study by the author, from multiple sources.

Mixed production landscapes exist in all South American countries, where the relative abundance of the different types of producers is the consequence of market forces, social phenomena and public policies that are unique to each country (Figure 3.13).

The presence of a diverse population of producer types opens the door for alternative business models that delink deforestation from the establishment of new oil palm plantations. Unlike industrial-scale operators, smallholders and independent producers are more likely to establish plantations on previously deforested landscapes. In the case of smallholders, many own properties that were homesteaded in previous decades and are now adopting oil palm cultivation because it is more lucrative than cultivating food crops. Independents are a diverse group: some are urban businessmen investing in a production with future growth potential, while others are successful smallholders who are purchasing and consolidating properties to benefit from the economies of scale. They are more likely than smallholders to deforest land, but they are also more likely to own properties that have been previously deforested, particularly cattle ranches.

The era of industrial-scale expansion of oil palm plantations at the expense of forest habitat may have come to an end in the Pan Amazon. There are multiple reasons that this is occurring, and they differ in each country, but the motivation is the same: producers are seeking to penetrate export markets, and the global demand for deforestation-free palm oil has created an opportunity they seek to exploit. The most astute companies have recognised that the best opportunities for growth are via partnerships with smallholders and independents who own or occupy most of the previously deforested land that can be converted to oil palm plantations.

Colombia

Colombia is Latin America's largest producer of palm oil, with nearly 450,000 hectares of oil palm plantations in 2020, with another 100,000 immature plantings that will expand production by ~ 20 per cent over the next few years. The sector generated approximately $US five billion in gross revenues in 2019* and contributes about 150,000 jobs to the national economy.[41] Colombia has a relatively diversified producer sector, and most of the industrial producers actively support independent and smallholder producers (Figure 3.13). The sector is self-organised via the *Federación Nacional de Cultivadores de Palma de Aceite* (FEDEPALMA) and its highly competent research and extension service, *Corporación Centro de Investigación en Palma de Aceite* (CENIPALMA).

---

* Nominal value based on the mean price of one ton of refined palm oil in international markets; smallholders and independent growers only receive a fraction of this amount.

Smallholders represented a small minority of plantation area in Colombia until 2000, when the government initiated the *Alianzas Productivas*, an initiative that supports collaboration between smallholder associations and industrial-scale producers. When the programme started in 1999, there were an estimated 390 farmers with oil palm groves smaller than twenty hectares; by 2015, almost 55,000 smallholder families were participating in the initiative.[42] This programme is expected to expand over the short term as part of Colombia's efforts to provide economic opportunity to displaced people who reside, or once resided, in conflict areas.

There are four major oil palm regions in Colombia, and none of them are in the Amazon.* Approximately fifty per cent of the nation's plantations have been established on landscapes that were transformed by human activity long ago. Deforestation did lead directly to the establishment of about five per cent of the nation's plantations on the Pacific Coast, while the conversion of natural savanna vegetation has preceded most plantations established on the *Llanos de Orinoco*. The Llanos region is immediately adjacent to the Amazon watershed, and the savanna habitat located in that ecoregion is broadly comparable to the Cerrado biome of Central Brazil. Colombia's palm oil sector promotes itself as 'deforestation-free'; however, this claim ignores the land-use change and environmental degradation that accompanies expansion into the *Llanos del Orinoco*, which now accounts for about forty per cent of palm oil produced in Colombia.

Only one industrial-scale oil plantation is located within the Colombian Amazon; a single plantation in Caquetá (HML #52) has struggled to survive in a region better known for cattle and as a centre of armed conflict and illicit drug production. Nonetheless, Caquetá and nearby landscapes in Putumayo (HML #51) and Guaviare (HML #54) have climates and landscapes appropriate for oil palm. In Caquetá alone, more than 1.2 million hectares of degraded pasture provide a unique opportunity to expand palm oil production with net positive environmental and social outcomes. Palm oil plantations established on degraded pastures sequester carbon and restore evapotranspiration functionality; they are also economically much more productive than cattle operations. Because relatively large landholdings exist in the region, they provide an opportunity for corporations seeking to establish new plantations. Investment in Caquetá has long been suppressed due to civil conflict, although the peace and reconciliation process provides an opening for expansion of the industry into the region.

Producers in high rainfall regions must deal with the constant threat from a serious plant disease, which affects plantations throughout South America (see Text Box 3.4). The disease has led to mass die-offs on large-scale

* Zona Norte (Caribbean Coast), Zona Central (Río Magdalena Valley), Zona Sur (Pacific Coast) and Zona Oriental (Llanos de Orinoco)

Text Box 3.4: Integrated pest management in oil palm plantations

Pudrición del Cogollo (Bud Rot) is a complex plant disease caused by an initial infection by a fungal pathogen (*Phytopthora palmivora*), which facilitates secondary infections by other fungi and bacteria that eventually overwhelm plant defence mechanisms. The disease is spread by the palm weevil (*Rhynchophorus palmarum*), which feeds on tender young palm leaves and acts as a vector for the initial fungus infection.

Prevention requires the use of pheromone-baited traps to control weevil populations, culling infected trees and adhering to best management practices, particularly soil nutrition, to ensure robust growth. Where infestations are severe and recurring, crop scientists recommend planting a resistant hybrid cultivar derived from a cross between the African oil palm (*Elaeis guineensi)* and its Amazonian cousin *(Elaeis oleifera)*. Hybrids must be pollinated manually because the plants' male flowers are sterile.

Corporate growers prefer to pursue prevention because best management practices bring collateral benefits in yield while manual pollination is labour-intensive and costly. In contrast, smallholders prefer to plant the hybrid, because it reduces risk and they can absorb the extra cost of pollination by using family labour. In addition, the dwarf stature of the hybrids facilitates harvesting and will prolong the lifespan of a plantations from twenty to thirty years, which will benefit smallholders in the future by lowering the need to replant at regular intervals to maintain peak production.

oil palm plantations on the Pacific Coast in both Colombia and Ecuador, which demonstrates the very real risk of catastrophic failure inherent in any monoculture production system where the plants under cultivation are genetically identical.

Ecuador

Ecuador has one of the most egalitarian palm oil sectors in the world, with more than 85 per cent of all oil palm plantations owned by independent and smallholder producers (Figure 3.13). This trend will become even more accentuated in the future because smallholders have increased the area under cultivation over the past decade, while the area managed by corporate producers has shrunk due to disease infestations (Figure 3.12). In 2020, there were about 6,800 oil palm farmers and about 50,000 jobs directly connected to the sector; an additional 100,000 indirectly benefit from palm oil production.[43] One of the most unusual aspects of Ecuador's palm oil sector is the diversity of its milling sector. Thirty independent mills are owned and operated by individuals who started as independent growers but have diversified their businesses by investing in medium-scale palm oil mills. Their combined milling capacity approximately equals that of the large corporations. Independent millers are, essentially, service providers to the independent growers and smallholder producers who dominate the sector in Ecuador.

Most of Ecuador's 410,000 hectares are located on the Pacific coastal lowlands, where transportation costs make producers more competitive in export markets and where banana growers have been diversifying their production systems. The Amazonian provinces of Sucumbíos and Orellana (HML #49) are important areas for palm oil production and plans for future expansion figure prominently in national development strategy documents. There are currently about 18,500 hectares of corporate plantations operated by two companies that were established via the standard model of deforestation in the 1980s. In 2018, there was one independent palm oil mill that was building partnerships with smallholders.* There are approximately 58,000 hectares of smallholder producers in Amazonian Ecuador; each farmer typically owns about forty hectares and pursues a diversified production model that includes coffee, cacao, livestock and subsistence crops, as well as forest remnants and fallow lands. These producers are increasingly adopting oil palm due to its overall profitability but also because of the economic stability it offers by producing throughout the year. In practical terms, this means farmers receive a cheque from the mill about every two weeks, which is a powerful incentive for a low-income family.

Peru

The first oil palm plantations in Peru were established in the 1970s by a state-owned palm oil company near the village of Tocache in the Upper Huallaga Valley in the Department of San Martin (HML #42); similar facilities were soon established near Pucallpa in Ucayali Province (HML #41) and between Tarapoto and Yurimaguas in the Lower Huallaga Valley (HML #43). Economic mismanagement combined with civil unrest eventually pushed these enterprises into bankruptcy; nonetheless, they left a legacy of oil palm plantations and mill infrastructure that are important parts of Peru's current palm oil supply chain. Each of these production landscapes is now home to farmer-owned companies whose shareholders received their equity stake as severance pay when the state-owned company declared bankruptcy in the mid-1980s.† Growth in the smallholder sector stagnated

* Most independent processing mills, known as extractors, compete on price to attract feedstock, but one company (Oleana) seeks to establish long-term partnerships with small farmers by providing technical expertise to generate loyalty (see https://www.oleana.ec/es/nosotros)

† In the Upper Huallaga Valley, workers organised as the Asociación Central de Palmiculltores de Tocache – (ACEPAT), while the industrial facilities were reconstituted as Oleaginosas del Perú S.A (OLPESA); the members of ACEPAT are the majority shareholders of OLPESA. This business model has been replicated

between 1990 and 2010, despite investments by international development agencies who viewed it as a viable alternative to illicit crops.[44]

A biofuel policy promulgated by the government in 2008 instigated a new phase in the palm oil sector in Peru. The policy motivated growers to expand production, as well as attract new farmers and investors to adopt the palm oil production system. Medium-scale independent growers, previously nonexistent, have established a presence on landscapes dominated by smallholders. As in Ecuador and Colombia, some of these middle-class producers are successful smallholders who have scaled-up production by purchasing more land; others are urban investors seeking a profitable long-term investment. In 2017, the members of grower associations linked to grower-owned companies operated about 33,000 hectares and provided about forty per cent of the country's palm oil supply.

At approximately the same time the state was introducing oil palm in Tocache, one of the country's largest private corporations, *Grupo Romero*, established a plantation and mill complex in Tocache. That enterprise also suffered from the economic and civil unrest of the 1980s and 1990s, but its owners persevered and are now Peru's largest producer of palm oil via its subsidiary corporation *Grupo Palmas*. This family-owned enterprise operates two industrial-scale oil palm plantations and associated industrial-scale processing facilities: *Palmas de Espino* in Tocache with about 12,000 hectares and *Palmas de Shanusi*, which has approximately 10,000 hectares located between Tarapoto and the Amazonian port city of Yurimaguas (HML #44).

Almost one hundred per cent of the original smallholder plantations were established via the direct clearing of primary forest during the 1980s and 90s; subsequently, they expanded via the conversion of pasture, idle cropland and secondary forest, as well as by clearing remnant natural forest patches.[45] The vast majority of the *Grupo Palmas'* 23,000 hectares of plantations have been established via the direct clearing of primary natural forest. Approximately 9,000 hectares were established in the 1980s and 1990s at Tocache; more recently, direct forest clearing also preceded the establishment *Palmas de Shanusi* between 2005 and 2013 (Figure 3.14). This last expansion generated considerable controversy and was challenged in court by environmental advocates. The company prevailed through the regulatory process, however, because they processed their permits via the Ministry of Agriculture and its land-use planning law, which the court determined had precedence over the guidelines established by the Ministry of the Environment (see Chapter 4).[46]

There is a fourth major player in Peru's palm oil industry, known as the *Grupo Melka*, a corporate entity that represents (or represented) a group

in Ucayali, which is home to three farmer-owned mills (OLAMSA-1, OLAMSA-2 and OLPASA) and in the Lower Huallaga Valley near the town of Tarapoto (INDUPALMA).

Source: Google Earth

Figure 3.14: Satellite images of the oil palm plantations on the border between San Martín and Loreto, Peru: (a) Regional perspective showing the Shanusi plantation (Grupo Palmas) and adjacent smallholder communities; (b) close-up of a heterogenous smallholder landscape with palm groves from 5 to 50 hectares; and (c) close-up within the homogeneous industrial plantation. The industrial plantation was established between 2005 and 2015 by the direct conversion of natural forest habitat; smallholder groves were planted on land adjacent to a trunk highway (PE-5NB) that was deforested in the late 1960s.

of investors with connections to Southeast Asia. This company attempted to establish two oil palm plantations in Coronel Portillo Province of Ucayali Region in 2013. In contrast to the *Grupo Romero*, these investors have had less success using the legal and regulatory processes in Peru to legitimise their plantations and in 2017 abandoned their investment.[47] The failure of the *Grupo Melka* represents a significant financial loss to its investors and may signal the end of attempts to create industrial-scale oil palm plantations at the expense of the forest estate in Amazonian Peru. The *Grupo Palmas* essentially ratified this new reality when it publicly embraced a no-deforestation policy in April of 2017. Presumably, the company adopted the zero-deforestation policy based on its evaluation of the business risk associated with deforestation and a desire to increase exports to North American and European markets.[48]

Brazil

The first cultivation of oil palm outside of Africa occurred in Bahía, where it was introduced as a food crop more than 400 years ago. Currently, there are about 54,000 hectares of oil palm groves scattered across the coastal municipalities of Bahía where they are cultivated on plots that are seldom larger than ten hectares. These producers supply most of the national consumption of *dendé*, which is the name for the crude red palm oil that is an ingredient in many traditional Brazilian recipes. Industrial-scale palm oil is concentrated in the Amazonian state of Pará, where a modern palm oil industry was created in the 1970s in response to Brazil's geopolitical strategy to populate and develop the Amazon (see Chapter 6). Production in Pará grew steadily over the next three decades, averaging about three per cent growth per year until 2007. Over the last ten years, however, the area under cultivation grew rapidly rising from about 55,000 hectares in 2010 to more than 165,000 hectares in 2019 (Figure 3.12)

Like other oil palm landscapes in South America, Pará has a combination of large, medium and small-scale farmers (Figure 3.13); however, large-scale producers dominate the sector and can be stratified into three subcategories: large (2,000–4,000 ha), very large (5,000–10,000 ha) and massive (> 35,000 ha).[49] Of the three massive producers, Agropalma is a long-established company that has grown organically since its founding in the 1980s. The other two, Belem Bioenergía Brasil and Biopalma da Amazônia were established in 2011 by two of Brazil's largest corporations (Petrobras and Vale) when the government was promoting biofuel as a component in its national energy and rural development strategies.

Agronomics and logistics plagued both operations and neither company started to produce significant volumes of palm until 2018; moreover, the economics of biofuels were undermined by price declines in the fossil-fuel market after 2015. Both parent companies divested their interests in 2019 and 2020 at a loss estimated at more than R$ 1 billion each.[50] Both plantations were acquired by Brasil Biofuels, a start-up company cultivating oil palm as a biodiesel supplement for thermal generation plants in Roraima, Acre and Rondônia and Amazonas. They are currently the largest oil palm producer in Latin America, with more than 63,000 hectares that supply vegetable oil to power plants in municipalities that are physically isolated from Brazil's integrated energy grid.[51]

Brazil's palm oil corporations have pursued a strategy of sourcing oil palm fruits from smallholders and independent producers, an effort that has been proactively supported by the Brazilian government via two programmes: *Programa Nacional de Fortalecimento da Agricultura Familiar* (PRONAF) a nationwide programme that provides technical assistance and concessionary loans to small farmers, and the *Produção Sustentável de Óleo de Palma* (PSOP), which has provided tax incentives to companies and concessionary loans to small and medium-scale farmers since 2010.[52] In spite of these programmes, efforts to expand smallholder production have faced many challenges; most are related to a general lack of technical capacity and the multiple problems linked to making credit available to small farmers.

Prior to 2000, almost all expansion occurred via forest clearing; however, all the producers in Pará have halted these practices and embraced a zero-deforestation policy.[53] *Agropalma* has reported that 45 per cent of its total plantation area was established via forest clearing. The change in practices stems, in part, from regulatory rules to ensure that palm oil production for biofuel avoids deforestation, but also because Brazilian producers seek to compete in overseas markets by providing certified, identity-preserved, deforestation-free palm oil.

Industrial infrastructure

Palm oil is different from most agricultural commodities because the raw harvested product, fresh fruit bunches, must be processed within 48 hours or it will spoil. This fact dictates that plantations and mills must be in close juxtaposition. In the case of soy and beef, the decision on where to locate a crushing mill or slaughterhouse is an option with considerable leeway, and its existence is not a prerequisite to the installation of a production system. In contrast, palm oil is a hundred per cent dependent upon the simultaneous creation of a mill and a plantation at the initiation of the development process. Palm oil mills require a significant capital investment and a large-scale facility capable of competing in the global market requires an investment of about $US 40 million.* To justify this capital outlay, insti-

* A plantation with 5,000 hectares of production oil palms would require a mill capable of processing 20 tonnes of fresh-fruit bunches per hour.

© Timothy J. Killeen

© First Resources / Shutterstock.com

Large-scale oil palm plantations require a sophisticated logistical operation and an industrial mill, which needs a large capital investment, typically in excess of $US 50 million.

© Timothy J. Killeen

© Timothy J. Killeen

The orange flesh (mesocarp) of oil palm fruits is the source of palm oil (left vial), while seeds (white) are the source of palm kernel oil (right vial). Palm oil is widely used for food, consumer products and as a biofuel feedstock; palm kernel oil is used in cosmetics and for other specialty purposes.

tutional investors require that the mill be accompanied by a plantation of at least 5,000 hectares to ensure the supply of sufficient feedstock to safeguard the viability of the mill. At about $US 10,000 per hectare, a 5,000-hectare plantation would require another $US 50 millions of investment capital.

Global versus national markets

Oil palm is cultivated for two basic products: palm oil, which is extracted from the fruit, and palm kernel oil, which is extracted from the seeds.* Palm oil is used in products ranging from cooking oil and ice cream to soap and toothpaste and is a feedstock for the biofuel and chemical industries. Palm kernel oil is similar to coconut oil and enjoys a market niche linked to cosmetics and personal care products. Thirty years ago, palm oil represented less than two per cent of global consumption of fats and oils; today that figure stands at 41 per cent. In 2020, the cultivated extent of oil palm reached 28 million hectares globally, with an annual growth rate of 5.5 per cent between 2000 and 2020, more than double the annual growth rate of soybeans (2.6 per cent). Palm oil displaced soy as the world's most important vegetable oil in 2006.[54] Its dominance as a feedstock for the consumer goods industry is due to its lower cost of production versus soy oil and the chemical characteristics of its constituent fatty acids, which make it a more attractive ingredient for many recipes and formulas.

Global supply chains for palm oil are dominated by producers from Southeast Asia because they have created a hyper-efficient production system based on access to state lands, low labour costs and strategic investments in technology and management systems. Producers in Latin America have missed out on this spectacular growth because their costs of production are significantly higher than their competitors in Indonesia and Malaysia. According to a recent study, the total cost per tonne of crude palm oil produced by an integrated Colombian or Brazilian producer was approximately double that incurred by similar companies in Southeast Asia.[55†] The difference was due largely to higher labour costs, but efficiency and superior yields also favour Southeast Asian producers. The cost differential makes it difficult for South American producers to compete in international markets and forces them to accept lower profit margins. This has caused them to focus on domestic markets, at least in the early stages of their development, but most companies are now focusing on export markets as a growth strategy. Each country has pursued different development strategies, which has

* In addition, two valuable byproducts are generated by the palm oil processing mill: palm kernel cake, an ingredient in animal feed, and biomass that is used to generate thermal energy to power the palm oil mill.

† Mean costs over several years calculated for Colombia ($US 477/ton), $US Brazil ($US 509/ton), Malaysia ($US 266) and Indonesia ($US 233) include plantation establishment and operations, harvest and transport and mill operations.

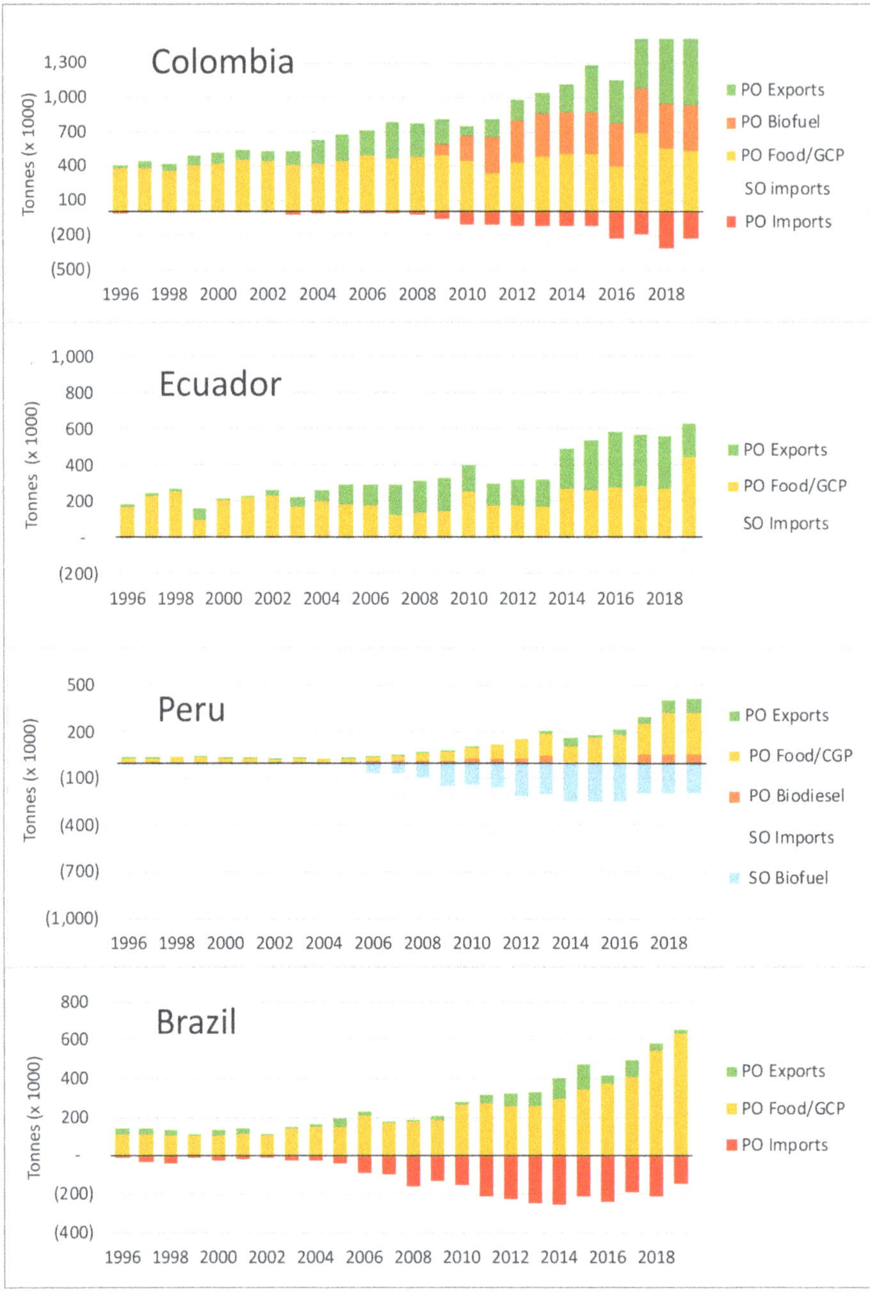

Figure 3.15: Trade flows of palm oil in four South American counties. Soy oil (SO) is shown for the Andean countries, where it is mixed with palm oil (PO) to ensure cooking oils remain liquid at the low temperatures common at high altitudes.

Data source: FAOSTAT

influenced how rapidly they have expanded and their ability to compete in national and global markets (Figure 3.15).

Colombian producers were making progress in penetrating overseas markets, but a disease outbreak in 2010 combined with drought conditions limited gains at a time when the sector was increasing plantation area in response to a national biofuel policy (Figure 3.14). The biofuel policy did promote the expansion of the industry and enabled the sector to export greater quantities overseas. Ecuador has a similar history, including periodic bouts with plant pathogens but, unlike Colombia, Ecuador's government did not embrace a biodiesel policy. Consequently, the palm oil sector has expanded by focusing exclusively on exports. Domestic consumption in both Colombia and Ecuador is flat; major export markets include Venezuela, the EU, Mexico, Chile and Brazil. Ecuador actually exports about thirty per cent of its production to Colombia.

Peruvian producers have not only failed to garner a significant export market but have also failed to reduce the country's dependence on imported soy oil. The macroeconomic incentives for expanding palm oil production in Peru would seem obvious to a casual observer since it would replace imported soy oil. The expansion of oil palm plantations in Peru is increasing and will be a significant driver of land-use change in Amazonian Peru over the medium term.

Brazil has enjoyed consistent growth of its palm oil sector, but domestic demand has far outpaced the ability of producers to meet supply (Figure 3.15). Brazil is the world's second-largest producer of soybeans, so palm oil must compete with soy oil for market share. For example, Brazil has a long-established policy of using biofuels as alternative energy sources, but very little palm oil has been allocated to the biodiesel market (see below).*

Brazil has a massive market for consumer goods, and many global brands manufacture their products in Brazil using derivatives from palm oil or palm kernel oil.† Apparently, the lack of domestic production combined with cheaper imports has motivated companies to source between twenty and forty per cent of Brazilian demand from overseas suppliers. Colombian and Ecuadorian imports represent about ten per cent of total imports, so the remainder must be coming from Southeast Asia.

The future Brazilian market for palm oil may be at a crossroads. The 100,000 hectares of new plantations established in Pará between 2010 and 2016 were intended originally to be used as feedstocks in an expanding biofuel industry, either for export to the EU (Belem Bioenergia Brasil) or to

* According to the Agência Nacional do Petróleo, Gás Natural e Biocombustíveis (ANP), Brazil consumed about 3.9 billion litres (3.5 million tons) of biodiesel in 2015, of which 80% was manufactured using soy oil. National production of soy oil in 2015 was approximately 7.5 million tons (FAOstat).

† Unilever, Procter & Gamble, Mars, Mondelez, Kraft-Heinz

defray the cost of diesel consumed by Vale's heavy machinery and railroad operations (Biopalma da Amazônia). However, the decline in demand for biodiesel may change this calculus, and this new production – which has yet to come fully online – might be targeted at Brazil's expanding consumption of traditional uses of palm oil.

Over the short term, there is considerable uncertainty regarding the future of the palm oil sector in the Pan Amazon. Production in Ecuador and Peru will probably continue to expand, but almost all this expansion will occur via smallholders and independent producers. In Colombia, the current government has changed the biofuel policies that contributed to the expansion of the sector over the last decade,[56] but a decline in domestic demand may be offset by increasing exports. Brazil has almost unlimited capacity for expansion, and government policy in the recent past has favoured the sector; however, recent expansion may have saturated domestic demand for the next several years. In all four countries, expansion will occur without deforestation by large-scale producers due to market pressure. Some small-scale deforestation will occur in Ecuador and Peru because smallholders are not subject to the same level of monitoring. That forest loss will occur via the loss of forest remnants or the gradual expansion of the agricultural frontier.

Biofuels

The spikes in commodity prices in the first decade of the twenty-first century coincided with a global boom in biofuels. The interest in biofuels was motivated allegedly by a concern for global warming but other, more mundane, reasons played a vital role. In advanced economies, this included a geopolitical urgency to lessen dependence on fossil-fuel imports from countries suffering political instability, combined with the political expediency of supporting domestic production of agricultural commodities. Well-known examples include the corn lobby in the United States and the rapeseed constituency in the European Union. Biofuel policies in both jurisdictions are now under pressure. Low oil prices have reduced demand for biofuels, while the unexpectedly rapid development of electric vehicles and renewable energy provide more attractive pathways for lowering greenhouse gases (GHG). The long-delayed operation of advanced biofuel production has caused several companies to declare bankruptcy and dampened investment in these once-promising technologies. Finally, concern about indirect land-use change and food security issues has motivated the EU to roll back subsidies and end policies that once favoured the biofuel industry.[57] The US is a net exporter of ethanol but imports about twenty per cent of its biodiesel consumption, which is largely based on soy.[58]

In South America, several countries have adopted biofuel policies. The motivation in these countries has little to do with climate change, being more obviously meant to advance rural development and boost export industries. Brazil, Colombia, Peru and Ecuador all have biofuel mandates that require distributors and retailers to blend a specified amount of biofuel with traditional fossil fuels. Gasoline is mixed with ethanol produced from sugar cane and maize, while biodiesel is a blend using derivatives of soy or palm oil or waste oil and fats collected from the food chain.

Brazil has a long history of promoting bioenergy, particularly ethanol from sugar cane, and the Brazilian sugar industry is renowned for its efficiency and low GHG footprint. Brazilian companies also lead efforts to produce cellulosic ethanol and most use waste biomass as a source of thermal energy to generate electrical energy that is monetised in domestic electricity markets.[59] Brazil is the world's largest producer and exporter of ethanol and has recently started to export biomass pellets from agricultural waste to North American and European utilities seeking to reduce emissions from coal-fired power plants.[60] An increasing number of corporations with landholdings in Mato Grosso are building thermal biomass plants with plans to install carbon-capture and sequestration technologies over the medium-term.[61]

The expansion in maize cultivation in Mato Grosso has catalysed its use as a feedstock for the production of ethanol. Manufactured in Brazil for the first time in 2012, maize-based ethanol has exploded as a business model with 1.4 billion litres produced in 2019/2020 and 3.2 billion litres projected for 2021/2022. Current production represents about four per cent of total national ethanol consumption, but is expected to reach twenty per cent by the end of the decade (8.0 billion litres).[62]

As of 2019, there were seven maize-based distilleries operating in Mato Grosso, three under construction and seven in different stages of planning.* A large plant, capable of manufacturing 500 million litres annually, requires a $US 100 million investment; it will generate gross revenues of ~ $US 200 million and create as many as 8,000 direct and indirect jobs.[63] The nominal value of maize-based ethanol production was approximately $US 1.3 billion in 2020/2021; about half can be attributed to Mato Grosso.

The economic impact of maize-based ethanol is maginified by fermentation byproducts with their own commercial value: corn oil and a solid residue known as DDGS (distiller's dried grains and solvents). Corn oil has chemical characteristics similar to soy and can be sold into vegetable oil markets; DDGS is even more lucrative because it is rich in protein

---

* As of 2020, there were four corporations operating in Mato Grosso: FS Fueling Sustinability (6 plants); USIMAT (2), INPASA (2) and ALD Bioenergia (1); source: União Nacional do Etanol de Milho, http://etanoldemilho.com.br/quem-somos/

and vitamins that makes it an excellent feed supplement for the livestock industry (see above). The development of a maize-based ethanol industry has resolved a crucial challenge for the industrial farm sector: surplus maize stocks. Unprocessed grain is non competitive in global markets due to high transportation costs, but as a biofuel feedstock it can be commerecialised in domestic energy markets, while corn oil and DDGS add value to meat products destined for overseas markets (see above).

The Brazilian government supports biodiesel production with a blend mandate of ten per cent. Most of the feedstock is soy oil, which in 2019 totalled 4.8 million cubic metres. This volume is equivalent to the oil content of ~18 per cent of the total national production and fifty per cent of the soy oil processed domestically within Brazil. Mato Grosso produced 26 per cent of that total. In contrast, < 2 per cent of the biodiesel mandate was supplied by palm oil, which represented less than six per cent of the total national harvest in 2019.[64] Most of that was produced by Brasil Biofuels, which blends it with conventional diesel to fuel its eighteen power plants located in Acre, Roraima, Rondônia and Amazonas.[65]

The decision to promote soy rather than palm oil as a biodiesel feedstock is noteworthy because the GHG footprint of Brazilian palm oil is significantly smaller than soy. The carbon emissions of palm oil are approximately fifty per cent less than soy when land-use change is excluded from the calculation; however, if palm oil is assumed to be deforestation-free and soy expansion occurs via the conversion of Cerrado landscapes, then the GHG footprint is about 140 per cent less for palm oil when compared to soy.* Palm oil produced on degraded pastures is carbon negative because both above and below ground biomass increases over time until it reaches an equilibrium after a couple of decades.

Presumably, the decision to give priority to soy over palm oil was driven by the pre-existing logistical infrastructure near the urban centres in Southeast Brazil and the limited capacity of palm oil companies to produce the volume of feedstock required by regulatory mandates.† Political

---

* The GHG footprint for palm oil versus soy, excluding land-use is 15 to 20 g CO2/MJ versus 27 to 59 g CO2/MJ. The GHG footprint for palm oil versus soy, assuming palm is cultivated in pastures and soy in converted Cerrado, is -35 to -60 g CO2/MJ versus 82 to 142 g CO2/MJ. Derived from emission factors in Stratton et al. 2010.

† Total Brazilian production of crude palm oil in Pará in 2014 was ~300,000 tons, while the total biodiesel demand was 3.4 million tons. Source: IBGE –Instituto Brasileiro de Geografia e Estatistica (2019) Sistema IBGE de Recuperação Automática – SIDRA, Produção Agrícola Municipal: https://sidra.ibge.gov.br/pesquisa/pam/tabelas and USDA-FAS / GAIN Report, Brazil Biofuel report (2016): https://apps.fas.usda.gov/newgainapi/api/report/downloadreportbyfilename?filename=Biofuels%20Annual_Sao%20Paulo%20ATO_Brazil_8-12-2016.pdf

influence may also have contributed to favouring the soy sector, which had gross revenues of about $US 40 billion in 2015 compared to about $350 million for palm oil.

Colombia designed its biofuel programme to support the domestic sugar cane and palm oil industries, two business sectors with influential constituencies that also generate tens of thousands of jobs in the rural economy. Biodiesel mandates increased production and now represents about a third of total production, which helped palm oil companies manage supplies impacted by disease outbreak and fluctuating export markets (see Figure 3.15). Producers hope to penetrate overseas biodiesel markets in the future by creating deforestation-free supply chains.

In Peru, the decision to adopt biofuel mandates was meant to fortify its rural economy and, simultaneously, reduce imports of vegetable oil, valued at about $US 500 million per year (see Figure 3.15). This did not have the desired effect, however. Instead, fossil-fuel companies imported soy oil from Argentina by ocean transport, which enjoyed a price advantage when compared to palm oil hauled over the Andes by truck. The price differential was so large that Grupo Palmas shuttered its biofuel refinery in Tocache in 2014 – only two years after its inauguration.

In Ecuador, the government elected to impose ethanol mandates, but declined to do the same for biodiesel. Apparently, the subsidised price for fossil-fuel diesel makes the implementation of a biodiesel mandate problematic because it would lead to an increase in the cost of diesel fuel at the pump. Producer groups have called for a policy that promotes biodiesel, but the enthusiasm for a biodiesel mandate varies among stakeholders: larger companies are active proponents, while the representatives of the smaller producers are less enthusiastic. The conventional logic for opposition to a biodiesel mandate is hard to understand given the obvious macro-economic benefits of a domestic biodiesel market and the example of Colombia, which has used the domestic biodiesel market as a mechanism to expand both production and exports.[66]

In Bolivia, the government of Evo Morales rejected all biofuel policies that would have diverted revenues from the state-controlled oil company to the agricultural sector, which is dominated by private companies linked to the political opposition. This may be about to change, however; in 2021, the newly elected government of Arce Catacora announced a plan to develop a biodiesel refinery. The proposed $US 250 million investment would alleviate the country's reliance on imported diesel.* Unlike the biofuel mandates in Colombia, Peru and Brazil, which are based on blends of fatty acid me-

* Bolivia imported approximately $US 500 million in conventional diesel in 2020 during a time of record low oil prices. Government subsidies of about 50% have been maintained in order to ensure civil peace among its conflictive transportation sector and to support its agrobusiness sector. See Laserna 2018.

thyl-ester (B5, B10, B20), the proposed refinery would convert vegetable oil to pure biodiesel (B100).[67] The project would require an additional 250,000 hectares of soybean plantations, which historically have been created by deforestation. The government expressed a commitment to use other feedstocks, such as waste animal fat, or to invest in the industrial production of a Macauba palm, a native species with commercial potential (see Chapter 7).

Coffee and Cacao*

Coffee and Cacao have much in common. Both are descended from understorey trees adapted to the low-light conditions of the forest floor. Each have multiple cultivated varieties that differ with respect to quality, as defined by the aromas and other phytochemicals that act as flavour enhancers and stimulants. Both are perennial cash crops that require unskilled labour at harvest and a certain amount of technical proficiency for post-harvest processing. The basic commodity of commerce is a seed, referred to as a bean, while the fruits are subject to fermentation to facilitate the collection of beans, which are washed, dried and bagged for transport and sale.

Unlike oil palm, the post-harvest processing of coffee and cacao do not require a large capital-intensive industrial facility. This is important because it allows small farmers who reside in remote villages to process their own production and transport it to the nearest logistical centre. Like all artisanal systems, there are cultivation practices and processing procedures that influence the quality of the product. A combination of these factors leads to the production of elite coffees and fine cacaos, which have niche markets that impact the prices paid to farmers.

Both cacao and coffee have been implicated in the loss of natural forest habitat, most of which is associated with smallholder production systems.[68] The proximate cause of this deforestation is a desire to expand production or a need to replace existing plantations that have become infested with pests, or have lost vigour due to age. These are crops common to the agricultural frontier, where farmers' *modus operandi* is to occupy and clear forest to establish new plantings. Several of the most popular varieties of both coffee and cacao are adapted to full sunlight; in these cases, the plot is completely deforested prior to planting. Some varieties of both crops grow better under shade, which motivates landholders to expand production into forest habitat to take advantage of the canopy trees. Although this is less damaging than clearing the forest of all its biomass, it is still a form

* The term cocoa is often used as a synonym for cacao, but cocoa actually refers to a processed product (cocoa powder) that is derived from the cacao bean. Using the term cacao also avoids confusion with coca, the plant whose leaves are used to make cocaine.

Both coffee and cacao are small trees adapted to the conditions of the forest understory. Coffee cherries (drupes, in botanical terminology) are harvested once a year by hand when ripe (top). Cacao pods (fleshy capsules) develop in an unusual manner, being born on the stems rather than on the branches, and maturing gradually over several months (bottom).

of cryptic deforestation and contributes to the loss of biodiversity and ecosystem services.[69] As trees mature and production declines, farmers tend to remove shade trees in order to maintain yields over the short term; eventually, the land is converted into some other land use, typically pasture for cattle and dairy

Cacao is a labour-intensive crop that requires an artisanal, post-harvest fermentation process that makes it a good option for small farmers. Both coffee and cacao are often promoted as alternatives to the cultivation of illicit coca, in part because they fetch a decent price, but also because their smallholder production system is similar to coca leaf. Numerous initiatives have sought to promote coffee and cacao as development options on conflictive landscapes, and most of these initiatives have also sought to avoid new deforestation by offering technical assistance. Unfortunately, these efforts have not been particularly successful, either in eradicating illicit coca or in avoiding deforestation.[70]

The cultivation of cacao and coffee has expanded on some, but not all, of their traditional landscapes in the Amazon; in some instances, production has declined (Figure 3.16 and Figure 3.17). Changes in crop area occur largely in response to market demand that is determined by conditions in other parts of the world, either by weather events or structural challenges that are motivating commodity traders to diversify their supply chains. In the case of declining coffee production in Ecuador, overseas competition has caused producers to abandon a long-established cultivation system in favour of other crops. The increase in supplies from South America has been accompanied by renewed attempts to delink the expansion of coffee and cacao from new deforestation;[71] these initiatives are taking advantage of new subsidies linked to climate-change programmes or improved pricing mechanisms linked to certification systems that support farmers who embrace the concepts of sustainability.[72]

Coffee

There are two major cultivated species of coffee: arabica (*Coffea arabica*) and robusta (*Coffea canephora*), each with a multitude of varieties adapted to a broad range of ecological conditions. Arabica represents seventy per cent of global production, while robusta represents about thirty per cent. Traditionally, arabica has been cultivated as 'shade coffee' grown at higher elevations, while robusta is 'sun coffee' cultivated at lower elevations. There are exceptions, including arabica varieties grown without shade at higher elevations and with shade at lower elevations. Both species of coffee are cultivated in the Amazon. Elite coffee varieties tend to come from arabica trees grown under shade at optimum altitudes, which vary by latitude but range between 750 and 1,500 metres above sea level. Harvesting and post-harvest processing practices are also important for maintaining quality in elite coffees.*

In the Andes, arabica is the predominant coffee cultivar. Colombia is third largest producer of coffee and is well known for its high-quality elite arabicas. However, most Colombian coffee is grown in the Magdalena watershed, while production in Venezuela is concentrated in the mountains overlooking the Caribbean coast. The largest producer of arabica in the Amazon is Peru (HML #37, #38, #42, #43, #45), followed by Bolivia (HML #33) and Ecuador (HML #48). Robusta was once widely cultivated in Amazonian Ecuador (HML #49 and #$50), but after about 2000, Ecuadorian smallholders began shifting production to oil palm or cacao.† Robusta is cultivated sporadically in lowland Peru (HML #40, #41, #44, #46) and Chiquitania Bolivia (HML #29).

In Peru, coffee is commonly cultivated on small farms that also produce food crops and livestock; more than 85 per cent of production originates on 150,000 family farms with coffee groves smaller than five hectares. Coffee is a cash crop harvested once a year over several weeks. Family labour is key to its success because it allows small farmers to absorb the fluctuations of price volatility; however, it also limits their ability to expand production. Commercial farms cultivating larger extensions obtain better yields, but since they rely on contract labour, they are also exposed to greater price risk from international markets. Net income varies depending upon production strategy but gross annual revenues between $US 1,000 and $US 1,500 are possible with yields of about 700 and 900 kilograms per hectare. The coffee sector exported about $US 670 million in 2016, roughly two per cent of total Peruvian exports.[73] According to the United States Department of Agriculture, coffee generates some 855,000 jobs in Peru, the vast majority of which are on-farm jobs performed by family members.[74]

The arabica coffee plant grows in a climate zone subject to weather fluctuations that can greatly impact yield and, in some cases, lead to large-scale die-off of plantations. When this type of event hits one of the major producing countries, particularly Brazil, price hikes can dramatically im-

---

* Management of the roasting and blending processes is essential to ensure consistent quality and these are done at the end of the supply chain, because roasted and ground coffee loses aroma and flavour over time. Consequently, the supply chain in producing countries tends to end with the commercialisation and export of green coffee beans.

† This collapse of the robusta coffee sector in Ecuador coincided with phenomenal growth of the Vietnamese coffee industry, which expanded by an order of magnitude (10 x) between 1990 and 2000; yields in Vietnam average about 1.5 tons per hectare versus about 300 kg per hectare in Amazonian Ecuador.

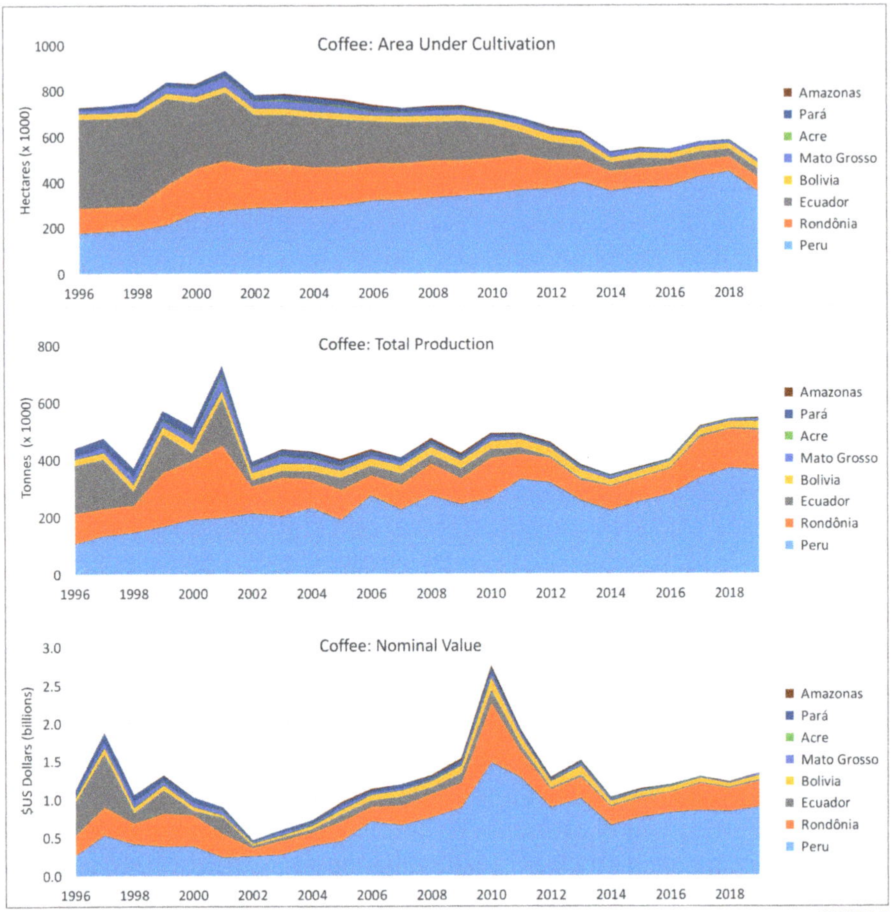

Figure 3.16: Coffee production trends in Amazonian territories of four countries. Cultivated area is decreasing in Rondônia, Brazil, but yield increases have kept production relatively stable. Ecuadorian producers have abandoned the cultivation of [robusta] coffee in Ecuador's Amazonian provinces, where growers are switching to cacao and oil palm. Price volatility linked to weather events in other regions (mainly Southern Brazil) causes periodic windfalls for Amazonian and Andean producers.

Data sources: FAOSTAT and IBGE/SIDRA.

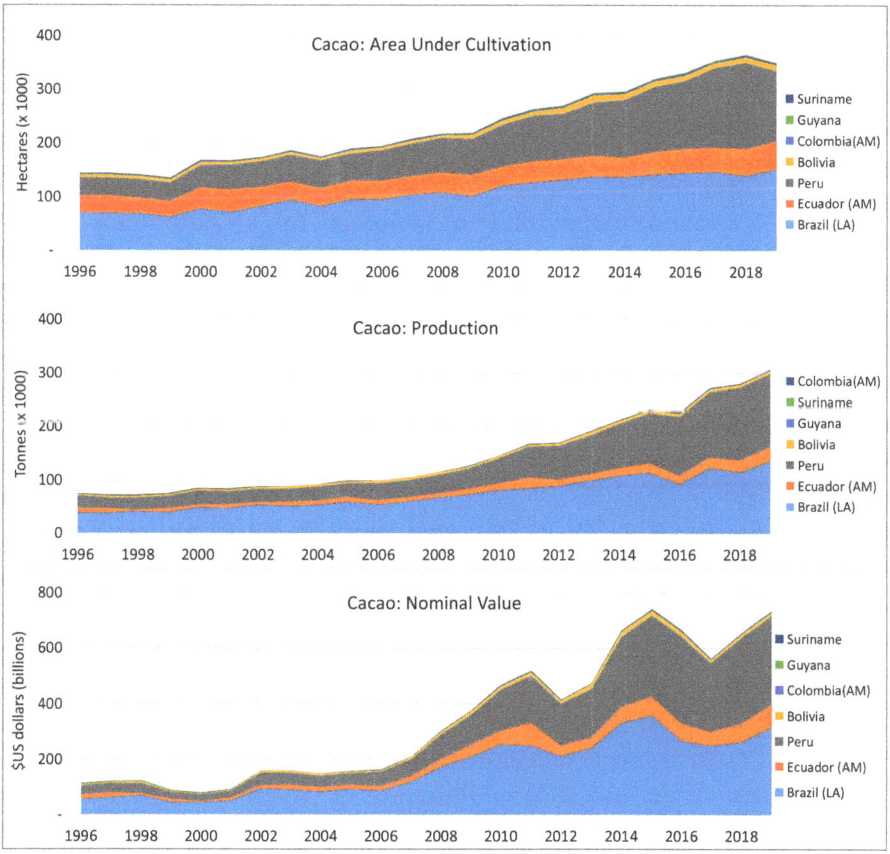

Figure 3.17: Cacao cultivation has been increasing in all jurisdictions; the lower production in Ecuador reflects Amazonian growers' predilection for low-yielding but high quality varieties. Variation in international commodity markets impacts annual revenues. The graphic excludes coastal producers in Ecuador (~90% of national production) and Bahia, Brazil (~75% of national production).

Data sources: FAOSTAT and IBGE/SIDRA.

prove incomes.* Many Peruvian coffee farmers go years with only marginal rates of return, but a price spike will provide a windfall that justifies what is essentially a long-term investment (Figure 3.16). Producers may lose their investment due to local weather events, such as drought or exceptionally humid years, or a disease outbreak. Coffee plantations across the Americas suffer from a coffee rust known as *la roya*, a fungal pathogen that can quickly devastate a plantation. Peru suffered from an outbreak of coffee rust in 2012/2013, which reduced the harvest and forced growers to replant infected groves with resistant varieties.†

Conventional and traditional producers in Peru represent about seventy per cent of total production and typically do not participate in certification programmes, choosing instead to emphasise yield, maximise income over the short term and reduce risk from plant pathogens. Entrepreneurial growers who adopt shade or organic practices tend to participate in growers' associations in order to improve market access via a certification programme. Shaded, organic, certified beans pay a premium of about ten per cent when compared to traditional coffee beans. These growers often obtain assistance from NGOs or alternative development programmes.[75]

The Brazilian coffee industry generates about 25 per cent of global coffee production. Arabica is the main coffee crop, with production in the states of Paraná, São Paolo and Minas Gerais contributing about eighty per cent of national production, all of which is outside the Amazon. In Amazonian Brazil, production is concentrated in Rondônia, where robusta varieties are cultivated by about 22,000 smallholders, each with plots between four and ten hectares in size. It is also cultivated in Acre and in the smallholder communities of Northern Mato Grosso. Referred to by Brazilians as *conilon*, this non-elite coffee is an important part of the smallholder production model in Rondônia, contributing about $US 150 million annually in gross revenues. The total area under cultivation has declined over the last couple of decades from more than 200,000 hectares in 2005 to less than 95,000 hectares in 2016; however, yields climbed from 550 kg/ha to 1.2 tonnes/ha over the same period (Figure 3.16).

According to agronomists with the Brazilian extension service, *Empresa Brasileira de Pesquisa Agropecuária* (EMBRAPA), the potential yield for robusta varieties in Rondônia is ~4 tonnes/ha when grown with irrigation and optimum nutrient management.[76] Depending upon the price, which

* Large scale frosts or extended drought have decimated coffee plantations in Southern Brazil on 10 different occasions between 1900 and 2000; the 2016 harvest was also impacted by an extended drought. Source: http://www.coffeeresearch.org/market/frosthistory.htm

† According to the Peruvian ministry of agriculture, replanting coffee plantations decimated by coffee rust has reduced infestation by 50% of all trees to only 17% between 2013 and 2016.

© hack78/Shutterstock.com

© Martin Nabert/Shutterstock.com

Processing coffee is an artisanal activity that is performed on the farm and requires only a moderate amount of infrastructure and know-how. Coffee beans are separated from the fruit (top) and then dried in ovens or in the sun (bottom).

varies greatly from year to year, a small farmer in Rondônia can generate revenues of between $US 1,100 per hectare to more than $US 6,300.* The feasibility of this production system depends partially on family labour, though less so than in the Andes.

The global coffee market has experienced significant changes over the last decade. Global demand for mass-market coffees has increased by an order of magnitude (x 10) due to changes in consumer preferences in traditionally tea-drinking nations such as China. Simultaneously, coffee-drinking counties, like the US, have increased consumption of elite coffees. Peru and the other Andean countries are focusing on the elite coffee market, and many producers are adding value to their production by embracing certification and organic production paradigms. These countries have large areas of idle land located on landscapes with current growing conditions capable of producing premium coffee beans that are highly competitive in global markets. Nonetheless, climate change threatens the long-term viability of existing plantation landscapes, and growers may be forced to 'migrate upward' into intact montane and cloud forest habitat.[77] If so, that would provoke widespread deforestation, the main driver of which would be climate change in combination with the ongoing demand for elite coffees from arabica varieties grown under ideal climate conditions.

Cacao

Cacao (*Theobroma cacao*) is native to the Amazon rainforest and has been cultivated and consumed throughout the Americas since before Columbus. Cacao can be broadly segregated into two contrasting types based on quality: bulk cacao is used for most candy and food products, and fine cacao is preferred for speciality chocolates. There are dozens of varieties, strains and hybrids, but these two major types have dominated production and trade for centuries. The supply of bulk cacao was largely diverted from Latin America to West Africa and Southeast Asia during the colonial period of the late nineteenth century. Fine cacao represents only about five per cent of global cacao consumption, but almost all of that originates in Latin America where the genetic diversity of the wild species has been used to improve the concentration of aromatic compounds in the cacao bean. A combination of events has stimulated a revitalisation of cacao production in South America, and the production of both bulk and fine coca has been increasing at about ten per cent annually over the last decade (Figure 3.17).

The traditional method for establishing a plantation is to clear the understorey of a natural forest and plant cacao seedlings that were germinated in a nursery or directly under an intact canopy. It takes about four years for young trees to flower and fruit, after which light management

* Yield = 0.5 t/ha @ $US 2/kg versus yield = 1.2 t/ha @ $US 2.6/kg.

is important for maximising production and quality: too much light and plants will suffer stress; too little and yields decline. An individual tree can live decades, but most commercial plantings are programmed to last about twenty years due to declining yields. One of the practices used to improve or prolong productivity is to open the canopy to increase light and stimulate photosynthesis, which will increase yields over the short term. Eventually, productivity will decline, however, and groves are converted to pasture or allowed to lapse into a forest fallow. Cacao farming as traditionally practised is a form of slow-motion deforestation.

Traditional cacao farming has contributed to the devastation of the tropical forests of West Africa, which provided about sixty per cent of global demand throughout the twentieth century. As forest habitat has become increasingly scarce in West Africa, cacao growers have adopted full-sun production systems to maintain yields. Many analysts predict that these practices will lead to a permanent decline in West African cacao production.[78] Not surprisingly, global commodity traders and food companies have been seeking alternative sources of cacao, and that is one reason production in the Amazon has been on the increase (Figure 3.17).

Brazil produced more than 275,000 tonnes of cacao in 2014, mainly from Bahía (58%) and Pará (42%), particularly in the municipalities located on the north bank of the Amazon River (HML # 1) and the Transamazônica highway (HML #10), where yields are almost double those of traditional cacao growers in Bahía. The popularity of cacao in smallholder landscapes is expected to grow over the short-term, and the *Associação Nacional das Indústrias Processadoras de Cacau* (AIPC) has pledged to double production over the next ten years.[79]

Ecuador produced approximately 160,00 tonnes in 2016 with about 85 per cent cultivated in the provinces of the coastal Pacific and the remainder originating in the Amazon (HML #49 and HML #50). Most of Colombia's production occurs in non-Amazonian watersheds, but its Amazonian provinces have the required climatic and edaphic characteristics. Caquetá has about 400 growers with about 1000 hectares (HML #52), which is less than two per cent of national production; efforts to increase production are linked to efforts to displace illicit coca. Venezuela produces cacao but not in the Amazon, while Bolivia's limited production is largely targeted at domestic consumption, except for an association of growers in the Yungas of La Paz (HML #33)

Ecuador has a special market niche because its growers supply about seventy per cent of the fine cacao commercialised in global markets. The source is a local variety known as *Nacional,* which has been cultivated on Ecuadorian farms for decades. It was recently christened *Sabor Arriba* by the Ecuadorian association of cacao exporters. Ecuador's position as a provider of fine cacao is being challenged, or at least modified, by the increasing popularity of a hybrid cacao variety known as CCN-51, which has gained market share because of its robust growth and high yields (see Text Box 3.5).

There is concern among chocolate connoisseurs and export companies that CCN-51 may displace *Nacional* and compromise Ecuador's dominance of the specialty cacao market. It is preferred by farmers because *Nacional* trees yield only about 300–400 kg/ha compared to 700–1,100 kg/ha for CCN-51, which can yield three tonnes per hectare when cultivated in experimental stations under ideal conditions. The increased productivity of CCN-51 is the result of its genetic makeup, which controls the size and number of the fruits per tree and seeds per pod. It can be cultivated in full sun, has a lower rate of disease infestation, and responds better to agrochemicals. These attributes cause concern among environmental advocates who criticise its genetic narrowness as a clone and cite its potential to become another monoculture production system.

Like all commodities, price varies enormously but, on a per hectare basis, cacao producers' revenues will vary from as low as $US 500 to as much as $3,000.* Smallholders in Amazonian Ecuador typically cultivate only between one and five hectares of cacao, while commercial producers on the Pacific Coast might have as much as fifty hectares. Certified production of fine cacao delivers a price premium to the farmer of about twenty per cent, but this is not sufficient to offset the lower yields of *Nacional* trees. Currently, CCN-51 contributes about 36 per cent to Ecuadorian national production compared to 37 per cent from *Nacional.* Amazonian producers largely produce a third type, known as *Cururay*, which is similar to *Nacional* in both yield and cocoa characteristics.[80] Production in Amazonian Ecuador is static (see Figure 3.17), but efforts to expand production are underway as part of initiatives to promote alternatives to deforestation-based agriculture.[81]

In Peru, a similar phenomenon is occurring with respect to CCN-51, which is making inroads among both established and new producers. The area under cultivation has approximately doubled between 2005 and 2015, with particularly strong growth on landscapes that are the focus of 'alternative development' programmes linked to the war on drugs. Apparently, these programmes have been proponents of the cultivation of CCN-51 due to its superior yield and the need to compete with coca cultivation (see below). In 2015, more than 53 per cent of all cacao groves were planted in CCN-51 with only about 44 per cent dedicated to fine cacao, which apparently includes a mixture of introduced *criollo* cultivars and native varieties derived from Amazonian wild populations.[82]

Production has grown annually by about eight per cent, reaching more than 85,000 tonnes in 2016 produced on about 110,00 hectares. The major production area is San Martin (HML #42 and HML #43) with 31 tonnes,

* Estimates based on 350 kg/ha @ $500/ton and 1000 kg/ha @ $300/ton.

© hack78 / Shutterstock.com

© William Borney / Shutterstock.com

Ecuador is renowned for its fine cacao, most of which is harvested from the variety known as Nacional (top). It has lost some market share, however, to CCN-51 (bottom), a hybrid developed by a plant breeder in the 1990s, which is inferior in quality but has better yield.

Text Box 3.5: The origin of CCN-51

CCN-51 was developed by an Ecuadorian horticulturist, Homer Castro, in the 1960s, when he was leading a plant-breeding programme to develop genetic resistance to a common plant pathogen, *mal de machete*, which causes plants to wilt when their vascular system becomes infested with fungal tissue.

Genomic studies of more than 900 accessions of cacao held by plant breeders revealed that the CCN-51 hybrid is a descendent of wild Amazonian seed collected near Iquitos (45% of the genome) and two common cultivars: Criollo (22% of the genome), which is the leading type of fine cacao in Central America, and Amelonado (21.5% of the genome), which is similar to Forastero and the source of about 80% of global bulk cacao. The remainder of CCN-15 genome is composed of other genotypes collected from wild populations in the upper Amazon, including Contamana (~4%), Purús (2.5%) and Marañon (~2%).

In other words, CCN-51 is a hybrid that combines the genetic attributes of both bulk and fine cacao; however, it is not considered to be a fine cacao due to the chemical properties of the beans. Many chocolate gourmets consider CCN-51 to be of extremely low quality and its flavor has been described as: 'weak basal cacao with thin fruit overlay; lead and wood shavings; astringent and acidic pulp; quite bitter'.

Source: Motamayor et al. 2008

which represents 44 per cent of total national volume; these tend to be new producers, and they are overwhelmingly adopting CCN-51. Other regions include landscapes near Pucallpa (HML #41), Huanuco and Junín (HML #40) and La Convención (HML #35). *Criollo* and native varieties predominate on these landscapes, but CCN-51 contributes between twenty and 35 per cent of the harvest.[83] Prior to 2000, most cacao production was dedicated to meeting domestic demand; subsequent growth can be attributed to exports, which accounted for more than two-thirds of total production in 2016 with a value of approximately $US 200 million. These proceeds are distributed among 26 growers' associations representing approximately 30,000 families, with mean gross revenues of between $1,000 and $1,500 per hectare, depending upon price.

An experiment in corporate cacao production may be underway courtesy of a controversial development located near Tamshiyacu, a village located about forty kilometres upriver from Iquitos, the capital city of Loreto Department in Amazonian Peru. This industrial-scale plantation was established by an entity knows as United Plantations, a subsidiary of the *Grupo Melka*, the same corporation that attempted to establish large-scale oil palm plantations in Ucayali Department (see above). The more than 2,400 hectares of primary forest that were cleared were intended, apparently, to

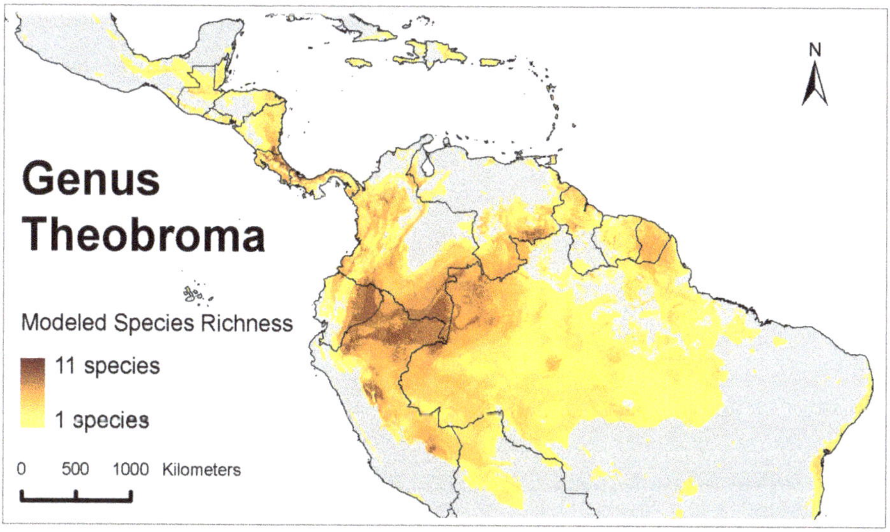

Source: Thomas et al. (2012) CC BY 4.0

Figure 3.18: The upper Amazon is a centre of diversity for the genus Theobroma. This genetic diversity is the source of the region's fine cacao, which is highly prized by chocolatiers.

establish what would be the first industrial-scale cacao plantation in the world.* Fortunately, the legality of this deforestation has been challenged in court and activities have been paralysed. The *Grupo Melka* has sold its assets and abandoned its investments in Peru.

Despite the predominance of CCN-51, the association of Peruvian cacao exporters is seeking to improve its market position by emphasising the quality of its fine cacao offerings. The motivation is driven in part by a commercial logic to provide a niche product with a price premium, but it is also influenced by a long-held desire to monetise the value of a biodiversity asset of the Amazonian biome. The Western Amazonian is a centre of genetic diversity of cacao and the evolutionary origin of the genus (Figure 3.18).[84] Ecuador and Peru hope to capitalise on the genetic diversity of their wild populations to develop new cultivars that combine both the improved aroma characteristics of fine cacao and the disease resistance and superior yield attributes that make CCN-51 popular with growers.

---

* This may not be an isolated case, however, as palm oil companies in Indonesia, which is the world's third largest producer of bulk cacao, are allegedly seeking to diversify their income stream by establishing cacao plantations. Key to the success of these initiatives is the prospect of using the CCN-51 cultivar, which under ideal conditions can produce up to 3 tons of cacao seed per hectare.

Local and National Food Crops

Many of our staple food products are highly dependent on industrial agriculture and global supply chains, but small farmers in the Amazon continue to produce foodstuffs for their families, local communities and national markets. This is particularly true in the Andean nations where poverty and a strong cultural tradition of subsistence farming influence both land use and food production. Most of the small farmers in the Andean Amazon are migrants from the Andean highlands where landholdings are often extremely small (see Chapter 6).

An economic production system based on small farms accompanied these migrant communities to the Amazon lowlands; although holdings are small, typically between ten and fifty hectares, they are an order of magnitude larger than what migrants were accustomed to in the villages of the High Andes. The opportunity to acquire land and grow food is the primary driver of deforestation in Andean countries; in all these countries, legal systems exist that allow individuals to homestead public lands and acquire tenure if they occupy and work the land. For many families, this is one of few viable pathways out of poverty.

Small farmers in the Andean Amazon grow a diversity of crops for household consumption and for sale to local, national and global markets. The use of technology varies but, like small farmers across the globe, they depend on family labour. This includes clearing the land, preparing the soil, sowing seeds or transplanting individual plants, weeding and pest control, as well as harvest and post-harvest activities. Some crops have labour-intensive stages when the family may contract outside labour, especially successful small farmers who have consolidated multiple small plots into a larger family landholding.[85]

Many food crops are annual species planted immediately after forest clearing, which may be natural forest on forest frontiers or second-growth forest in areas with a longer history of settlement. Common annual species are rice, cassava, maize, yams, beans and peanuts; farmers also plant fast-growing perennial crops such as sugar cane, plantains, bananas and papaya, while investing in longer-lived tree species such as mangoes, avocados and citrus. Sometimes these are planted simultaneously with perennial cash crops targeted at overseas markets, such as cacao, coffee and oil palm (see above). Over the short term, food crops produce essential resources that families need to survive, but it requires a constant effort to maintain production using the slash-and-burn/forest-fallow system. In contrast, perennial plantations produce for ten to twenty years, or longer, although they require up to five years to start producing revenue. Importantly, neither the perennial nor annual production models are profitable without the benefit of family labour.[86]

© guentermanaus/Shutterstock.com © nomadkate/Shutterstock.com

© Olga Kot Photo/Shutterstock.com © Sven Schermer/Shutterstock.com

Small farmers tend to grow foodstuffs for domestic markets, including manihot (top left), plantains (top right), rice (bottom left) and tropical fruits (bottom right).

For a recent arrival to the frontier, land clearing provides a positive cash flow from the production of food crops that can be consumed and/or sold into national or local markets. Maize production may be used to raise swine and poultry, which adds value to the farmer's primary production while contributing protein to the family diet. The effort to create a long-term cash flow from cacao, coffee or palm oil is cost-effective, but these will not turn cash-flow positive until the fourth or fifth year. More importantly, perennial plantations generate a windfall if they are producing during a periodic price spike, which tends to occur in decade-scale intervals (Figure 3.16 and Figure 3.17).

Established farmers have greater flexibility than migrants because they can discount the capital costs of land acquisition and deforestation as 'sunk costs', which makes the perennial production models more attractive on long-established landholdings. Moreover, an established farmer can use contract labour to expand his or her landholdings, since the return from an annual food crop will largely offset the cost of expansion. The most logical option is to pursue both strategies: invest in existing properties via perennial cash crops while growing food crops by expanding into unoccupied or inexpensive land (see Chapter 4). Combining family labour with contract labour can maintain a positive cash flow over the short term while acquiring a real estate asset that will accrue value over time. This includes more land, cleared land and land with a perennial production system.

For both the recent arrivals and the long-established small farmer, speculation in land represents the primary means for generating wealth. More wealth is created via perennial production systems when compared to annual crops, but annual crops initiate the cycle by allowing new immigrants the opportunity to acquire land, while providing established farmers the opportunity to expand their holdings. This logic and cycle hold true in most of the major smallholding landscapes in the Andean Amazon, although the relative mix of annual and perennial crops varies among geographies.

Coca – The Anti-Development Crop

The most lucrative agricultural system in the Amazon is neither soy nor palm oil, but coca leaf, which is cultivated for both legal and illegal markets. There are two species, *Erythroxylum coca*, which is grown at higher elevations and is preferred for the legal market, and *Erythroxylum novogratenensis*, which is cultivated at lower elevations and is the primary feedstock for the manufacture of illicit cocaine. Historically, coca was cultivated in the montane forests of the Eastern Andes in Bolivia and Peru, where the leaf is consumed as a mild stimulant via infusions or mastication. The consumption of legal coca leaf has grown steadily over the past several decades, as it has

Coca leaf (top) is cultivated to produce a variety of consumer products as well as illicit cocaine. It is cultivated legally in the Coroico municipality of Bolivia (bottom).

become a habit adopted by consumers inside and outside of the Andes. The consumption of illegal cocaine has boomed since the 1970s when it became a popular drug among the urban elite in North America and Europe, a habit which become democratised and globalised as cocaine consumption spread to populations in other economic strata and social groups across the world.

Coca farmers are the smallest of the commercial farmers in the Amazon; a legal coca plantation in Bolivia is a 40 x 40 metre plot of land referred to as a *cato de coca*. Individual plantings can produce for years, if not decades but, since the overwhelming majority of plantations are illegal and subject to eradication efforts, most coca plantations are probably younger than five years old. A well-managed coca plantation can produce up to two tonnes of dried coca leaf per year; as with any crop, harvesting a significant amount of biomass will deplete the nutrient content of the soils, which is another incentive to constantly move and renew coca plantations.

Starting in the mid 1990s, Colombia surpassed Peru and Bolivia as the primary source of coca leaf, a transition that coincided with an increase in the civil unrest in Colombia and the initiation of interdiction efforts in Peru and Bolivia that were financed by the United States and the European Community (Figure 3.19).[87] Subsequently, Colombia increased its efforts to combat cocaine trafficking; according to recent reports, the sum total of coca was reduced to about 50,000 hectares by 2014, of which about 25,000 hectares were located in the four departments located with the Amazon (Table 3.2).

Cultivating coca is legal in both Bolivia and Peru due to the traditional market for coca leaf; however, its conversion to cocaine or its precursors is widespread. Efforts to promote alternative production schemes have met with only limited success, mainly because of the economic advantages of cultivating coca. One hectare of coca leaf will produce about $US 5,000 to $US 7,000 of income based on the mean yield and market value of the leaf as paid to the farmer; a similar area of any of the other cash crops generates revenues of about $US 1,000 to $US 1,500. The economics of coca production reflect market demand, which ensures that somewhere, somebody will grow coca leaf for conversion to illicit cocaine.

If the reported area under cultivation is accurate, this translates into between $US 300 to $US 500 million in Peru, with slightly lower values in Bolivia and slightly higher values in Amazonian Colombia. However, these are only the farm-gate proceeds that flow directly to the farmer, and the total value to the national economy is many times greater when the post-harvest processing and commerce are considered. According to the UNODC,* the full

* United Nations Office on Drugs and Crime.

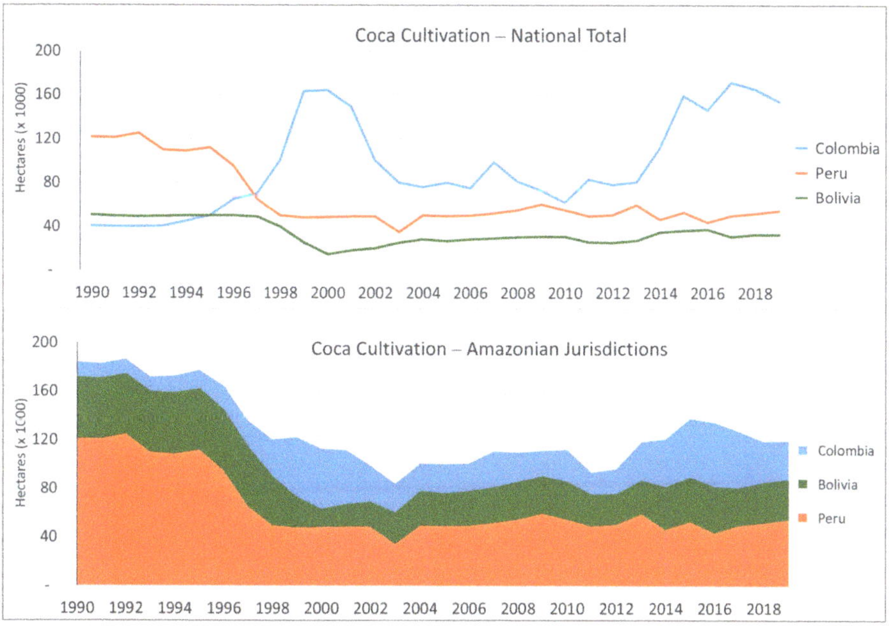

Figure 3.19: Coca cultivation between 1990 and 2019. There was a large displacement of coca cultivation in the 1990s, following the defeat of the Marxist guerillas in Peru and the decision by Colombia's armed militias to embrace illicit drugs as a source of revenue. The spike in coca cultivation after the beginning of the peace process in 2015 was more pronounced in the extra-Amazonian provinces of Colombia than within the Colombian Amazon.

Data source: United Nations Office on Drugs and Crime (UNODC).

coca-cocaine supply chain contributes about 0.9 per cent to Bolivia's GDP, which would place the total value of the supply chain in Bolivia at about $US 4 billion in 2019, with estimates for Colombia and Peru at around $US 8 billion each. Presumably, this does not include the proceeds from 'money laundering', which acts as a subsidy to other sectors of the economy. For example, in Bolivia the construction sector is used to turn illicit proceeds into real estate assets, because buildings can be constructed with cash and sold through banks via mortgages. In Bolivia and Colombia, unusually large investments in cattle ranches and industrial farms are commonly assumed to be financed in part by financial resources of dubious provenance.

Table 3.2: The major coca producing regions in the Andean countries.

HML #*	UNODC hotspot	Annual deforestation (ha)	Total anthropic area (ha)	Coca fields (ha)	% Coca/ anthropic	% Coca/ deforestation	Revenues ($US million) @ $6k/ha	@ $10k/ha
32	Chapare	15,000	392,100	8,720	2.2%	93%	52.3	87.2
33	La Paz Yungas†	7,007	276,900	17,390	6.3%	248%	104.3	173.9
35	Interoceanic Corridor	16,786	691,900	5,514	0.8%	33%	33.1	55.1
35	La Convención-Lares*	--	--	13,090	--	--	62.7	104.5
35	VRAEM	1,757	45,500	19,247	42.3%	1095%	115.5	192.5
40	Selva Central	23,514	970,300	3,235	0.3%	14%	19.4	32.4
44	San Martin & Yurimaguas	23,993	1,583,300	5,777	0.4%	24%	34.7	57.8
45	Marañon	3,457	596,300	3,250	0.5%	94%	19.5	32.5
51	Putumayo	3,400	330,800	8,432	2.5%	248%	50.6	84.3
51	Caquetá	20,486	1,726,400	4,093	0.2%	20%	24.6	40.9
54	La Macarena	9,779	639,400	3,339	0.5%	34%	20.0	33.4
54	Guaviare	8,621	364,700	5,355	1.5%	62%	32.1	53.5

*HML: human modified landscape (see Chapter 1, Figure 1.2)
† Coca plantings are legal in these regions.
Source: UNODC (Bolivia, 2019; Colombia, 2019; Peru, 2017) [88]

Comparing the deforestation data with the coca monitoring data compiled by the United Nations reveals multiple patterns of coca production. In some, coca plantations are located in areas with a centuries-long tradition of coca cultivation; these include the La Paz Yungas (HML #33) of Bolivia and the La Convención-Lares region near Cuzco, Peru (HML #35). Coca farming in these regions seems to be stable and practised openly on landscapes that have been settled for decades or centuries. They are legal because they are assumed to be producing coca leaf for domestic consumption; nonetheless, the vast majority of coca leaf is channeled into the illegal production of cocaine. In other areas, such as the Chapare of Bolivia (HML #32), the VRAEM (HML #35b) and all the regions in Colombia (HML #51 and #54), the cultivation of coca is a more recent phenomenon and is entirely illegal. In all localities, however, new deforestation patches can be observed in the remote areas and, typically, are less than one hectare in size and are isolated from villages and roads (Figure 3.20).

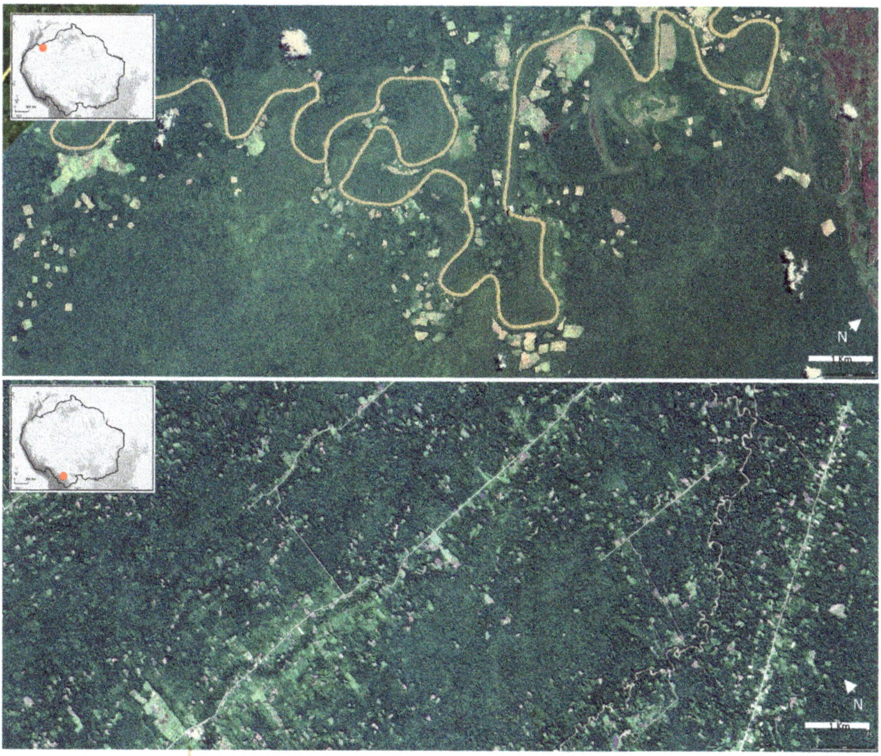

Source: Google Earth

Figure 3.20: Deforestation pattern typical of coca-producing landscapes: (a) Guaviare, Colombia, where rivers are the primary source of transit; and (b) the Chapare region of Bolivia where successive governments have supported or tolerated coca cultivation.

Recent press reports indicate that coca cultivation in Colombia has skyrocketed since 2016 from a historic low of about 48,000 hectares to more than 170,000 hectares because of the peace process and the policy of tolerance that was implemented during the negotiations. In the Caquetá- Putumayo region (HML #51) the coca cultivation area increased by an average of forty per cent per year between 2014 and 2017. Ironically, this would make the peace process a driver of deforestation.

Roundtables and Certification Schemes

Sustainability initiatives have been organised for most of the agricultural commodities of the Pan Amazon, including palm oil, soy and beef, but also for coffee and cacao. Several of these initiatives have adopted the term 'roundtable' in their names because it conveys the notion of inclusiveness that is a core concept in these multi-stakeholder initiatives. Typically, the stakeholders include all the participants in a supply chain, from the farmer to the retailer, but also commodity traders, consumer goods manufacturers, banks and service supplies, as well as civil society groups. Their shared goal is to identify effective solutions to the social and environmental challenges associated with conventional production systems. The mechanism used to reform supply chains is typically a voluntary certification system that verifies that the production, trade and transformation of a commodity has complied with a set of best practices that have been agreed to by all the parties.* The search for consensus is important, because it means all of the stakeholders have agreed to accept this package of solutions and commit to supporting the commercialisation of the goods that have been certified as 'sustainable'.

Some environmental activists view these initiatives as a form of green-wash and have questioned their efficacy. Participating companies certify the production within their own supply chain, but roundtable initiatives have not succeeded in transforming their respective sectors.[89] Demand for certified commodities has failed to attract a critical mass of producers that would actually transform the market and change the economic drivers of deforestation. Adoption is highest for coffee (40%) and cocoa (22%), while commodities linked to industrial plantations tend to be lower: palm oil (20%), sugar (3%) soy (2%) and beef (<1%). Part of the explanation for the slow uptake of the voluntary standards is the lack of demand; typically, only about fifty per cent of certified production is actually sold as a certified commodity.[90]

The lack of uptake is yet another manifestation of the dilemma of allocating the cost of environmental protection and social justice. Sustainability protocols cost money, which either adds to the price of a consumer good or reduces the profit margin of commodity producers. Although North American and European consumers are concerned about deforestation, most still choose a lower-cost product, while those in Asia, Latin America and the Middle East are overwhelmingly focused on price.† Moreover, global commodity markets are dominated by producers on landscapes that were transformed by agriculture decades or centuries in the past, and these

* The proliferation of voluntary standards and certification protocols led to the creation of an initiative known as the Committee on Sustainability Assessment (COSA) that seeks to standardise evaluation criteria across platforms and demonstrate that these initiatives lower production costs, increase yields and improve profit margins, as well as work to ensure that these positive benefits are reaching small farmers.

† The emphasis on price is superseded only by health concerns. In advanced economies, consumers seek 'healthy' products, while in developing countries there is a greater focus on quality and, in the case of China, an ingrained fear of contaminated products.

farmers operate without fear of being accused of environmental crimes. Consequently, traders are not motivated to pay a premium to farmers on the agricultural frontier.*

A few producers seek to differentiate their products as organic, deforestation-free, fair-trade or antibiotic-free because they are selling their products into a differentiated market and receive a premium for their production in compensation for the extra cost and reduced yields that these systems [allegedly] entail. Others participate because it guarantees them market access. Most producers opt to circumvent the voluntary guidelines or sell to traders unconcerned about sustainability or just ignore the whole process entirely.

Social advocates have questioned the economic benefits of certification because they tend to discriminate against small-scale producers who cannot meet the record-keeping and logistical demands of a certification process. These protocols are negotiated by large-scale producers that dominate the roundtable initiatives and tailor the certification criteria to their supply chains.† As formalisation spreads throughout national and international markets, smallholders could be increasingly marginalised within regional and even local markets in contradiction with the stated social objectives of these certification schemes.[91]

Rural Finance

The agricultural producers of the Amazon have access to radically different levels of credit depending upon national policies, the willingness of each country's financial services industry to engage rural populations and, most importantly, the scale of their production system. Brazil has the most sophisticated agricultural sector and, not surprisingly, the most generous and far-reaching system to support its producers. Industrial-scale farmers have access to multiple forms of credit, which they access to pay operational costs, acquire technology and invest in on-farm infrastructure. If they are entrepreneurial, and many are, they borrow money to acquire land and expand production. Small family farms have fewer options, but the federal government has programmes to provide them with affordable short-term credit. Regardless, the cash economy predominates on forest frontiers and within smallholder landscapes where producers must overcome barriers imposed by physical isolation and subsistence livelihoods. Financial cred-

* Rainforest Alliance, UTZ-Certified, Fair Trade are the best-known organisations offering certification protocols that cover multiple commodities.

† Roundtable for Sustainable Palm Oil (RSPO), Round Table Responsible Soy (RTRS), Global Roundtable for Sustainable Beef (GRSB), BONSUCRO (the global sugarcane platform for sustainability), 4C-Association for a better coffee world.

it to support production is largely absent in the Andean Amazon, where small farmers operate within an informal economy with limited access to financial services.

Brazil's financial system operates on two tracks: the *Sistema Nacional de Crédito Rural* (SNCR), which is managed by the financial industry according to rules established by the federal government; and an independent system managed by multinational trading companies designed to capture commodities for their competing supply chains. The latter includes the four well-known western giants: ADM, Cargill, Bunge and Louis Dreyfus, as well as second-tier companies based in Brazil (Amaggi), Japan (Gavilon), Europe (Sodrugestvo) and China (COFCO). Within the Amazon, the SNCR provides most of the credit used by the livestock sector, while the region's grain farmers depend upon credit obtained from the SNCR, loans from commercial banks and, most importantly, short-term credit provided by commodity traders.

The SNCR was established in 1965 in conjunction with government policies to promote settlement and investment in the agricultural frontiers of the Southern Amazon. Its main objective is to provide producers with working capital at below-market interest rates so they can plant and harvest a crop or raise a herd of cattle for slaughter. The national rural finance plan (*Plano Safra*) of 2020/2021 provided $R 236 billion (~$US 45 billion) in loans to the livestock, farm and plantation sectors; 75% was used for short-term credit, and 25% was allocated for medium to long-term investments. Small producers had access to R$ 33 billion with interest between 2.75% and 4%, while medium-sized producers received a similar sum at 5%. Large-scale producers, who receive the bulk of the finance, were charged between 6 and 7% annual interest.[92]

The SNCR programme has been, and remains, an important element in national development strategies and has catalysed the impressive growth of Brazilian agriculture. The success of the SNCR rests on its ability to leverage the domestic savings of the Brazilian people with the technical capacity of Brazil's commercial banking sector. Its genius was to provide low-cost credit to strategically important producers in an economy characterised by high interest rates. The lion's share of the SNCR's financial resources is generated by a regulatory requirement that obligates commercial and savings banks to either: (a) transfer 34 per cent of their deposits to the *Banco Central do Brasil* or (b) use those resources to fund loan portfolios in agriculture and forestry.[93] Attractive interest rates are a magnet for investors, especially when combined with an easily understood business model based on conventional economics. Brazil's abundant soil and water resources are the foundation of its rural economy, but the SNCR shares much of the credit for creating an agricultural powerhouse. It also shares responsibility for the conversion of approximately eighty million hectares of Amazonian rainforest and an approximately equivalent area of Cerrado savannas.[94]

The SNCR channeled hundreds of millions of dollars into the Southern Amazon during the 1970s to establish a cattle industry on land being distributed to influential families and corporations.[95] In the 1980s, the programme loaned money during a period of hyperinflation at interest rates well below the rate of inflation, an untenable situation that led to its near-collapse in the early 1990s. The SNCR was revitalised following the stabilisation of the Brazilian economy in the administration Fernando Henrique Cardoso, who introduced two additional programmes managed by the national development bank: PRONAF, which is targeted at smallholders, and PRONAMP, which provides finance to medium-scale producers.* Regional development banks, known as *Fundos Constitucionais de Financiamento*, also have credit programmes targeted at their rural constituents.[96]

A recent review revealed that between $US 9 and 12 billion dollars were loaned annually via the SNCR to Amazonian producers during the last decade. Of this amount, approximately 44% went to Mato Grosso, followed by Tocantins (18%), Pará (13%), Maranhão (13%), and Rondônia (9%).[97] This study looked only at four commodities considered to be the main drivers of deforestation and reported that 64% of the loans went to cattle ranchers and 35% to soy farmers, with smaller amounts to timber (0.5%) and oil palm plantations (0.7%). Data for PRONAF were reported only at the national level, but small farmers cultivating soy, beef and palm oil received only two percent of the resources channeled via the SNCR.†

The contribution of the trading companies is difficult to know because they do not break out those numbers in their annual reports. They can be estimated, however, using bottom-up methods and government reported statistics. In the 2019/2020 crop-year, soy and maize were cultivated on approximately eighteen million hectares in the states of the Legal Amazon, where the leading extension agency reported the cost of seed, fuel, agrochemicals and labour at $US 650 per hectare.[98] Assuming that a hundred per cent of crop is planted and harvested using short-term credit, then agribusiness would need approximately $US 13 billion to plant and harvest a crop. Since only $US 1.6 billion was obtained via the SNCR,[99] the remaining $US 11.4 billion was probably supplied by the commodity trading companies. That may be an overestimate, however, because large-scale

* *Banco Nacional de Desenvolvimento Econômico e Social* (BNDES); *Programa Nacional de Agricultura Familiar* (PRONAF); *Programa Nacional de Apoio ao Médio Produtor Rural* (PRONAMP); in total, there are 19 separate financial programmes that fall within the confines of the SNRC. Source: Souza et al. 2020.

† According to the *Banco Central*, those two programmes receive about 10% each of total rural credit, but most is lent to small farmers in the South and Southwest. Source: Capellesso et al. 2018.

producers, who control approximately 46 per cent of the agricultural land in Brazil, are often subsidiaries of diversified corporations. As such, they have access to multiple forms of credit, including domestic bond markets and overseas private equity.[100]

Bond markets include corporate bonds, which are 'debentures' backed by the reputation of the company and *Certificados de Recebíveis do Agronegócio* (CRA), which are securities that place a lien on a physical or contractual asset. Both are used by agribusinesses and banks to finance medium to long-term investments (two to twelve years). If the CRA is issued by a bank, it is likely to be a basket of loans to family farmers, while corporations use them to fund individual projects or activities. The Brazilian bond market has attracted international attention over the last several years (post-2015) because it is viewed as a venue for sustainable finance that seeks to minimise risk from 'environmental, social and governance' (ESG) factors that harm society and, presumably, increase the risk of losing money.

The most common types, and the largest by volume ($US 10 billion) have been 'green bonds' issued by corporations accessing capital markets without the intermediation of banks.* In the Amazon, companies are committing to consume (self-generated) renewable energy, increase productivity, sequester soil carbon and, allegedly, conserve biodiversity and water resources. One of the largest offerings is FS Bioenergia ($US 639 million) a maize-based ethanol producer that is a joint venture between the Iowa-based Summit Holdings and a Tapajos Participaes S/A, a Brazilian subsidiary of a Chinese holding company (Hunan Dakang). Brazilian agribusiness giants are likewise accessing the green bond market, including SLC Agrícola ($US 480 million) that farms 150,000 hectares in Mato Grosso and Maranhão; and Amaggi S/A ($US 750 million), which operates an integrated supply chain spanning 259,000 hectares in Mato Grosso and includes logistics and processing facilities in Rondônia, Amazonas and Pará.[101]

One of the most controversial offerings, is a 'transitional loan' to Marfrig Alimentos S/A ($430 billion), a beef packing company committed to eliminating illegal deforestation and unfair labour practices from its supply chain. The tender is classified as a loan rather than a bond due to the contractual terms of the offering; it is controversial because most of the resources will be used to support their dedicated supply chain (Marfrig Club) without adequate guarantees to reform or exclude calves originating from independent producers who are not in compliance with the Forest Code.[102] The criteria for evaluating the ESG performance will rely on Key

* The forest products sector has issued the most green debt, but none operate within the Legal Amazon; they inlcude Suzano ($US 5.1 billion), Klabar ($US 1.7 billion) and Irani Cellulose ($US 131 million). Source: Climate Bond Initiatives (2021) https://www.climatebonds.net/files/reports/cbi-brazil-agri-sotm-eng.pdf

Performance Indicators (KPIs), metrics that are specified in the prospectus of individual bond tenders that must be validated by independent third party review.*

There is only limited potential for finance to change agricultural practices in the Andean Amazon, because landscapes are largely populated by smallholders who are notoriously risk-adverse in how they manage their finances and cropping systems. They are cautious because the consequences of a crop failure are catastrophic for their families; consequently, they are less likely to adopt novel production systems that require a capital investment that would have to be financed by debt. Nearly all understand the value of credit and its potential to transform their lives; however, the options available to them are neither friendly nor fair.[103]

Andean governments have launched multiple efforts over several decades trying to create mechanisms and institutions to provide financial credit for rural communities, but they have failed to change the calculus that impedes investment on smallholder landscapes. One manifestation of the challenge is the high proportion of families that are unbanked, a term economists use to describe individuals who, by choice or circumstance, do not use financial services. Another is the role of microfinance institutions that provide credit to individuals who do not qualify for loans from conventional banks; instead, they borrow money based on 'good faith' and reputational integrity.† Their presence has materially benefitted the lives of tens of thousands of individuals, many of them female, by allowing them to participate in the informal market economy that characterises commerce in these nations. In addition, they offer savings accounts and provide families with a digital identity for interacting with government agencies and utility companies. Unfortunately, these institutions lend money at interest rates that are out of reach for small farmers.

The microfinance business model was born in the marginalised neighbourhoods of major cities, but these institutions are now present in most mid-sized cities where they also cater to the needs of surrounding rural communities. Interest rates, which range from twenty to eighty per cent, reflect the risk of default associated with their clientele and the high transaction costs associated with administering tens of thousands of small loans.[104] Most microfinance entities operate with capital obtained from conventional banks and investment funds and pay those institutions

* External reviews known as Second Party Opinions (SPO) and Certification under the Climate Bonds Standard are two methodologies recognized by the Climate Bond Intiatitive

† Loans are given to anybody with valid ID and an explicit promise to lend more money in the future if the current loan is serviced and amortised. Many impoverished individuals cherish their personal reputation because it is the only thing of value they own.

standard commercial interest rates (five to eight per cent). Microfinance, which is marketed as a pro-poor public service, is also a highly lucrative business model.[105]

Presumably, a farmer would be a low-risk debtor when compared to an individual engaged in speculative commerce, but the financial sector considers small family farms as high-risk creditors due to weather and pests. In Peru, inflation-adjusted interest rates for small farmers are between twenty and thirty per cent, a value that is out of reach for all agricultural production systems, much less smallholders living on the edge of poverty. Large- and medium-scale producers have access to conventional forms of credit because they can meet the conditions required by lending agencies, particularly legal title to their land and a documented history of production and sales. Even these numbers are disappointing, however. In 2019, government agencies reported that $US 33 million were loaned to 4,199 beneficiaries in a country with an estimated 2.2 million farmers.[106]

Peru has attempted various models to channel funds via savings and loan cooperatives (COOPACS), privately-owned rural savings banks (Caja Rurales), mixed associations of private capital and local government (Caja Munciaples) and a specialised state-owned development bank (Agrobank). None have succeeded in providing affordable credit to small farmers. The most recent attempt, a special fund (FAE-Agro) that is capitalised by the national development bank (COFIDES)[*] is doomed to failure because recipients are required to show legal title for their properties, a condition enjoyed by only fifteen per cent of the farmers of Peru (see Chapter 4).[107]

Civil society has had better success working with grower's associations that simultaneously provide technical support in agronomics, pest management and business administration for individual growers and their umbrella organisations. Successful initiatives are characterised by a long-term commitment on the part of civil society organisations and specialty buyers willing to invest in programs that guarantee a supply of coffee and cacao that is certified as deforestation-free, organic, indigenous and/or gender positive.[†]

Bolivia's agricultural sector is similar to Brazil's but also quite different. There is a limited number of large-scale producers, quite a few (upper)

---

* FAE-Agro: *Financiamiento Agrario Empresarial*; COFIDE: *Corporacion Financiera de Desarrollo S.A*

† Root Capital works with the APROCAM coffee and cacao cooperative of Awajún communities in Amazonas Region; APROCASSI, a woman's coffee cooperative in Cajamarca region; CAC Pangoa, a coffee cooperative of Asheninka growers in the Selva Central in Junin region. Solidaridad supports 1,200 coffee producers in San Martin region; Rabobank works with the Norandoino, a consortium of three coffee cooperatives with 8,250 members and 9 regional offices in Northern Peru. Sources: https://rootcapital.org; https://www.rabobank.nl/over-ons/rabofoundation; https://www.solidaridadsouthamerica.org/es.

middle-class landowners and a large, dynamic small-farm constituency. It lacks, however, a state sponsored rural credit programme that obligates the financial industry to channel money to its producers. Large and mid-scale farmers access capital via the commodity trading companies, as well as from an informal credit market best described as a normalised system of loan sharks.* Ranchers rely on family capital, personal savings or cash flow generated by urban business ventures (medical services, real estate, commerce).

The smallholders of Bolivia are also active participants in the national foodstuffs market, and quite a few have evolved into successful soybean farmers. Many are descendants of Andean indigenous migrants with a strong cultural tradition of savings and investment, traits shared by a large Mennonite community. These groups also have an informal credit market they use for short-term finance. Microfinance institutions are present and, as in Peru, have opened offices in regional cities. Government policies to distribute cash income to elderly and school-age children have motivated tens (hundreds) of thousands of rural families to open savings accounts. High interest rates, however, preclude their ability to borrow money to invest in agricultural technology.

Ecuador's microfinance industry is dominated by savings and loan associations that serve both urban and rural populations; interest rates range between 25 and 28 per cent. The traditional banking system treats agricultural credit as one of several types of 'productive enterprise', all of which have annual interest rates between eight and twelve per cent.[108] The government hopes to support its agricultural sector via a newly constituted public bank, *BanEcuador*, which offers loans specifically designed for, and marketed to, producers of coffee, cacao and oil palm.† Producers can borrow up to $US 150,000 for both short-term credit and to improve productive capacity (plantations); for the latter, terms of up to fifteen years are available, with a grace period of between three to five years.[109] Loans greater than $US 20,000 require a mortgage guarantee.

The *BanEcuador* programmes show an understanding of the needs of their potential clientele, including payment schedules based on the cash flow of individual production strategies (monthly, quarterly, or annually).

* Individuals with a reputation for probity collect capital from individuals seeking investment opportunities with a good rate of return. They lend money to established farmers, who pay off the loan after each harvest cycle, which is typically four to six months. This system thrives because the financial system pays extraordinarily low interest rates on savings accounts and depositary certificates (<1%); Bolivia's bond and equity markets are closed to retail investors.

† The administration of Rafael Correa created *BanEcaudor* in 2015; it inherited the assets and liabilities of the *Banco de Fomento de Ecuador*. Its board of directors comprises five cabinet ministers. Source: https://www.banecuador.fin.ec/historia-banecuador/

Annual interest rates range between 9.75 and 16.5 per cent, in line with business loans from private banks and significantly lower than those available from microfinance institutions. Regardless, interest rates at this level are not likely to catalyse a wave of much-needed investment in agricultural production. In 2019, BanEcuador loaned $US 3 million to 700 producers, a relatively small amount that would translate into only about 500 hectares of new oil palm plantations.

Colombia has a programme similar to the SNCR of Brazil. It is administered by FINAGRO (*Fondo Para el Financiamiento del Sector Agropecuario*), a public agency that operates as a second-tier lender to private institutions and a guarantor for a variety of financial products, including short and long-term credit and crop insurance. The FINAGRO system establishes standard terms and rates for a diversified portfolio of credit products specifically designed for the needs of producers in agriculture, livestock and plantation forestry. Programmes span the landholder spectrum and include special initiatives for producer associations. Interest rates range from three to ten per cent above a base rate set by the central bank, which has fluctuated between three and five per cent since 2010. FINAGRO also offers discounts to the intermediary institution to make the loan more accessible to the retail client.[110] For a commission, FINAGRO will guarantee the loan for the producer, which is essentially a form of crop insurance; it also offers conventional crop insurance to protect the producer and the lending agency from climate risk and pests.

At the national scale, FINAGRO facilitated financial credit worth approximately $US 7.1 billion in 2020, up from $US 2.9 billion in 2011, increasing by an impressive twenty per cent annually over the same period. Although the Colombian programme is well designed and considers both market reality and the special needs of producers, it operates only on landscapes where the state has imposed the rule of law.[*] Unfortunately, Amazonian landscapes are characterised by the absence of the state, either because they are remote or because they are under the control of armed criminals. In Amazonian departments, FINAGRO facilitated only about $US 80 million in 2020, a number that has remained essentially flat since 2010.[111] Approximately half of that was in Caquetá and, presumably, was lent to the cattle sector, which is also the largest single recipient of agricultural credit within the FINAGRO system.

Harnessing finance to change behaviour

Rural finance has enormous potential to reform conventional agricultural production systems. Consequently, it is a common component of policy

* Approximately, 60% of FINAGRO credits are issued in Bogota (D.C.), Cundinamarca, Antioquía and Valle de Cauca.

proposals to combat deforestation where it is viewed as a 'carrot' to accompany the 'sticks' that seek to coerce landholders to reform land use practices (see Chapter 7).

The experience of the Cattle Agreement and the Soy Moratorium show the potential when commercial intermediaries are used as pressure points to eliminate illegal deforestation. These initiatives, which focus on excluding transgressors from supply chains, could be expanded by conditioning access to the billions of dollars of short-term loans provided annually by international commodity traders and meat packing companies. As a driver of sustainability, credit might be even more effective if these same companies offered long-term loans with concessionary rates that motivated their suppliers to restore forest that had been converted illegally in the recent past.

Similar changes to the *Sistem Nacional de Credito Rural* (SNRC) could likewise catalyse widespread change, particularly within the Brazilian cattle industry where decades of overgrazing have degraded millions of hectares of pasture. Pasture restoration begins and ends with soil conservation, which relies on management practices to increase soil organic matter and, in the process, create a long-term carbon sink (see Chapter 4). This is essentially the goal of Brazil's *Agricultura de Baixo Carbono* (ABC) programme, a subcomponent of the SNCR with attractive interest rates and an extended pay-back period.[112] Supported technologies include reduced tillage, pasture renovation, integrated crop and livestock management, and the restoration of riparian habitat (see Chapter 4 and Chapter 7). The ABC programme has enjoyed modest success – in 2020, the programme lent approximately $R 2 billion ($US 400 million) – nonetheless, that is less than one per cent of the total channelled through the SNCR in 2020. The potential, given Brazil's history of using the SNCR to subsidise its agricultural producers, is massive and eminently practical.

Green bonds and similar types of ESG finance are the fastest growing component of Brazil's financial sector. International capital markets are frenetically seeking viable projects to satisfy massive global demand for ESG investment. Brazil's potential to satisfy this demand by reducing GHG emissions caused by deforestation can be leveraged by an equally massive capacity to sequester carbon via economically advantageous technologies to make conventional agriculture more sustainable. This type of risk-limited green investment will be a magnet for global investors. The country's attractiveness is reinforced by the nation's cultural commitment to a market economy, its openness to international capital and the abundant natural resources that are the foundation of its rural economy.

The performance of green bonds in Brazil is being closely watched by policy analysts, because of their potential to drive climate change action 'at scale'. Nonetheless, these instruments, and others like them, risk being labled as 'greenwash' if they succeed in improving the performance of participating companies but fail to resolve the deforestation crisis. That outcome will depend, in large part, on the ability of the Brazilian state – and its private sector partners – to incorporate smallholders in a revitalised and reformed rural economy. Brazil has created the institutional mechanisms for pursuing that goal (INCRA, EMBRAPA, PRONAF, SNCR), but its track record for dealing equitably with its own citizens is not particularly encouraging (see Chapter 6)

In the Andean Amazon, the potential to link finance, including short and long-term credit, with effective policies to transform their agricultural sector is even more challenging. No nation has succeeded in delivering affordable credit to their Amazonian populations, nor developing an extension system capable of ensuring those resources are invested in productive enterprises that are globally competitive and environmentally sustainable. If they have any advantage, compared to Brazil, it is the preponderance of smallholder systems that creates a precondition that favours social equity. That advantage, however, is a double-edged sword. It may ensure that a reformed system will be socially sustainable, but it also makes it enormously more difficult to implement.

If the Amazon Forest is to be saved, deforestation must end. Full stop. Global and national markets, however, will continue to demand more commodities from the farmers and ranchers of the Southern Amazon and Andean Piedmont. They will respond by increasing production. Full stop.

Producers can expand by investing in technology to intensify land use, or they can purchase more land to enlarge the area under cultivation. Left to their own devices, they will pursue both options because that is the logical pathway to maximise the return on their investments. Producers do not operate in a vacuum, however. Farmers and ranchers, large and small, allocate their resources in response to regulatory and market forces that govern the agricultural economy. Among the most important are the constraints and incentives in rural real estate markets (see Chapter 4). When the forest frontier ceases to be a source of inexpensive land, the agricultural sector will be forced to invest in the land under production – and only the land under production. Making that happen sooner, rather than later, is essential for saving the Amazon.

Bibliography

ABIEC – Associação Brasileira das Indústrias Exportadoras de Carnes. 6 April 2021. 'Estatístcas, Exportações': http://abiec.com.br/en/exports-consult/

Alves, R. 2021. Agribusiness faces turning point as fire outbreaks soar. The Brazilian report: https://brazilian.report/environment/2021/01/10/agribusiness-faces-turning-point-as-fire-outbreaks-soar-brazil/

Assunção, J., C. Gandour and R. Rocha. 2015. 'Deforestation slowdown in the Brazilian Amazon: prices or policies?' *Environment and Development Economics* **20** (6): 697–722.

Barrientos-Fuentes, J.C. and J.C. Torrico-Albino. 2014. 'Socio-economic perspectives of family farming in South America: cases of Bolivia, Colombia and Peru'. *Agronomía Colombiana* **32** (2): 266–275.

Barthel, K., V. Cespedes, B. Salazar, R. Torres and M. Varón. 2016. *Land and Rural Development Policy Reforms in Colombia: The Path to Peace*, USAID/Colombia Land and Rural Development Program (LRDP): https://www.globalcommunities.org/publications/2016-Colombia-Rural-Development-Policy.pdf

CONAB – Companhia Nacional de Abastecimento. 2021. 'Safra Brasileira de Grãos': https://www.conab.gov.br/info-agro/safras/graos

De Beenhouwer, M., R. Aerts and O. Honnay. 2013. 'A global meta-analysis of the biodiversity and ecosystem service benefits of coffee and cacao agroforestry'. *Agriculture, Ecosystems & Environment* 175: 1–7.

Benami, E., L.M. Curran, M. Cochrane, A. Venturieri, R. Franco, J. Kneipp and A. Swartos. 2018. 'Oil palm land conversion in Pará, Brazil, from 2006–2014: evaluating the 2010 Brazilian Sustainable Palm Oil Production Program'. *Environmental Research Letters* 13 (3): 034037.

Bowman, M.S., B.S. Soares-Filho, F.D. Merry, D.C. Nepstad, H. Rodrigues and O.T. Almeida, 2012. 'Persistence of cattle ranching in the Brazilian Amazon: A spatial analysis of the rationale for beef production'. *Land Use Policy* **29** (3): 558–568.

Brandão, F. and G. Schoneveld. 2015. *The State of Oil Palm Development in the Brazilian Amazon.* Working Paper 198. Bogor, Indonesia: CIFOR.

Capellesso, A.J., A.A. Cazella and F.L. Búrigo. 2018. 'Evolução do Pronaf Crédito no Período 1996-2013: redimensionando o acesso pelos cadastros de pessoa física'. *Revista de Economia e Sociologia Rural* **56** (3): 437–450.

Collins, Ley. 2017. 'The road beyond 2020, EU Biofuels is facing the threat of a "lost decade"'. *Biofuels International*, March/April. https://issuu.com/horseshoemedialtd/docs/biofuels_march-apr_17_2

Conecta – Parcerias de Agropecuária Responável. 25 May 2021. *Uma plataforma digital que valoriza o trabalho do pecuarista legal*, Safe Trace (ST), Amigos da Terra – Amazônia Brasileira (AdT) e The Nature Conservancy (TNC), Partnerships for Forests (P4F): http://agropecuariaresponsavel.com.br/#sobre

Dammert, J.L. 2013. 'Expansión de palma aceitera en la Amazonía: en las puertas del escándalo'. *Revista Agraria* 153.

ESRI – Environmental Systems Research Institute. 2020. 'Esri 10-Meter Land Cover': https://livingatlas.arcgis.com/landcover/

FAOSTAT – Food and Agriculture Organization of the United Nations. 2021: http://www.fao.org/faostat/en/#data/TP

Fehlenberg, V., M. Baumann, N.I. Gasparri, M. Piquer-Rodriguez, G. Gavier-Pizarro and T. Kuemmerle. 2017. 'The role of soybean production as an underlying driver of deforestation in the South American Chaco'. *Global Environmental Change* **45**: 24–34.

Finer, M. and T. Olexy. 2016. Oil Palm Deforestation in the central Peruvian Amazon. MAAP: 48: http://maaproject.org/2016/oil-palm-tocache/

Garcia, Andrea Santos and Maria Victoria Ramos Ballester. 2016. 'Land cover and land use changes in a Brazilian Cerrado landscape: drivers, processes, and patterns'. *Journal of Land Use Science*: 1–22.

Hunke, P.P. 2015. *The Brazilian Cerrado: Ecohydrological Assessment of Water and Soil Degradation in Heavily Modified Meso-scale Catchments*. Geoökologie/Ökohydrologie, Universität Potsdam: https://publishup.uni-potsdam.de/frontdoor/index/index/docId/8511

IBCE – Instituto Boliviano de Comercio Exterior. 2021. 'Estadísticas de Bolivia': https://ibce.org.bo/informacion-estadisticas-bolivia.php

IBGE – Instituto Brasileiro de Geografia e Estatística. 2016. 'Mapa da Cobertura e Uso da Terra do Mato Grosso': https://www.ibge.gov.br/geociencias/informacoes-ambientais/cobertura-e-uso-da-terra/15833-uso-da-terra.html

IBGE/SIDRA – Instituto Brasileiro de Geografia e Estatística / Sistema IBGE de Recuperacao Automatica – SIDRA. 2021. https://sidra.ibge.gov.br/pesquisa/pam/tabelas

IMEA – Instituto Mato-Grossense de Economia Agropecuária. 2021. 'Custo de Produção de Soja GMO, Mato Grosso': https://www.imea.com.br/imea-site/relatorios-mercado-detalhe?c=4&s=3

Jha, S., C.M. Bacon, S.M. Philpott, V. Ernesto Mendez, P. Läderach and R.A. Rice. 2014. 'Shade coffee: update on a disappearing refuge for biodiversity'. *BioScience* **64** (5): 416–428.

Killeen, T.J., A. Guerra, M. Calzada, L. Correa, V. Calderon, L. Soria, B. Quezada and M.K. Steininger. 2008. 'Total historical land-use change in eastern Bolivia: who, where, when, and how much'. *Ecology and Society* **13** (1): 36.

Killeen, T.J. and G. Harper. 2015. Coffee in the twenty first century: Will global warming and increased demand lead to new deforestation? Conservation International: http://www.conservation.org/publications/Documents/CI-Coffee-Report.pdf

Klingler, M., P.D. Richards and R. Ossner. 2018. 'Cattle vaccination records question the impact of recent zero-deforestation agreements in the Amazon'. *Regional Environmental Change* **18** (1): 33–46.

Lambin, E.F., H.K. Gibbs…. and C. Nolte. 2018. 'The role of supply-chain initiatives in reducing deforestation'. *Nature Climate Change* **8** (2): 109–116.

Laserna R. 2018. *Energy Dividends in Bolivia: Are There Any Alternatives to Price Subsidies?* Center for Global Development: https://www.cgdev.org/sites/default/files/energy-dividends-bolivia-are-there-any-alternatives-price-subsidies.pdf

MAAP – Monitoring of the Andes Amazon Project. 7 Apr 2021. 'MAAP #136: Amazon Deforestation 2020': https://maaproject.org/2021/amazon-hotspots-2020-final/1

Marques, G.H.F., E. De Stefano, C.P. Ribeiro, L.H.A Turissi, R.A. Dias, J. Naranjo, P.S. Pozzetti, J.F. Costa, and E.M. Pituco. 2016. 'A experiência brasileira na erradicação da febre aftosa e o emprego do sistema I-ELISA 3ABC/EITB para certificação sanitária de bovinos e bubalinos'. *Arquivos do Instituto Biológico* **82**: 1–11.

McGrath, D.G., L. Castello, O.T. Almeidaand G.M. Estupiñán. 2015. 'Market formalization, governance, and the integration of community fisheries in the Brazilian Amazon'. *Society & Natural Resources* **28** (5): 513–529.

Modernel, P., W.A. Rossing, M. Corbeels, S. Dogliotti, V. Picasso, and P. Tittonell. 2016. 'Land use change and ecosystem service provision in Pampas and Campos grasslands of southern South America'. *Environmental Research Letters* **11** (11): 113002.

Motamayor, J.C., P. Lachenaud, J.W.D.S. e Mota, R. Loor, D.N. Kuhn, J.S. Brown and R.J. Schnell. 2008. 'Geographic and genetic population differentiation of the Amazonian chocolate tree (*Theobroma cacao L*)'. *PLoS One* **3** (10): e3311.

Parra, E.M. 2014. 'Desarrollo alternativo en el Perú: treinta años de aciertos y desaciertos'. *Perspectivas Rurales Nueva Época* **23**: 85–104.

Pinzon, A. 2019. Redefining finance for agriculture: Green agricultural credit for smallholders in Peru. Global Canopy: https://www.meda.org/innovate/innovate-resources/822-partner-publication-redefining-finance-for-agriculture-green-agricultural-credit-for-smallholders-in-peru-eng/file/

Potts, J. et al. 2014. State of Sustainability Initiatives Review – 2014,International Institute for Sustainable Development: https://www. iisd. org/sites/default/files/pdf/2014/ssi_2014. pdf

PRODES. 2020. 'Desflorestamento nos Municípios da Amazônia Legal Projeto Prodes – Monitoramento Da Floresta Amazônica Brasileira Por Satélite': http://www.obt.inpe.br/OBT/assuntos/programas/amazonia/prodes

Sengupta, R. and C.P. Aubuchon. 2008. 'The microfinance revolution: An overview'. *Federal Reserve Bank of St. Louis Review* 90 (Jan./Feb.)

Souza, P., S. Herschmann and J. Assunção. 2020. *Política de Crédito Rural no Brasil: Agropecuária, Proteção Ambiental e Desenvolvimento Econômico*. Climate Policy Initiative, Rio de Janeiro: https://www.climatepolicyinitiative.org/pt-br/publication/politica-de-credito-rural-no-brasil-agropecuaria-protecao-ambiental-e-desenvolvimento-economico/

Spera, S.A., G.L. Galford, M.T. Coe, M.N. Macedo and J.F. Mustard. 2016. 'Land-use change affects water recycling in Brazil's last agricultural frontier'. *Global Change Biology* **22** (10): 3405–3413.

Strassburg, B.B., T. Brooks, R. Feltran-Barbieri, A. Iribarrem, R. Crouzeilles, R. Loyola… and A. Balmford. 2017. 'Moment of truth for the Cerrado hotspot'. *Nature Ecology & Evolution* **1** (4): 1–3.

Stratton, R.W. et al. 2010. Life Cycle Greenhouse Gas Emissions from Alternative Jet Fuels, Partnership for Air Transportation Noise and Emissions Reduction, Project 28: http://web.mit.edu/aeroastro/partner/reports/proj28/partner-proj28-2010-001.pdf

Tarouco, J.U., J.B.S. Ferraz, R.D. Gomes, P.R. Leme and E.A.Navajas. 2012. 'Prediction of retail beef yield, trim fat and proportion of high-valued cuts in Nellore cattle using ultrasound live measurements'. *Revista Brasileira de Zootecnia* **41** (9): 2025–2031.

Thomas, E., M. van Zonneveld, J. Loo, T. Hodgkin, G. Galluzzi and J. van Etten. 2012. 'Present spatial diversity patterns of *Theobroma cacao L.* in the neotropics reflect genetic differentiation in Pleistocene refugia followed by human-influenced dispersal'. *PLoS One* **7** (10): e47676.

Tondoh, J.E., F.N.G. Kouamé, A.M. Guéi, B. Sey, A. Koné and N. Gnessougou. 2015. 'Ecological changes induced by full-sun cacao farming in Côte d'Ivoire'. *Global Ecology and Conservation* **3**: 575–595.

Torrijos Rivera, R. 2019. Cifras de Contexto Ganadero Caquetá 2019. Comité Departmental de Ganaderos del Caquetá, Florencia, Caquetá, Colombia: https://issuu.com/rafaeltorrijos/docs/contexto_2019_con_portada_publicable

USDA/FAS – United States Department of Agriculture/Foreign Agricultural Service: https://www.fas.usda.gov/data

Voila, M. and D. Triches. 2015. 'A cadeia de carne de frango: uma análise dos mercados brasileiro e mundial de 2002 a 2012'. *Revista Teoria e Evidência Econômica* 21 (44). http://seer.upf.br/index.php/rtee/article/download/5357/3477

Wasserstrom R. and D. Southgate. 2013. Deforestación, reforma agraria y desarrollo petrolero en Ecuador, 1964–1994, *Natural Resources* **4**: 34–44, http://www.scirp.org/journal/nr

Yuri Ramos, S. and G. Bueno Martha Jr. 2010. Evoluçcão da Politica de Créditop Rural Brasileria, EMBRAPA Cerrado, Brasilia.

Zalles, V., M.C. Hansen, P.V. Potapov, S.V. Stehman, A. Tyukavina, A. Pickens, A… and S. Chavez. 2019. 'Near doubling of Brazil's intensive row crop area since 2000'. *Proceedings of the National Academy of Sciences* **116** (2): 428–435.

Zero Deforestation Working Group (2017). *A Pathway to Zero Deforestation in the Amazon, Greenpeace, IPAM, IMAZON, IMAFLORA, WWF, ICVC, ISA,* https://ipam.org.br/wp-content/uploads/2017/11/A-Pathway-to-Zero-Deforestation-in-the-Brazilian-Amazon-full-report.pdf

Notes to Chapter 3

1. Assunção, J., C. Gandour and R. Rocha. 2015. 'Deforestation slowdown in the Brazilian Amazon: prices or policies?' *Environment and Development Economics* **20** (6): 697–722.
2. Bowman, M.S., B.S. Soares-Filho, F.D. Merry, D.C. Nepstad, H. Rodrigues and O.T. Almeida. 2012. 'Persistence of cattle ranching in the Brazilian Amazon: A spatial analysis of the rationale for beef production'. *Land Use Policy* **29** (3): 558–568.
3. Tarouco, J.U., J.B.S. Ferraz, R.D. Gomes, P.R. Leme and E.A.Navajas. 2012. 'Prediction of retail beef yield, trim fat and proportion of high-valued cuts in Nellore cattle using ultrasound live measurements'. *Revista Brasileira de Zootecnia* **41** (9): 2025–2031.
4. ABIEC - Associação Brasileira das Indústrias Exportadoras de Carnes 2016. Brazilian Livestock Profile 2016. http://abiec.com.br/en/catpub/printed/
5. Ibid.
6. LAPIG – *Laboratório de Processamento de Imagens e Geoprocessamento*. 2020. Instituto de Estudos Socioambientais (IESA) da Universidade Federal de Goiás (UFG): https://lapig.iesa.ufg.br; ABIEC – Associação Brasileira das Indústrias Exportadoras de Carnes. 2020. Slaughterhouse locations: http://abiec.com.br/en/industries-location/
7. PRODES. 2020. Desflorestamento nos Municípios da Amazônia Legal Projeto Prodes Monitoramento Da Floresta Amazônica Brasileira Por Satélite: http://www.obt.inpe.br/OBT/assuntos/programas/amazonia/prodes
8. Beef Point. 2016. Confinamentos de bovinos deverão cair em 24,37% nesse ano no MT: https://www.beefpoint.com.br/confinamentos-de-bovinos-deverao-cair-em-2437-nesse-ano-no-mt/; Canal Rural. 6 Apr. 2021. Brasil bateu recorde de bois confinados em 2020, aponta censo: https://www.girodoboi.com.br/destaques/brasil-bateu-recorde-de-bois-confinados-em-2020-aponta-censo/
9. SóNoticias. 6 Apr. 2021. Quantidade de gado em confinamento em Mato Grosso diminui: https://www.sonoticias.com.br/agronoticias/quantidade-de-gado-em-confinamento-em-mato-grosso-diminui/
10. ABIEC – *Associação Brasileira das Indústrias Exportadoras de Carnes*. 6 Apr. 2021. Estatístcas, Exportações: http://abiec.com.br/en/exports-consult/
11. Milhorance, M. 8 Oct. 2020. The murky process of licensing Amazonian meat plants, Mongabay.com: https://news.mongabay.com/2020/10/the-murky-process-of-licensing-amazonian-meat-plants/
12. Killeen, T.J., A. Guerra, M. Calzada, L. Correa, V. Calderon, L. Soria, B. Quezada and M.K. Steininger. 2008. 'Total historical land-use change in eastern Bolivia: who, where, when, and how much'. *Ecology and Society* **13** (1): 36.
13. AgroMeat. 31 May 2016. Bolivia negoció en París exportación de carne de res a Medio Oriente y otros países: https://www.agromeat.com/184370/bolivia-negocio-en-paris-exportacion-de-carne-de-res-a-medio-oriente-y-otros-paises
14. IBCE – *Instituo Boliviano de Comercio Exterior*. 10 Nov. 2020. Bolivia: Exportaciones de carne de la especial bovina: https://fegasacruz.org/bolivia-hasta-abril-se-triplica-la-exportacion-de-carne/

15. La Razón Digital. 4 Jan. 2014. Bolivia puede exportar hasta $US 200 MM en carne bovina: http://boliviaemprende.com/noticias/bolivia-puede-exportar-hasta-us-200-mm-en-carne-bovina
16. Mendez, C. 21 Feb. 2021. Ganadería en Bolivia: se amplía la exportación, se reducen los bosques, Mongabay LATAM: https://es.mongabay.com/2021/02/ganaderia-bolivia-deforestacion-bosques-china-incendios-forestales/
17. MAAP – Monitoring of the Andes Amazon Project. 7 Apr. 2021. MAAP #136: Amazon Deforestation 2020: https://maaproject.org/2021/amazon-hotspots-2020-final/1
18. Amazon Muncipal Production and Land-use Model. 2016. T. Killeen and Frank Hajek, Unpublished analysis.
19. Wasserstrom R. and D. Southgate. 2013. Deforestación, reforma agraria y desarrollo petrolero en Ecuador, 1964–1994. *Natural Resources* **4**: 34–44, http://www.scirp.org/journal/nr
20. FINKEROS. 15 Oct. 2013. El problema de la ganadería en Colombia: http://abc.finkeros.com/el-problema-de-la-ganaderia-en-colombia/
21. FEDEGAN – *Federación Colombiana de Ganaderos*. 1 May 2021. Ganadería Colombiana Sostenible: http://www.fedegan.org.co/programas/ganaderia-colombiana-sostenible
22. Torrijos Rivera, R. 2019. Cifras de Contexto Ganadero Caquetá 2019. Comité Departmental de Ganaderos del Caquetá, Florencia, Caquetá, Colombia: https://issuu.com/rafaeltorrijos/docs/contexto_2019_con_portada_publicable
23. PID-Amazonia. 11 Oct. 2018. Pacto Caquetá: Cero Deforestación y Reconciliación Ganadera. Retos y Avances del Programa. Plataforma de Informacifon y Dialogo para la Amazonia Colombiana: https://pidamazonia.com/content/pacto-caquet%C3%A1-cero-deforestaci%C3%B3n-y-reconciliaci%C3%B3n-ganadera-retos-y-avances-del-programa
24. MAAP – Monitoring of the Andes Amazon Project. 7 Apr. 2021. MAAP #136.
25. Barthel, K., V. Cespedes, B. Salazar, R. Torres and M. Varón. 2016. *Land and Rural Development Policy Reforms in Colombia: The Path to Peace*, USAID/Colombia Land and Rural Development Program (LRDP): https://www.globalcommunities.org/publications/2016-Colombia-Rural-Development-Policy.pdf
26. INFOBAE. 1 Feb. 2020. Venezuela: La producción ganadera está quebrada y los venezolanos consumen apenas 7 kilos de carne al año: https://www.infobae.com/america/venezuela/2020/02/01/la-produccion-ganadera-esta-quebrada-y-los-venezolanos-consumen-apenas-7-kilos-de-carne-al-ano/
27. Roraima em Foco. 29 Jan. 2020. Pecuária: Roraima Avança e Rebanho Já Chega a Quase 900 Mil Cabeças: https://roraimaemfoco.com.br/acervo/pecuaria-roraima-avanca-e-rebanho-ja-chega-a-quase-900-mil-cabecas/
28. IMEA – *Instituto Mato-grossense de Economia Agropecuária*. 2021. Custo de Produção: http://www.imea.com.br/imea-site/relatorios-mercado-detalhe?c=4&s=3
29. FAOSTAT – Food and Agriculture Organization. 2021. Data: http://www.fao.org/faostat/en/#data/TP
30. Garcia, Andrea Santos and Maria Victoria Ramos Ballester. 2016. 'Land cover and land use changes in a Brazilian Cerrado landscape: drivers, processes, and patterns'. *Journal of Land Use Science*: 1–22.

31. Zalles, V., M.C. Hansen, P.V. Potapov, S.V. Stehman, A. Tyukavina, A. Pickens, A… and S. Chavez. 2019. 'Near doubling of Brazil's intensive row crop area since 2000'. *Proceedings of the National Academy of Sciences* **116** (2): 428–435.
32. Hunke, P.P. 2015. *The Brazilian Cerrado: Ecohydrological Assessment of Water and Soil Degradation in Heavily Modified Meso-scale Catchments*. Geoökologie / Ökohydrologie, Universität Potsdam: https://publishup.uni-potsdam.de/frontdoor/index/index/docId/8511
33. Fehlenberg, V., M. Baumann, N.I. Gasparri, M. Piquer-Rodriguez, G. Gavier-Pizarro and T. Kuemmerle. 2017. 'The role of soybean production as an underlying driver of deforestation in the South American Chaco'. *Global Environmental Change* **45**: 24–34.
34. MPD – *Ministerio de Planificación del Desarrollo*. 2013. Agenda Patriótica 2025, 13 Pilares De La Bolivia Digna Y Soberana: http://www.planificacion.gob.bo/uploads/AGENDA_PATRIOTICA2025_MPD.pdf
35. Canal Rural. 2014. http://www.canalrural.com.br/noticias/agricultura/produtores-mato-grosso-investem-construcao-silos-individuais-10040
36. Ibid.
37. WWF – World Wide Fund for Nature. 2014. The Growth of Soy: Impacts and Solutions. WWF International, Gland, Switzerland: http://awsassets.wwfdk.panda.org/downloads/wwf_soy_report_final_jan_10.pdf
38. USDA / FAS – United States Department of Agriculture / Foreign Agricultural Service. 2016. Ukraine Oilseeds Annual Production Growth Continues, GAIN-Global Agricultural Information Network: https://gain.fas.usda.gov/Recent%20GAIN%20Publications/Oilseeds%20and%20Products%20Annual_Kiev_Ukraine_3-22-2016.pdf
39. IBGE – *Instituto Brasileiro de Geografia e Estatística*. 2021. Pesquisa Pecuária Municipal 2021 Sistema IBGE de Recuperacao Automatica – SIDRA: https://sidra.ibge.gov.br/pesquisa/pam/tabelas
40. Voila, M. and D. Triches. 2015. 'A cadeia de carne de frango: uma análise dos mercados brasileiro e mundial de 2002 a 2012'. *Revista Teoria e Evidência Econômica* **21** (44). http://seer.upf.br/index.php/rtee/article/download/5357/3477
41. FEDEPALMA. 2015. Análisis financiero del sector Palmicultor y la industria relacionada: http://web.fedepalma.org/sites/default/files/files/08012016_An%C3%A1lisis%20financiero_2010_2014_fin.pdf
42. MINIAGRICULTURA – *Ministerio de Agricultura y Desarrollo Rural*. 2015. Sistematización de una experiencia exitosa: http://alianzasproductivas.minagricultura.gov.co:81/tutorial/archivos/sistematizacionpaap-versionimprimir-25_11_2015.pdf
43. Potter, L. 2011. Palm oil industry in Ecuador. Good business for small farmers? Eutopia: http://repositorio.flacsoandes.edu.ec/bitstream/10469/3980/1/RFLACSO-E02-02-Potter.pdf ; El Universo. 30 May 2020. Unos 6800 palmicultores esperan la apoprbación de la Ley de la Palma Aceitera: https://www.eluniverso.com/noticias/2020/05/30/nota/7855980/6800-palmicultores-esperan-aprobacion-ley-palma-aceitera/

44. UNODC. 2013. Perú: Desempeño Comercial de las Empresas Promovidas por el Desarrollo Alternativo / 2012. Oficina de las Naciones Unidas contra la Droga y el Delito para Perú y Ecuador.
45. Finer, M. and T. Olexy. 2016. Oil Palm Deforestation in the central Peruvian Amazon. MAAP: 48: http://maaproject.org/2016/oil-palm-tocache/
46. Dammert, J.L. 2013. 'Expansión de palma aceitera en la Amazonía: en las puertas del escándalo'. Revista Agraria 153.
47. EFEverde. 24 Feb. 2017. Empresario acusado de deforestar la Amazonía peruana evadió pago: http://www.efeverde.com/noticias/empresario-acusado-deforestar-la-amazonia-evadio-pago-justicia-dice-ong/
48. Chain Reaction Research. 2017. Grupo Palmas: First Peruvian NDPE Policy Creates Business Opportunities but Strands Land: https://chainreactionresearch.files.wordpress.com/2017/04/grupo-palmas-first-peruvian-ndpe-policy-170406-final.pdf
49. Brandão, F. and G. Schoneveld. 2015. *The State of Oil Palm Development in the Brazilian Amazon*. Working Paper 198. Bogor, Indonesia: CIFOR.
50. BrasilAgro. 21 Jan. 2020. Petrobras torra centenas de milhões em projeto de biocombustível de dendê: https://www.brasilagro.com.br/conteudo/petrobras-torra-centenas-de-milhoes-em-projeto-de-biocombustivel-de-dende.html ; BNamericas. 23 Oct. 2020. Regulador autoriza a Vale a vender mayor productor de aceite de palma de Brasil: https://www.bnamericas.com/es/noticias/regulador-autoriza-a-vale-a-vender-mayor-productor-de-aceite-de-palma-de-brasil
51. Brazil Biofuels. 16 Apr 2021. Agribusiness: https://www.brasilbiofuels.com.br/en/agribusiness/
52. INCRA – *Instituto Nacional de Colonização e Reforma Agrária*. 2010. Governo Federal lança Programa de Produção Sustentável de Óleo de Palma: http://www.incra.gov.br/governo-federal-lanca-programa-de-producao-sustentavel-de-oleo-de-palma
53. Benami, E., L.M. Curran, M. Cochrane, A. Venturieri, R. Franco, J. Kneipp and A. Swartos. 2018. 'Oil palm land conversion in Pará, Brazil, from 2006–2014: evaluating the 2010 Brazilian Sustainable Palm Oil Production Program'. *Environmental Research Letters* **13** (3): 034037.
54. FAOSTAT. 2016.
55. Lia Guterman. 2013. Medición de costos de Producción e indicadores de productividad laboral en la agroindustria de la Palma de aceite enColombia 2011 – 2012. XLII Asamblea Geneal de Fedepalma: http://web.fedepalma.org/sites/default/files/files/Fedepalma/CongresoPalmero_LiaGutterman.pdf
56. El Espectador. 6 Oct. 2016 – 6:21. Medidas del Gobierno afectan de nuevo producción nacional de biocombustibles y palma de aceite: http://www.elespectador.com/noticias/economia/medidas-del-gobierno-afectan-de-nuevo-produccion-nacion-articulo-658988
57. Collins, Ley. 2017. 'The road beyond 2020, EU Biofuels is facing the threat of a "lost decade"'. *Biofuels International*, March/April. https://issuu.com/horseshoemedialtd/docs/biofuels_march-apr_17_2
58. US Energy Information Administration. 24 May 2017. Monthly Energy Review: https://www.eia.gov/totalenergy/data/monthly/index.php#renewable

59. Secretaria de Energia e Mineração. 27 July 2016. Biomassa representa 8,8% da matriz elétrica do Brasil: http://www.energia.sp.gov.br/2016/07/biomassa-representa-88-da-matriz-eletrica-do-brasil/
60. Bloomberg. 10 Mar. 2016. Brazil's ailing sugar cane growers get boost from anti-coal push: https://www.bloomberg.com/news/articles/2016-03-11/brazil-s-ailing-sugar-cane-growers-get-boost-from-anti-coal-push
61. Bioenergy International. 21 June 2021. FS plans South America's first BECCS project at FS Lucas do Rio Verde in Brazil: https://bioenergyinternational.com/biofuels-oils/fs-plans-south-americas-first-beccs-project-at-fs-lucas-do-rio-verde-in-brazil
62. RPAnews 4 Mar 2020. Etanol de milh: https://bioenergyinternational.com/biofuels-oils/fs-plans-south-americas-first-beccs-project-at-fs-lucas-do-rio-verde-in-brazil o avança no Brasil https://revistarpanews.com.br/etanol-de-milho-avanca-no-brasil/
63. CanalRural 26 Aug 2020. Etanol de milho: nova usina entra em operação em Mato Grosso: https://www.canalrural.com.br/agronegocio/etanol-milho-usina-mato-grosso/
64. EPE – *Empresa de Pesquisa Energetica*. 2019. Análise de Conjuntura dos Biocombustíveis, Ano 2019, Ministerio de Minas e Energia: https://www.epe.gov.br/sites-pt/publicacoes-dados-abertos/publicacoes/PublicacoesArquivos/publicacao-489/Analise_de_Conjuntura_Ano_2019.pdf
65. Brasil Biofuels. 19 Apr. 2021. Power Generation: https://www.brasilbiofuels.com.br/en/power-generation/
66. USDA/FAS – United States Department of Agriculture / Foreign Agricultural Service. 2018. Ecuador: Ethanol Industry Emerging in Ecuador: https://www.fas.usda.gov/data/ecuador-ethanol-industry-emerging-ecuador
67. World News Today. 25 Dec. 2020. Government bets on biodiesel, but resist the use of more transgenics: https://www.world-today-news.com/government-bets-on-biodiesel-but-resists-the-use-of-more-transgenics/
68. De Beenhouwer, M., R. Aerts and O. Honnay. 2013. 'A global meta-analysis of the biodiversity and ecosystem service benefits of coffee and cacao agroforestry'. *Agriculture, Ecosystems & Environment* **175**: 1–7.
69. Jha, S., C.M. Bacon, S.M. Philpott, V. Ernesto Mendez, P. Läderach and R.A. Rice. 2014. 'Shade coffee: update on a disappearing refuge for biodiversity'. *BioScience* **64** (5): 416–428.
70. Parra, E.M. 2014. 'Desarrollo alternativo en el Perú: treinta años de aciertos y desaciertos'. *Perspectivas Rurales Nueva Época* **23**: 85–104.
71. Athelia Ecosphere Fund. 2017. Tambopata-Bahuaja Redd+ and Agroforestry Project: https://althelia.com/investment/tambopata-bahuaja-redd-and-agroforestry-project/
72. Rainforest Alliance. 2015. Climate-smart coffee and cacao in the Peruvian Amazon, http://www.rainforest-alliance.org/videos/icaa-megantoni-ra
73. Gestión. 17 May 2017. Producción peruana de café aumentará 8% en el 2017, proyecta el Scotiabank: http://gestion.pe/economia/produccion-peruana-cafe-aumentara-8-2017-reporto-scotiabank-2176971

74. USDA / FAS – United States Department of Agriculture / Foreign Agricultural Service. 2016. Coffee, Annual: https://gain.fas.usda.gov/Recent%20GAIN%20Publications/Coffee%20Annual_Lima_Peru_4-30-2015.pdf

75. Rainforest Alliance, 'Climate-smart coffee and cacao'.

76. EMBRAPA – Empresa Brasileira de Pesquisa Agropecuaria. 13 July 2015. Agricultura familiar: https://www.embrapa.br/web/mobile/noticias/-/noticia/3609412/investimento-em-tecnologia-impulsiona-producao-de-cafe-em-rondonia

77. Killeen, T.J. and G. Harper. 2015. Coffee in the twenty first century: Will global warming and increased demand lead to new deforestation? Conservation International: http://www.conservation.org/publications/Documents/CI-Coffee-Report.pdf

78. Tondoh, J.E., F.N.G. Kouamé, A.M. Guéi, B. Sey, A. Koné and N. Gnessougou. 2015. 'Ecological changes induced by full-sun cacao farming in Côte d'Ivoire'. *Global Ecology and Conservation* **3**: 575–595.

79. Alves. R. 4 Apr. 2021. Chocoholic Brazil seeks to redeem seat as top cocoa producer, The Brazil report: https://brazilian.report/business/2021/04/04/brazil-top-cocoa-producer/

80. Motamayor, J.C., P. Lachenaud, J.W.D.S. e Mota, R. Loor, D.N. Kuhn, J.S. Brown and R.J. Schnell, R. J. 2008. 'Geographic and genetic population differentiation of the Amazonian chocolate tree (*Theobroma cacao L*)'. *PLoS One* **3** (10): e3311.

81. PNUD – *Programa de las Naciones Unidos para el Desarrollo*. 2017. Ecuador fortalece sus esfuerzos para enfrentar la deforestación en la cuenca amazónica: http://www.undp.org/content/undp/es/home/presscenter/pressreleases/2017/03/17/ecuador-fortalece-sus-esfuerzos-para-enfrentar-la-deforesta-ci-n-en-la-cuenca-amaz-nica.html

82. MINAGRI. 2015. Estudio del Cacao en el Perú y el Mundo, Situación Actual y Perspectivas en el Mercado Nacional e Internacional al 2015: http://www.minagri.gob.pe/portal/analisis-economico/analisis-2016?download=10169:estudio-del-cacao-en-el-peru-y-en-el-mundo

83. Ibid.

84. Thomas, E., M. van Zonneveld, J. Loo, T. Hodgkin, G. Galluzzi and J. van Etten. 2012. 'Present spatial diversity patterns of *Theobroma cacao L.* in the neotropics reflect genetic differentiation in Pleistocene refugia followed by human-influenced dispersal'. *PLoS One* **7** (10): e47676.

85. Barrientos-Fuentes, J.C. and J.C. Torrico-Albino. 2014. 'Socio-economic perspectives of family farming in South America: cases of Bolivia, Colombia and Peru'. *Agronomía Colombiana* **32** (2): 266–275.

86. World Bank. 2017. Gaining Momentum in Peruvian Agriculture: Opportunities to Increase Productivity and Enhance Competitiveness: http://documents1.worldbank.org/curated/en/107451498513689693/pdf/P162084-06-26-2017-1498513685623.pdf

87. UNODC – United Nations Office on Drugs and Crime. 2010. World Drug Report 2010, Chapter 1.3 The global cocaine market: https://www.unodc.org/documents/wdr/WDR_2010/1.3_The_globa_cocaine_market.pdf ;

UNODC – United Nations Office on Drugs and Crime. 2019. Bolivia, Monitoreo de Cultivos de Coca 2019: https://www.unodc.org/documents/crop-monitoring/Bolivia/Bolivia_Informe_Monitoreo_Coca_2019.pdf ;

UNODC – United Nations Office on Drugs and Crime. 2020. Colombia, Coca cultivation survey 2019: https://www.unodc.org/documents/crop-monitoring/Colombia/Colombia_Monitoreo_Cultivos_Ilicitos_2019.pdf ;

UNODC – Oficina de las Naciones Unidas contra la Droga y el Delito. 2017. Perú, Monitoreo de Cultivos de Coca 2017: https://www.unodc.org/documents/crop-monitoring/Peru/Peru_Monitoreo_de_Cultivos_de_Coca_2017_web.pdf

88. Ibid.
89. Lambin, E.F., H.K. Gibbs.... and C. Nolte. 2018. 'The role of supply-chain initiatives in reducing deforestation'. *Nature Climate Change* **8** (2): 109–116.
90. Potts, J. et al. 2014. State of Sustainability Initiatives Review – 2014,International Institute for Sustainable Development: https://www. iisd. org/sites/default/files/pdf/2014/ssi_2014. pdf
91. McGrath, D.G., L. Castello, O.T. Almeidaand G.M. Estupiñán. 2015. 'Market formalization, governance, and the integration of community fisheries in the Brazilian Amazon'. *Society & Natural Resources* **28** (5): 513–529.
92. MAPA – *Ministério da Agricultura, Pecuária e Abastecimento*. 2020. Política Agrícola, Plano Safra 2020/2021 entra em vigor nesta quarta-feira Plano Dafra: https://www.gov.br/agricultura/pt-br/assuntos/noticias/plano-safra-2020-2021-entra-em-vigor-nesta-quarta-feira
93. Ibid.
94. Strassburg, B.B., T. Brooks, R. Feltran-Barbieri, A. Iribarrem, R. Crouzeilles, R. Loyola… and A. Balmford. 2017. 'Moment of truth for the Cerrado hotspot'. *Nature Ecology & Evolution* **1** (4): 1–3.
95. Yuri Ramos, S. and G. Bueno Martha Jr. 2010. Evoluçcão da Politica de Créditop Rural Brasileria, EMBRAPA Cerrado, Brasilia.
96. Souza, P., S. Herschmann and J. Assunção. 2020. *Política de Crédito Rural no Brasil: Agropecuária, Proteção Ambiental e Desenvolvimento Econômico*. Climate Policy Initiative, Rio de Janeiro: https://www.climatepolicyinitiative.org/pt-br/publication/politica-de-credito-rural-no-brasil-agropecuaria-protecao-ambiental-e-desenvolvimento-economico/
97. Forests & Finance Coalition. 2021. Finance data, Explore the data: https://forestsandfinance.org/data/
98. IMEA – *Instituto Mato-Grossense de Economia Agropecuária*. 2021. CUSTO DE PRODUÇÃO, DSoja GMO: https://www.imea.com.br/imea-site/relatorios-mercado-detalhe?c=4&s=3
99. Forests & Finance Coalition, Finance data.
100. OXFAM – BRASIL. 2016. Terrenos da Desigualdade: https://www. oxfam. org. br/publicacoes/terrenos-da-desigualdade-terra-agricultura-e-desigualdade-no-brasil-rural
101. Climate Bond Initiatives. 2021. https://www.climatebonds.net/files/reports/cbi-brazil-agri-sotm-eng.pdf

102. FAIRR Farm Animal Investment Risk and Return Initiative (2019) https://www.fairr.org/article/marfrigs-transition-bond

103. Pinzon, A. 2019. Redefining finance for agriculture: Green agricultural credit for smallholders in Peru. Global Canopy: https://www.meda.org/innovate/innovate-resources/822-partner-publication-redefining-finance-for-agriculture-green-agricultural-credit-for-smallholders-in-peru-eng/file/

104. SBS – *Superintendencia de Banca, Seguros y AFP*. 2021. Tasa de interés promedio del sistema de cajas rurales de ahorro y crédito: https://www.sbs.gob.pe/app/pp/EstadisticasSAEEPortal/Paginas/TIActivaTipoCreditoEmpresa.aspx?tip=C

105. Sengupta, R. and C.P. Aubuchon. 2008. 'The microfinance revolution: An overview'. *Federal Reserve Bank of St. Louis Review* 90 (Jan./Feb.)

106. Gestión. 26 Nov. 2020. El FAE-AGRO Es Un Completo Fracaso, Grupo Propuesta Ciudadana: https://gestion.pe/blog/propuesta-ciudadana/2020/11/el-fae-agro-es-un-completo-fracaso.html/?ref=gesr

107. Ibid.

108. ASOBANCO – *Asociación de Bancos del Ecuador*. 2019. Informe Técnico: Tasas de Interés, Volumen N°1: https://www.asobanca.org.ec/file/2286/download?token=IHmybA6t

109. BanEcuador. 29 May 2021. Credito Productivo Palma, Microempresas: https://www.banecuador.fin.ec/productos-ciudadanos/credito-micro/productos-microempresas/credito-palma/

110. FINAGRO – *Fondo para el Financiamiento del Sector Agropecuario*. 2 June 2021. Productos y Servicios, Portfolio de Servicios: https://www.finagro.com.co/productos-y-servicios

111. FINAGRO – *Fondo para el Financiamiento del Sector Agropecuario*. 2 June 2021. Estadísticas: https://www.finagro.com.co/estad%C3%ADsticas

112. Banco do Brasil (26 July 2021) Agricultura de Baixo Carbono (ABC), https://www.bb.com.br/pbb/pagina-inicial/agronegocios/agronegocio---produtos-e-servicos/credito/investir-em-sua-atividade/agricultura-de-baixo-carbono-(abc)#/

Chapter 4

Land: The Ultimate Commodity

New roads open wilderness landscapes to development, and commodity markets drive the expansion of the agricultural frontier. These two causes of deforestation are at the centre of deforestation policy discussions. A third factor – land values and their tendency to appreciate over time – is a synergistic product of these two phenomena. Understanding the dynamics of rural real estate markets is essential in devising policies to halt the advance of the conventional economy into the forest wilderness.

The agricultural frontier in the Pan Amazon is the product of centuries of cultural tradition and decades of economic policy. This phenomenon, which is central to the history of the Western Hemisphere, became a major disruptive force in the Pan Amazon only in the 1960s, when governments implemented programmes to occupy and develop their Amazonian hinterlands (Chapter 6). Unlike previous colonisation periods, such as the rubber boom of the nineteenth century, this latter period included initiatives to promote the mass migration of families into the region, which were combined with strategies to attract investment in market-based production systems. These policies were contingent on the offer of free, or nearly free, public land.

Access to land was conditional, however, and pioneers had to install a productive enterprise, which obligated them to replace natural vegetation with cultivated plants. Official policies have changed, but this practice continues to motivate individuals on the forest frontier, where people clear forest as a strategy to project ownership of land they view, rightly or wrongly, as their own. Most believe they are acting in the best interest of their families and their country by generating economic activity. They are aided and abetted by functionaries in agricultural ministries who implement outdated policies that facilitate the transfer of public lands to private individuals. Layered on top of (or underneath) this dysfunctional regulatory framework is a culture of graft, impunity and entitlement.

doi: 10.3197/9781912186228.ch04

The forest frontier continues to be invaded by migrant settlers and land grabbers (grileiros or traficantes de tierra), who use legal, extra-legal and illegal mechanisms to appropriate public land. The practice, which was once organised by the state, is still tolerated by some local and national authorities.

Rural real estate markets regard land as part commodity and part capital asset. As a commodity, its price is mediated by supply and demand: Parcels near to the forest frontier are less expensive because there is an available supply that can be acquired at low cost. As the forest frontier recedes, land appreciates in value because it becomes a more limited commodity. As a capital asset, properties increase in value with investment in on-farm infrastructure and perennial crops that generate cashflow over the short-term, such as coffee, cacao and oil palm, as well as timber species that pay a substantial dividend over the medium term.

Other considerations influence the price of land. If the soil is arable, land has additional value because farming is more lucrative than ranching. Forest remnants may or may not have commercial value, depending on whether they retain stands of hardwood timber. Despite their intrinsic value, degraded forests are viewed as 'unproductive' – unless they have been converted into 'productive land' dedicated to conventional agriculture.

All too frequently, landholders will first monetise the value of their timber and then use that capital to finance the conversion of the degraded forest into pasture or farmland.

The economics are straightforward: a pasture can support cattle and generate cash flow of ~ $US 200 per hectare annually, or $US 2,000 over ten years. This is a reasonable return on an investment that requires a rancher to clear the forest, build fencing and construct a water impoundment at a cost of about $US 500 per hectare. More importantly, the value of land itself will appreciate over time, reflecting both the improvement of on-farm infrastructure and the generally upward direction of real estate markets (see below). Similar economic calculations drive investment decisions on smallholder landscapes, where properties can experience a step-change in value with the establishment of a perennial crop like coffee, cacao or oil palm.

Pioneer families are active participants in rural real estate markets. They use their knowledge of soil, water and natural vegetation to develop additional landholdings that they sell to investors and newly arrived migrants. Some become frontier entrepreneurs who specialise in the acquisition and development of properties. Many are businesspeople who are 'improving' properties deforested during previous cycles of settlement. One of their main marketing tools – and a core service – is to complete the titling process. A certified legal title significantly enhances the market price of a property.

Unfortunately, legitimate real estate investors share the marketplace with unscrupulous individuals who invade public lands or displace families who have informally occupied them. Referred to as 'land grabbers' in the English-language media, in Brazil they are known as *grileiros* and in Spanish speaking countries as *traficantes de tierra.*[*]

The Distribution of Public Lands

Public lands have been, and continue to be, distributed via a variety of legal, quasi-legal and blatantly illegal mechanisms.[1] These mechanisms have evolved over time, but they can be broadly organised into four main categories:

1. *State-sponsored colonisation schemes:* This policy was predominant during the 1970s and 1980s and was managed by agencies with various names and acronyms (Table 4.1). Approaches varied among countries, but all targeted the rural poor and distributed landholdings between forty and 100 hectares. Some were organised via a communal tenancy regime while others ceded plots to individual families. Only Brazil

* They are called *grileiros* because they would place forged titles and deeds in a box with a few crickets (*grilo* in Portuguese) to make them appear old.

Source: Google Earth

The settlement of the alluvial plain of Santa Cruz began in the 1960s with a state-sponsored scheme that settled Andean migrants in villages with radially organised landholdings (a). In the 1980s, they were joined by additional migrants, organised in sindicatos, who established their own settlements by building roads and appropriating public forest (b). Immigrants from Canada and Mexico purchased large estates from local intermediaries to create Mennonite colonies (c). All three types of settler now pursue intensive agriculture using technologies similar to industrial-scale corporate farms (e), while new migrants continue to settle a floodplain once zoned for conservation and forest management (d).

continues to distribute land among its citizens via projects organised by a national agency, or it did until 2017 when an audit led to a temporary suspension of its activities (see Chapter 6).

2. *Direct land grants or sales by the state:* This mechanism was widely used in Brazil over several decades but was most prominent in the

1970s when Amazonian development was a core policy of the military government.* The distribution of large landholdings led to the development of the agro-industrial model that dominates the economy of Mato Grosso, Eastern Pará and Tocantins. A similar phenomenon occurred in Bolivia, where military governments distributed land to influential families using the agrarian reform institution originally created to address land tenure inequality. Large land grants in Ecuador led to the establishment of two large-scale oil palm plantations in the early 1980s.† The most recent example comes from Peru, where an influential corporation obtained large tracts of natural forest in 2005 to establish that country's largest oil palm plantation.‡

3. *Privately sponsored colonisation schemes:* This type of land distribution is a variant of the previous mechanism in which the state would grant a concession to a private company or cooperative that would subdivide and resell plots to settlers (see Chapter 6). This method promoted a middle-class farm model based on properties that range from a few hundred to several thousand hectares. It was a common business model in central Mato Grosso between the late 1950s and the early 1980s. Mennonite immigrants have employed a variant of this scheme in Bolivia, where a group of families collectively purchase a large private property, which they subdivide among themselves to create a 'colony' of 100-hectare family farms. This system is being replicated in Peru and Colombia, where Mennonite immigrants have been accused of clearing forest on landscapes zoned for forest management.[2] Mennonites are not known to invade public lands, choosing instead to purchase land from intermediaries, a tactic that improves the probability they will obtain legal title.

4. *Spontaneous settlement and land grabbing*: The appropriation of public lands via informal and blatantly illegal processes is common on all the forest frontiers in both the Andean republics and Brazil. It can occur as a land rush when a new trunk highway is created through a pristine forest landscape, but more often it occurs over decades as

* The infrastructure development initiative, *Plano de Integração Naciona* (PIN), incorporated within its framework the *Programa de Redistribuição de Terras e de Estímulo à Agroindústria do Norte e Nordeste* (PROTERRA), the *Programa de Desenvolvimento do Centro-Oeste* (PRODOESTE) and the *Programa de Pólos Agropecuários e Agrominerais da Amazônia* (POLOAMAZÔNIA). Source: Girardi (2015).

† Palmeras del Río (10,000 ha) near Coca and Palmeras del Ecuador (9,500 ha) near Shushufindi.

‡ Palmeras de Shanuzi is a subdivision of the Grupo Palmas, which is a subsidiary of the Grupo Romero; apparently, the original land grant consisted of 7,000 hectares, but the company has acquired adjacent land to establish a plantation covering 17,000 hectares. Source: Dourojeanni (2013).

secondary road networks expand outward from a trunk highway. In the 1980s, governments facilitated this process via special initiatives created to respond to petitions from interest groups and regional governments.* Depending upon the social and political environment, it can lead to the proliferation of large properties or small landholdings or a mixture of both.

Obtaining a Certified Legal Title

Occupying a plot of land is the first, and perhaps easiest, step in the process of creating a legally constituted private property. In all eight Amazonian nations, a title or a certification of a title must be issued by an agency of the central government, which in most cases is a lineal descendant of the colonisation agencies of the 1960s and 1970s (Table 4.1). At the time, these agencies issued provisional titles because full tenure was contingent upon establishing a successful homestead. This negative legacy grew over decades as the rural economy expanded and the number of landholdings multiplied.† One key responsibility of these agencies was the compilation of a land registry, known as a 'cadaster', which functions as a documentary reference point for all legal transactions involving rural property.

The decision to delegate the task of title certification to a national rather than a local agency was a logical consequence of the distribution of public lands by the central government.‡ A national solution probably appealed to central planners who doubted the capacity of local (frontier) governments to manage a large and technically complex undertaking. Individual landholdings are incorporated into the national rural cadaster, but only after their spatial attributes and legal providence have been validated by public servants.

The failure to complete this process and consolidate national cadasters is a major driver of the lawlessness that defines frontier society. These agencies, whether by design or happenstance, oversee a chaotic system where fraud and graft facilitate the misappropriation of public lands. As such, it is a fundamental driver of deforestation. It is also a massive moral failure because the system has failed to provide millions of smallholders with legal title to their most important financial asset. Successive governments and multilateral agencies have organised multiple initiatives to reform

---

* Examples include the *Grupo Executivo das Terras do Araguaia/Tocantins* (GETAT) and *Grupo Executivo para a Região do Baixo Amazonas* (GEBAM). Source: Hecht and Cochburn (2010).

† Urban properties are managed by local governments, typically the municipality.

‡ The decision to include properties from long-settled landscapes in the national registry was driven in part by the agrarian reform process where central governments expropriated and redistributed properties.

and modernise these agencies and, most importantly, complete the task of determining land ownership in the Pan Amazon. They have all failed.

Historically, transactions involving land are recorded with a notary public, known as a *Cartório* in Brazil and as *Notaría de Fe Pública* in Spanish-speaking countries. These legal offices provide a more substantive service than their counterparts in than United States because they keep a legal copy of all contracts and transactions as well as validate certain legal principles common to contract law. This type of documentation provides the primary legal basis for most rural landholdings.

The agencies compiling national cadasters have protocols for validating landholdings and incorporating them into the national cadaster using a property's '*historial*', essentially a paper trail that documents its origin and previous transfers or subdivisions. These protocols open a door for fraud because land grabbers use them to invent a legal history or to clone another property's past with forged documents. Since land registries have massive backlogs of unprocessed land claims, it is often necessary to provide a cash payment to 'expedite' legitimate transactions. The practice of paying a bribe to process a legal transaction provides cover for land grabbers processing illegitimate documents.

The regulatory framework is further complicated by two distinct levels of land tenure: ownership and possession. As the terms imply, an owner (*proprietário*) holds a legal title to a property, while a possessor (*poseiduero*) lacks a legal document validating ownership but is occupying the property and using it for his or her economic benefit. Logically, ownership has more rights than possession, but a possessor is not devoid of legal protection, including the right to not be evicted from the property if he or she is utilising it according to principles referred to as a 'social and economic function'. There is an implicit assumption that possession will eventually be transformed into ownership; nonetheless, the lack of a clear legal title impacts a property's value in real estate transactions.

Insecure land title and corrupt systems also impact legitimate landholders. In Brazil, there is a long history of ranchers dispossessing smallholders and forest dwellers by inventing documents and then using violence to evict them from their homes. In Bolivia, squatters will invade a property if its owner is incapable of demonstrating clear title and lacks the economic resources to physically defend the property. On occasion, squatters are paid agents acting on behalf of a land grabber who is preying on a family perceived to be weak. Insecure land tenure is an invitation for bad actors to use force to obtain what does not belong to them.

Violence and Land

The adage 'possession is nine-tenths of the law' is not legally true, but the concept reigns supreme on frontier landscapes in the Pan Amazon. Land grabbers and peasant pioneers share a *modus operandi*: they occupy land that does not belong to them. Historically, this process was condoned by the state, and conflict occurred only when the two groups competed for the same territory – or when either group sought to steal land from forest communities. Smallholders have the advantage of numbers, while land grabbers use their political connections to formalise their claims and label their competitors as 'squatters'. In Brazil and Bolivia, ranchers use force to clear landholdings, usually by hiring thugs to beat the smallholders and destroy their belongings. The smallholders resist by organising themselves into peasant syndicates associated with the *Movimento Sim Terra* (MST) and the *Confederación Sindical Única de Trabajadores Campesinos de Bolivia* (CSUTCB). Resistance leads to an escalation of violence.

In Brazil, criminal land grabbers contract *pistoleiros* to murder *posseiros** who stand in their way. The most famous incidents have involved activists who were assassinated for defending the rights of forest people and smallholder peasants, most notably Francisco Alves (Chico) Mendes, who was ambushed at his home in Xapuri, Acre, in 1988; and Dorothy Stang, who was executed in 2005 on a remote road near Anapú, Pará. These crimes led to high-profile public prosecutions and the incarceration of the men who pulled the trigger, as well as the ranchers who contracted their services. Unfortunately, it is more common for these mafia-style murders to remain unsolved and, even when identified, most perpetrators escape justice – the exact definition of impunity.

The *Comissão Pastoral da Terra* (CPT) has monitored rural violence in the Brazilian Amazon for more than four decades and has compiled a gruesome historical archive: *Massacres no Campo* lists 47 incidents and details the murder of 341 individuals.[3] More than half are identified as 'leaders' and more than seventy per cent are linked to disputes over land tenure between large-scale landholders and landless peasants. That total vastly underestimates rural violence, however, because it only includes clashes where at least three people died. Since 2011, the CPT has compiled more precise statistics that reveal that little has changed and the situation may be getting worse. In 2017, there were more than 980 separate incidents impacting more than 98,000 families; 56 people were killed, mostly landless *posseiros* occupying ranches deemed vulnerable to an organised occupation.[4]

---

* A less derogatory term for squatter used by Brazilians to identify individuals and families that may have a legal right to occupy public lands. Source: Ferreira (1986).

The ongoing closure of the forest frontier has increased the pressure on consolidated landscapes in Southeast Pará and Southern Rondônia, where a new militant organisation, the *Liga de Camponeses Pobres* (LPT), has tapped into the discontent engendered by the inequities of land distribution. Large landholders increasingly use private security forces and police to enforce judicial evictions. The involvement of police, however, is no guarantee of a just or orderly process, as revealed in 2017 at the Fazenda Santa Lucia in the municipality of Pau D'Arco (Pará) where seventeen police officers were accused of executing ten *posseiros* in a court-ordered eviction process.* The most lethal municipalities are Anapú, Pará (16 dead), Vilhena, Rondônia (13 dead), Colniza, Mato Grosso (11 dead), Pau D'Arco, Pará (10 dead) and Porto Velho, Rondônia (10 dead).[5]

Although *posseiros* suffer the most violence, indigenous communities continue to be attacked on frontier landscapes where land grabbing is most prevalent, particularly along BR-163 in Pará and BR-230 in Amazonas. Communities suffering the highest levels of violence are in the heavily deforested regions of Maranhão, where nineteen indigenous men and women died while protecting their reserves from timber thieves.[6] Not even remote indigenous reserves are immune from violence, particularly the Munduruku and Yanomami communities, which must contend with the notoriously violent wildcat gold miners (see Chapters 5 and 11).

The plight of lowland indigenous communities in Peru and Bolivia is both more acute and very different when compared to Brazil. Although their national governments profess to support the territorial claims of native people, they have deployed security forces to violently suppress indigenous groups when they protest policies that threaten their territories. In 2009, the administration of Alain Garcia enacted laws that would have created a pathway for the privatisation of native landholdings. The resulting campaign of civil disobedience ended in a violent confrontation and the deaths of 33 individuals.† A similar skirmish occurred in Bolivia in 2011 when the government of Evo Morales attempted to obstruct a march protesting the

---

* The disputed landholding was first occupied in 2010 by families associated with *Liga de Camponeses Pobres* (LCP). According to witnesses, the unarmed victims were tortured prior to their execution; subsequently, three witnesses were murdered. The policemen were jailed briefly but, as of September 2020, none had been formally accused of murder. Source: EJA – *Environmental Justice Atlas*. 2020. Land-grabbing and disputed cattle ranch in Pau-d'Arco, Pará, Brazil. https://ejatlas.org/conflict/land-grabbing-and-disputed-cattle-ranch-in-pau-darco-para-brazil

† The incident is known as the *Baguazo* (after the locality of Bagua). Fifty-two protesters were accused of plotting to kidnap and kill police officers; all were found innocent at trial. The Peruvian Congress rescinded the law that threatened communal landholdings, but the conflict over mineral rights remains unresolved (see Ch. 11).

construction of a highway through the *Territorio Indígena y Parque Nacional Isiboro Securé* (TIPNIS). Nobody died, but police beat and arrested dozens of men and women in a flagrant violation of their civil rights.*

Less newsworthy but more insidious are the invasions of communal landholdings adjacent to colonisation zones, usually by highland indigenous migrants who enjoy the tacit support of their national and regional governments. In Bolivia, this is portrayed as agrarian reform by the central government, which is distributing public land to settlers and land grabbers in the forests of Chiquitania and Guarayos (see below). In Peru, native communities are struggling to protect their communal landholdings from wildcat gold miners and illegal loggers operating with the collusion of regional authorities. At least 22 indigenous leaders have been assassinated since 2013; more than half were threatened prior to their murder and had requested protection from police.[7] The forest frontier in Ucayali and Huánuco (HML #40) is particularly dangerous due to the presence of criminal gangs dedicated to the production of cocaine, who have targeted leaders of the Cacataibo and Shinobo–Conibo ethnic groups.† As of March 2021, none of the assassins had been apprehended by the police, allegedly due to the complicity of local authorities in the production and commercialisation of illicit drugs.[8]

Violence and murder are endemic to the Colombian Amazon due to decades of civil war and an economy based on the production of illicit drugs. The peace process has brought an end neither to armed conflict nor to the scramble for land. Criminal gangs composed of ex-combatants now fight for control of the borderlands between Caquetá, Meta and Guaviare. The central government has been unable to assert control, and competing bands recruit peasants to clear the forest to establish coca fields and cattle ranches.[9] There are no specific statistics on land-related violence but, pre-

---

* The incident, known as *Chaparina*, was a public relations disaster for the government, in part because citizens of Rurrenabaque overwhelmed the police escorting protesters for transport to their arraignment. The freed protesters completed their march to La Paz and the Brazilian construction company vacated the contract when the government failed to obtain the free, prior and informed consent of the communities that would be impacted by the highway (see below and Ch. 11).

† Five members of the Shinobo-Conibo were murdered (24 Apr. 2018), including an 81-year-old woman: Olivia Arévalo Lomas (source: https://www.frontlinedefenders.org/en); at least ten Cacataibo men have been murdered since 2020, including Arbildo Meléndez (12 Apr. 2020), Herasmo Gracia (25 Feb. 2021) and Yenes Ríos (20 Feb. 2021). Source: https://elpais.com/planeta-futuro/2021-03-16/que-hay-detras-de-los-asesinatos-de-lideres-indigenas-en-la-amazonia-peruana.html

sumably, it is a major cause of death in a region where the overall murder rate (32 per 100,000) is among the highest in the Americas.*

Rural Real Estate Markets

Mark Twain once said, 'Buy land, they're not making it anymore'. Samuel Clemens was a literary genius, but he was a notoriously poor investor.† Nonetheless, his observation on the intrinsic value of land is inherently true and explains, in part, the scramble for land in the Pan Amazon. Settlers, investors and politicians all know that the distribution of public lands will eventually end. The appropriation of public land no longer occurs on the consolidated frontiers, but it continues to plague the margins of agricultural frontiers and is the defining characteristic of forest frontiers. The ongoing creation of new landholdings, legal and otherwise, on the forest frontier impacts the price of land in more settled landscapes. Simultaneously, the demand for arable land in consolidated municipalities inflates the value of holdings on adjacent landscapes. Rural real estate markets reflect the dynamic of supply and demand across the entire development frontier.

In remote corners of the forest frontier, newly created homesteads and their associated land claims are typically hard to sell. Transactions are loaded with risk due to the dubious nature of deeds and the potential for squatters to invade properties. Risk is amplified by the threat of violence because settlers and land grabbers employ force to protect their claims. Both sell their land to risk-tolerant investors and later migrants and, in the process, create the market for rural real estate. Prices are low and rise slowly, but early-stage participants are confident that the region will eventually evolve into an agricultural frontier and reward them for their audacity and disregard for the law.

Properties on agricultural and consolidated frontiers are more valuable because of improved infrastructure and better access to markets. They also are located on landscapes with a more mature legal status, where holdings have been transformed into 'safe' investments. The jeopardy from bad papers has not disappeared, but due diligence and preventative legal action

---

* The highest murder rate (per 100,000) in the Americas is El Salvador (49) and the lowest is Canada (1.6); the most lethal department in Colombia is the Valle de Cauca (51); in Brazil it is Pará (46). Sources: United Nations Office on Drugs and Crime, DANE and IBGE.

† Mark Twain (Samuel Clemens) made a fortune writing books, but lost a fortune on investments in an automated engraving process, a magnetic telegraph, a steam pulley and a watchmaker, as well as in a turn-of-the-century railroad stock bubble. He declared personal bankruptcy in 1894 and, although he eventually paid all his creditors, he did so by writing and not by investing. Source: *Encyclopedia Britannica.*

can mitigate the risk. More importantly, conventional production systems generate cashflow and a decent return on investment.*

Land investment must be viewed in the context of the domestic economies: all Amazonian nations have suffered severe bouts of hyperinflation within the living memory of anybody older than fifty.† Capital invested in landholdings might be illiquid during times of crisis, but it always recovers its value. The same cannot be said for savings held in bank accounts or stock markets subject to erratic, often confiscatory, government policies. The return on real estate is the best option for most Latin American investors

In Iberian cultures, a predilection for land also has strong cultural appeal, which motivates urban professionals to invest in rural properties. In Brazil and Bolivia, this is manifest in an avocation for *estancias* or *fazendas* that raise cattle, while in Peru, Ecuador and Colombia urban investors are attracted to *fincas* that grow coffee, cacao or oil palm. Absentee landlords, including physicians, lawyers and other professionals, are attracted by the appreciation and the preservation of capital, but they also bask in the prestige of being a farmer or rancher.

The connection between city and countryside includes working-class families whose forebears settled smallholdings in Rondônia and Pará or one of the colonisation zones in the foothills of the Andes. As they have in rural families everywhere, younger generations have migrated to the cities while keeping their attachment to the family homestead. Money flows in both directions: towards urban dwellers attending school or seeking medical attention but also back to the farm as a remittance that can be used to invest in land, livestock and plantations. Successful families expand their holdings by buying adjacent parcels or by appropriating more land from the forest estate. Land tenure maps show hundreds of thousands of small plots (see Annex 4.1 to 4.11), but an individual family often owns multiple parcels. Small farms tend to be unviable, at least with conventional production models, and consolidation is a market-based cure for unviable settlement policies.

Savings and investment by professional and working-class families is a factor in the appreciation of rural real estate and, indirectly, a driver of deforestation. A more immediate economic force causing the appreciation

* Return on investment (ROI) is a standard metric that investors use when evaluating investment options; it includes both net revenues (profits) and capital appreciation (land value). Small differences (1–2%) will translate into very large differentials when compounded over several years.

† Bolivia (1984: 23,000%); Brazil (1994: 2,086%); Colombia (1985–1995: >25% annually); Ecuador (1985–2000: > 20% annually); Guyana (1991: 100%); Peru (1990: 7,481%); Suriname (1994: 125%); Venezuela (2018: 66,000%). Source: https://knoema.com/

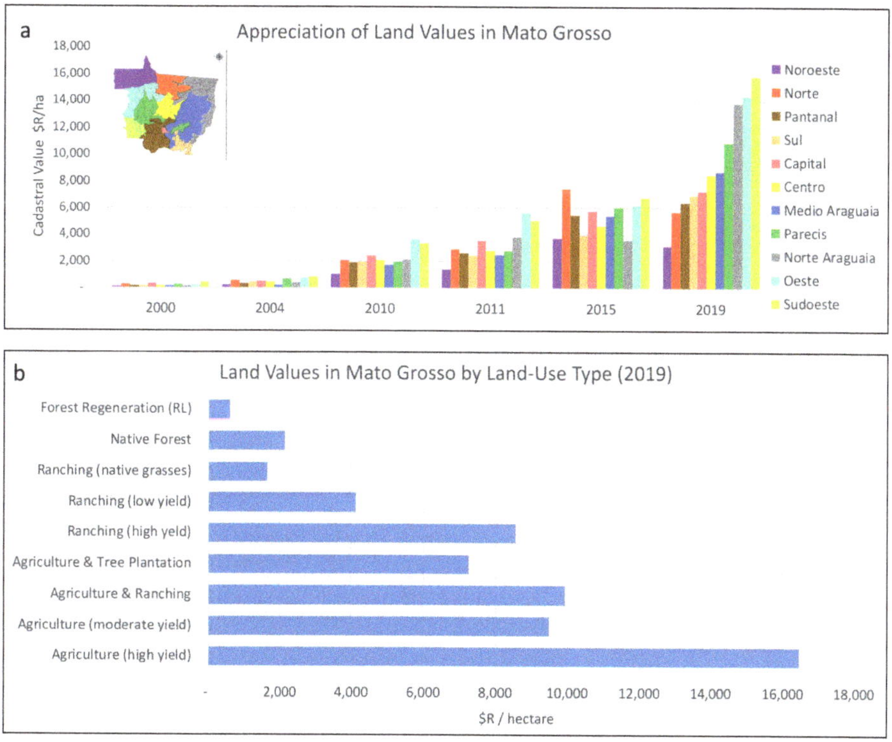

Figure 4.1: The appreciation of land value in Mato Grosso reflects the expansion and profitability of its agro-industrial sector. (a) Values increased by about 10% annually between 2000 and 2012, and 25% and 50% between 2012 and 2020. (b) The most valuable land has soil and topography appropriate for industrial agriculture; the lowest market value is assigned to land set aside to comply with the Forest Code (RL: Reserva Legal).
Data source: INCRA (2019).

of land values is the extraordinarily lucrative business model pursued by industrial agriculture.

The impact of this type of agricultural production on land values is most evident in Mato Grosso (Figure 4.1). In 2019, mean land values in the municipalities dominated by massive corporate farms were about R$ 12,000 per hectare (*Parecis*). In contrast, properties in the northwest corner (*Noroeste*), where timber extraction and cattle raising predominate, had a mean value of R$ 3,100 per hectare. In both regions, however, the price of land has exploded over the last twenty years with reported increases in market value between 2,500% (*Noroeste*) and 3,500% (*Parecis*).*

This level of asset appreciation is equivalent to the increase in the average value of farmland in the US Midwest between 1900 and 2000 – a century of growth in only twenty years! Increases in valuations at this rate are often indicative of a market bubble.† Perhaps. The recently reported gains in three of the sample regions (*Sudeste, Oeste* and *Norte Araguaia*) are occurring on landscapes considered to be expansion zones and are overpriced, at least when compared to the farmland in the original soy belt (*Parecis* and *Centro*). Market corrections are evident in the decrease in valuations between 2015 and 2019 in the northern expansion zone (*Norte*), which experienced a surge in prices simultaneous with the paving of BR-163 (see Chapter 2). Regardless, land in central Mato Grosso ($US 3,000 per hectare) is still affordable when compared to other regions that produce soy and maize, such as Paraná ($US 8,000 per hectare)[10] and Iowa ($US 18,000 per hectare).[11]

The appreciation of land is a core component of the business model of cattle ranchers across the Southern Amazon. Many operate on relatively thin margins that cause them to overgraze pastures and degrade soils; many have expanded operations by clearing small patches of forest annually over many years. The opportunity to sell can be a windfall. For example, a middle-class rancher in Alta Floresta with a 1,000-hectare property valued in 2000 at approximately R$ 300,000 ($US 190,000) could potentially sell that property in 2020 for R$ 5.7 million (~$US 1.1 million). The capital gains would be roughly equivalent to his net earnings over that same twenty-year period (see Chapter 3). After paying a capital gains tax, the rancher would have ample resources for a comfortable retirement or could avoid paying capital gains tax by purchasing another landholding. One option might be to buy a ranch on a forest frontier where land values remain affordable.

The appreciation of land creates positive feedback loops that benefit large-scale agriculture. Consider an agribusiness corporation with landholdings of 100,000 hectares operating in central Mato Grosso. The capital appreciation between 2000 and 2019 would surpass $US 250 million. Although the increase in value might not be monetised via a sale, its

* Property values are reported by the regional office of INCRA based on a standardised methodology used to compensate parties involved in legal or administrative actions pursued by the state. They are allegedly significantly below (25–50%) actual commercial values.

† Farmland in the US Midwest saw bubbles in the early 1920s (post WWI) and 1970 (post farm reform) when mean farmland values increased about 10% per year for a decade then fell in the 1980s by about 25%. Source: USDA. https://www.nass.usda.gov/Publications/Trends_in_U.S._Agriculture/Land_Values/index.php

© Paralaxis/Shutterstock.com

Deforestation not only occurs on the forest frontier but also on long-established landholdings, whose owners progressively create new pastures to expand their herds or to replace pastures degraded by overgrazing.

book value would be incorporated into a corporate balance sheet. Strong balance sheets are at the core of corporate finance because they reduce the cost of credit and attract new equity investors. Approximately ten per cent of the private landholdings in Mato Grosso (15,000 properties) encompass seventy per cent (46 million hectares) of the total area allocated to private landholdings (67 million hectares).[12] The appreciation in the value of those properties would sum to between $US 83 and 100 billion; that value, however, is dwarfed by the capital gains enjoyed by the plutocrats who acquired their properties at virtually no cost in the 1970s.

The increase in the cost of land has motivated agroindustry to develop alternative financial models for accessing land. Joint ventures between farmer-entrepreneurs and landholding ranchers are now common. The most common type of joint venture is a lease negotiated in terms of *sacos*

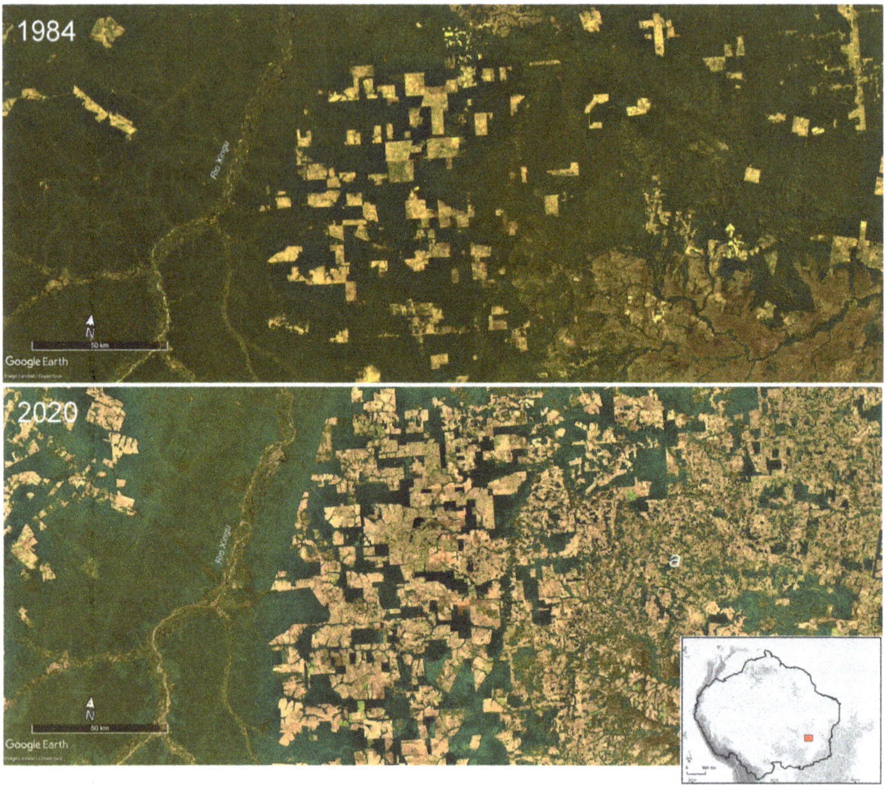

Source: Google Earth

Large-scale ranchers were well established by the 1980s on land acquired in northeast Mato Grosso state in the late 1970s. Influential investors acquired landholdings between 5,000 and 15,000 hectares at virtually no cost; several have consolidated these properties into even larger estates. Most retain significant areas of forest habitat but very few are in full compliance with the Forest Code. Small- and medium-scale ranchers settled the more heavily deforested landscape located to the East (a). The intact forest corridor along the Xingu River is the Parque Indigena de Xingu (PIX), the first large-scale indigenous reserve in the Amazon.

de soja (soy bags).* This stratagem mitigates risk from volatile commodity markets and exchange-rate fluctuations that can wreak havoc on a business enterprise with fixed costs measured in local currency. If the price of soy falls or the Brazilian currency weakens, the farmer is not locked into a contract based on a fixed monetary amount but instead shares the reduc-

* A *saco* (bag) is a traditional measure of bulk in Portuguese; its conversion to metric values (tonnes) depends upon the commodity, but in the case of soy is 60 kg.

tion in diminished revenues with the landholder. Ranchers can afford to be flexible because even reduced revenues are better than the proceeds from conventional beef cattle operations.

Ranchland is at a premium because sustainability protocols adopted following the Soy Moratorium (Chapter 3) limit the ability of landholders to convert native forest. This has inflated the value of pastures, including those adjacent to existing production landscapes, as well as those in more remote areas or along transportation corridors. Even highly degraded soils, the product of decades of overgrazing, can be attractive to a soybean farmer because the application of limestone ($CaCo_3$) or gypsum ($CaSO_4$), used to ameliorate soil acidity, also resolves the loss of fertility that limits the stocking rates of degraded pastures. Sophisticated farmers deploy technology to micromanage plant nutrient levels and use minimum tillage technology to rebuild soil organic matter; consequently, they view topography, soil texture and previous land use to be more important than the nutrient status of potential farmland. Essentially, ranchers are being paid to restore their degraded soils.

The interactions between ranchers and farmers are pulling more ranchland into the soy-maize production system, either permanently or periodically. Simultaneously, industrial farming is expanding onto municipalities in Rondônia, Pará and Tocantins. Even remote landscapes are being impacted, including in southeast Amazonas, Roraima and Amapá, where soy is being cultivated on savanna landscapes and previously deforested landholdings (Chapter 3). The economic boom in agroindustry is impacting the value of land across the entire region.

Agrarian Reform Agencies and National Land Registry Systems

Rural real estate markets in the Pan Amazon are regulated by institutions that are a legacy of the agrarian reform movements that played a prominent role in domestic politics during the last half of the twentieth century. Prior to World War II, the region was characterised by a quasi-feudal land tenure system, with ownership concentrated among affluent families of European extraction. In Bolivia, Peru and Ecuador, large estates were dependent on the labour of indigenous peasants (*campesinos*) with ancestral ties to the land, while in Brazil, Colombia and Venezuela, the rural labour force was composed of individuals with a contractual relationship with the landowner. The states of the Guiana coast were in the early stages of post-colonial rule, and the relationship between landlord and tenant was in a state of flux, but landless peasants were the majority in an economic system that was overwhelmingly rural.

Table 4.1: Land agencies that oversaw the distribution of public lands in the 1960s and 1970s (left column) and their descendants, now responsible for compiling national rural land registries.

Country	Land Reform Agency / Colonisation Institute	Current day successors
Brazil:	INCRA - Instituto Nacional de Colonização e Reforma Agrária (1971) INTERPA – Instituto de Terras do Pará (1975 INTERMAT– Instituto de Terras do Mato Grosso (1977) ITERAM – Instituto de Terras e Colonização do Amazonas (1979); ITEAM – Instituto de Terras do Amazonas (2003); SECT – Secretaria de Estado das Cidades e Territórios de Amazonas (2016) ITERTINS – Instituto de Terras do Tocantins (1989) AMAPÁ – TERRAS - Instituto de Terras do Estado do Amapá (2019) ITRON – Instituto de Terras e Colonização de Rondônia (2019) ITERACRE – Instituto de Terras do Acre (2001)	INCRA; Instituto Nacional de Colonização e Reforma Agrária Núcleos Municipais de Regularização Fundiária (NMRF) & Secretaria das Cidades e Territórios (SECT) State land agencies: INTERMAT, INTERPA, SECT, ITERTINS, AMAPÁ-TERRAS. ITRON, ITERACRE
Bolivia:	INRA: Instituto Nacional de Reforma Agrária (1954–present) INC: Instituto Nacional de Colonzación (1965–1992)	INRA: Instituto Nacional de Reforma Agrária Ministerio de Desarrollo Rural y Tierras
Ecuador:	IERAC: Instituto Ecuatoriano de Reforma Agrária y Colonización (1964–1992) INDA: Instituto Nacional de Desarrollo Agrario (1992–2010)	SSTA: Subsecretaría de Tierras y Reforma Agrária (STRA) Ministerio de Agricultura y Ganadería Regional Governments
Colombia:	INCORA: Instituto Colombiano de la Reforma Agrária (1992–2007) INCODR: Instituto Colombiano de Desarrollo Rural (2007–2015)	Agencia Nacional de Tierras (ANT), Ministerio de Agricultura y Desarrollo Rural
Guyana:	The Land Registry	The Land Registry
Peru:	IRAC: Instituto de Reforma Agrária y Colonización (1963–1992) PETT: Proyecto Especial Titulación de Tierras (1992–2016) COFOPRI (2007) Organismo de Formalización de la Propiedad Informal	(MIDAGRI= Ministerio de Desarrollo Agrario y Riego Catastro Rural es parte del trabajo del MIDAGRI Regional governments Superintendencia Nacional de los Registros Públicos (SUNARP)
Suriname:		Management Instituut voor Grondregistratie en Land Informatie Systeem (MI-GLIS)
Venezuela:	Municipal land registries	Instituto Geográfico de Venezuela Simón Bolívar (IGVSB)

This inherent inequality was a political tinderbox that was exacerbated by the expanded influence of Marxist philosophies and the explosion of radical movements after Fidel Castro consolidated the Cuban Revolution. Governments throughout the region responded by enacting agrarian reform legislation. Unsurprisingly, these policies were unpopular with conservative elites seeking to protect their financial patrimony. The decades following the Cuban Revolution were dominated by military governments; these governments varied in their adhesion to the principles of genuine agrarian reform, but all seized upon a solution originally championed by Abraham Lincoln: colonise public lands on the frontier.*

Distributing public lands in wilderness areas was popular; better yet, it avoided the politically perilous measure of violating the property rights of the landowning elite. Governments created agrarian reform agencies as a response to claims for social justice, but they simultaneously delegated to these agencies the task of dispensing public lands in their Amazonian provinces (Table 4.1). The United States supported these initiatives via the newly created the United States Agency for International Development (USAID) and the Alliance for Progress, a programme launched by John F. Kennedy in 1961.[13] Ironically, legitimate concerns about social inequality in Latin America catalysed one of the great social and environmental disasters of the twentieth century: the invasion of indigenous lands and the deforestation of millions of hectares of tropical forest.

Brazil

Agrarian reform in Brazil was initiated by the *Estatuto da Terra* in 1964, a law that created two entities: the *Instituto Brasileiro de Reforma Agrária* to address the inequal distribution of land and the *Instituto Nacional de Desenvolvimento Agrário* to manage colonisation processes then getting underway.[14] In 1971, these two institutions were fused to create the *Instituto Nacional de Colonização y Reforma Agraria* (INCRA) as an autonomous entity within the Ministry of Agriculture. INCRA's administrative functions can be divided into three main categories: (1) the redistribution of land by agrarian reform, (2) the allocation of public lands through settlement programmes and (3) the creation and management of a national rural land registry. The first category has always been politically difficult, while the second has been beset with inefficiency and corruption. The third is INCRA's most important function because rural real estate markets, which mediate investment in agricultural production, depend on a functional land tenure

* The Homestead Act of 1862 was a policy enacted to win the support of the population during the Civil War; it provided settlers with 160 acres of public land if they completed five years of continuous residence and made basic improvements to the landholding.

system that guarantees property rights. A dysfunctional registry not only impedes investment, it undermines efforts to promote sustainable land use and combat land grabbing.

INCRA as an agrarian reform programme

INCRA was created in response to Brazil's long-standing inequality in the ownership of land. Statisticians use a metric known as the 'Gini Coefficient' to measure inequality. Usually, it is employed to evaluate wealth, but it can be applied to land ownership. In Brazil, the Gini Land Coefficient is 0.87, well above the regional average and among the highest in the world.* Despite INCRA's efforts to redistribute land and to populate the Amazon with small farmers, the concentration of land in Brazil has increased over the last half-century. This inequality, combined with rural poverty, nurtured peasant movements throughout the mid-decades of the twentieth century; these were consolidated in 1984 as a national organisation: *Movimento dos Trabalhadores Rurais Sem Terra* (MST). The MST currently has 1.5 million members, representing 370,000 families residing on approximately seven million hectares of encampments acquired by a combination of non-violent civil disobedience and legal combat.[15]

The demand for land and the political power of the MST has motivated successive administrations to embrace the first leg of INCRA's institutional mission. Since its founding, INCRA has redistributed ~4.3 million hectares, benefitting about 130,000 families in the consolidated rural landscapes in the South, Southeast and Central-West regions.[16] Those numbers are not large in the context of Brazil's rural land assets, however, and have not materially alleviated inequality of land ownership. The limited impact of these policies, which are largely achieved by the purchase or expropriation of private estates, explains the political importance of INCRA's second institutional pillar, which is largely dependent on the forest landscapes of the Legal Amazon.†

---

* The Gini coefficient is a measure of equality that varies between 0.0 (total equality among all economic sectors) and 1.0 (total concentration of wealth in the highest economic strata). Gini coefficients on land tenure for the major regions of the World are: Latin America (0.79), Europe (0.57), Africa (0.56) and Asia (0.55); there are no published statistics for North America. See Guereña (2016).

† INCRA also has programmes in the Northeast region, where about 8 million hectares have been distributed to 127,000 families. Source: INCRA Instituto Nacional De Colonização e Reforma Agrária – INCRA (2020) Acervo Fundiario: http://acervofundiario.incra.gov.br/acervo/acv.php

INCRA as a colonisation institute

INCRA's approach to distributing public land has changed over time. Known as *terras devolutas*,* they were largely the domain of state governments until 1971, when the military dictatorship decreed that state land situated 100 kilometres either side of a national highway was the domain of INCRA.† This was the era of *Programa de Integração Nacional – PIN* when thousands of kilometres of roads were under construction (see Chapter 6).

The original law was based not on the highways that were under construction but on the proposed national highway system, including hundreds of kilometres of roads in remote regions that were never actually built. Pará, Mata Grosso, Amapá and Roraima relinquished about seventy per cent of their surface area, Acre lost about ninety per cent, and Rondônia and Tocantins‡ literally ceded all their territory to the central government. Only the state of Amazonas retained control over significant parts of its territory.[17] The newly obtained federal land bank was divided into subunits referred to as *glebas*,§ which are periodically opened for settlement, sold or allocated to a specific public category based on ecological, social and economic criteria.

In the 1970s, INCRA initiated its Amazonian settlement programme by organising *Projetos de Colonização* (PC) as part of the POLOAMAZONIA programme.¶ Between eight and twelve million hectares were allocated for distribution as fifty to 100-hectare holdings adjacent to highways under construction in Rondônia and Acre (BR-364), Roraima (BR-175), Mata Grosso (BR-163), Pará (BR-230) and Maranhão (BR-316).[18] The colonisation

---

* The term *terra devolutas* roughly translates as 'returned land'; usage comes from the colonial period when the crown issued hereditary land grants to individuals with the stipulation that they be turned into productive enterprises within a given period of time. Most lands reverted to the state, as represented by the crown, which was succeeded by the Brazilian Republic and the Federal Union. Source: *Dictonário Ambiental*, https://www.oeco.org.br/dicionario-ambiental/27510-o-que-sao-terras-devolutas/

† Decreto-Lei n° 1.164, de 1° de Abril de 1971; it was modified by the Decreto-Lei n° 2.375, de 24 de novembro de 1987, which returned tuition over some of the previous land to back the states, particularly in the remote areas of Amazonas state where 'projected' highways have never been constructed.

‡ Tocantins was separated from the state of Goiás in 1988.

§ The term *gleba* has multiple definitions depending on context; it is derived from Latin for soil and in Portuguese can signify arable land, church land, state land and, in this context, land that has not been adjudicated for a specific purpose.

¶ There were three colonisation categories: *Projeto de Asentamentos Conjuntos* (PAC), *Projetos de Asentamentos Dirigidas* (PAD) and *Projetos de Asentamentso Rápidos* (PAR); the data from these settlements is excluded from most INCRA databases because land was deeded (albeit imperfectly) to individual families rather than members of an INCRA-sponsored settlement.

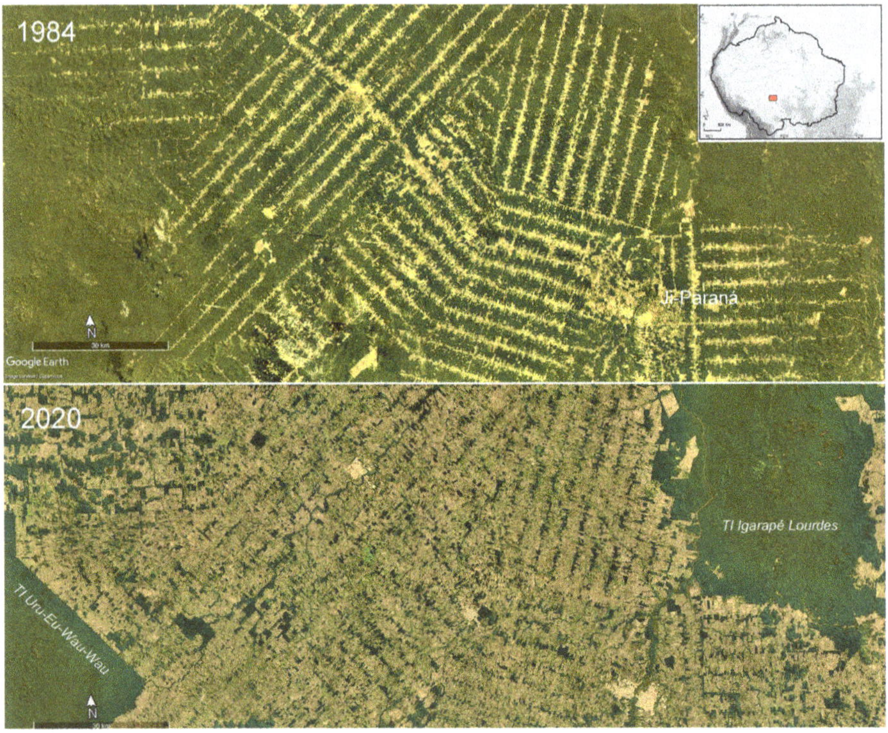

Source: Google Earth

Central Rondônia was settled by tens of thousands of settlers in the 1970s and 1980s on 50-hectare parcels distributed by INCRA in Projetos de Colonização (PC). The vast majority are small-scale livestock producers who retain the lowest proportion of remnant forest within the Brazilian Amazon (see Annex 4.7). The large blocks of remnant forest are indigenous territories.

programme was widely criticised because settlers were encouraged to migrate to remote landscapes and then left to fend for themselves. The land bank available to smallholders during the PC era had the capacity to accommodate about 120,000 families but INCRA succeeded in attracting only about 25,000 participants in the early stages of the programme.*

* The precise numbers are unknown because publicly available shapefiles that purport to show landscapes covered by the *Projetos de Colonização* are incomplete, particularly landscapes adjacent to BR-230 (Pará) and BR-364 (Rondônia). Most of these landscapes were not occupied during the first wave of settlement, but were settled in the subsequent decades with or without the intervention of INCRA. Source of data: INCRA (2020).

INCRA modified its procedures in 1984 and began taking a more coordinated approach to building pioneer communities, which were now referred to as *Projetos de Assentamento (PA)*. Like the previous policy, these explicitly favoured landless peasants, but INCRA now provided extension support and subsidised credit, while facilitating the delivery of public services by federal, state and municipal authorities (Figure 4.2 a).

Over time, the system evolved to include state and municipal settlement projects.* The PA system remained in place until 2000, allocating ~25 million hectares that currently benefits ~433,000 families (Table 4.2).[19] Within these territories, each family was granted a provisional right-of-use contract (*Contrato de Concessão de Uso* [CCU]) for a fifty-hectare plot; after five years, these are converted into a permanent right-of-use contract (*Contrato de Concessão de Direito Real de Uso* [CCDRU]) and, eventually, a title deed (*Título de Domínio* [TD]).

PA landholders can be transformed into owners (*proprietários*) after they have paid INCRA a nominal sum for their land and liquidated outstanding debts from credit programmes. Theoretically, the entire settlement can be '*emancipada*' if fifty per cent of the inhabitants opt for title deeds and vote to dissolve their settlement. This requires them to set aside land for public utilities (schools, clinics. etc.) and comply with norms dictated by the Forest Code (see Chapter 7); it also ends their access to INCRA-subsidised credit programmes and technical assistance. A fast-track emancipation process was approved in 2018 and the option is being promoted by the Bolsonaro administration as part of its policy of privatising public assets and promoting a market economy.[20]

Following the shift in environmental and development policies at the turn of the millennium, INCRA modified its land allocation paradigm to create *Projetos de Assentamento Ambientalmente Diferenciado* (PAAD). Unlike their agriculturally-oriented predecessors, these settlements are predicated on the sustainable exploitation of timber and non-timber forest products, fish and wildlife.† The difference in management philosophy has led INCRA to create larger land units with less dense human populations. Shifting agriculture is tolerated, but the emphasis is on sustainable production models informed (theoretically) by a management plan based on technical criteria elaborated via a consensual process.[21] As of 2020, INCRA had accommo-

---

* This category includes *Projeto de Assentamento Federal* (PA), *Projeto de Assentamento Estadual* (PE), *Projeto de Assentamento Municipal* (PAM), *Projeto de Assentamento Casulo* (PAC), *Reassentamento de Barragem* (PRB).

† This category includes *Projeto de Assentamento Agroextrativista* (PAE), *Projeto de Desenvolvimento Sustentável* (PDS), *Projeto de Assentamento Florestal* (PAF), *Projesto de Assentamenyo Colletivo* (PCC), *Projeto Descentralizado de Assentamento Sustentável* (PDAS), *Território Remanescentes Quilombola* (TRQ), *Reconhecimento de Assentamento de Fundo de Pasto* (PFP).

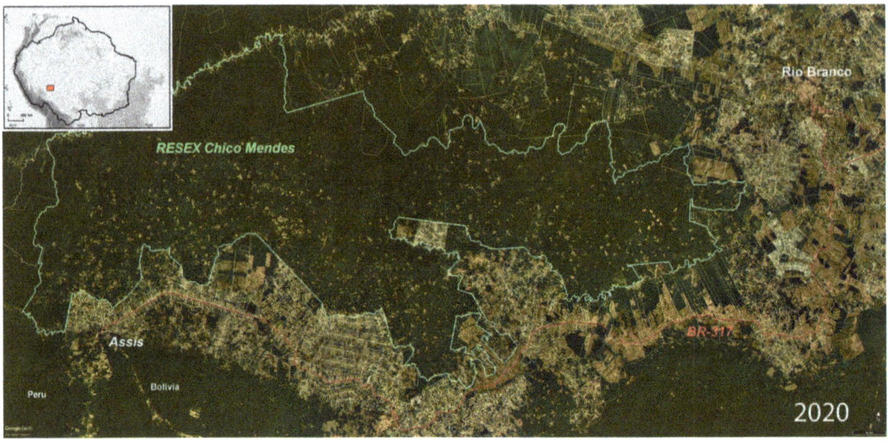

Source: Google Earth

Southern Acre was settled in the 1970s and 1980s by pioneers who established homesteads along BR-317. Most are small- to medium-scale ranchers (white polygons) and many reside in INCRA sponsored PA-type settlements. The settlement zone is bounded to the north by the Reserva Extrativa Chico Mendes (blue polygon), a sustainable-use conservation unit populated by families reliant on the annual harvest of Brazil nuts; most have cleared small patches of forest to generate cash income from micro-scale livestock operations.

Data source: Google Earth and INCRA (2020).

dated ~127,000 families within PAAD settlements covering ~13.5 million hectares (Table 4.2). Unlike the individual plots allocated to residents in the PA settlements, however, these have a communal tenancy regime. In most cases, residents are immediately granted a permanent long-term concession (CCDRU) because INCRA is essentially recognising the prior use-rights of established communities. Beneficiaries are never granted a full legal title, although they may sell their long-term concession to individuals who meet the legal conditions for participating in INCRA sponsored land projects.* Concessions within both PA and PAAD programmes can be passed on to heirs at the death of the beneficiary.

The PAAD settlements are similar to multiple-use protected areas managed by the Instituto Chico Mendes de Conservação da Biodiversidade (ICMBio), an agency within the Environmental Ministry. INCRA 'recognises'

* Eligibility is defined in negative terms; petitioners cannot (i) be public servants; (ii) have participated in a previous land reform, land regularisation or land credit programme; (iii) own a rural property unless that property is insufficient to support a family; (iv) own or be a shareholder in a private company; (v) be under eighteen years of age; and (vi) have income exceeding three monthly minimum wages. Source: INCRA (2021), https://www.gov.br/incra/pt-br/assuntos/reforma-agraria/acesso-a-terra

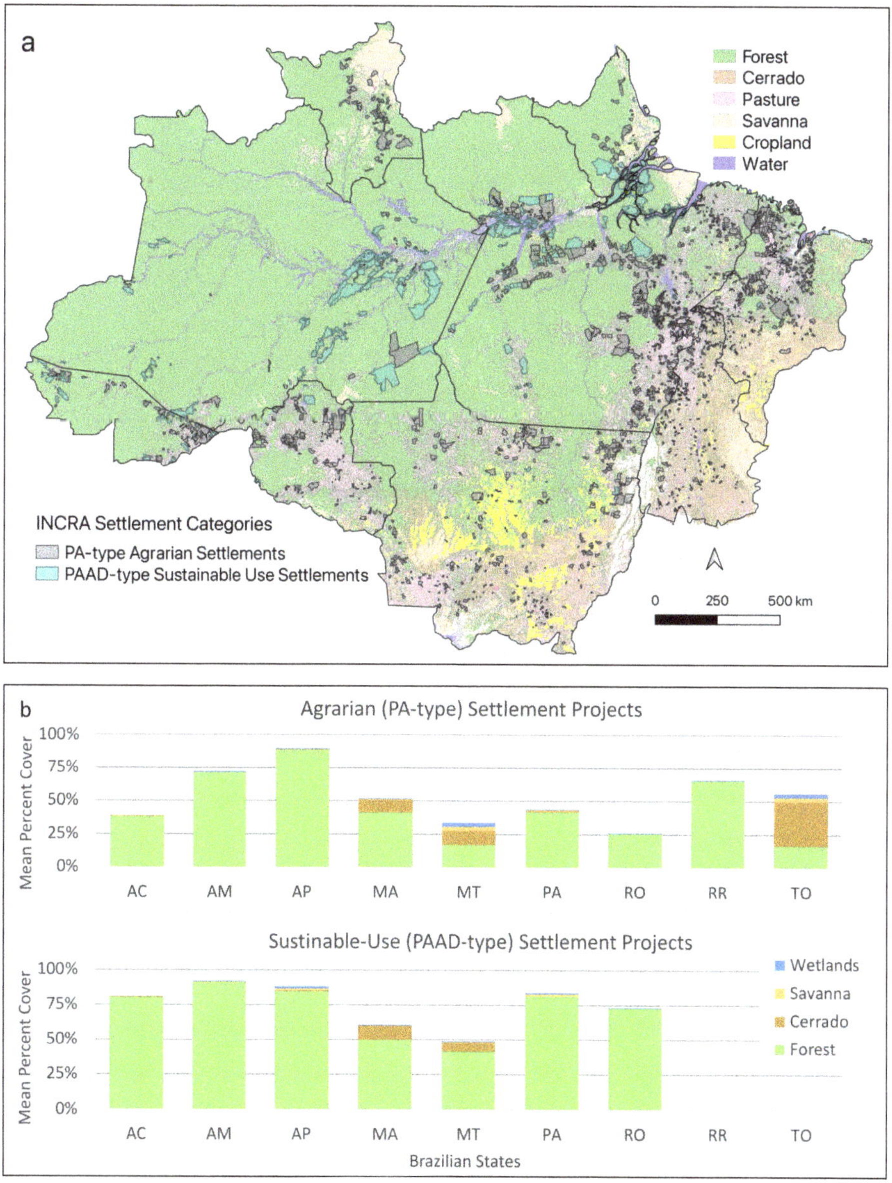

Figure 4.2: (a) The Instituto Nacional de Colonização y Reforma Agraria (INCRA) oversees two broad categories of settlement: agrarian communities (PA-type) established prior to 2000, and environmentally differentiated (PAAD-type) communities with sustainable-use management plans established after 2000. (b) Forest cover reflects category type but also geographic location.

Data sources: INCRA (2020) and MapBiomas (2021).

these conservation units within its institutional mission, which ensures their residents enjoy the same legal rights as beneficiaries in the agrarian reform settlements and have access to subsidised credit and key public services.* Because they are part of the protected-area system, they are subject to a greater level of scrutiny and, by many accounts, more institutional support.[22] They also enjoy a larger forest area to support their livelihoods, averaging about 500 hectares per family compared to only 100 hectares within the PAAD system. The difference in population density will be an important factor in determining whether these sustainable-use land-management units succeed in conserving the forest estate within their borders.

Forest monitoring programmes have identified the INCRA settlements as a significant source of deforestation. The earliest PC-landscapes in Rondônia and Mato Grosso have a mean forest cover of less than ten per cent although settlements with a similar history in Acre, Roraima and Pará retain between twenty and forty per cent. Similarly, PA settlements in eastern Pará have retained only vestigial areas of remnant forest (< 5%), while those in remote landscapes of Amapá and Amazonas retain as much as ninety per cent of their forest cover (Figure 4.2 b). Deforestation in PAAD settlements has been limited (0–10%), but is not insignificant. Forest conservation in both PA and PAAD landscapes is not necessarily a consequence of management criteria: remoteness, isolation and history also determine their fate as forest reserves. The annual deforestation rate within all the INCRA settlements fell from about 450,000 hectares between 2003 and 2005 to less than 70,000 hectares by 2015.[23]

INCRA as a regulatory agency

The third pillar of INCRA's institutional mission encompasses both administrative and legal aspects of land tenure and, as such, is the most important agency regulating rural real estate markets. Administratively, the institution is charged with collecting and organising the records of all rural properties in Brazil, including their creation and all subsequent sales, subdivisions and unifications. Legally, INCRA functionaries must review and verify that documents are legitimate and validate the spatial attributes of individual land parcels. This is a gargantuan task that would test the governance capacity

* Not all multiple-use categories qualify for INCRA accredited support – those that do include: *Reserva Extrativa* (RESEX), *Reserva de Desenvolvimento Sustentável* (RDS), *Floresta Nacional* (FLONA) and *Floresta Estadual* (FLOTU). Residents of *Área de Proteção Ambiental* (APA), a category used when there is a conflict concerning mineral exploitation or with private properties, are not eligible for the subsidised credit and services that come with INCRA's accreditation.

Table 4.2: The distribution of land via the settlement modes of INCRA and ICMBio within the Legal Amazon.

	PA system			**PAAD system**			**ICMBio system**		
INCRA Regional Office	Units	Families	Area (1,000 ha)	Units	Families	Area (1,000 ha)	Units	Families	Area (1,000 ha)
Pará-Belem	113	21,744	964	268	52,729	2,251	15	24,056	882
Pará-Marabá	511	71,806	4,656	18	27,439	27	-	-	-
Pará-Santarem	61	15,272	1,383	-	5,951	2,566	2	4,955	859
Pará-Altamira	39	12,787	1,252	-		921	4	2,252	2,728
Maranhão	1,007	125,868	4,117	14	1,349	61	4	1,603	214
Mato Grosso	528	75,380	5,214	13	1,539	112			
Acre	103	13,215	657	39	5,002	678	12	5,513	3,641
Amazonas	37	16,395	1,396	71	26,542	6,484	35	14,301	19,394
Rondônia	189	35,562	1,904	6	801	179,775	22	536	1,532
Amapá	32	8,248	1,567	21	5,042	175,698	1	1,409	502
Roraima	65	13,983	1,196	3	-	-	-	-	-
Tocantins	379	23,452	1,242	78	378	-	-	-	-
Total	**3,064**	**433,712**	**25,554**	**531**	**126,772**	**13,457**	**95**	**54,625**	**29,754**

Source: INCRA 2017.

of any country but is particularly challenging in a nation of continental dimensions undergoing a massive distribution of land.*

The decision to organise rural properties into a national land registry, *Sistema Nacional de Cadastro Rural* (SNCR), coincided with policies to transform the Amazon via migration and settlement. That task might have been completed automatically if the smallholder programmes, which distributed about twelve million hectares, had accurately and precisely recorded those transactions. Unfortunately, that did not happen. That missed opportunity

* In Europe, land registries are managed by national institutions, but countries are a fraction of the size of Brazil, and most properties enjoy centuries of historical records. In North America, the process is decentralised to the state or province for legal guidelines, while local jurisdictions manage the administrative tasks of record keeping and verification. At the national level, public and private databases are bottom-up compilations derived from thousands of systems.

was confounded by a collateral decision to facilitate a land rush that was occuring organically across the Southern and Eastern Amazon.

After about 1978, the military government became disenchanted by the smallholder settlement framework due to high overhead costs, low economic return and terrible public relations. Instead, they decided to expedite the transfer of public lands to corporations and influential families with the capacity to invest in productive enterprises at economies of scale. Over the next two decades, more than 100 million hectares of public land were transferred to large-scale landholders via a variety of legal and extra-legal operations.

The easiest way was to obtain a land grant from a government agency.* Sometimes these were disguised as a concession to organise a private colonisation project but were converted into a corporate estate. Another gambit was to cycle a small landholding through a series of transactions and to enlarge its dimensions at each stage. Many landholdings were manufactured out of whole cloth.[24] Questionable deeds were laundered by the *Fundo para Investimento Privado no Desenvolvimento da Amazônia* (FIDAM), a subsidiary of the *Superintendência do Desenvolvimento da Amazônia* (SUDAM),† which loaned money to corporate ranchers.[25] FIDAM required creditors to obtain documentation verifying their property rights from INCRA's regional offices or state agencies, all of which were staffed by individuals eager to facilitate the infusion of money into their jurisdictions.

The open collaboration of multiple state and federal agencies created a permissive environment that was exploited by speculators, who appropriated land that was sparsely populated by rubber tappers and indigenous communities. Tacit approval for fraudulent real estate transactions had been formalised in 1976 when the military government promulgated a land regularisation law that included a provision for conferring titles for properties that had been created via extra-legal procedures, if the current owners had purchased them in 'good faith'.[26] Each subsequent transfer of a property, or bank-mediated financial transaction, provided a layer of judicial security.

INCRA did not begin a serious effort to catalogue and review land tenure claims until about 1993, after which it launched periodic initiatives to consolidate the SNCR with increasing levels of electronic sophistication.[27] Landholdings deemed to be legitimate were incorporated into the SNCR

---

* The institutions with the most power (and land) were the *Companhia Desenvolvimento do Estado de Mato Grosso* (CODEMAT), which was succeeded by the *Instituto de Terras do Mato Grosso* (INTERMAT); the *Grupo Executivo das Terras do Araguaia/ Tocantins* (GETAT); and the *Instituto de Terras do Pará* (ITERPA). Source: Hecht and Cochburn (2010).

† SUDAM is a semi-autonomous federal agency (autarchy) which has existed in various permutations since 1944, and oversees development funds and programmes in Amazonia. See Ch. 6.

database, and owners were issued a *Certificado de Cadastro de Imóvel Rural* (see Text Box 4.1).

Text Box 4.1: Certificado de Cadastro de Imóvel Rural

> As the name *Rural Property Registration Certificate* implies, the CCIR is not a legal title residing within the *Cartorío* notary system; rather, the emission of a CCIR means that a landholding has passed through a due diligence process referred to as *regularização fundiária.* Brazilian law requires that a CCIR accompany legal contracts that 'transfer, lease, mortgage and dismember' a real estate asset; a CCIR is also routinely required by banks when approving a loan. A CCIR endows a property with *segurança jurídica* (legal certainty) and, in spite of claims to the contrary, effectively functions as legal title.

Following the restoration of democracy in 1985, civil society clamoured for legislative and legal action to combat land grabbing.[28] An exposé published by the national news magazine *Veja* in 1999 motivated the Cardoso administration to review the legality of unusually large Amazonian estates.[29] The INCRA secretariate in Manaus conducted an audit of landholdings greater than 10,000 hectares, identifying more than 2,900 holdings covering ~87 million hectares. Putative owners were required to provide documentation supporting their claims. The audit caused INCRA to rescind title for 63 million hectares.[30]

The scandal also motivated the creation of a congressional commission (*Comissão Parlamentar de Inquérito*), which investigated the illegal transfer of public lands in seven of the nine states of the Legal Amazon.[*] Referred to as the *CPI do Grilajem*, the probe identified an additional 37 million hectares of public forest that had been fraudulently obtained via transactions involving landholdings between 1,000 and 1.6 million hectares.[31] The report, which was published in 2003, provides a detailed account of the mechanics of land grabbing, the collusion of state functionaries and the complicity of magistrates who validated 24 million hectares in judicial hearings, including twelve million hectares in the name of a single individual.[†]

* Tocantins and Maranhão were excluded.

† In Pará, the name Carlos Medeiros was used in dozens of timber-related transactions totalling 12 million hectares; it was apparently a fictious name used by at least five different individuals acting with a power of attorney, most commonly Flávio Augusto Titan Viegas. In Acre and Amazonas, a later-day rubber baron named Falb Saraiva de Farias amassed forest estates totalling 7 million hectares. Source: CPI – Comissão Parlamentar de Inquérito. 2003. Relatório da Comissão Parlamentar de Inquérito destinada a investigar a ocupação ae terras públicas na região Amazônica: https://www2.camara.leg.br/atividade-legislativa/comissoes/comissoes-temporarias/parlamentar-de-inquerito/51-legislatura/cpiamazo/relatoriofinal.pdf

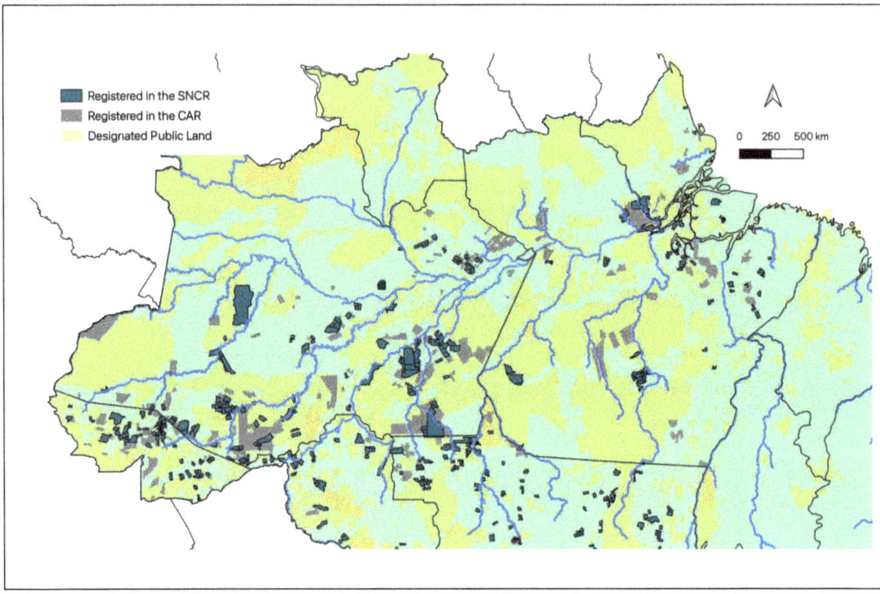

Figure 4.3: In the early 2000s, the Brazilian state clawed back approximately 100 million hectares of illegitimate forest holdings following a congressional review of land claims in the Legal Amazon. There are still ~12 million hectares of private forest holdings greater than 10,000 hectares registered within the Sistema Nacional de Cadastro Rural (SNCR) and an additional 16 million hectares of claims registered in the Cadastro Ambiental Rural (CAR) that have yet to be adjudicated.

Data sources: INCRA (2021) and IMAFLORA (2019).

Administrative action by INCRA reverted some, but not all, of those landholdings. Legal action on the part of aggrieved landholders delayed resolution, particularly in Pará where timber companies continued to exploit landholdings while the judicial system evaluated their claims.[32] As of 2021, INCRA databases continued to list multiple forest properties larger than 100,000 hectares, including a 913,000 hectare estate that was impounded in 2004 following the congressional investigation (Figure 4.3).*

* INCRA revoked the title (CCIR) of the *Gleba Santa Rosa do Tenque* in 2004, based on a reevaluation of the documents submitted on behalf of the corporate owner: APLUB *Agro Floresta Amazônia S/A*. The company successfully appealed the decision in 2005 and went on to develop a timber management plan in 2016 that was approved by IBAMA and certified by the Forest Stewardship Council (FSC). That venture failed to attract investors, however, and the property was recently listed on the Brazilian real estate market for R$ 60 million. Sources: CPI – Comissão Parlamentar de Inquérito. 2003. Relatório da Comissão Par-

The findings of the commission led to investment in INCRA's capacity to manage the SNRC, including the development of a geospatially precise database: the *Sistema Nacional de Certifcação de Imóveis* (SNCI). Launched in 2003, the SNCI was a cumbersome system that relied on public servants to verify information and then incorporate the property into its digital catalogue. It was replaced in 2013 by the *Sistema de Gestão Fundiária* (SIGEF), which relies on the landholder (or consultant) to upload data via a web application that is subsequently verified by INCRA staff.

For some reason, presumably technical, the two systems have never been merged. The SNCI incorporated 15,330 properties and, as of June 2021, SIGEF housed ~142,000 (up from 73,000 in 2017).[33] To put this in perspective, the number of certified properties in 2020 represented only fifteen per cent of the rural landholdings enumerated by the IBGE agricultural census of 2017.[*] Nonetheless, the properties registered within the SNRC encompass sixty per cent of the total spatial footprint of all agrarian landholdings, another data point highlighting the unequal distribution of public land in the Legal Amazon.[†]

The deficiencies in the SNCR led to the development of parrallel cadasters.[34] The national tax authority relies on the *Cadastro Fiscal de Imóveis Rurais* (CAFIR). This database does not incorporate spatial attributes but does include large and mid-scale landholders who register in the system to pay taxes and, in the process, further legitimatise their holdings. In 2015, the government moved to unite the SNRC and CAFIR into a single registry: the *Cadastro Nacional de Imóveis Rurais* (CNIR).[35] Presumably, this is part of a broader strategy to improve tax collection, but like the CAFIR, the CNIR will include both *proprietários* and *poseidueros*. The CNIR will not issue a certificate of title regularisation (CCIR), which will remain the responsibility of INCRA; however, the CNIR will generate a document to be required for future property transactions which, if so, will effectively function as a type

lamentar de Inquérito destinada a investigar a ocupação ae terras públicas na região Amazônica: https://www2.camara.leg.br/atividade-legislativa/comissoes/comissoes-temporarias/parlamentar-de-inquerito/51-legislatura/cpiamazo/relatoriofinal.pdf; *Tribunal Regional Federal da 1ª Região* (AMS 5665 AM 2004.32.00.005665-1) and OLX: https://am.olx.com.br/regiao-de-manaus/terrenos/fazenda-na-amazonia-913-000-hectares-644885176#

* The Instituto Brasileiro de Geografia e Estatística (IBGE) survey records holdings in 18 size classes ranging from 0 to > 10,000 hectares; Source of data: SIDRA – Sistema IBGE de Recuperação Automática (2021) *CENSO Agropecuario 2017*: https://sidra.ibge.gov.br/tabela/6780

† The estimate is derived from a model of land tenure compiled from 14 overlapping databases organised in the *Atlas Agropecuária – A Geografia da Agropecuária Brasileira*. Source: IMAFLORA (2019).

of deed.* Although authorised by legislation in 2001, the consolidation of the CNIR has taken on new impetus since the election of Jair Bolsonaro, and all landholders have been instructed to register by the end of 2022.[36]

Another parallel land registry is the *Cadastro Ambiental Rural* (CAR), created in 2009 as part of *Plano de Ação para a Prevenção e Controle do Desmatamento na Amazônia Legal* (PPCDAm), a highly successful cross-sectoral strategy to combat illegal deforestation (see Chapter 7). Registration in the CAR is obligatory but, in order to ensure its success, authorities and private sector stakeholders created multiple incentives to promote participation. Positive incentives include access to subsidised credit and the provision of technical assistance. Negative incentives include barriers for the commercialisation of crops and livestock that are enforced by commodity traders and meatpackers.† Companies use the CAR to monitor deforestation and (allegedly) exclude producers who illegally clear forest from their supply chains. Agrobusiness has aggressively promoted the CAR as a key component in its strategy to protect Brazil's overseas markets from consumer boycotts (see Chapter 3). Unfortunately, land grabbers are attempting to use the CAR to establish documentary history to support fraudulent claims, a strategy that may succeed, considering the Bolsonaro administration's support for expanding the agricultural frontier.

As a cadaster, the CAR has avoided the pitfalls of SNCR by accepting registration of all landholdings regardless of legal status and by ignoring conflicting land claims. Participants are expected to conform to environmental regulations; however, inscription provides a flexible (open-ended) pathway for coming into compliance with the Forest Code (see Chapter 7). Consequently, the response of landholders has been overwhelming, and the CAR provides an alternative depiction of the number and location of all land claims (Table 4.3 and Table 4.4).

The massive gap between the CAR (755,000 landholdings) and SNCR (135,000 landholdings) shines a spotlight on both the dysfunction and inequity in INCRA's programmes to formalise land tenure. The technical tasks associated with verifying the legal and geospatial attributes of a landholding require the services of a professional surveyor. Large-scale producers have self-financed this process because they can, but smallholders of limited means must wait until INCRA organises a campaign in their municipality. The dysfunction is evident in the municipality of Ariquemes (Rondônia), where hundreds of landholders lack CCIRs, even though the region was settled in the 1970s and 1980s (Figure 4.4).

---

* A 'title' is declaration of ownership; a 'deed' is the registry of a real estate transaction; both are legal documents.

† Grain traders: ADM, Bunge, Cargill, Louis Dreyfus, Amaggi group and a few others from China and the EU (see Ch. 3); Meatpackers: JBS, Marfrig, Minerva.

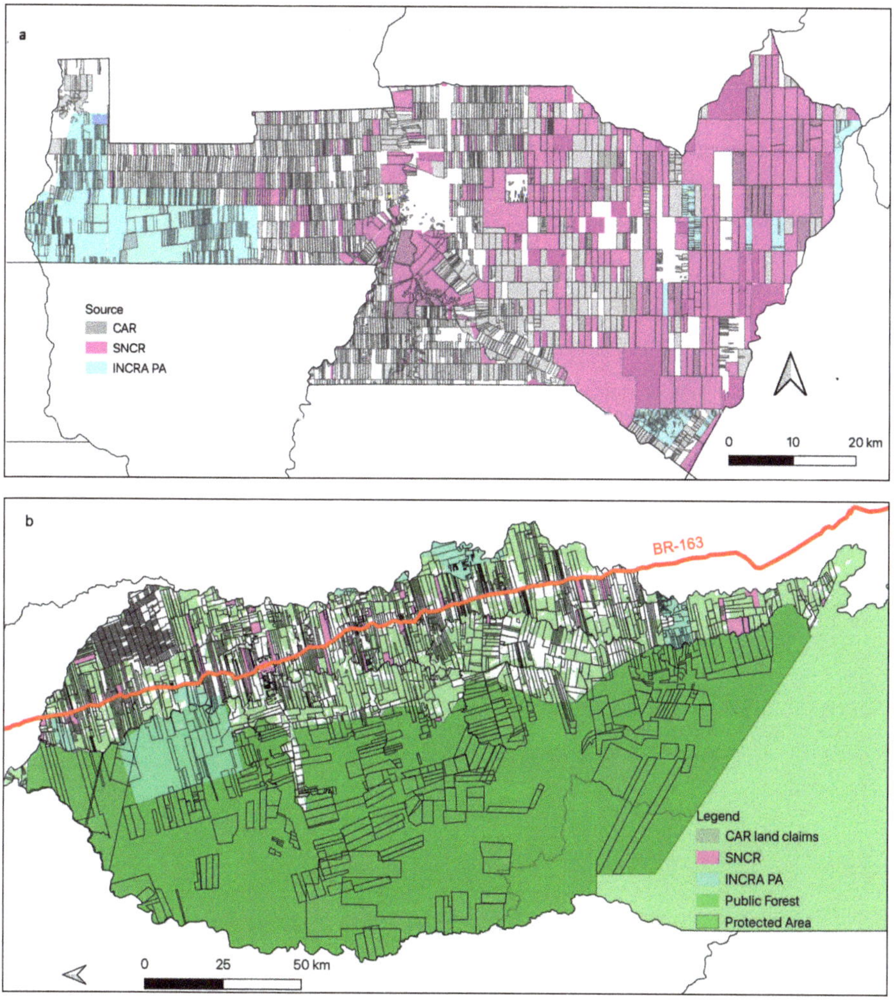

Figure 4.4: The Cadastro Ambiental Rural (CAR) has allowed hundreds of smallholders to obtain an official registration for their properties in Arequimes municipality of Rondônia, which was settled in the 1970s (a). Unfortunately, the CAR also provides opportunities for land grabbers to document dubious land claims on the public forests adjacent to BR-163 in Novo Progresso municipality in Pará (b).

Data source: INCRA (2020) and SFB (2020).

Table 4.3: Summary statistics on rural landholdings in the Legal Amazon of Brazil. The IBGE data is from the agricultural census of 2017 and excludes forest properties; the CAR incorporates data for forest and agrarian landholdings, including many of dubious provenance; the INCRA-SNCR incorporates only those that have been verified for their spatial and administrative attributes, while excluding landholdings within INCRA PA-type settlements, which are included in both the IBGE and CAR databases.

	IBGE		**CAR**		**INCRA SNCR**		**INCRA Settlements**	
	number (x1,000)	hectares (x1,000)	number (x1,000)	hectares (x1,000)	number (x1,000)	hectares (x1,000)	number (x1,000)	hectares (x1,000)
Acre	37	5,211	36	6,172	1	3,156	20	1,173
Amapá	9	1,244	6	2,323	1	572	16	1,337
Amazonas	81	4,467	50	26,122	5	6,905	8	1,557
Maranhão	220	14,166	102	16,302	22	10,124	102	3,779
Mato Grosso	119	59,669	142	73,033	56	51,436	66	4,285
Pará	282	32,324	213	44,498	13	13,420	124	11,762
Rondônia	91	10,589	123	12,003	9	4,024	37	1,897
Roraima	17	2,986	10	4,259	2	1,069	15	1,259
Tocantins	64	17,155	73	17,762	25	13,470	24	1,283
Total	**919**	**147,810**	**755**	**202,474**	**134**	**104,176**	**413**	**28,333**

Sources: SIDRA (2021); CAR from IMAFLORA (2019); INCRA-SNCR and INCRA-PA from INCRA (2020).

The need to fast-track the *regularização** of smallholder titles motivated the *Terra Legal* programme, which sent teams of surveyors to selected municipalities to accelerate the process for landholdings established prior to 2004.† The initial goal was to review and certify 300,000 smallholdings in 463 municipalities; [37] however, the programme collected data on only 117,000 landholdings and issued less than 23,000 CCIRs.[38] As of June 2021,

* This is the term used in Brazil to refer to the process whereby land titles are reviewed and certified; in Spanish-speaking countries, the preferred term is *saneamiento;* both terms reflect the perception that the *status quo* is 'irregular' or 'unclean'.

† Eligibility was limited to landholdings smaller than 15 '*modulos fiscales*', a standardised unit-area defined by Brazilian law that varies regionally – in the Amazon it is nearly always 100 hectares.

Table 4.4: Summary statistics on rural landholdings in the Legal Amazon of Brazil. The IBGE , CAR and INCRA–SNCR data as in Table 4.3. The IMAFLOR model combined data from the CAR and SNCR database, but excludes landholdings within the INCRA–PA system, thus undercounts the number of small holdings by about 413,000 and 28.3 million hectares (see Annex 4.1 to 4.9)

	IBGE		**CAR**		**INCRA–SNCR**		**IMAFLOR Model**	
Size class (hectares)	number	hectares (x1,000)	number	hectares (x1,000)	number	hectares (x1,000)	number	hectares (x1,000)
< 100	735,863	18,365	534,064	20,714	66,263	2,792	396,530	6,069
100–500	118,558	26,573	158,229	32,569	58,900	16,839	311,225	12,908
500–1,000	16,813	12,610	26,641	19,144	16,370	13,230	135,969	27,673
1,000–10,000	19,770	64,631	34,893	81,998	19,969	50,159	69,085	80,553
10,000–50,000	1,318	25,632	1,409	26,287	791	14,781	2,282	16,046
50,000–100,000	-	-	92	6,375	44	2,953	1,230	22,803
> 100,000	-	-	69	15,387	17	3,526	158	21,022
Total		**147,810**	**755,397**	**202,474**	**162,354**	**104,280**	**916,479**	**187,073**

Sources: SIDRA (2021); CAR and IMAFLOR model from IMAFLORA (2019); INCRA (2020).

none of these recently registered properties has been incorporated into the SIGEF databases available via INCRA's public portal.*

Although the *Terra Legal* system failed significantly to increase the inscription of smallholders in the SNCR, it demonstrated how a wall-to-wall effort can resolve potential conflicts among neighbours and achieve impacts at scale by engaging an entire community. That experience will be replicated in *Titula Brasil,* an initiative launched in 2021 by the Bolsonaro

* Terra Legal was organised outside of INCRA via the *Secretaria Extraordinária de Regularização Fundiária na Amazônia Legal* (SERFAL) and the *Secretaria Especial de Agricultura Familiar e do Desenvolvimento Agrário* (SEAD)/ *Casa Civil* (i.e., the president's office). The programme was funded by the German government and terminated in 2019; its functions and databases were transferred to INCRA. Source: A. Borges (27 Nov. 2020), TCU aponta lentidão na legalização de terras, *Terra, Noticias*: https://www.terra.com.br/noticias/brasil/politica/tcu-aponta-lentidao-na-legalizacao-de-terras,0cd955783b4bcb176a885243b150216dpzo8skel.html

Source: Google Earth

The forest landscape adjacent to BR-319 south of the Amazon River near Manaus is largely intact forty years after the construction of this federal highway; nonetheless, numerous land claims have been regularised (white polygons) and more have been inscribed into the Cadastro Ambiental Rural (red polygons).

Data source: Google Earth and INCRA (2020).

administration, that will delegate most of the administrative and technical tasks of property mensuration to the newly created *Núcleos Municipais de Regularização Fundiária (NMRF)*. These offices are meant to function as decentralised units of INCRA and, like *Terra Legal*, prioritise assistance for smallholders.

The *Titula Brasil* programme will first target the approximately 430,000 households that are resident in the INCRA's 3,000 PA-type settlements; however, these municipal offices will be open to other small and mid-sized landholders. According to the IBGE agricultural census, there are at least 680,000 landholdings smaller than 100 hectares in the Legal Amazon,[39] while data derived from the CAR indicates the number located outside the PA system might be as large as 500,000.[40]

The Lei de Grilagem

The effort to resolve the backlog in the regularisation of small farms collides, unfortunately, with the fight to end land grabbing. Congressional representatives affiliated with conservative groups often referred to as the *banca ruralista** have consistently pushed for a regulatory approach that would

* Also known as the *bancada BBB*, which stands for *bala* (bullets), *boi* (cattle) and *biblia* (bible) – see Ch. 6.

issue CCIRs to thousands of medium- and large-scale ranchers with problematic land tenure documents. Previous policy initiatives, particularly a land law passed in 2009, included measures that would recognise the legality of landholdings settled in the 1990s and 1980s, when land acquisition rules were laxly enforced.[41] Environmental and social advocates characterised the law as an amnesty, however, and insisted that it incorporate a *quid pro quo.* Consequently, the law included measures to limit the size of landholdings eligible for an expedited process to 1,500 hectares and set cut-off dates to exclude lands illegally occupied after 2004.

In 2016, an executive order by the Temer administration modified the regularisation protocols by moving the cut-off date to 2009 and expanding the size of the landholding eligible for an expedited process to 2,500 hectares.[42] The rules were modified again in 2019 by the Bolsonaro administration, first by an executive order that evolved into a legislative act known by its critics as the *Ley da Grilagem.*[*] Critics contend that the recent (and proposed) changes represent another amnesty for past infractions and open the door for another round of land grabbing.[43] Like all legislative proposals, the final version will depend upon last-minute negotiations but, as of August 2021, opponents point out multiple deficiencies: [44]

1. Extends the cut-off date for the expedited resolution of land claims to 2014 (rather than 2009).
2. Includes provisions to auction illegal landholdings that allow rejected applicants to participate and, in certain cases, make bids prior to the public auction.
3. Limits on-sight verification for environmental compliance for landholdings greater than 1,000 hectares (rather than 400 hectares).
4. Condones illegal deforestation by relying on (seldom enforced) future commitments to remediate past infractions (*Termo de Ajustamento de Conduta – TAC*).[†]
5. Weakens the ability for INCRA to recover (claw-back) landholdings that fail to comply with environmental regulations (see #4 above).

---

* *Projeto de Lei nº 510/2021,* https://www25.senado.leg.br/web/atividade/materias/-/materia/146639

† *Termo de ajustamento de conduta* (TAC) is an agreement between a public entity (prosecutor's office or regulatory agency) and a person or legal entity to correct a violation of a collective right protected by law (e.g., environmental or consumer). It is an extrajudicial measure used to resolve conflicts without resorting to legal action. In case of non-compliance, the public entity can demand the execution of the TAC without the need to file a public civil action to seek remediation or compensation. Source: https://comunicacao.mppr.mp.br/2020/08/21443/Termo-de-Ajustamento-de-Conduta.html

6. Facilitates land grabbing as a business model by allowing individuals to submit multiple applications to INCRA to regularise a landholding.
7. Inappropriately rewards speculators by extending discounts (ranging from fifty to ninety per cent of the appraised value of the land) that were originally intended only for residents within INCRA PA-like settlements.
8. Creates a mechanism for the ongoing distribution of public lands by sale or auction that would open a door for the further privatisation of public lands.

Proponents of the reorganisation of INCRA protocols argue that it is necessary to impose order on the chaos of the land-tenure system while providing economic justice to hundreds of thousands of rural families. Opponents contend that the law represents (another) amnesty for past illegal activity that will foster future abuse. Moreover, they contend that none of the proposed changes are needed to expedite the regularisation of smallholder properties and, instead, suggest investing in the capacitation of INCRA staff and the provision of a budget commensurate with the size of the task – which all parties agree is very large and long overdue.

Underlying the debate are two opposing philosophies about the future of development in Amazonian Brazil. On the right, economists and political scientists view land as a financial asset and believe the regularisation of private property will stimulate investment and create economic growth. On the left, social and environmental advocates view access to land as a human right, and seek to ameliorate the inequality that defines Brazilian society and conserve biodiversity and protect indigenous and traditional cultures in the Amazon (Annex 4.1 to 4.10).

Bolivia

Bolivia was a leader in the agrarian reform movement in South America. A defining moment in its modern history was the national revolution of 1952, which started as an uprising against the feudal system that bound indigenous communities to estates owned by wealthy families. The revolutionary government created the *Instituto Nacional de Reforma Agraria* (INRA) in 1958 to provide legal status to the lands occupied and claimed by indigenous peasants. The revolution largely occurred in the Andean highlands and eventually led to the proliferation of extremely small (micro) landholdings that motivated many *campesinos*[*] to migrate to urban areas or the eastern lowlands. Large estates in the Bolivian Amazon avoided confiscation but

* In Andean countries, indigenous peasants from the highlands self-identify as *campesinos*.

their owners were forced to bequeath a fraction of their properties to the indigenous communities upon which they depended for labour.*

In 1965, Bolivia established the *Instituto Nacional de Colonización* (INC) to foster the migration to the lowlands and, in the process, created a parallel and overlapping bureaucracy for granting land titles. Both agencies distributed land in the Bolivian Amazon to the growing stream of indigenous migrants from the Andean highlands. Organised colonisation projects in the 1970s created smallholder landscapes in the Chapare, Cochabamba (HML # 32); Alto Beni, La Paz (HML #33); and San Julián, Santa Cruz (HML #31).

Japanese immigrants also arrived in the 1960s and established colonies in Santa Cruz at Yapacaní (HML #32) and Okinawa (HML #31), landscapes with unusually fertile soils uniquely suited for the cultivation of irrigated rice. Mennonites settled south of the city of Santa Cruz in the 1970s, initiating a process of colonisation on the alluvial plain of the Río Grande (HML #31) that eventually extended to Chiquitania (HML #29) and the Guarayos regions (HML #30). These foreign migrants were welcomed by both military and civilian governments because they brought practical knowledge that aligned with government policy to develop the agricultural economy of Santa Cruz. The main beneficiaries of that policy, however, were *Cruceño* families who used their influence to acquire millions of hectares of public forest.[45]

The 1980s and 1990s were characterised by the adoption of neoliberal economic policies imposed on Bolivia by multilateral agencies.† One of the most far-reaching decisions was the closure of unprofitable state-owned mines that led to another round of mass migration, this time by indigenous miners who joined their *campesino* peers in the lowlands to start a new life as small farmers. About half moved to the Chapare region and took up the cultivation of coca.[46] They brought with them a tradition of union activism that would define the political struggles of the first decade of the twenty-first century (see Chapter 6).

In 1992, the INC was merged into INRA, which was reformed to protect property rights as part of the ongoing programme to create a market economy. Codified by the *Ley INRA* of 1996, the reformed land-tenure system included provisions for the regularisation (*saneamiento*) of land titles

* An estancia with an indigenous *rancho* within its boundaries would excise that village and its associated slash-and-burn farmland and pass ownership to the village as a communal holding. The estancia would continue, however, to rely for manual labour on its inhabitants, who continue to view the owner as their *Patrón.*

† 'Structural adjustment' was the euphemism that the World Bank and International Monetary Fund used when obligating developing countries to adopt the economic reforms – including trade liberalisation, deregulation, fiscal discipline and privatisation – embodied in the so-called Washington Consensus.

and the compilation of a national land registry.* Like most of the structural adjustment policies of the 1990s, it included provisions to safeguard the ancestral claims of indigenous communities (see Chapter 11).

The World Bank and other multilateral agencies supported the land-titling process† while financing investments in infrastructure (see Chapter 2) and land-use planning (see below). International commodity traders created local subsidiaries, provided affordable credit and incorporated the expanding agricultural frontier into their global supply chains (see Chapter 3). Inexpensive land, fertile soils and an accommodating government attracted investors from North America, Argentina and Brazil. The soybean boom was well under way by the year 2000, which fostered a rise in real estate values similar to that experienced by the soybean landscapes of central Mato Grosso. On the alluvial plain near San Julián (HML #31), mean annual prices for farmland increased by six per cent between 1990 and 2000, by fifteen per cent between 2000 and 2010 and by seven per cent between 2010 and 2020.[47]

While the boom in agriculture was getting underway, Bolivia was also transforming its forest sector following a playbook designed by forest ecologists eager to implement sustainable forest management via the private sector. The centrepiece of that strategy consisted of thirty-year concessions covering approximately 6.5 million hectares awarded to companies that promised to pursue sustainable forest management (see Chapter 7).[48] Bolivia was viewed as an experiment where democratic reform and a market economy would promote sustainable development and social justice. The citizens of Bolivia, however, had a different vision for the future of their country.

Estado Plurinacional de Bolivia

In 2005 Bolivia elected Evo Morales in a landslide victory that gave his political party, the *Movimiento al Socialismo* (MAS), the first congressional majority since the restoration of democracy in 1986. Key to his electoral success was a platform based on indigenous rights and their demand for *tierra y*

* The process of *saneamiento* should: (1) reconcile conflicting claims to a landholding; (2) issue title to a valid claim; (3) annul invalid claims; (4) validate imputed properties that meet criteria of social and economic function (FES, by its Spanish acronym); (5) issue a certification that a property meets FES criteria; (6) revert to the state land claims that do not meet the FES. Properties that have been *saneada* are registered in the national rural cadaster. Source: http://www.inra.gob.bo/InraPb/paginaController?cmd=contenido&id=6561

† The World Bank: National Land Administration Project (P0061907) $US 20 million in 1995; the IDB: the Land Regularization and Legal Cadaster project (BO0221) $US 22 million; USAID contributed funds for land tenure review for the Chapare region. However, most costs have been borne by the Bolivian state with an annual budget of about $US 10 million. See *Fundación Tierra*: http://www.ftierra.org/

Source: Google Earth

The Bolivian government completed the verification and validation (saneamiento) of tens of thousands of small farms (white polygons) in the coca-growing region of the Chapare in a five year period between 2007 and 2012, thus demonstrating that technical competence and political will can resolve long-standing backlogs of tenure review.
Data source: Google Earth and INRA (2018).

territorio, which attracted the overwhelming support of lowland indigenous nations (see Chapter 11) and the Quechua- and Aymara-speaking people of the Andean highlands. Evo Morales and his indigenous allies rewrote the constitution, changed the name of the country and began to dismantle the institutional framework imposed on Bolivia by the Washington Consensus.[*] One of the first items on the legislative agenda was a reform of the *Ley INRA* with an emphasis on the rights of communities and smallholders.[†]

Communal land tenure is common to indigenous cultures and the previous regime had started the process of recognising *Territorios Comunitarios de Origen* (TCO).[‡] This tenure category largely benefitted ethnic groups that inhabited remote lowland landscapes in the tropics and pastoral communities on the arid plains of the Altiplano. However, scant progress had been made in formalising titles for communal landholdings on agrarian landscapes where most indigenous families actually lived. The Morales administration made agrarian populations a priority and proceeded to demarcate the boundaries around thousands of village landholdings in the

---

* The first administration of Evo Morales also renationalised the oil and gas industry, telecommunications and the electrical utility sector. Source: Farthing (2019).

† Ley N° 3545 *de Reconducción de la Reforma Agraria* (28 de noviembre de 2006).

‡ The Constitution of 2006 changed the official designation of this category of communal reserves to *Territorio Indígena Originario Campesino* (TIOC), but the original term persists in both the academic and popular press.

Andean highlands. In Chiquitania (Santa Cruz), this consisted of formalising the land grants conferred by large-scale landholders to their indigenous tenants in the 1960s and 1970s.* In Northern Bolivia, INRA used the communal (*campesino*) system to distribute land to communities of forest-dwelling families who had settled the region during the rubber booms of the 1890s and 1940s (See Annex 4.11).

Concurrently, the Morales administration responded to the claims of hundreds of thousands of indigenous families who had migrated to the lowlands over the previous forty years. Although they had voted for a socialist government, these families wanted full legal title to their small farms. INRA began to review and approve land titles at an unprecedented pace; between 2006 and 2015, INRA processed and validated hundreds of thousands of small farms, vastly exceeding the dismal record of the consulting companies that had been contracted during the first phase of the land tenure regularisation process (Table 4.5).†

Despite its anti-capitalist rhetoric, the socialist government did not attempt to impose a far-reaching agrarian reform, although there were a few high-profile attempts to confiscate large-scale estates. Resistance from civil society in Santa Cruz and an (alleged) agreement with business magnates muted attempts to change the land tenure regime in Bolivia's most productive and valuable landscapes. Agribusiness is too important for the health of the domestic economy.[49]

The Ley INRA of 2009 include a limit on properties larger than 5,000 hectares and provisions that allow the state to claw-back properties that do not meet the criteria of having a '*función económico - social*'(FES). In other words, owners must 'use the land or lose the land'. Large-scale owners manage these requirements by subdividing their landholdings while hiring agronomists, foresters and lawyers to maintain the documents required to demonstrate FES. Medium-scale producers, however, can fall prey to predatory functionaries seeking to extort a bribe, or unscrupulous land grabbers who invade properties with significant forest assets or problematic documents.

* Between 1996 and 2005, INRA recognised 550 communal landholdings (2 million hectares), compared to ~8,700 (7.2 million hectares) between 2006 and 2015. Source: Colque et al. (2016).

† Between 1996 and 2005, INRA validated 19,500 properties < 500 hectares (67,000 hectares); in contrast, between 2006 and 2015, the agency certified more than 358,000 (3.1 million hectares). Source: Colque et al. (2016).

Table 4.5: The distribution of land in the Bolivian Amazon.

	Number	Area (ha)	%	% Agrarian
Private Properties				
Small-scale (< 100 ha)	377,802	3,852,050	3.6%	9%
Medium-scale (100–2,500)	3,148	1,667,651	1.6%	4%
Large-scale (>2,500)	1,374	3,091,530	2.9%	7%
Communal Lands				
Agrarian properties	8,921	9,176,971	8.6%	21%
Amazonian estates	901	2,977,144	2.8%	
Indigenous territories - TCO (Amazonian)	440	22,998,273	21.5%	
Public Lands				
Protected Areas	35	15,522,327	14.5%	
Forest Estate (ex-concessions)	93	11,327,956	10.6%	
Unknown				
Under review (agrarian and private)		26,867,679	25.2%	60%
Undefined		14,555,048	13.6%	
Total Agrarian (farm and rangeland)		44,655,881	41.8%	
Total		106,751,722		

Data source: Colque et al. 2016.

A coalition created by a demand for land is splintered by a competition for territory

The political movement that brought Evo Morales to power incorporated a latent conflict between highland and lowland indigenous communities. The lowland nations are intent on recuperating their ancestral territories, which had been appropriated by families of European extraction or, more recently, allocated to timber companies as long-term forest concessions. The promise of recovering these lands was the reason lowland indigenous groups overwhelmingly supported Evo Morales in 2005. In contrast, highland indigenous groups believe they have a constitutional right – as Bolivian citizens – to settle unoccupied public lands, particularly the forest concessions that were rescinded in the early days of the Morales' administration. The highland and lowland indigenous groups are competing for the same land.

This conflict is manifest in the evolving self-identity of the Andean migrants, who for decades referred to themselves as *colonizadores.*[*] Since about 2000, however, they have self-identified as *interculturales,* a term that recognises their status as indigenous people who have left their ancestral homeland. They are politically powerful, in part because they maintain familial and commercial ties with a large population of urban migrants, but also because they have organised militant syndicates skilled in the tactics of economic blockade. They exercise their electoral power by demanding that INRA, which is controlled by the central government, distribute land via settlement associations affiliated with the *Confederación Sindical de Comunidades Interculturales Originarios de Bolivia* (CSCIOB) or the *Confederación Sindical Única de Trabajadores Campesinos de Bolivia* (CSUTCB).

The alliance among the highland and lowland indigenous groups was fractured in 2011 when the Morales administration announced its intention to build a highway through the *Territorio Indígena y Parque Nacional Isiboro – Securé* (TIPNIS). The proposed highway ignored several legal precepts, including the requirement to obtain 'free prior and informed consent' (FPIC) for the highway project from the Moxeño, Trinitario, Yuracaré and Tsimane peoples who have held communal title to the reserve since 1990 (see Chapter 11).[50] The official justification for building the road was to integrate two regions of the country, but the inhabitants of TIPNIS know that it would also trigger a land rush by the coca-growing farmers of the Chapare, who have already colonised the southern sector of their reserve (HML #32).

The attempt to build the highway revealed that Evo Morales would not honour his campaign promises to lowland indigenous groups when it conflicted with the interests of the more numerous and politically assertive *interculturales.*[†] The *Confederación de Pueblos Indígenas del Oriente Boliviano* (CIDOB) expressed solidarity for the tribes native to the TIPNIS, an action that coincided with a slow-down in the titling process for the TCO reserves for ethnic groups affiliated with CIDOB (Figure 4.5).[51] Simultaneously, INRA administrators ignored requests by Chiquitano and Guarayos organisations for the restitution of ancestral territories that had been incorporated into

---

* The association of smallholder farmers in the Bolivian lowlands was founded in 1971 as the *Confederación Sindical de Colonizadores es de Bolivia* (CSCB); the organisation changed its name in 2008 to *Confederación Sindical de Comunidades Interculturales de Bolivia* (CSCIB) and again in 2013 to the *Confederación Sindical de Comunidades Interculturales Originarios de Bolivia* (CSCIOB). It represents approximately 2.5 million small farmers. Source: García Yapur et al. (2014).

† Evo Morales is himself a member of the *interculturales;* he was born in the Altiplano community of Orinoca (Oruro), but moved to the Chapare region as a boy, where he became active in the campesino syndicates that represent the interests of the region's coca farmers. Source: Harten (2011).

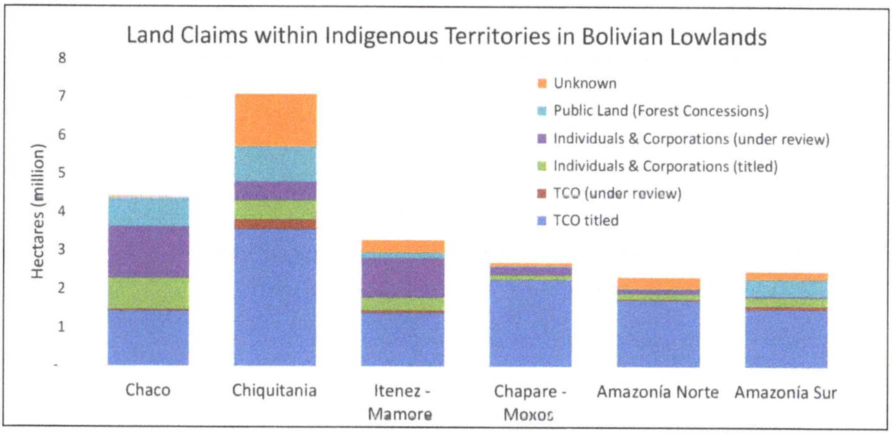

Figure 4.5: The Instituto Nacional de la Reforma Agraria (INRA) of Bolivia must adjudicate conflicting claims within the territories claimed by lowland Indigenous nations in 1996. The most conflictive are in Chiquitania and Chaco regions and the flooded savannas of Iténez–Moxos, where ranchers of European descent have long-standing land claims that are recognised by the county's judicial system. The subregions used in the analysis correspond to cultural regions rather than administrative jurisdictions: Santa Cruz (Chaco and Chiquitania), Beni (Iténez–Mamore) Beni/ Cochabamba (Chapare–Moxos), Pando (Amazonia Norte), La Paz/Beni (Amazonia Sur).

Data sources: Colque et al. (2016) and INRA (2018).

forest concessions in the 1990s (Annex 4.11). Instead, they began distributing land to associations affiliated with the *interculturales*.*

The conflict between the indigenous groups is part of the shifting political coalitions that have defined recent elections. Although they lost power at the national level, *Cruceño* elites still dominate local and regional governments and now support the territorial demands of Chiquitano and Guarayos communities. This regional coalition is advocating for the creation of multiple-use protected areas controlled by local jurisdictions, which would allow timber extraction by the region's indigenous and non-indigenous inhabitants. Implicit in this political manoeuvring are strategies focused on the demographics of Santa Cruz and the fear (aspiration) that an expanding population of *interculturales* will lead to the electoral success of the political party associated with Evo Morales.[52]

* Between 2011 and 2020, INRA approved between 1,400 and 2,000 settlement communities in northeast Santa Cruz; allegedly, the distribution has favoured intercultural migrants rather than the Chiquitano inhabitants. Source: Zegada (2019).

Overlaying the ethnic and political conflicts is the ever-present spectre of corruption, which permeates almost all land transactions and involves unscrupulous individuals within every stakeholder group. This includes land grabbers, military officials and leaders of campesino syndicates that use their political connections to obtain large-scale landholdings for resale to corporate farmers and Mennonites.[53] Even lowland indigenous leaders have been tempted to participate in the political melee, most notably when the government created a parallel slate of indigenous leaders within CIDOB that supported their attempt to violate the TIPNIS.[54]

The expansion of the agricultural frontier

INRA has done a fairly competent job of processing the huge backlog of land claims, but there is no indication that any government will end the distribution of public land. Over the last twenty years, INRA has issued title to thousands of landholdings within two forest reserves specifically created to ensure the long-term management of timber resources. The first to be dismembered was El Choré and the same process is underway within the Guarayos Reserve, even though it enjoys dual status as a TCO and forest reserve. A third forest reserve, Bajo Paraguá, is at the centre of the competition between Chiquitanos, *Interculturales* and local politicians. Recent statements by INRA functionaries indicate they view land claims by settlers as having precedence over efforts to create municipal protected areas within forest reserves.

The jockeying for land reflects a broad consensus that expanding the agricultural frontier is in Bolivia's national interest. This includes all major political parties, the central and regional governments, the agribusiness sector, ranchers and intercultural settlers.[55] These policies originated in the administration of Evo Morales (2005–2019), which approved five laws between 2013 and 2019 that facilitated access to public lands, legalised landholdings appropriated during previous administrations and opened the door to deforestation and the use of fire.* These policies were embraced by the transitional government of Jeanine Añez and the administration

* *Ley 337, de Apoyo a la Producción de Alimentos y Restitución de Bosques* (2013). This law provided amnesty to landholders who did not file forest-clearing permits between 1996 and 2011. The beneficiaries were largely smallholders (76%) but also medium-scale properties (11%), agroindustry (7%) and even lowland indigenous communities (4%). *Ley 741, Ley de Autorización de Desmonte hasta 20 hectáreas* (2015). The law authorises landholders to clear up to 20 hectares of forest per year; clearing land is essential to establish property rights for new land claims. *Ley 1098, de Aditivos de Origen Vegetal* (2018). A biofuel measure that proponents claim will add $US 480 million in revenue and create 27,000 direct and indirect jobs. *Ley 1171 de Uso y Manejo Racional de Quemas* (2019). This measure gave smallholders the legal right to use fire as a management tool without seeking a permit from local authorities. *Decreto Supremo N° 3973* (2019). This 2001 executive decree eliminated the need to obtain a forest-clearing permit on

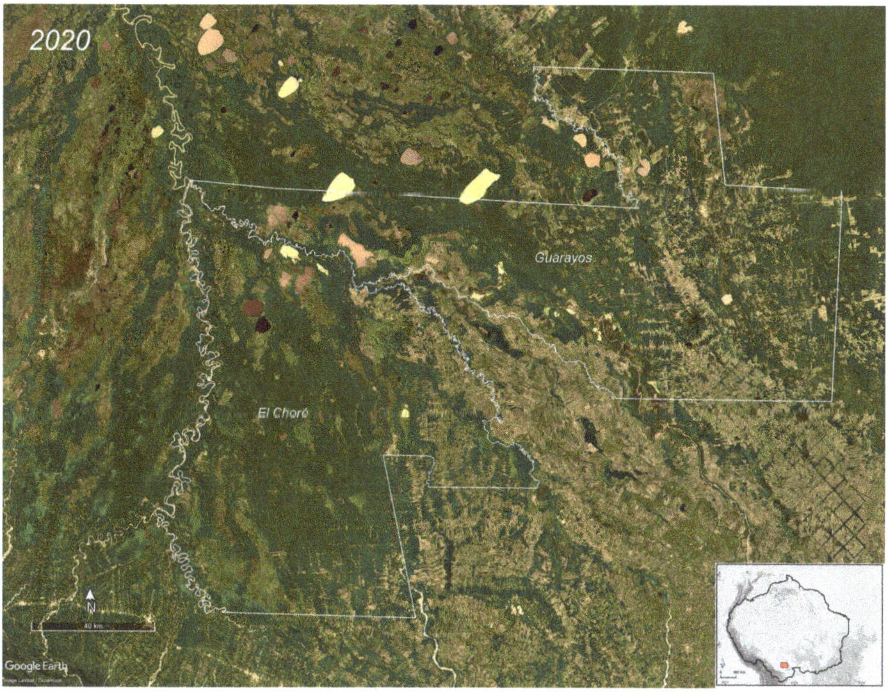

Source: Google Earth

In 2000, the Bolivian state reaffirmed the status of two forest reserves, El Choré and Guarayos, created in 1969 in areas once rich in mahogany. The most remote sections remain intact due to heavy flooding, but the scramble for land has overwhelmed all attempts to protect the reserves, including large areas that were claimed by the Guarayos indigenous people as part of their TCO.

of Luis Arce Catacora, who was elected in October 2020 as the candidate endorsed by Evo Morales.

The controversial policies led to a spike in wildfires in 2018 that coincided with a review of the Interamerican Development Bank (IDB) project that financed INRA's land tenure programme.* Among its findings were: (1) the agency had issued no new titles for indigenous communities; (2) the ongoing distribution of public lands had generated new social and environmental conflicts; and (3) the disregard for national laws and environmental regulation violated the IDB's policies.[56] The IDB halted disbursements of

landscapes located within 41 million hectares of Permanent Forest Reserves (Decreto Supremo 26075).

* IDB Project BO-LT113, *Programa de Saneamiento de Tierras II*; $US 60 million loan + $US 40 million in matching funds from the Bolivian government. Source: https://www.iadb.org/en/project/BO-L1113

funds in 2018 and is awaiting actions by INRA to address the concerns documented in the monitoring report.[57]

Peru

The agrarian reform process in Peru began in 1964. Originally a cautious effort targeting the more egregious examples of peasant exploitation, it was dramatically expanded by a left-wing military government in 1969. Between 1970 and 1975, more than 15,800 landholdings, covering slightly more than nine million hectares, were confiscated and redistributed to more than 370,000 *campesino* families.[58] The original owners were to be compensated by the sale of sovereign bonds, but hyperinflation in the 1980s forced the government into default and the outstanding debt continues to be the object of legal action.* The original plan by the military government was to form producer-owned collectives that empowered communities to assume control of their land while preserving the economies of scale. This idea was not embraced by the peasants, however, who divided the land among themselves while managing tenure communally according to traditional highland customs.

The military regime ended in 1980 with the election of Fernando Belaunde, an advocate of Amazonian development and the original proponent of the *Carretera Marginal de la Selva* (see Chapter 2). Among his first actions was to create the *Instituto Nacional de Desarrollo* (INADE), an autonomous agency affiliated with the Ministry of Agriculture; INADE organised *Proyectos Especiales,* which included settlements in the tropical lowlands, irrigation systems on the coast and mechanised agriculture in the highlands.[59] The six lowland projects were analogous to the colonisation projects in Brazil that distributed land adjacent to highways then under construction, largely, as in Bolivia, to indigenous migrants from the Andean highlands.

The government reportedly deforested 615,000 hectares in anticipation of the arrival of settlers; however, only ~125,000 hectares were occupied by the first wave of colonists.[60] Over the next decade, a steady stream of migrants flowed into the tropical forest of the foothills (*Selva Alta*) and piedmont (*Selva Baja*). The estimated rate of deforestation between 1980 and 1990 exceeded 250,000 per year,[61] approximately twice the annual rate

---

* The Tribunal Constitutional (TC) in 2001 determined that the bonds should be paid based on a calculation of their 'real value' but did not indicate a method of calculating that value until 2013, when it dictated that an inflation index selected by the Ministry of Economy and Finance should be used. Bondholders rejected the settlement, arguing it was equivalent to less than 1% of the bonds' real value and continue to litigate their cause in both US and Peruvian courts. Source: APJBA – *Alianza por el Pago Justo de los Bonos Agrarios* (30 Aug. 2021): https://bonosagrarios.pe/preguntas-frecuentes-sobre-el-pago-de-bonos-agrarios/

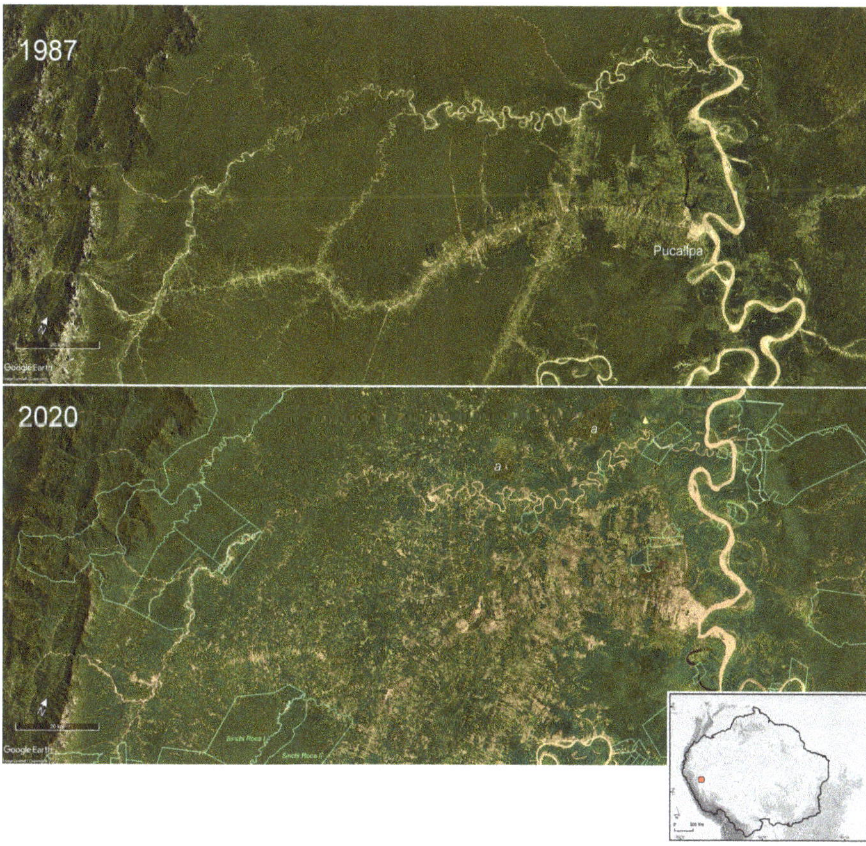

Source: Google Earth

The piedmont between the Andean foothills and the port city of Pucallpa on the Ucayali River was opened for colonisation with construction of the Carretera Federico Basadre in the 1980s. Most landholdings have yet to be validated (saneamiento) and incorporated into the national cadaster, including two large oil palm plantations established in 2013 and 2014 (a). Communal landholdings of indigenous communities (green polygons) are under pressure from migrants and land grabbers.

documented between 2000 and 2020.[62] The rural population of Amazonian jurisdictions surged from three to four million inhabitants, approximately double the rate of growth of previous and subsequent decades.[63] These years also saw an explosion in the cultivation of coca across the Peruvian Amazon, as well as the rise of the *Sendero Luminosa* and the *Movimiento Revolucionario Túpac Amaru* (see Chapter 6).

In 1992, the government of Alberto Fujimori created the *Proyecto Especial de Titulación de Tierras* (PETT) to regularise the titles for all Peruvian landholders. As in Bolivia, this coincided with policies emanating from multilateral agencies to foster a market economy and provide *seguridad* *jurídica* to landholders.* The PETT included safeguards to protect and recognise the communal lands of indigenous people, which in the vernacular of Peru included both highland groups (*comunidades campesinas*) and lowland tribes (*comunidades nativas*). The process was coordinated with other programmes to create a system of forest concessions (see Chapter 7)† and organise a protected-area system (see Chapter 12). The objective was to allocate the public lands among the various stakeholders of the Peruvian nation (see Annex 4.12).

The land tenure regularisation (*saneamiento*) process has been subject to periodic administrative reforms. This included a decentralisation decree in 2003, which passed the implementation to regional governments (*Gobiernos Regionales* – GORE), and an anti-corruption drive that transferred the programme back to the central government in 2007. In 2008, the PETT was fused into the *Comisión de Formalización de la Propiedad Informal* (COFOPRI),‡ a high-profile programme created to formalise property rights in the country's volatile urban barrios. This agency had the technical capacity to compile a digital cadaster, but devolved the administrative responsibilities back to the *Ministerio de Agricultura y Riego* (MINAGRI).[64] In 2014, MINAGRI delegated field operations to regional governments in a renewed effort to decentralise the administrative functions of the state (see Chapter 6).[65] In 2018, MINAGRI assumed full responsibility from COFOPRI for the compilation and management of a national rural land cadaster, which is known as the *Sistema Catastral para Predios Rurales* (SICAR) and managed by the *Dirección General de Saneamiento de la Propiedad Agraria y Catastro Rural* (DIGESPACR).

Despite the administrative shuffling, the land tenure project maintained a level of institutional continuity because the government leveraged its investments with loans from the IDB via the *Proyecto de Catastro, Titulación y Registro de Tierras Rurales* (PTRT).§ Executed as three consecutive projects

* Unlike in Bolivia and Brazil, Peruvian land law does not condition long-term tenure on the demonstration of a social and economic function, instead endowing owners with property rights typical of a market-oriented liberal democracy. Source: Fort (2007).

† Concessions have been granted for timber extraction (75%), conservation (12%), NTFP/wildlife, ecotourism (1.2%) and reforestation (1%); concessions are governed by 20-year or 40-year contracts, which include a clause renewal. Source: SERFOR – Servicio Nacional Forestal y de Fauna Silvestre (2017) Las Concesiones Forestales, http://www.serfor.gob.pe/portal/wp-content/uploads/2017/07/mapa-concesiones-24-07-2017.pdf

‡ Now known as the *Organismo de Formalización de la Propiedad Informal* (COFOPRI), it was established in 1996 to address the problem of urban property titles; it is now part of the *Ministerio de Vivienda, Construcción y Saneamineto.*

§ PTRT1, $US 36 million, approved in 1996 (PE0037); PTRT2, $US 46.7 million; approved in 2005 (PE0107); and PTRT3, approved in 2014, $US 81 million (PEover 25 years, the communal component has benefitted from strong oversight from civil society and the participation of indigenous organisations.* In spite of numerous setbacks, the PTRT has succeeded in establishing a nationwide system to register rural properties, certify their titles and incorporate them into a cadaster.†

Predios comunales (indigenous landholdings)

The demarcation and regularisation of communal landholdings is well advanced; nonetheless, significant hurdles impede the completion of the process.[66] As of August 2021, the ministry had registered the landholdings of more than 5,680 *comunidades campesinas*, covering 21 million hectares on the coast and in the highlands. Unfortunately, 25 per cent have yet to be fully validated by DIGESPACR, apparently due to litigation stemming from the agrarian reform of the 1970s.

In the lowlands, SICAR has registered the claims of 1,950 *comunidades nativas* covering ~13 million hectares; about two-thirds have received a validated title while the remainder are awaiting resolution of bureaucratic or legal obstacles (Figure 4.6a). Organisations representing indigenous communities report that there are at least 500 additional villages seeking 'recognition', an administrative stage that is a prerequisite for soliciting a title for communal lands.[67] Progress has been stymied by the conflicting land claims of other stakeholders. A survey conducted in 2017 enumerated 2,703 communities of which 808 (30%) reported some kind of land conflict. These included conflicts with other communities (45%), private landholdings (27%) or individuals within their own community (24%), as well as with timber (14%), petroleum (7.3%) and mining companies (5%), and with wildcat miners (1.6%).[68]

The process of distributing land is further complicated because the state must resolve conflicts among its own institutions. For example, approximately twelve million hectares of the forest estate are unavailable to communities because they have been leased as a concession for a determined period of time (see Chapter 7). Similarly, protected areas created in

---

L1026); IDB – Interamerican Development Bank (2014) Land Regularization and Administration Projects: https://publications.iadb.org/publications/english/document/Land-Regularization-and-Administration-Projects-A-Comparative-Evaluation.pdf

* *Instituto del Bien Común* (IBC), *Asociación Interétnica para el Desarrollo de la Selva Peruana* (AIDSP) and *Confederación Campesina del Perú* (CCP); *Confederación Nacional Agraria* (CAN).

† Peru has created a decentralised system of public registries: *Superintendencia Nacional de los Registros Públicos (SUNARP)* that manages a nationwide database covering four legal categories: properties, legal entities (e.g., corporations), natural persons (e.g., powers of attorney), and contracts and assets (such as vehicles). Source: https://www.gob.pe/sunarp

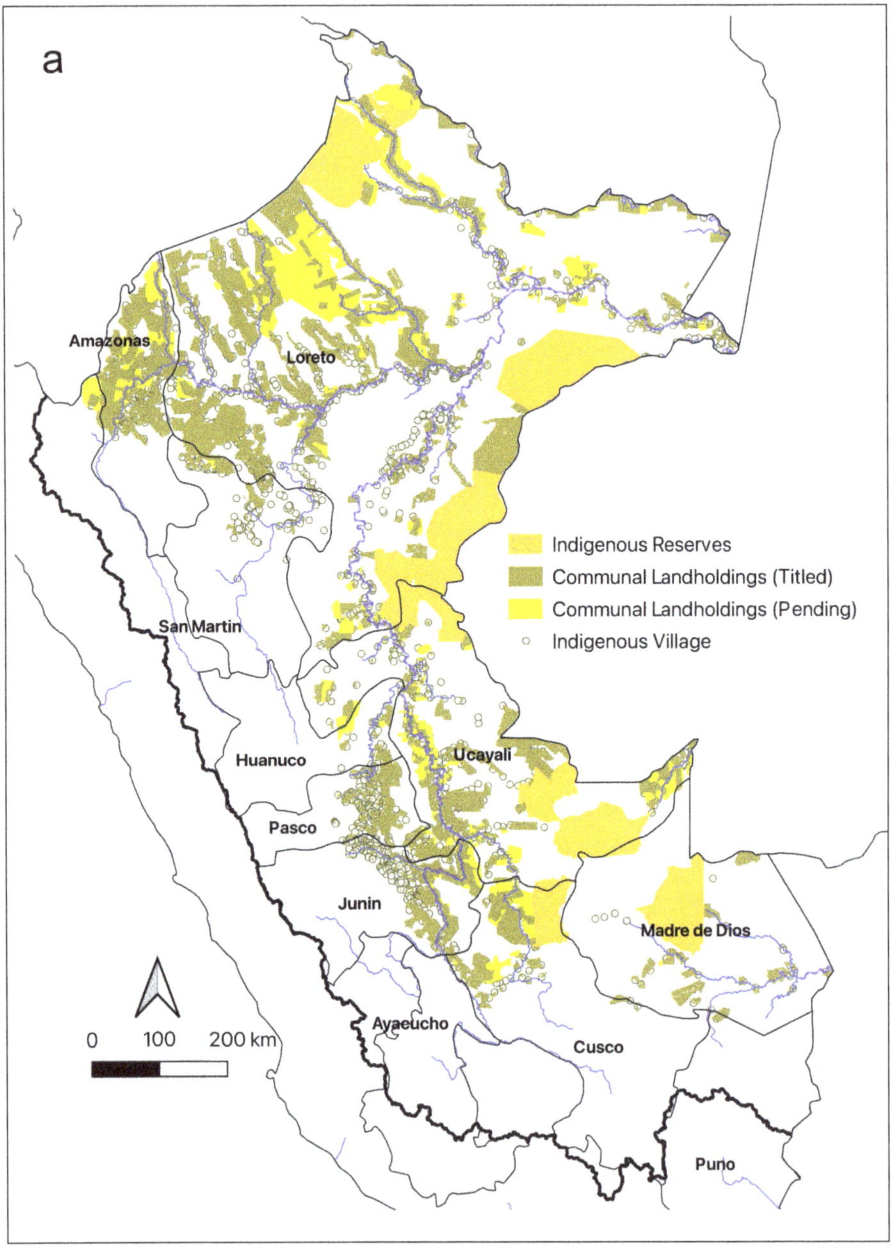

Figure 4.6a: Peru has recognised the territorial rights of ethnic indigenous groups via landholdings linked to individual communities. Many villages are yet to have their land claims adjudicated. The state has also created several Indigenous Reserves to protect groups living in voluntary isolation.

Data source: IBC (2020).

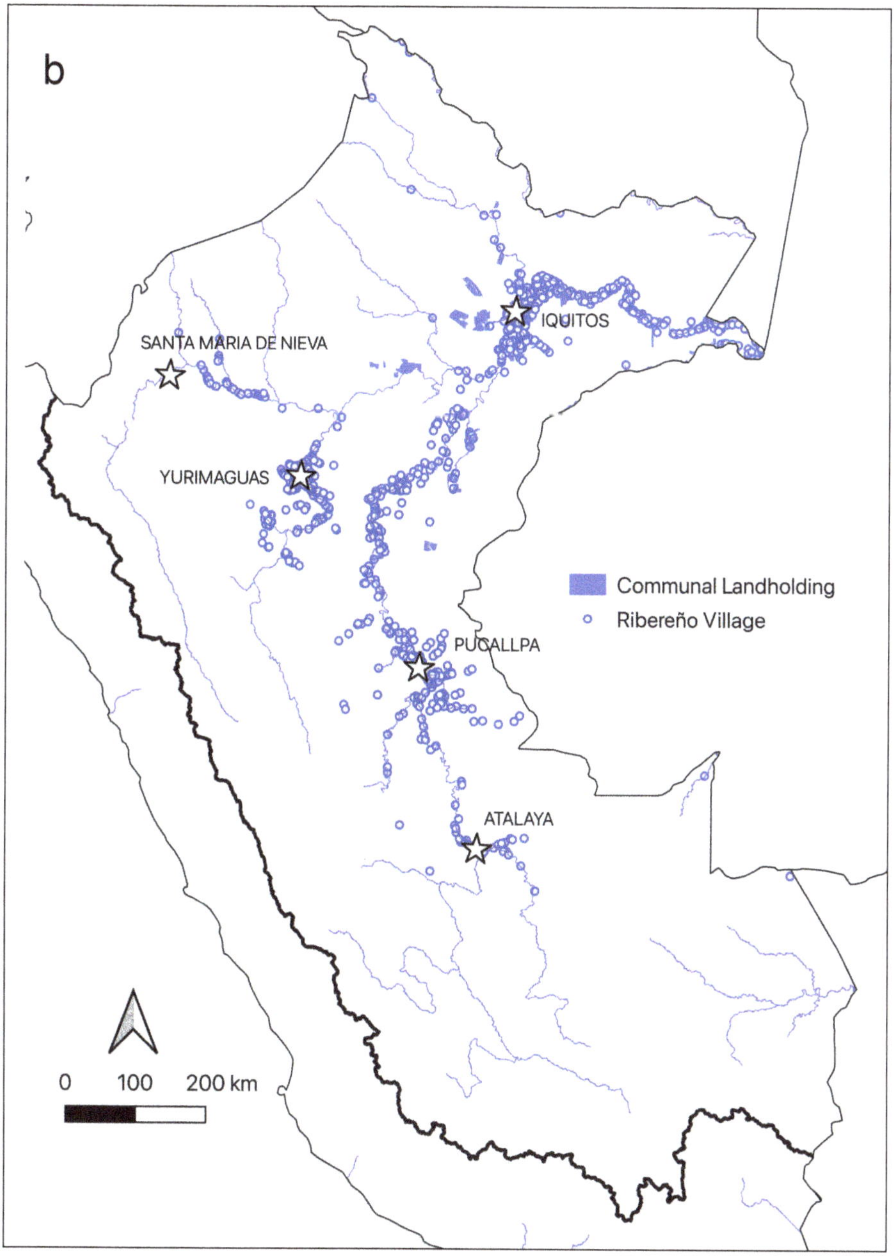

Figure 4.6b: Hundreds of villages on the Amazon, Ucayali and Marañon rivers are populated by communities with no specific ethnic affinity but who self-identify as Ribereños; the Peruvian state has only recently begun to recognise their right to the land adjacent to their riverside communities.

Data source: IBC (2020).

the 1980s and 1990s, before there was clarity as to the priority of indigenous claims, impede the state's ability to grant legal title to long-established communities. Management guidelines guarantee inhabitants' access to natural resources, but unlike their peers on nearby landscapes, indigenous inhabitants of a national protected area do not have communal property rights.Below-ground mineral resources are a major source of contention: while they legally belong to the state, their exploitation is contingent upon the consent of local indigenous communities (see Chapter 6).

These limitations are particularly vexing for the approximately 750 villages inhabited by about 35,000 families who self-identify as *comunidades ribereñas* (Figure 4.6b). These forest-dwelling people have a mixed heritage that includes an indigenous legacy but lacks an ethnic affinity due to intermarriage and deculturisation (see Chapter 6). They reside along all the major rivers but are most densely settled near Iquitos. Often, *Ribereños* coexist and share resources with ethnic communities, particularly along the southern border of the *Reserva Nacional Pacaya Simiria*. Recently, the regional government of Loreto (GOREL) used the *comunidad campesina* protocol to formalise the status for 64 landholdings covering ~376,000 hectares.

Karol Moraes/ Shutterstock.com

Hundreds of riverside villages in the Peruvian Amazon are inhabited by families who self-identify as Ribereñas. Most are descended from immigrants and survivors of indigenous communities that were disrupted during the rubber boom of the nineteenth century. Because they lack a specific ethnic heritage, they have not benefitted from the state's programme to allocate communal landholdings to native communities.

Near Pucallpa (HML #41) and Yurimaguas (HML #44), both *comunidades nativas* and *comunidades ribereñas* must compete for land and resources with an expanding population who self-identify as *colonos*. As in Bolivia. they come from a highland culture where communal landholdings are the norm, but on the forest frontier of the Peruvian Amazon, they have embraced private property as their pathway to prosperity.

Predios rurales (private landholdings)

Regularising land tenure on landscapes where private property predominates is more challenging than on landscapes occupied by communal landholdings. In part, this is due to their greater number, but the task is further complicated by the limited resources of their owners and the chaotic nature of frontier landscapes.[69] The *Censo Nacional Agropecuaria* of 2012 enumerated about 3.7 million private properties in all of Peru and the first two phases of the PTRT programme registered approximately two million of these landholdings into what would eventually become the SICAR database[70] The overwhelming majority are located on the coast or in the highlands, where PTRT technicians and regional authorities have succeeded in the *saneamiento* of about 75 per cent of all private properties.[71]

Unfortunately, the limited technical capacity of the regional offices in lowland provinces, exacerbated by administrative reshuffling that preceded the implementation of PTRT3, has impeded progress in Amazonian jurisdictions.[72] A comparison of data compiled by the agriculture ministry (MINAGRI) and the census (INEI) are broadly similar (Table 4.6); however, an inspection of spatial data available in the public domain reveals that tens of thousands of farmsteads have not been incorporated into either database (See Annex 4.12). It is difficult to know with any level of precision how many smallholdings actually exist in the region, but 'back of the envelope' estimates suggest that the number of farmsteads in the region is well over 500,000, implying that the process of *saneamiento* is less than 25 per cent complete. When the other departments with tropical provinces are considered, that number might approach one million.*

Unfortunately, many farmsteads are destined to persist as illegal or informal holdings for the foreseeable future. In the Huallaga Valley (HML #42 and HML #43), settlers have invaded forest concessions on the upper slopes of both the upper and lower valley. These cannot be legally regularised without a modification of the legislative and regulatory framework governing the forest estate. The most conflictive landscapes are the agri-

---

* In addition to the five regions that are wholly part of the lowland tropics (see Table 4.6), there are four regions composed of both the montane and lowland tropics (Cuzco, Huánuco, Junín, Pasco) and three that are largely covered by highland ecosystems with a small area with tropical climates (Ayacucho, Cajamarca, Puno).

Table 4.6: The number of private landholdings in five Amazonian departments of Peru.

Region	CNA (2012)[i]	MINAGRI (2016)[ii]	SICAR (2020)[iii]	Land Use Model [iv]
Amazonas	69,125	60,044	17,450	75,000
Loreto	67,003	24,044	5,079	138,000
Madre de Dios	6,602	14,249	11,036	44,000
San Martin	91,137	157,668	64,451	158,000
Ucayali	25,325	30,804	23,774	99,000
Totals	259,192	286,809	121,790	505,000

i IV Censo Nacional Agropecuario (2012) Instituto Nacional de Estadísticas e Informática, Cuadro No. 051, *Unidades agropecuarias y superficie de las parcelas.* http://censos.inei.gob.pe/cenagro/tabulados/

ii MINAGRI – Ministerio de Desarrollo Agrario y Riego (2016). MINAGRI asume administración de la información nacional referida a saneamiento y titulación de predios rurales, y de comunidades campesinas y nativas. https://www.midagri.gob.pe/portal/present-catastro-rural

iii SICAR (2020). Summary data from shape file downloaded from https://www.geogpsperu.com/2020/10/mapa-de-predios-rurales-descargar.html

iv Total historical deforestation reported by *Geobosque* divided by an assumed mean size of 10 hectares of deforestation per landholding. http://geobosques.minam.gob.pe/geobosque/view/descargas.php

cultural frontiers that surround the eastern terminus of Peru's three major Amazonian highways: the *Interoceanico Norte* (HML #44), the *Interoceanico Central* (HML #40 and HML #41) and *the Interoceanico Sur* (HML #27). On all three landscapes, settlers are expanding outward from long-established agrarian landscapes and, in the process, invading both indigenous lands and forest concessions. The SICAR system was specifically designed to exclude this type of blatant illegality, and these landholdings should be excluded from the cadaster regardless of (corrupt) attempts by local authorities to include them.

Land grabbers are using the SICAR system to launder unallocated forest lands from the state. The most egregious examples are large-scale oil-palm plantations in San Martin, Loreto and Ucayali (see Chapter 3). Some of these plantations have passed through a legal adjudication while others have been declared illegal. Regardless, the perpetrators have not suffered any significant penalty via the criminal justice system, while the plantations continue to operate and expand.

The SICAR system is being used to create smallholdings carved out of unallocated public lands via deliberately planned development projects designed to appeal to local constituencies. For example, the system shows a string of (~50) contiguous land parcels on several tributaries of the Ucayali River. Hopefully, this is an effort to recognise the property rights

Source: Google Earth

The national rural cadaster shows two types of land claim on the east bank of the Ucayali River, near Pucallpa in the Peruvian Amazon: (a) irregularly shaped holdings along a tributary with long-established communities and (b) blocks of uniformly-sized parcels on upland landscapes not associated with any specific village or community.

of *Ribereña* families and not the actions of *traficantes de tierra*. In the Madre de Dios region, the SICAR system shows ~250 identical plots adjacent to two regional highways that transect the gold mining landscapes west of Puerto Maldonado.

The west bank of the Ucayali Valley has attracted land speculators and immigrant settlers. The region has long been viewed as an expansion zone for agriculture and several blocks of forest have been claimed and registered in the SICAR system. The region was the focus of a proposed investment by Grupo Palmas, Peru's largest operator of industrial oil palm plantations (see Chapter 3).[73] The company abandoned its plans in 2017 following a legal battle and public relations scandal;[74] however, the fate of these landholdings has yet to be resolved and they were not included in a corporate programme to support forest conservation announced in 2021.*

Evidence of accelerating change was highlighted by the arrival of Mennonite farmers in 2020, establishing the first colony of this type in Peru

* In 2017, the Grupo Palmas abandoned plans to develop oil palm plantations at two localities: Tierra Blanca and Santa Catalina (14,000 ha at Sarayacu, Ucayali, Loreto) and the Manatí and Santa Cecilia (10,000 hectares at Indiana, Maynas, Loreto). As of 2021, only the Indiana properties had been allocated to a forest conservation initiative. Forest Conservation Fund (2021) Maniti Promise Forest, https://www.fundforests.org/maniti-promise

Source: Google Earth

The west bank of the Ucayali Valley has attracted land speculators and immigrant settlers. Several blocks of forest have been claimed and registered in the national rural land register (a; b; c), while Mennonite farmers have purchased land from intermediaries whose holdings do not (yet) appear in the national rural cadaster (d; f). Access to the area is being facilitated by logging roads that connect to the port cities of Orellana and Sarayacu (arrows); eventually, they will link to the national road system via Huimbayoc. The region includes two forest blocks (g; h) ceded in 2013 to one of Peru's largest corporate entities (Grupo Romero), which abandoned plans to establish oil palm plantations in 2017 (see text).

Data source: Google Earth and SICAR (2020).

and providing further evidence of the disfunction of the SICAR system.[75] Mennonites are astute and experienced in the dark arts of rural real estate markets in Latin America; they are unlikely to risk their investment capital without a deed documenting the legality of the landholding. Environmental journalists have reported that these types of legally dubious transactions are being approved by local authorities but not reported to DIGESPAC, the agency within MINAGRI charged with updating the SICAR system.[76]

The piedmont landscapes located west of the Ucayali River will eventually connect with the national highway system, which will trigger more speculation in land and deforestation in previously remote areas. This ongoing development demonstrates the potential for local governments to expand the agricultural frontier by approving timber contracts, facilitating road construction and issuing land grants without the intervention (or knowledge) of central authorities.

Ecuador

The sharecropping system that defined land tenure in the Ecuadorian highlands prior to agrarian reform was known as the *'huasipungo'*, a Quechua word that describes the relationship between landlords and tenant farmers. The end of this feudal system had a radically different outcome when compared to Peru and Bolivia, however, because landowners preempted the confiscation of their lands by expelling tenant farmers.[77] Owners mechanised farm operations and turned to contract labour, while thousands of peasant families were evicted from their homes. Some moved to urban centres, but many chose to migrate to the agricultural frontiers in the tropical lowlands of the Amazon and the Pacific coast.[78]

The official effort to promote settlements in the Ecuadorian Amazon began in 1957 when the democratically elected government created the *Instituto Nacional de Colonización* (INC). In 1964, a military government enacted the *Ley de Reforma Agraria y Colonización*, which merged the *INC* into the newly created *Instituto Ecuatoriano de Reforma Agraria y Colonización (IERAC)*.[79] Between 1964 and 1994, IERAC distributed about five million hectares of land with support from USAID and the Alliance for Progress; about 1.8 million hectares were located in the five Amazonian provinces.[80] Land was distributed in forty-hectare plots, which suggests that about 45,000 families acquired plots in the Amazon during this thirty-year period.

Unfortunately, the IERAC did not provide homesteaders with documents that were equivalent to a legal title because ownership was contingent upon residency and evidence of development. Settlers were provided with a provisional deed that required future administrative action to be converted into a full legal title. The IERAC did not incorporate this information in a national archive; instead, the information was preserved in 'folders' housed at their regional offices.

In 1994, a new law replaced IERAC with the Instituto Nacional de Desarrollo Agrario (INDA), and, as in Peru and Bolivia, a major objective was to introduce market economics into the rural economy as part of 'structural readjustment' policies. The law centralised the land titling process in

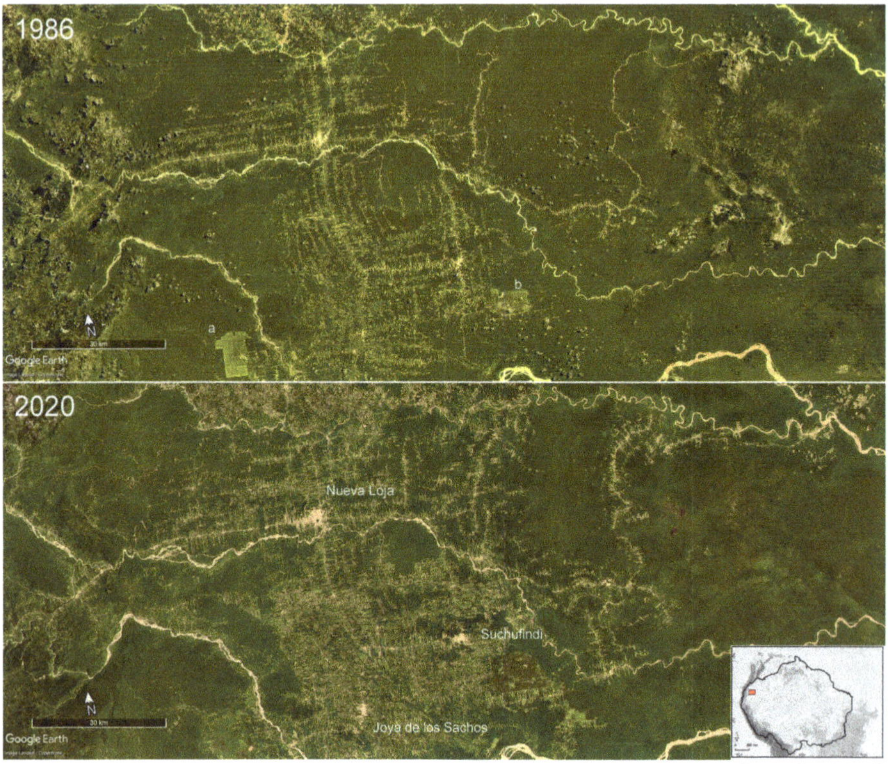

Source: Google Earth

The Ecuadorian provinces of Succumbios and Orellana were opened to settlement in the 1960s with the discovery of oil. By the 1980s, the distribution of 40-hectare landholdings was well advanced, as was the establishment of two large-scale oil palm plantations (a; b). The frontier has essentially been closed, but deforestation continues around the margins of the settlement zone and within individual landholdings.

Quito and in 2002 was supported by an IDB-financed initiative to create a digital database.* This initiative had only limited success, however, and by 2010 a total of 700,000 'folders' were waiting to be processed when INDA was dissolved and its functions were transferred to the *Subsecretaria para Tierras y la Reforma Agraria* at the Ministry for Agriculture.[81]

The responsibilities of the new agency had been fundamentally changed, however, because the constitution of 2008 devolved administrative

* IDB – Interamerican Development Bank; Project EC-0191: Land Titling and Registration in 2001 @ $15 million; it was not actually a fully funded initiative but a pilot project to design and test a digital cadaster.

authority over land tenure back to municipal governments.* Nonetheless, the law tasked the national government with the responsibility of compiling and maintaining a digital database, which led to the creation of a national cadaster known as SIGTIERRAS. The IDB supported the effort with another loan that incorporated the experiences of the previous decade's pilot project.†

The land tenure process is now managed by the *Autoridad Agraria Nacional* (AAN), a new entity with an expanded portfolio of obligations that was created by the *Ley de Tierras* of 2016.‡ This far-reaching legislative act seeks to ameliorate inequality of land tenure in Ecuador and establishes limits on the maximum dimensions for properties: 200 hectares for the highlands, 500 hectares for the coast and 1,000 hectares for the Amazon. It also provides the ANN with the power to confiscate properties that are larger than these dimensions or do not meet criteria regarding social and economic function.[82] In spite of its populist appeal, the land law generated criticism because it did not incorporate specific protocols for resolving the claims of lowland indigenous communities.[83]

Although their territorial rights are enshrined in the constitution and codified by the *Ley de Tierras*, there are only a handful of fully demarcated indigenous entities. As in other countries, these can be organised into two broad categories: communal landholdings associated with one (or a few) villages on frontier landscapes and large reserves extending across wilderness landscapes with several isolated villages. The indigenous territories shown in maps prepared by civil society organisations show both types of tenure categories (Annex 4.13). Most territories have been established by presidential decree, but only a few of the village landholdings have been formalised and demarcated. Most represent claims presented to the government. Their final size and exact boundaries are awaiting the land tenure review process that has been underway for at least two decades.

As in Peru and Brazil, the land regularisation (*saneamiento*) process in Ecuador is being organised using field campaigns that target specific municipalities in order to maximise the participation of property owners and achieve wall-to-wall coverage. As of October 2017, the AAN had registered 1.4 million rural properties in 59 municipalities, a significant number but only a fraction of those awaiting regularisation in the nation's 221 municipalities.[84] Incomplete as it may be, this number dramatically

---

* In Ecuador, municipalities are referred to traditionally as *cantóns*, but in the constitution they are referred to as *Gobiernos Autónomos Descentralizados Municipales*.

† IDB – Interamerican Development Bank: Project EC-L1071: National System for Rural Land Information and Management and Technology in 2010 @ $US 90 million, plus $US 38 million in matching funds. The project was closed in 2018.

‡ *Ley Organica de Tierras Rurales y Territorios Ancestrales*, 14-mar.-2016: https://www.gob.ec/regulaciones/ley-organica-tierras-rurales-territorios-ancestrales-1

overturns previous estimates about the number, size and distribution of rural properties in the country.

The agricultural census of 2000 enumerated a total of 850,000 *Unidades de Producción Agropecuaria* and reported that about seventy per cent had obtained legal title. Although they are more than twenty years old, these statistics have been reproduced in subsequent reports and used to guide policy.[85] Preliminary results from the IDB pilot project registered about 2.7 million parcels and suggested that about sixty per cent were lacking secure title.[86] The results from the second IDB project would place the number of landholdings between four and five million.[87] The most recent survey found that ~75% of the landholders hold some kind of legal document that supports their possession, although fewer than a quarter of them had registered their property with regional land offices.[88]

The regularisation process, which has been carried out in three of 41 municipalities in the Ecuadorian Amazon, likewise shows that previous assumptions underestimate their number and overestimate their dimension (Table 4.7). For example, the number of landholdings registered in a single municipality was greater than the number reported by the census for the entire province.* The discrepancy may be caused, in part, by an expansion of the agricultural frontier; however, most of the difference can be explained by the subdivision of existing properties. The original distribution in the 1970s averaged between forty and sixty hectares; in contrast, the *predios* registered in SIGTIERRAS averaged between ten and twenty hectares, indicating that many have been legally subdivided, probably via inheritance, into smaller units.†

The census of 2000 reported a total of 46,000 farmsteads in Amazonian Ecuador covering a total of 2.5 million hectares (~27% of the total area). Nonetheless, the total area identified as human modified landscapes (see Chapters 1 and 2) spans approximately 3.9 million hectares. Assuming the mean size of a private landholding lies between twenty and forty hectares, then there should be between 75,000 and 150,000 landholdings that need to be registered, validated and incorporated into SIGTIERRAS.

The resolution of indigenous lands is likewise unfinished. Their claims for communal landholdings sum to ~2.5 million hectares distributed across more than 4,000 communal landholdings. Only 85 have actually been

---

* Succumbios Dept. (7,300) versus Lago Agria Munic. (16,578); Orellana Dept. (4,948) versus Coca Munic. (9,239); Morona Santiago Dept. (4247) versus Sucúa Muncip. (5,410). Source of Departmental data: INEC – Instituto Nacional de Estadística y Censos 2002. Censo Nacional Agropecuario 2000: https://www.ecuadorencifras.gob.ec/censo-nacional-agropecuario/

† The census also records functional enterprises, enumerated as *unidades productivas agropecuarios*, which could be composed of multiple properties managed under a common enterprise or operator.

Table 4.7: Land tenure in three municipalities in Amazonian Ecuador.

	Lago Agrio Succumbios		**F. Orellana (Coca) Orellana**		**Sucúa Morona - Santiago**	
	number	area (ha)	number	area (ha)	number	area (ha)
Private Properties	16,578	227,217	9,239	219,531	5,410	53,279
Communal Land-holding	21	25,960	19	305,509	45	16,109
Indigenous Reserve	1	10,000	1	75,000	1	5,000
Protected Areas	1	15,000	1	90,204	1	15,000
Other		37,549		13,000		
Total		**315,726**		**703,247**		**89,388**

Source: SIGTIERRAS (2017) *Catastro Rural en el Ecuador*. Ministerio de Agricultura y Ganadería del Ecuador (MAG) Quito: http://www.sigtierras.gob.ec/publicaciones/

demarcated and issued a legal title, all within the three municipalities that benefitted from the SIGTIERRAS project, which confirmed that indigenous claims often overlap with private properties. This was particularly the case in Morona-Santiago, where Shuar families have laid claim to individual landholdings using the legal options available to them in the decades before the state started recognising communal landholdings.

The state has the infrastructure in place to resolve this longstanding administrative obligation to its rural citizens. The ANN has offices in all five provincial capitals and an online application where property owners can self-register their landholdings. Hopefully, the IDB will finance a third phase of the SIGTIERRAS project that will allow the government to complete the task.*

Colombia

The unequal distribution of land in Colombia is the root cause of that nation's violent history. Multiple policy initiatives spanning decades have failed to resolve the problem. The first agrarian reform law was promulgated in 1936, but it only motivated landowners to protect their assets by converting tenant farmers into contract labour.[89] A backlash to land reform eventually led to a civil war between 1948 and 1958 when the two major political parties battled for power during a period referred to as *La Violencia*. Subsequently, a coalition government pursued a renewed effort at agrarian reform with

* The SIGTIERRAs project received a favorable review, and the IDB funded a technical support consultancy to prepare a phase 2 proposal, which was completed in 2021. See https://www.iadb.org/en/project/EC-T1382

the creation of the *Instituto Colombiano de la Reforma Agraria* (INCORA) in 1961. This initiative established clear criteria for the expropriation of land and instituted mechanisms to indemnify landowners. As in other countries, it had the support of the Alliance for Progress and promoted colonisation programmes within the Amazon. This effort also failed and contributed to the formation of the guerilla armies and decades of violent conflict.

A third agrarian reform in 1994 was based on a market-based approach for redistributing land by providing subsidies so peasant farmers could purchase land from large estates. This followed the precepts of the Constitutional reform of 1991 and coincided with the legal decrees in 1995 that recognised the rights of indigenous and traditional people. INCORA was replaced in 2003 by *Instituto Colombiano de Desarrollo Rural* (INCODER), which diversified its mission by sponsoring the sustainable development of *campesino*, indigenous and Afro-Colombian communities. These initiatives also failed to resolve the long-standing grievances linked to land tenure and rural poverty, a task that was essentially rendered impossible by the violence that consumed the country for another 25 years.[90]

The competition for territory between leftist guerrillas and their equally violent paramilitary competitors has enormously compounded the problem of land tenure. Both sides dispossessed legitimate landowners, either by direct confiscation or forced sale at gunpoint. Land theft created a legacy that plagues the national economy because investors are unwilling to commit resources to a productive enterprise if there is the risk of forfeiture due to illegitimate title. The most conspicuous attribute of this legacy, however, is the massive number of displaced people, estimated at five million in 2020.[91] Small farmers were particularly vulnerable, and the violence greatly aggravated the inequity in the distribution of land.[92] In 2015, civil society organisations estimated that seventy per cent of the country's small farmers hold only 2.7 per cent of its arable land while 68 per cent was controlled by only 0.5 per cent of all landholders.[93]

This legacy was supposed to be addressed via the Colombian Peace Process. The final agreement is a long and complex document that covers a multitude of complex and thorny issues. The first chapter deals with land, and the first section of that chapter outlines a pathway for providing fair and equitable access to land.[*] Land issues were treated first because unequal access to land sparked the conflict, and fifty years of war magnified that injustice. The agreement goes further, however, because it also recognises that resolving land-related discord and uncertainty of land tenure is essential for closing the agricultural frontier and conserving the natural patrimony of Colombia.

---

* *Acuerdo Final para la Terminación del Conflicto y la Construcción de Una Paz Estable y Duradera.* See full text at: http://www.eltiempo.com/contenido/politica/proceso-de-paz/ARCHIVO/ARCHIVO-16682558-0.pdf

The agreement created a process entitled the *Reforma Rural Integral (RRI)*, which is to be implemented by two institutions: *ANT*, a clearing-house for all issues related to land tenure, and *Agencia de Desarrollo Rural* (ADR), which will foster investment and provide technical support.* The RRI has four major components:[94]

1. Provide land to displaced families using land seized from criminals or acquired by purchase.
2. Formalise rural land tenure and grant free land to low-income families via a territorial-based process.
3. Establish an agrarian judicial system to resolve all property disputes.
4. Organise and execute a modern land registry (cadaster).†

Resolving land tenure in the Colombian Amazon is essential for the success of the peace process. The region was at the centre of the conflict and one of the last bastions of the *Fuerzas Armadas Revolucionarias de Colombia* (FARC). During the war, FARC maintained a logistical corridor that spanned three national parks in the foothills of the Andes (Los Picachos, Tinigua, Macarena) and the crown jewel of Colombia's protected area system in the Amazon lowlands, Serranía de Chiribiqueta. The landscapes surrounding the three montane reserves attracted tens of thousands of peasant farmers who cultivated coca under the auspices of FARC. The government tried to assert control via police action but made no real attempt to control land use in the buffer zones (*Distritos de Manejo Integral*)‡ that surrounded the four protected areas.

The peace process has stimulated long-suppressed investment in adjacent agricultural landscapes in Meta, Caquetá and Guaviare, which has stimulated a scramble for land across the forest frontier that separates these agrarian landscapes from the wilderness areas of the Colombian Amazon. The area is now riddled with roads where land grabbers collude with ex-combatants who employ settlers to clear forest to establish both coca farms and cattle pastures across an 'arc of deforestation' more than 500 kilometres long.[95] The short-term cash flow is being driven by illicit drugs,

---

* See http://www.adr.gov.co/index.php and http://www.agenciadetierras.gov.co/

† There are an estimated 3.7 million rural properties in Colombia, and less than 5% have clear legal title. Although 48% of the known landholdings have been incorporated into a preliminary version of the national cadaster, an unknown number has yet to be enumerated. The World Bank has financed P162594, Multipurpose Cadaster Project @ $US 100 million (2019) and P172972, Additional Financing for the Multipurpose Cadaster Project @ $US 42.9 million (2020): https://projects.worldbank.org/en/projects-operations/projects-home

‡ DMI Ariari Guayabero, DMI Macarena Norte and DMI Macarena Sur.

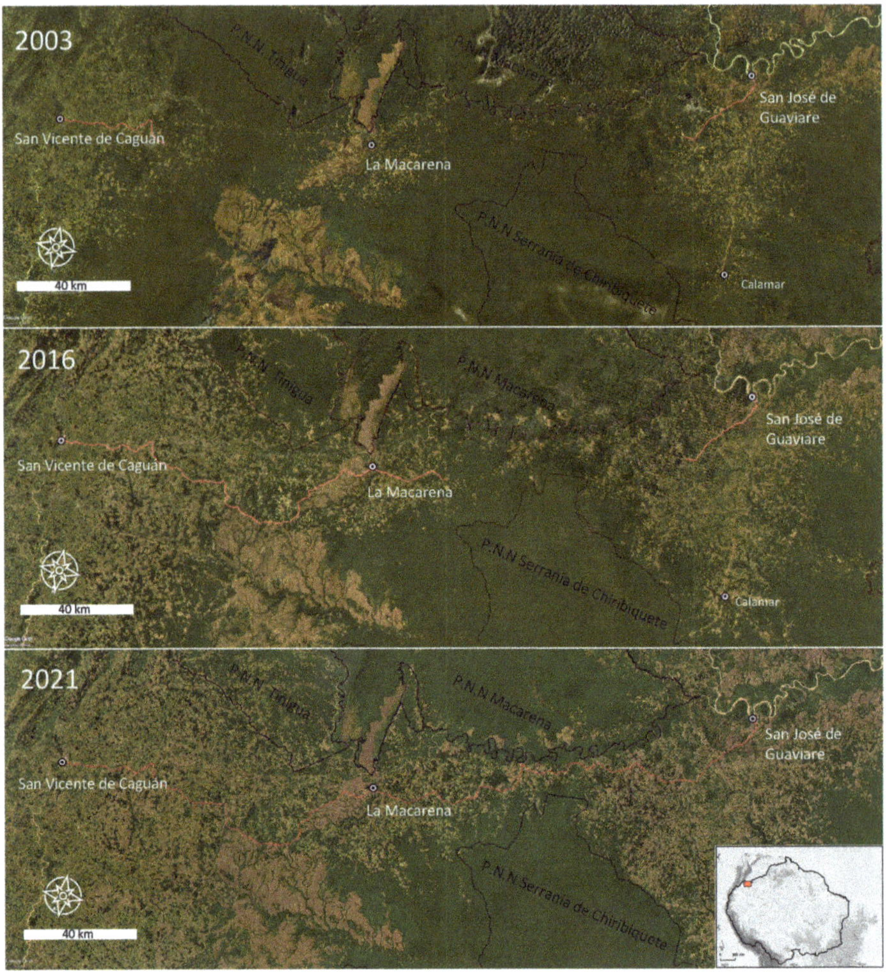

Source: Google Earth

The forest corridor between Chiribiquete and Macarena national parks in the Colombian Amazon was gradually deforested and fragmented between 2003 and 2021. Settlers and land speculators originally accessed their claims via the river system while developing an informal road network, which eventually linked the towns of San Vicente de Caguán, La Macarena and San José de Guaviare.

but the medium-term speculation is now quite clearly focused on land and the rapidly expanding cattle industry (see Chapter 4).

This dynamic will persist until the central government or regional authorities establish the rule of law and the presence of the administrative state. Until that happens, land grabbers and campesino settlers will continue to appropriate state lands within the last habitat corridor connecting the lowland forests of the Amazon with the montane forests of the Andean Cordillera.

Venezuela and the Guianas

Historically, agrarian reform was never a major political issue in any of the countries on the Guiana Shield. Because of its oil wealth, the rural poor of Venezuela flocked to the cities to enjoy the benefits of subsidised housing, transport and food. Agrarian reform became a priority only when the government of Hugo Chavez sought to transform the nation via a socialist revolution. A new land tenure regime in 2010 led to the confiscation of several million hectares of private estates. Most of these actions occurred in non-Amazonian regions, and colonisation of Amazonian wilderness has never been pursued as a deliberate policy (Annex 4.15).

Land tenure in Guyana and Suriname reflects their shared colonial history and the legacy of Crown lands, which were transferred to the republican governments upon independence in the 1960s.[96] Agrarian landscapes are restricted to the coastal provinces where tenure is a combination of freeholders and leaseholders on public lands.[97] The former are few in number and include both family farms and plantation estates, while the latter include cooperative societies of small farmers who operate as independent producers. Away from the coast, both governments enjoy a near-monopoly on land tenure, managed via concessionaire systems governing both minerals (Chapter 5) and timber (Chapter 8).

In Guyana, the state owns approximately 73 per cent of the national territory, freeholders control twelve per cent and indigenous villages hold communal title to about fifteen per cent of the country, mostly in the interior.[98] In Suriname, the state holds title to more than 95 per cent of all land, despite demands by Maroon* and indigenous communities for the recognition of their territorial rights (Annex 4.16). Failure to accede to these requests was one of several causes of a civil war that plagued the nation between 1986 and 1991, which was followed by an extended period of political stagnation that allowed successive governments to ignore their demands – despite multiple rulings by the Inter American Court of Human Rights (see Chapter 11). In 2016, the government finally made a commitment to resolve all outstanding issues; however, as of January 2022, the final details had yet to be finalised.

Land Use Planning: An Aspirational Tool with Mixed Results

Regulating land tenure is not the only power available to the state for influencing how people use land. Land-use planning and land-use zoning are two closely related mechanisms that Pan Amazonian nations wield to foster sustainable development on their forest and agricultural frontiers. Like policies governing infrastructure, agriculture and land tenure, these

* Maroons are a traditional people of mixed African and indigenous heritage living in Suriname and French Guiana.

technical programmes have evolved in response to the shifting economic and social forces within countries, as well as to the prescriptions from multilateral agencies and civil society groups seeking to protect the biodiversity of the Amazon Forest.

In the 1970s and 1980s, most land-use planning programmes used a methodology developed by the United States Department of Agriculture (USDA) that identifies optimum land use based on climate and soil and stratifies regions and landscapes into categories ranging from full protection to intensive agriculture. Known in the United States as Land Capability Classification, in Latin America it has been promoted by USAID as *Capacidad de Uso Mayor de la Tierra* (CUMAT). A similar system developed by the *Instituto Interamericano de Cooperación para la Agricultura* (IICA) and sponsored the Food and Agriculture Organisation (FAO) was known as a *Zonificación Agro Ecológica* (ZAE). The technical details and output from these studies were of very high quality, but they suffered from one fundamental defect: they did not include a participatory process, which caused them to overlook economic trends already underway and customary uses that might not coincide with the best technological option for land use.

These limitations quickly became apparent, and the ZAE framework was modified and renamed as *Zonificación Ecológica Económica* (ZEE), which uses technical analysis as a baseline but incorporates additional social and economic criteria. Most importantly, it included a participatory process to ensure the aspirations of different stakeholder groups are considered, including indigenous and traditional communities, but also small farmers and agroindustry. All the Pan Amazonian countries have embraced some variant of the ZEE methodology and have enacted it into their regulatory processes to govern land-use planning (recommendations) and regulatory frameworks (zoning).*

The effectiveness of these studies is decidedly mixed. Settlers and corporate farmers have used the technical components to inform their investments, but most deforestation is driven by infrastructure development (see Chapter 2), demand for commodities (see Chapter 3) and land speculation. Nonetheless, the ZEE process coincided with programmes to create protected area systems (see Chapter 12) and has supported territorial claims by indigenous communities (see Chapter 11). Governments, NGOs and multilateral institutions continue to invest in these studies, arguing they are essential for discovering a path towards truly sustainable development.

---

* Bolivia: *Plan de Uso del Suelo* (PLUS); Brazil: *Zoneamento Ecológico e Econômico* (ZEE); Colombia: *Plan de Ordenamiento Territorial* (POT); Ecuador: *Planes de Desarrollo Ordenamiento Territorial* (PDOT); Guyana: National Land Use Plan (NLUP); Peru: *Zonificación Ecológica Económica* (ZEE); Suriname: *Bestemmingsplannen*; Venezuela: *Plan Nacional de Ordenación del Territorio* (PNOT).

The ZEE in the Brazilian Amazon

The history of the ZEE in Brazil began in 1981 when Congress passed the National Environmental Policy Act, which recognised 'environmental zoning' as a regulatory tool for promoting the rational use of soil and the protection of ecosystems. This was followed in 1990 by the formation of a working group to review the different methodologies and establish a standard approach for the Legal Amazon. The responsibility was transferred to the states in 1994 and incorporated as a key component of the *Programa Piloto para Proteção das Florestas Tropicais (PPG7).*[*] The methodology was formalised as a regulatory procedure via presidential decree in 2002, at which time the government established a federal commission to coordinate the process (*Comissão Coordenadora do Zoneamento Ecológico-Econômico do Território Nacional* – CCZEE) and convened a working group to accelerate its implementation (*Consórcio ZEE Brasil*). In 2000, the ZEE was incorporated into the constitutionally mandated four-year, state-level, strategic planning process (*Plano Plurianual* – PPA).[†]

In 2010, the Ministry of the Environment published a Macro ZEE (1:1,000,000) of the Legal Amazon derived from preliminary state-level studies that provided the first official vision of the future of the Legal Amazon. (Figure 4.7).[99] The Forest Code of 2012 reinforced the importance of the ZEE by stipulating its use for the implementation of key provisions and obligated the state to produce a more detailed version (1:250,000).

As of October 2021, Acre, Pará and Rondônia had completed final versions that have been approved by federal authorities, while Maranhão, Tocantins and Roraima had draft versions under review. Amazonas and Amapá have completed studies for selected sub-regions that are most exposed to land-use change and land grabbing.[‡] The classification criteria

* The acronym PPG7 was used because the $US 450 million programme was financed by seven advanced economies: Germany (43%), EU (24%), UK (7%), USA (5%), Japan, Netherlands and France. See: Kohlhepp (2018).

† CCZEE (*Comissão Coordenadora do Zoneamento Ecológico-Econômico do Território Nacional*) is composed of twelve cabinet ministries. See: http://www.mma.gov.br/informma/item/7596. *Consórcio ZEE Brasil* (*Grupo de Trabalho Permanente para a Execução do Zoneamento Ecológico-Econômico)* is composed of two ministries and fifteen autonomous agencies and technical institutes. See: http://www.mma.gov.br/informma/item/10407. PPA *(Plano Plurianual)* is a medium- to long-range strategic planning process for activities and investments and establishes the budget for government actions and investments in infrastructure and productive activities. http://www.planejamento.gov.br/servicos/faq/planejamento-governamental/plano-plurianual-ppa/o-que-eacute-o-ppa

‡ Amazonas: Purus, Madeira, Baixo Amazonas; Amapá: Sul do Amapá. MINAM – Ministerio do Meio Ambiente 2016. O Zoneamento ecológico – económico na Amazônia Legal, Trilhando o caminho do Futuro: https://antigo.mma.gov.br/images/arquivo/80253/ZEE_amazonia_legal.pdf

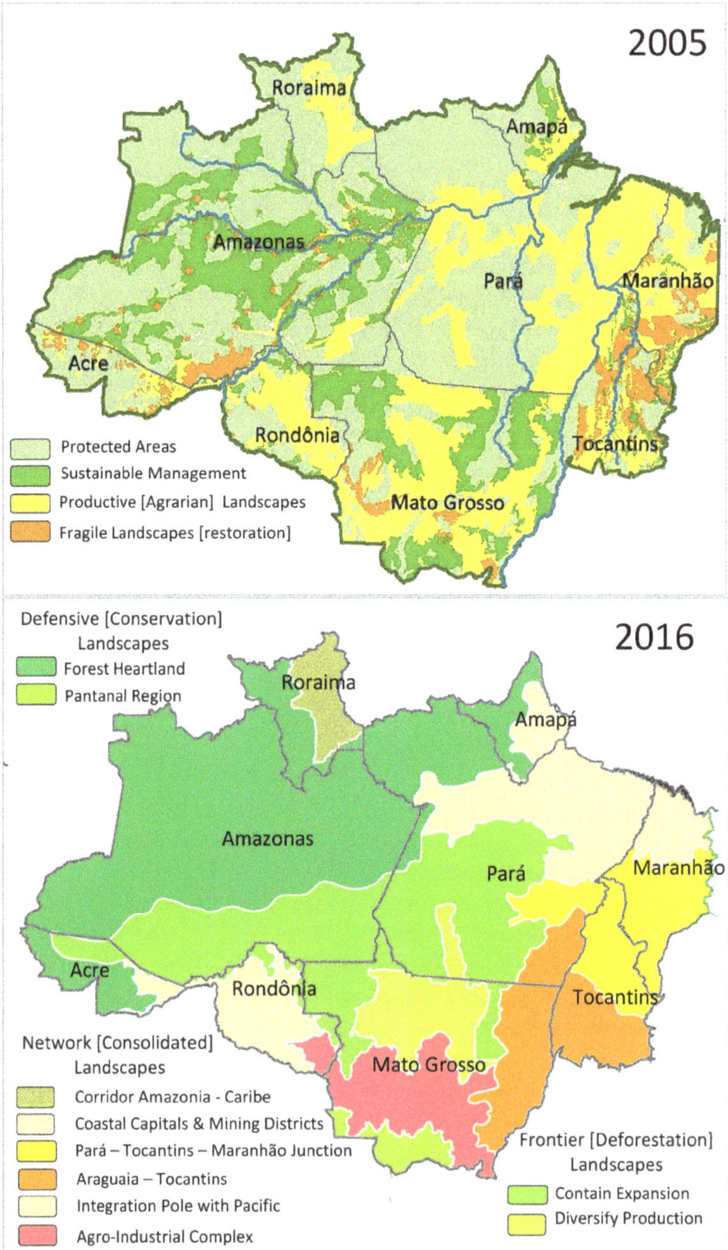

Figure 4.7: (Top) A draft of the Macro ZEE for the Legal Amazon prepared from a harmonised version of state-level ZEE (1:1,000,000). (Bottom) The final official version promulgated by the Temer administration in 2016, stratifying the region into three major (Defensive, Frontier and Network) and ten minor categories.

Data sources (both maps): MMA (2009) and MMA (2016), with additional material from SIAGEO Amazônia (2021).

generally fall into one of three categories: (1) consolidation of existing production landscapes; (2) sustainable use of natural resources; and (3) protected areas and indigenous land.* The first category always includes landscapes where large-scale agriculture and ranching predominate, but can include small-scale farms (Maranhão, Rondônia and Pará) or forest-based livelihoods near major highways (Acre and Amazonas).

The second category typically contains landscapes that support forest-based livelihoods, including those within PAAD-type INCRA settlements, but also privately held forest estates (Amazonas and Roraima) and smallholding communities (Mato Grosso). The third category includes conservation units in all jurisdictions, including those that support forest-based livelihoods and, in some cases, cattle ranching.† Several versions also recognise fragile areas requiring special management (Mato Grosso, Amazonas) and provide for an accelerated process to review and resolve issues related to land tenure (Acre, Roraima). The differences reflect the idiosyncrasies of individual states and the social and economic heterogeneity of the Brazilian Amazon.

The ZEE process is viewed favourably in Brazil, where it impacts both federal and state planning, such as the PPA investment process and environmental review overseen by the environmental protection agency. The first iteration of the ZEE coincided with a parallel effort to protect large swathes of the Amazon and provided technical criteria and legal support for the creation of dozens of conservation units and indigenous territories. For example, fourteen conservation units and indigenous territories were created in Acre after the completion of its preliminary ZEE, while a total of 44 such entities were set aside in Pará.[100] Conservation initiatives would have occurred independently but, by integrating them in a multi-sector analysis with explicit considerations for alternative land uses, the Brazilian state has avoided many future conflicts.[101]

The ZEE documents support efforts to halt or slow deforestation by providing geographic clarity as to which landscapes are off-limits for agricultural development while acting as a legal benchmark that reduces opportunities for land grabbing (Figure 4.8). Public sector financial entities, such as the *Banco do Brasil*, are obligated to review investment projects and

---

* Both Tocantins and Maranhão have adopted a different framework that stratifies their state into geographic subregions based on their biophysical attributes, which are then characterised by development potential. Source: Maranhão: http://www.zee.ma.gov.br/ and Tocantins; https://zee.seplan.to.gov.br/

† There are 13 categories of conservation units, including forest management for non-timber forest products: *Reserva Extrativa* (RESEX), *Reserva de Desenvolvimento Sustentável* (RDS), *Floresta Nacional* (FLONA) and other areas that include a broad range of land use and allow inholdings: *Área de Proteção Ambiental* (APA). See Chs 8 and 12).

ensure they comply with the provisions of the regional ZEE. These plans have widespread public support – except in the state of Mato Grosso – because the public consultation incorporated stakeholder aspirations.

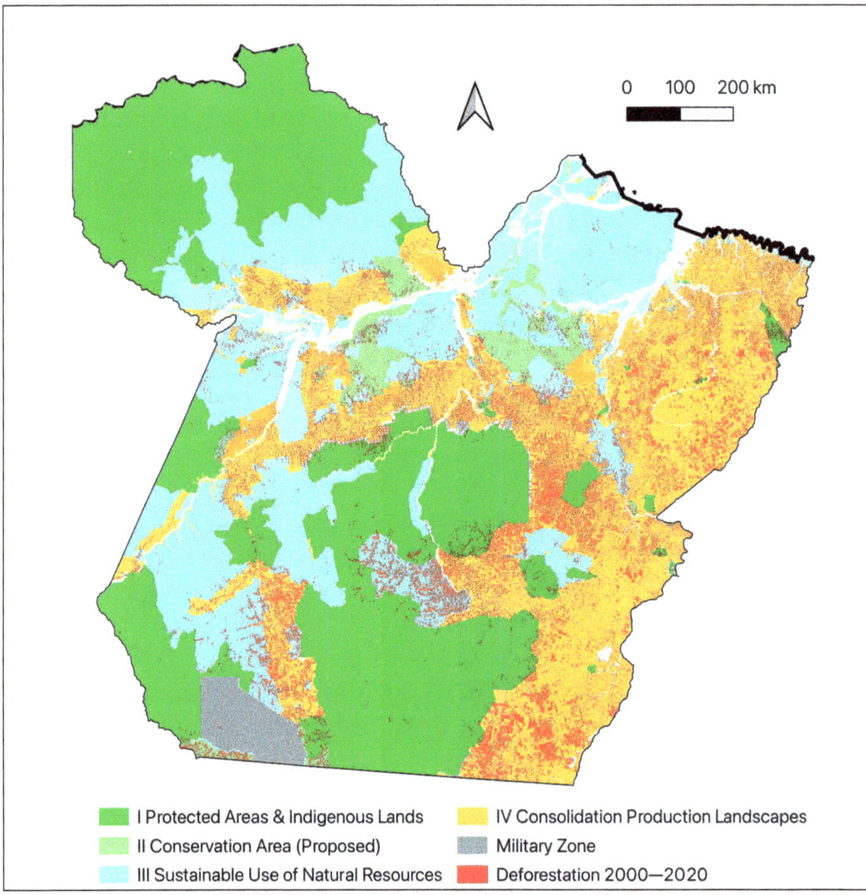

Figure 4.8: Deforestation in Pará state since 2000 in relation to the Macro ZEE of 2012. Category I covers conservation units and indigenous lands that severely restrict economic activities. Category II areas were identified as appropriate for sustainable use but were listed in 2012 as undesignated public land. Category III includes PAAD-type (forest and riparian) INCRA settlements and sustainable use conservation units, including several that allow private inholdings. Category IV includes PA-type (agrarian) INCRA settlements, private landholdings and undesignated public lands. Although much of the documented deforestation is illegal, it has been largely restricted to landscapes zoned for development (consolidation of productive activities).

Data sources: SEMAS (2012) and RAISG (2021).

Table 4.8: The major categories and subcategories of the ZEE of Mato Grosso proposed in 2018.

1. Areas for Agricultural Uses with Protection of Water Resources	
1.1. Mechanised Agriculture	Upland (flat) landscapes, appropriate for large-scale agriculture; important for aquifer recharge; private forest reserves (*Areas Permanente de Proteção* (APP) & *Reserva Legal* (RL) subject to environmental regulation).
1.2. Agriculture and Livestock	Upland (variable) landscapes; aptitude from mechanised agriculture to cultivated pasture in areas susceptible to erosion; important for aquifer recharge; extensive forest remnants that require management; potential for mineral extraction.
2. Areas for Diversified Use	
2.1. Family Agriculture	Characterised mainly by tenure and size of landholding; upland (variable) landscapes; aptitude from agriculture to cultivated pasture in areas susceptible to erosion; important for aquifer recharge; extensive forest remnants that require management; potential for mineral extraction.
2.2. Forestry and Agriculture in the Forest Landscapes	Predominantly forest cover; appropriate for exploitation of timber and non-timber products with management plans; includes area with pasture/livestock, family agriculture and agroforestry; potential for mineral extraction.
2.3. Extensive Livestock, Tourism and Fishing in Wetlands	Wetlands subject to seasonal flooding in three watersheds (Araguaia, Paraguai, Guaporé); aptitude for low-density grazing on natural grasslands; important wild fisheries, biodiversity and scenic beauty; potential for a diversity of tourist markets.
2.4. Livestock and Reforestation in Fragile Environments	Upland landscapes dedicated to cattle ranching based on cultivated pasture, soils often rocky and susceptible to erosion; scenic beauty; mineral potential, including limestone and precious stones.
3. Protected Areas	
3.1. Existing Protected Areas	Indigenous lands: Traditionally occupied by ethnic people over many years, dependent upon a forest livelihood. Quilombo lands: Home to long-established quilombo communities. Conservation Units: territorial entities established by federal, state and municipal jurisdictions recognised for biodiversity and ecological importance.
3.2. Proposed Protected Areas	The creation of new or the modification of existing protected areas, including indigenous territories and quilombo landholdings and conservation units.

Source: SEPLAG – Secretaria de Estado de Planejamento e Gestão (2018). *Dispõe sobre o Zoneamento Socioeconômico Ecológico do Estado de Mato Grosso—ZSEE/MT, e dá outras* providências: http://seplag.mt.gov.br/index.php?pg=ver&id=6304&c=117&sub=true

The environmental secretariat of Mato Gross completed a detailed ZEE in 2008, but its provisions were vehemently opposed by agribusiness. The release of the ZEE coincided with international boycotts that targeted the state for its deforestation-linked production systems (see Chapter 3). The zoning plan would have further complicated the sector's image by labelling the farms established in the previous decade as unsustainable,

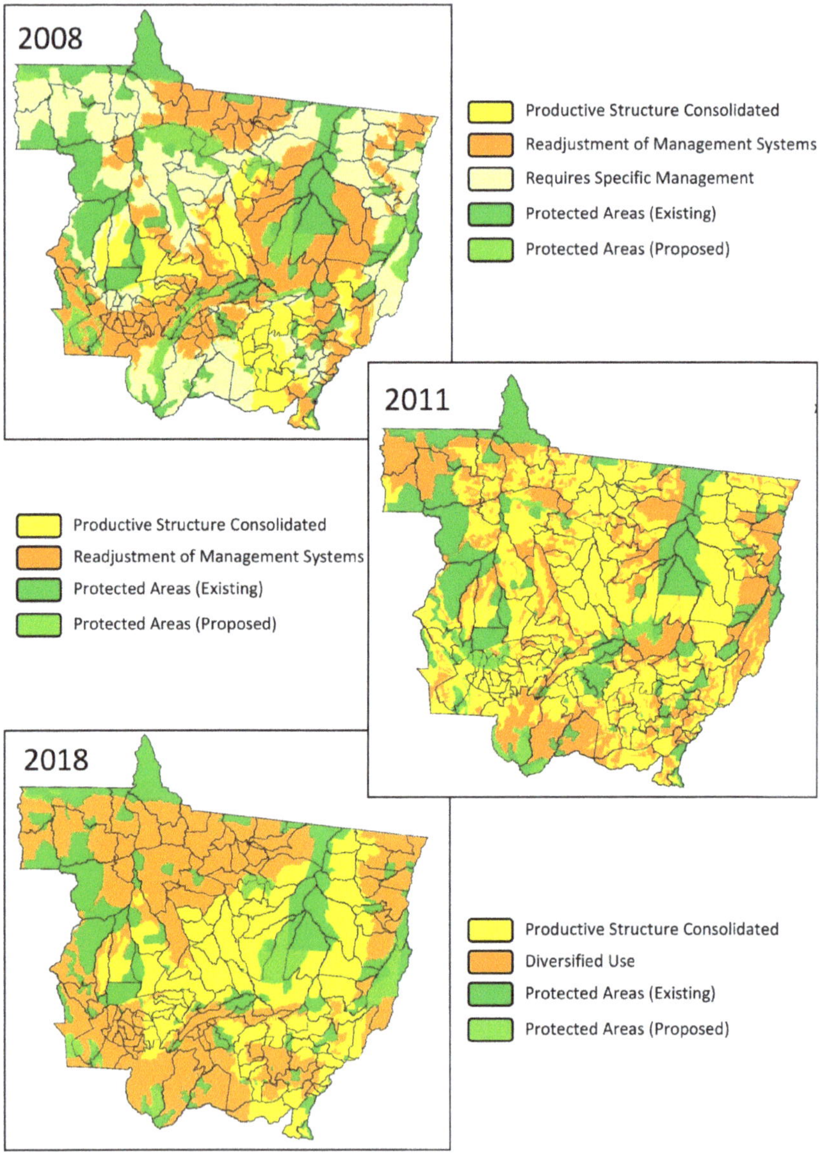

Figure 4.9: Three versions of the Zoneamento Socioeconômico e Ecológico (ZSEE) for the state of Mato Grosso: 2008, the version prepared by the state environmental agency but rejected by the state legislature; 2011, the revised version approved by the state legislature but declared invalid by the Supreme Court; 2018, another revised version prepared by state authorities but opposed by the state's agricultural sector.

Data sources: SEPLAN (2018), with additional information from Schönenberg et al. (2015) and the Fórum Mato-Grossense da Agropecuária (2021).

particularly those on the upper watershed of the Xingu River. It also would have curtailed the future expansion of the soy-maize production model onto previously deforested pastures on the northern tier of municipalities and in the valley of the Araguaia near the border with Pará (Figure 4.9).

The state legislature commissioned an alternate study and approved a radically different version in 2011. The revised study did not adhere to federal guidelines, however. It was challenged in court by the public prosecutor and was rejected by the CCZEE in 2012.[102] The state government, which is obligated by law to promulgate a ZEE, initiated another study that produced a third version in 2018 that essentially split the difference between the two previous iterations (Table 4.8).

The third version has been rejected by institutions representing farmers, ranchers, lumber companies and manufacturers.* Critics contend that the zoning provisions would threaten the livelihoods of thousands of rural families because they would: (1) label ~20% of the existing farmland as nonsustainable; (2) limit the potential of ~ 69% of existing pastureland to be converted to intensive agriculture; and (3) create environmental obstacles for ~78% of the proposed bulk transport systems.[103] The state legislature created a special commission in June of 2021 to review the proposal.[104]

The Andean Amazon

Land-use maps and their explicit recommendations are most relevant on pioneer landscapes that are in the flux of change. Recommendations can provide sound information and support an expanding agricultural production system; more often, however, they are ignored in a frenzy of land speculation. This is, unfortunately, the case in Bolivia, Peru and Colombia.

Bolivia

One of the most notable examples of zoning with positive and negative outcomes is the *Plan de Uso de Suelos* (PLUS) of Santa Cruz, Bolivia.† The PLUS identified the productive capacity of the alluvial plain located east of the Río Grande, which was legally deforested over the following decade to create a soybean production landscape known as the 'eastern expansion zone'.[105] That same document classified a similarly flat alluvial landscape located to the north and west of the Río Grande as inappropriate for intensive agriculture due to poor drainage. Nonetheless, this seasonally flooded

---

* *Fórum Agro MT, Federação das Indústrias de Mato Grosso* (FIEMT), the *Centro das Indústrias Produtoras e Exportadoras de Madeira do Estado de Mato Grosso* (CIPEM) and *Aprosoja Mato Grosso*. Source: *AproSoja/MT*: http://www.aprosoja.com.br/comunicacao/release/setor-produtivo-pede-a-al-novo-estudo-do-zoneamento--socioeconomico-e-ecologico

† The PLUS was a component in the Eastern Lowlands Project of the World Bank; P006152 @ $US 35 million + $US 20 million in counterpart funds.

wetland was settled by farmers who drained the marshes to create a second soy production district known as the 'northern expansion zone'. These two landscapes have made Bolivia into the ninth largest soybean producer in the world (HML #31).

The PLUS was part of a larger strategy to promote sustainable development in the neoliberal phase of Bolivia's recent history (1986–2005). It incorporated an element of *quid pro quo,* with multilateral agencies supporting the expansion and diversification of the rural economy by promoting intensive agriculture via deforestation on arable soils and forest conservation by creating protected areas and indigenous reserves. In between these two extremes were land-use classifications that could be managed for cattle farming (via deforestation) or timber management (via logging). Land adjacent to roads was zoned for *Uso Agrosilvopastoral* while more remote areas were zoned for *Uso Forestal Ganadero Reglamentado,* both of which are different versions that mixed agriculture, cattle ranching and forest exploitation. Landholders ignored the PLUS and developed their properties according to their capacity to mobilise financial capital. In Chiquitania (HML #29) most have adopted the Brazilian beef production model, while those in Guarayos (HML #30) are cultivating field crops.

Laws enacted in the 1990s obligated Bolivian municipalities to downscale the PLUS recommendations via a *Plan Municipal de Ordenamiento Territorial* (PMOT). Revenue sharing and decentralisation policies promoted the compilation of PMOTs; most were abandoned prior to completion, although some have led to the creation of municipal protected areas.[106] The information and recommendation from the PLUS / PMOT regulatory system were meant to be implemented on individual landholdings via a *Plan de Ordenamiento Predial* (POP). The original objective of the POP protocol was to ensure that forest corridors and river margins were protected as conservation easements. Landholders were motivated to complete the study because it is required to regularise land tenure (see above). Most landholders contracted consultants who provided desktop documents that met the administrative requirements of the forest authority; however, the implementation of conservation measures remained at the discretion of the property owner.

The lack of government commitment to forest conservation on private property was revealed in 2013 with the *Sembrando Bolivia* programme, which is central to the government's goal of expanding the agricultural footprint from three to ten million hectares. As part of that process, the agrarian reform agency (INRA) used the POP system to fast-track the regularisation of land tenure on properties deforested between 1996 and 2013 (later extended to 2017). The Bolivian forest authority* approved POPs

---

* *Autoridad de Fiscalización y Control Social de Bosques y Tierra* (ABT).

Source: Google Earth

The potential for intensive agriculture is limited on the seasonally flooded savannas south east of Trinidad, the capital of Beni Department, but upland forests are being converted to agriculture by Mennonites (a) and indigenous migrants who self-identify as Interculturales (b, c, d).

covering 850,000 hectares and issued new forest-clearing permits for 154,000 hectares.[107] Ironically, this land planning instrument, originally intended to foster forest conservation, was used to promote deforestation to expand agricultural production in the Bolivian Amazon.

Another example of the government's use of land-use zoning regulations to promote agricultural expansion is the recent modification of the PLUS for the Department of Beni.[108] The original version (PLUS Beni 2002) reflected the traditions of cattle ranching on the Llanos de Moxos and the forest livelihoods of its indigenous people. The revised version (PLUS Beni

Mennonite immigrants first came into Bolivia in the 1970s. As their population expanded, their colonies cleared approximately 900,000 hectares of forest to pursue a variety of production models including intensive agriculture, dairy and livestock. Their holdings, typically about 100 hectares, are characterised by the absence of remnant forest. These tight-knit communities pool their capital resources to purchase lands from intermediaries and are attracted to frontier landscapes where land is inexpensive.

2019) has made several substantive changes, including the recognition of the agro-industrial production zone along the highway from Santa Cruz (HML #30) and a new deforestation frontier being settled by Mennonites and *Interculturales* east of Trinidad.[109] These landscapes were zoned for forest management in the 2002 version but have been reclassified for a type of agroforestry (*agrosilvopastoril)* in the revised plan. If history is a guide, however, these settler groups will soon be cultivating row crops.

Several lowland ethic groups will be impacted by the 2019 version. Sirionó and Baure communities inhabit the forest landscapes adjacent to the new settlement zones east of Trinidad, while the reclassification of 500,000 hectares to allow '*agrosilvopastoril*' will impact dozens of Moxeño and Movima communities on the highway west of Trinidad.[110] The PLUS Beni 2019 also changes the classification of approximately two million hectares in the north, where Cerrado savannas are deemed amenable for intensive cattle management. This previously remote area is now accessible by an IIRSA sponsored trunk highway that connects its ranchlands with urban markets in La Paz (see Chapter 2). This area is surrounded by Caviñeno, Cayubaba, Chacobó and Tacana communities that have large territories (TCOs) in the adjacent forest landscapes.

Peru

Peru embraced the ZEE in 1996 as a pillar in its strategy to manage national development. This was soon followed by a pilot project funded by IDB and USAID in 2000 to develop a ZEE for Madre de Dios. Implementation guidelines published in 2004 included a mandate to develop them in coordination with regional (macro-ZEE) and local (meso-ZEE) governments. A flurry of studies was completed between 2005 and 2015, but lack of financial support has left the task unfinished. As of 2021, seventeen of 24 regional governments have developed and published a Macro ZEE, but only one has been completed since 2015. Fortunately, this includes most Amazonian jurisdictions (Amazonas, Cuzco, Huánuco, Madre de Dios, San Martin and Ucayali). Loreto has yet to complete a macro-ZEE, but has developed detailed meso-ZEEs for its most populated provinces.*

The Peruvian system, like Brazil's and Bolivia's, groups land-use into several major zones: (a) productive, (b) protected, (c) recuperation, (d) special and (e) urban/industrial.[111] It differs from the Bolivian and Brazilian systems by placing less emphasis on tenure or land-use, and more on underlying biophysical attributes. For example, long-settled agrarian landscapes in the Andean foothills (*Selva Alta*) are zoned for recuperation, reflecting their degradation by erosion caused by steep inclines and extreme precipitation.[112] Similarly, drainage is major determining factor in restricting development on riparian landscapes, regardless of whether the land has been cleared or not.[113]

Another major difference is the treatment of indigenous lands. The most common type, *comunidades nativas*, are zoned for forest management, agroforestry and subsistence agriculture (a), rather than for protection (b). This reflects their status as communal landholdings, which are open to development, rather than as territorial reserves, which are classified as a protected zone; these include *Reservas Comunales*† that were created as dual-status protected areas and *Reservas Territoriales*‡ that have been created to protect indigenous groups in voluntary isolation.

* The Amazonian jurisdictions have enjoyed the support of the *Instituto de Investigaciones de la Amazonia peruana* (IIAP), which has led the development of three macro-ZEES: Amazonas, Madre de Dios, San Martín; seven Meso-ZEEs: Aguaytia (Ucayali), Nanay, Pastaza-Motona, Alta Amazonas (Loreto), Tocache, Alto Mayo (San Martín), Tahuamanu (Madre de Dios), Satipo (Junín), Selva Huániuco (Huanuco) and the Valle del Río Apurimac (Ayacucho); and two Micro-ZEES: Shabilo (Ucayali) and Iquitos Nauta (Loreto). Source: IIAP (2021): http://terra.iiap.gob.pe/macrozee-mdd.html

† Yanesha, Chaya Nain, Tuntanain, Amarakaeri, Ashininka Machiguena, El Sira, Purús, Sierra del Divisor. Source: RAISG (2021).

‡ Madre de Dios, Kugapakori, Nahua, Nanti, Yavarí-Tapiche, Yavarí Mirim, Napo Tigre. Source: RAISG (2021).

The ZEE is a technical document that provides information and recommendations, but it is not a legally binding land-use plan. Rather, it is the first step in the labyrinthian process of developing a *Plan de Ordenamiento Territorial* (POT), which requires seven additional *Estudios Especiales*: (1) disaster and climate change risk analysis, (2) documentation of past and ongoing land-use change, (3) a description of natural ecosystems, (4) an assessment of land tenure, (5) an analysis of the regional economic dynamic, (6) an evaluation of the nature and status of ecosystem services and (7) an assessment of the institutional capacity of the pertinent jurisdiction.[114]

All of this information is synthesised in yet another study entitled *Diagnostico Integral del Territorio* (DIT) prior to the promulgation of the POT, which is a binding regulatory document that can constrain (or foment) specific types of land use. As of October 2021, no ZEE had been used to initiate a process to formulate a POT in any part of Peru.[115]

The compilation of the ZEEs has improved the potential for sustainable development of the Peruvian Amazon. The information is of very high quality and is readily available to most stakeholders via government websites. The public consultation process would appear to have been fairly comprehensive and democratically organised. Nonetheless, their impact on guiding development and conservation has been limited.

The Peruvian ZEEs were not used to design the protected area systems, which largely occurred independently and, in most cases, prior to the compilation of the regional ZEE. Neither have they been used to regulate mineral exploitation or investments in infrastructure, although they have undoubtedly had a positive influence on the preparation of the environmental impact studies (see Chapter 6). The ZEE documents show the chaotic nature of land-use on private properties while providing a snapshot of the ongoing scramble for public lands. A comparison of maps prepared in the mid-2000s with recent satellite images in Aguaytía (Ucayali) reveals that land zoned for forest management has been converted into an industrial-scale oil palm plantation surrounded by dozens of small agricultural fields.

It is, perhaps, more accurate to think of the Peruvian ZEEs as a depiction of the *status quo* combined with the aspirational recommendations of technocrats trained in the methodologies of sustainable development. The actual decisions are made by local politicians in control of the regional offices of the forest service, the land tenure agency and the environmental agency, who routinely ignore the recommendations of the ZEE as they promote conventional development initiatives in their jurisdictions (Figure 4.10).

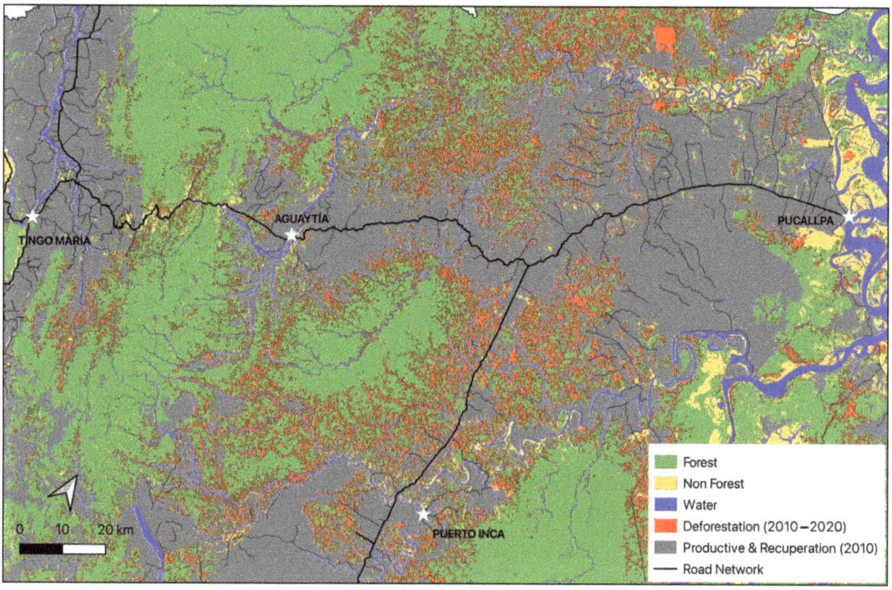

Figure 4.10: The ZEE prepared for Ucayali and Huánuco Regions in Peru sought to freeze the agricultural frontier at its maximum extent in 2010 (grey), but deforestation (red) continued to expand at the expense of land that was zoned for conservation and forest management.

Data sources: IIAP (2010) and RAISG (2021).

Ecuador

The territorial planning framework in Ecuador is known as the *Plan de Desarrollo y Ordenamiento Territorial* (PDOT). It is one component in an ambitious effort to articulate a *Plan Nacional de Desarrollo* (PND), which is being implemented via a constitutionally mandated policy to decentralise the administrative functions of the state.* The PDOT is a process that is highly participatory and multi-sectoral, as well as re-iterative, since it is designed to be updated in future decades. The goal is to create a legally-binding framework that will promote (or constrain) the economic activities supported by an autonomous regional government at three scales: Provincial, Cantonal and Parochial. The PDOT incorporates three main elements:

1. Strategic Diagnosis: an analysis of the current situation using biophysical and socio-economic information. The PDOT has all the compo-

* The decentralisation and planning process is coordinated by the *Secretaría Nacional de Planificación* (SENPLADES), a high-level executive agency chaired by the vice president of Ecuador. Source: https://www.planificacion.gob.ec/

nents of a ZEE but also includes data on demographics (migration, stakeholder groups), finance and economics (GDP/sector), land tenure, infrastructure (energy, transport, communications), threats from climate change and governing capacity.

2. Proposal: a vision of the future based on strategies implemented over the short, medium and long-term, which is built around an aspirational land-use plan (*Categorías de Ordenamiento Territorial*) and policies to achieve sustainable growth and conservation outcomes.

3. Management Model: programmes and proposals to be implemented by the autonomous government, including a participation plan and a monitoring component to evaluate progress and set the stage for the next iteration of the PDOT.[116]

The PDOT planning framework borrows concepts from various planning methodologies, such as a strategic environmental evaluation, and, as such, provides one of the most comprehensive frameworks for guiding a sustainable development in the Pan Amazon.[117] The process, formally initiated in 2016, has compiled an impressive amount of information at all three scales that is freely available via government portals. It is not without its shortcomings, however. The *Plan Nacional de Desarrollo* is organised around a jurisdictional scheme that assigns the six Amazonian provinces into five different autonomous regions. Logic and tradition argue for a legal and administrative framework that coordinates governance among Amazon jurisdictions. Fortunately, that approach continues to prevail in other programmes, such as the *PROAmazonía*[*] and *Fondo Común del CETA (Circumscription Territorial Especial de la Amazonia)*.[†]

PROAmazonía is a joint project of the ministries of agriculture and the environment and has coordinated the recent development of the PDOTs. More importantly, it provides technical assistance to promote sustainable production and biocommerce.[118] Land-use planning is important, but it must be accompanied by programmes that motivate landholders to reform

* The project is jointly administered by the ministries of the environment and agriculture with financing from the Green Climate Fund ($42 million), the Global Environment Facility ($US 12 million) and the Ecuadorian state ($US 35,000), with contributions from civil society, academia and the private sector ($US 10 million). Source: UNDP (2020) 'Sustainable Development of the Ecuadorian Amazon: integrated management of multiple use landscapes and high value conservation forests': https://erc.undp.org/evaluation/evaluations/detail/12449

† The trust fund receives contributions from mineral royalties and revenues from state-owned enterprises operating in the Amazon; between 2018 and 2020, it distributed $US 500 million dollars to the local and regional governments of the region. Source: *Secretaria del CTA* (2021): https://www.secretariadelamazonia.gob.ec/mas-de-500-millones-de-la-ley-amazonica-desde-su-promulgacion/

their business models and reward forest communities. The *Fondo Común del CTEA*, a trust fund that distributes royalties, is the principal source of capital for public investment in physical and social infrastructure.[119]

Colombia

The government of Colombia is renowned for its capacity to develop conceptually coherent programmes for the challenges that beset the country. The programme to implement a modern land-use plan was adopted as a national strategy in 1997, and by 2003 all of the country's municipalities had a POT.* The relatively rapid compilation of these studies is due to Colombia's unique system of environmental management, which depends upon an institution known as a *Corporación Autónomo Regional*. The CAR are regional regulatory agencies, broadly organised around watersheds, that advise sub-national departmental and municipal governments on natural resource management (see Chapter 7). Most were established and consolidated in the 1970s and over decades have compiled a knowledge base and institutional capacity unsurpassed in the Pan Amazon.

There are three CARs in the Colombian Amazon,† and all have conducted dozens of POTs to support the regulation of land use within strategic watersheds and wetlands that provide socially and economically important ecosystem services to urban areas. They have also completed similar documents at the department scale, referred to as *Agenda Ambiental* (Caquetá, Putumayo, Amazonas) that are similar in content and format to a ZEE (see Annex 4.14). More recently, the government of Caquetá invested in a strategic evaluation entitled *Directrices de Ordenamiento Territorial*, which is similar to the three-part scheme used in Ecuador (Diagnosis/ Proposal/ Management).[120] As commented previously, the challenge in Colombia is not access to information or capacity but the inability to establish the presence of the state on lawless landscapes.

---

* These are prepared at three scales. In addition to the POT, there are also *Planes de Ordenamiento Departamental (POD)* and *Planes Estratégicos Metropolitanos de Ordenamiento Territorial* (PMOT). Source: DNP - *Departamento Nacional de Planeación* (7 june 2016) *A partir de hoy, 100 municipios y 25 departamentos le apuestan a ser territorios modernos: DNP*. https://www.dnp.gov.co/DNP/Paginas/acerca-de-la-entidad.aspx

† CORPOAMAZONIA: *Corporación para el desarrollo sostenible del sur de la Amazonía* (Putumayo, Caquetá and Amazonas); CORPOMACARENA: *Corporación para el desarrollo sostenible de la Macarena* (Caquetá, Meta and Guaviare); CDA: *Corporación para el Desarrollo Sostenible del Norte y Oriente Amazónico* (Guainía, Vaupes and Guaviare) https://www.minambiente.gov.co/entidades-adscritas-al-ministerio/

The Countries of the Guiana Shield

While the Andean republics have invested in land-use planning with limited success, the nations of the Guiana Shield were latecomers in the effort to plan development of their forested hinterlands. Fortunately, the agricultural and infrastructure drivers of deforestation have been weak or absent historically.

In 2013, Guyana completed a National Land Use Plan (NLUP), the first comprehensive review of the nation's renewable natural resources since independence. It is essentially a national-scale ZEE and combines information from multiple sources to identify development options based on potential land use. It incorporates an explicit effort to consider future climate change and was an integral part of the agreement between Guyana and Norway to develop a Low Carbon Development Strategy (LCDS).* The NLUP includes a section dedicated to a REDD+ mitigation programme† via forest management and identifies the need to shift agricultural production and populations away from the coastal plain due to rising sea levels.

Part of the motivation for the NLUP was a strategy to link its port facilities with the agricultural landscapes in Roraima, Brazil, via a highway,‡ and a need to revitalise the economy to reduce the loss of human resources from emigration (see Chapter 5). Many of the issues that motivated the LCDS have become less germane, however, because of the discovery of offshore oil and gas reserves. The future influx of royalty revenues and investment should lessen fiscal pressures that might motivate a future government to unsustainably exploit the nation's hardwood stocks or convert marginal soils to plantation agriculture.[121]

Suriname has a land-use history similar to Guyana's, with development concentrated on the coast, but it has yet to conduct a comprehensive

---

* The LCDS was launched by a Memorandum of Understanding (MoU) that committed Norway to provide Guyana up to $US 250 million by 2015 for avoided deforestation, contingent upon certain performance indicators, one of which was the NLUP. While generally considered a successful initiative, the LCDS is essentially irrelevant since the discovery of oil offshore in 2015. See: https://www.lcds.gov.gy/ and http://www.worldoil.com/magazine/2017/june-2017/columns/offshore-in-depth

† Reduced Emissions from Deforestation and Forest Degradation (REDD+) is a UN-backed framework to mitigate climate change; the "+" refers to the emission reductions from sustainable management of forests and carbon sequestration via reforestation. REDD+ finance includes direct payments to countries or via the exchange of 'carbon credits' monetized on international carbon markets. The concept was formalised in 2009 and has functioned for more than a decade via voluntary carbon markets.

‡ IIRSA / COSIPLAN, API Project: Boa Vista - Bonfim - Lethem - Linden - Georgetown Road, Guianese Shield Hub: GUY09 (Lethem - Linden Road); GUY42 (Boa Vista - Bonfim Road and GUY43 (Linden - Georgetown ROAD). See: http://www.iirsa.org/proyectos/detalle_proyecto_api.aspx?h=15

study of its land resources.[122] This will soon change due to an ambitious new effort to reform the country's environmental legislation being led by the Ministry of Spatial Planning, Soil and Forest Management: *Project Onze Natuur op 1* (Our Nature is One), a development initiative that will consider the value of natural capital when considering development options.[123]

The most difficult challenge in both Guyana and Suriname are the small and medium-scale gold miners exploiting selected landscapes with gold-bearing rocks (see Chapter 5). Although it is unlikely that a land-use planning document will change the nature of those activities, which are regulated by different agencies, the public forum in which land issues are discussed is often dominated by debates about the mining sector.

Land-use on the Venezuelan sector of the Guayana Shield exists in two broad categories: (1) Areas Under a Special Administrative Regime (ABRAE) and (2) everything else (see Annex 4.15). The ABRAE system was established in 1984 and includes the national protected area system (national parks, monuments, biosphere reserves and wildlife reserves/ sanctuaries/refuges), forest reserves, cultural monuments, select hydrological basins, tourist attractions, frontier zones and even farm land.[124] About 55 per cent of the Venezuelan Amazon has been designated as an ABRAE, mainly as national parks (6) or natural monuments (22).* In addition, there are a massive hydrological reserve and six large forest reserves, only one of which is being exploited for timber.† Deforestation linked to agriculture is essentially nonexistent, and there are no conspicuous reports of land grabbing for agricultural development, although gold miners appropriate state lands with the consent of military authorities who have administrative authority over mining landscapes.

Presumably, some type of study preceded the construction of the Guri Hydropower facility in the 1960s; however, the first formal land-use study wasn't completed until 2004.[125] That plan was narrowly focused on biophysical features of the watershed and ignored the mining sector; it also lacked a participatory process.[126] Spurred by several power-management crises linked to water levels in the lake, the government initiated an evaluation and planning process in 2008.[127] Like most recent government initiatives, there is no evidence this project ever advanced beyond the planning stage. Any effort to improve land-use zoning in the Caroni watershed will be forced to contend with the massive gold rush that has overwhelmed the region, a development that highlights the real challenge of any land-use plan: it will not be effective if there is not the political will to enforce it.

---

* The national monuments are all *tepúis* (tepuyes), table mountains renowned for their endemism.

† Reserva Forstal: Río Caura, Sigapo, La Paragua; Reserva de Biosfera Alto Orinoco- Casiquiare; Reserva Productor Sur de Bolivar. Only the Reserva Forestal Imataca has active timber concessions.

Undesignated Public Land

One of the objectives of the ZEE process was to assist the nations of the Pan Amazon to allocate their public lands among different constituencies and stakeholder groups. The group with the highest public profile, at least in recent years, is the indigenous people who have organised a highly successful campaign to assert their territorial rights and formalise their claims to their ancestral territories. They are joined in their quest for land rights by tens of thousands of local communities, known variously as *Caboclos, Ribereñas, Quilombolas, Maroons, Seringueros* and *Castañeros*, that also rely on the forest and aquatic habitats for their livelihoods. They are competing for land with other societal groups that have economic, demographic and political power, including the ranching sector, large and small farmers and the timber industry. The competition for land is influenced by the interests of mining companies and the oil and gas industry, who have separate rights to below-ground resources, but are concerned that access to those natural resources can be constrained by whomever controls the surface rights.

The multi-decade campaign to prepare ZEEs and formally designate the precise physical borders of public land has succeeded in limiting the expansion of agriculture, particularly in Brazil and Ecuador, and to a lesser extent in Bolivia and Peru. The sharp forest boundaries between indigenous territories and adjacent agrarian landscapes (with several notable exceptions) demonstrates that settlers and land grabbers will not occupy territory they cannot eventually claim as private property.* The ongoing scramble for land is largely occurring on landscapes that have been tacitly identified as expansion zones and highway corridors, many of which were assigned a category of land-use in a ZEE that is purposely vague (see Table 4.8 and Figure 4.8)

Environmental advocates are factually correct when they (a) state that deforestation on these landscapes is illegal and (b) accuse the individuals involved as misappropriating state lands.[128] Regardless, elected officials and government functionaries, either by action or inaction, facilitate settlement on these landscapes which have been zoned for development. It is widely assumed that these lands will be occupied by somebody using some kind of legal or extralegal mechanism. Public forest must be formally designated and managed — or it will pass to the private sector, which increases the probability they will be cleared or degraded.

How much undesignated land is left in the Pan Amazon? Government agencies maintain a running tally by compiling the various land tenure categories managed by their agencies.[129] Those estimates are imprecise,

---

* This behaviour does not hold true for timber thieves or wildcat gold miners who have no long-term interest in acquiring title since their activities are inherently short-term.

however, because of the deficiencies in land tenure registries and the ongoing appropriation of public land. Moreover, what is perceived as public land varies, particularly when it comes to indigenous or communal territory, which may or may not be considered as a protected area. Likewise, the degree of protection differs depending upon the type of economic activity, which may or may not be allowed, and some protected areas coexist with private inholdings.

The amount of land that remains to be formally designated provides an approximate estimate of the land available for conservation and development (Figure 4.11). The mathematical exercise used to make that estimate also provides a snapshot of the existing distribution of land among major stakeholder groups (Figure 4.12).

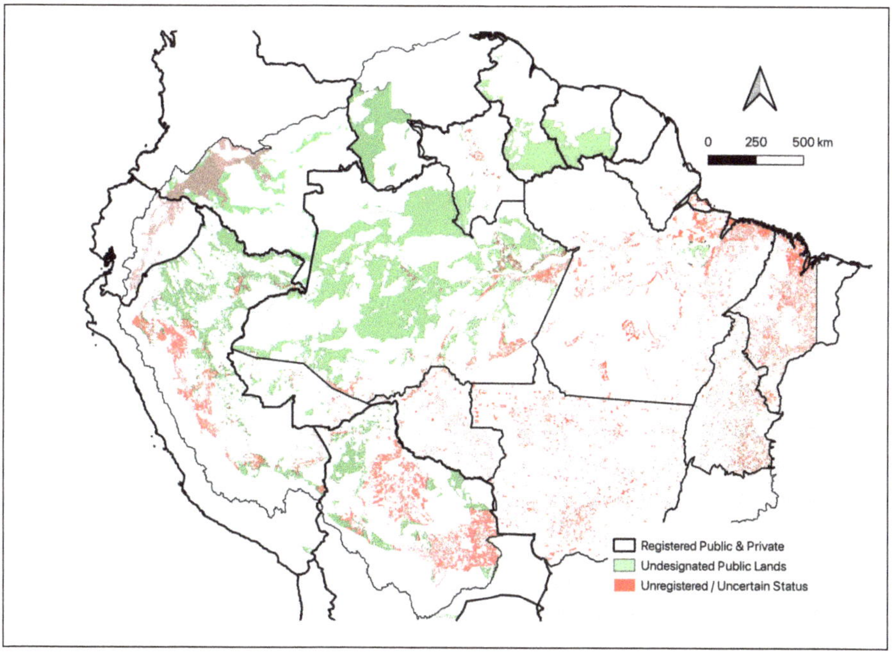

Figure 4.11: The approximate distribution of undesignated public lands in the Pan Amazon. The green areas are ~92 million hectares of public lands not yet legally encumbered or physically settled by pioneers. Pink areas are ~47 million hectares of putatively state land located on landscapes with large backlogs of unregularised land holdings. The white space represents land deeded as private property or formally designated as a protected area, indigenous land or sustainable use management unit. See Annex 4.1 to 4.16 for jurisdiction-specific graphics, maps and sources.

Figure 4.12: The relative proportion of nine major land tenure categories in the Pan Amazon, stratified by jurisdiction based on a harmonised comparison of classification systems unique to each country: Brazil: Acre (AC), Amapá (AP), Amazonas (AM), Maranhão (MA), Mato Grosso (MT), Pará (PA), Rondônia (RO), Roraima (RR), Tocantins (TO); Bolivia (BO); Colombia (CO); Ecuador (EC); French Guiana (FG); Guyana (GY); Peru (PE); Suriname (RN); and Venezuela (VE). See text for explanation of classes. See Annex 4.1 to 4.16 for jurisdiction-specific graphics, maps and sources.

Private denotes large and small landholdings registered in a national cadaster (Bolivia, Brazil, Peru, Ecuador) or landscapes zoned for agricultural activities (Colombia, Guyana, Suriname, French Guiana). It includes large-scale forest estates (Amazonas and Acre) included in the national cadaster of private properties (SNCR) but excludes forest holdings in the *Cadastro Ambiental Rural* (CAR). This category is stratified into landholdings that have been titled (regularised) and claimed (title pending review).

Communal denotes private lands held by a communal title and public lands where tenure or the permanent right-of-use has been deeded to non-ethnic communities in Brazil, Bolivia and Peru. This includes: (a) INCRA-sponsored settlements that benefit forest dwelling communities (PAAD-type) and (b) pioneer farmers (PA-type) in Brazil; (c) *Castañera* forest communities in Northern Bolivia and *Campesino/ Interculturales* pioneer farmers in Santa Cruz, Beni and La Paz; and (d) *Ribereña* communities in Peru.*

Indigenous denotes communal landholdings deeded to specific communities, as well as state lands where permanent use-rights have been granted by law or decree to one or more ethnic peoples. These can be small or large, but are specific for indigenous groups with a specific ethnic heritage. It includes protected areas that enjoy a dual status as indigenous reserves and those created to protect indigenous groups living in voluntary isolation.

Protection denotes national and regional protected areas established with the primary goal of conserving biodiversity and natural ecosystems, typically referred to as 'indirect use' (IUCN Categories I, II and II). Those with dual status as indigenous territories are excluded to avoid double accounting.

Sustainable-Use denotes public lands allocated to the sustainable management of timber and non-timber resources and include both national and regional protected areas, as well as forest concessions in Peru, Bolivia, Venezuela, Guyana and Suriname. Excluded from this category are multiple-use protected areas with privately held inholdings (e.g, *Areas Natural de Manejo Integrado* and *Áreas de Proteção Ambienteis*) and communal landholdings dedicated to sustainable management.

Other denotes urban areas and public lands that have been registered within cadasters; they include military properties, infrastructure and their associated rights-of-way, rivers, lakes and water ways and urban areas.

Encumbered denotes putatively state lands in areas with unresolved land tenure; many (perhaps most) are occupied by a possessor (*poseiduero*) who has yet to have his or her title validated and registered in the national ca-

* Although the agrarian settlements are legally communal, they are functionally private smallholdings and may (probably) eventually be split up and distributed amongst their inhabitants.

daster. These were identified by subtracting the total area from registered properties in cadasters from the total area within polygons defined as 'Human Modified Landscapes' (see Chapter 1).

Undesignated denotes all other public lands, estimated by subtracting the above categories from the total area of each jurisdiction as defined by administrative criteria (Bolivia, Brazil, Colombia, Venezuela) or by the approximate tree-line on the eastern slope of the Andes (Bolivia, Colombia, Ecuador, Peru).

This large-scale accounting shows there are still significant areas of public land that await to be allocated as protected area, indigenous reserve or open to some type of sustainable development. It also highlights the dimension of the challenges in resolving land tenure on frontier landscapes. Insecure and uncertain land tenure are directly linked to the deforestation crisis. The ability of land grabbers and settlers to appropriate state lands is made possible by the incomplete nature of land registries. The forest frontier will only be closed when all legal properties enjoy fully certified titles and all public lands are clearly demarcated and assigned a management category.

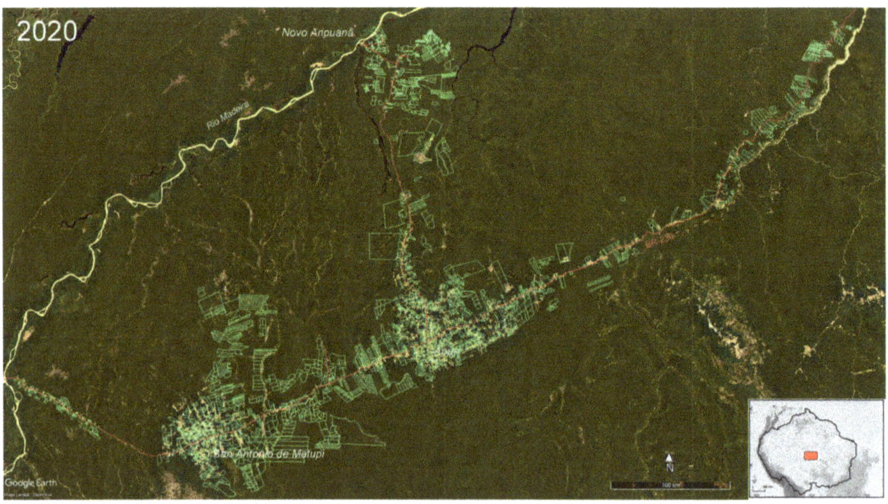

Source: Google Earth

The land claims (green polygons) registered in the Cadastro Ambiental Rural (CAR) along BR-230 in Southeast Amazonas state reflect the ongoing speculation for land along Brazil's most active settlement and deforestation frontier.

Land Sparing Versus Land Sharing

The creation of protected areas and indigenous reserves offers the best hope for conserving the biodiversity of the Amazon; however, the management of the human modified landscapes will determine whether society protects the ecosystem services essential for the economic health of the continent. Models predict that an ecological tipping point will be crossed when about 25 per cent of the region's forests have been converted to agriculture – just a few percentage points above the current level of eighteen per cent.[130] When (if) that tipping point is crossed, the decline in atmospheric water recycling will lead to a catastrophic decline in rainfall across the farmlands of South America, including those in the Southern Amazon, but also in Central Brazil, Paraguay, Bolivia and Northern Argentina (see Chapter 1 and Chapter 10).

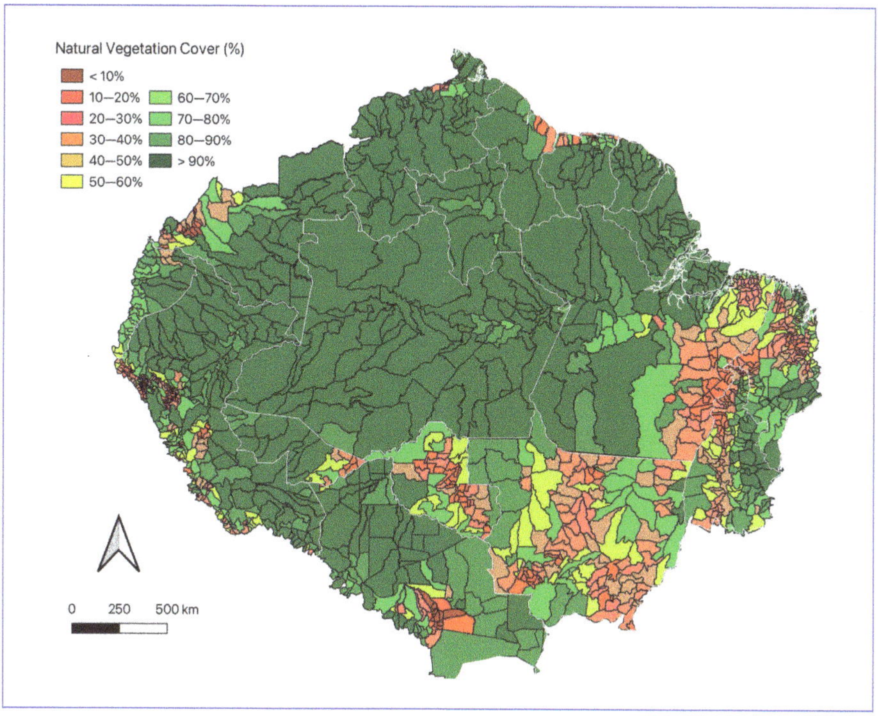

Figure 4.13: The loss of native vegetation, mainly forest but also Cerrado scrubland, has degraded atmospheric recycling of the Southern Amazon. Models forecast that precipitation sufficient to support a rainforest ecosystem declines dramatically once forest cover falls below 60%, a situation that has already impacted more than 600 municipalities in the Pan Amazon (yellow and red polygons).

Data sources: MapBiomas (2021) and RAISG (2021).

The predicted tipping point at ~25% deforestation is a basin-wide metric; however, large parts of the Southern Amazon passed that metric approximately twenty years ago. Dozens of municipalities in Pará, Mato Grosso and Rondônia have lost more than forty per cent of their original forest cover (Figure 4.13). Those landscapes are now both hotter and drier.[131] It could be worse. Producers still benefit from water recycled in the Central Amazon and, as more upwind landscapes are deforested, these will cease to provide this precipitation subsidy. When that happens, the farmers and ranchers of Mato Grosso will be forced to adapt to a new reality.

Some producers will migrate into landscapes less susceptible to precipitation declines, a process already underway as farmers expand northward, attracted by cheap land and lower logistical costs (see Chapter 3). Most will use new drought-resistant cultivars and adopt management practices that conserve soil moisture. Some will seek to use irrigation technology (see below). There will also be pressure – and incentives – to change how they use the land.

Some academics advocate for a 'land-sparing' approach that relies on technology to intensify production on existing production landscapes to reduce the demand for new cropland. Others contend that a 'land-sharing' approach that diversifies production systems is needed to conserve ecosystem services. Both tactics have a place in a coherent development strategy, but their social, economic and environmental impacts vary depending on the perspective of the observer and the scale of the evaluation.

The growing recognition in financial markets that climate change is an existential threat to global society has created a demand for investments that comply with criteria defined as Environmental, Social and Governance (ESG). Among the most common are 'green bonds' that purport to fund business ventures that reduce greenhouse gas emissions, sequester carbon and conserve forest and biodiversity. Simultaneously, public and private commitments to eradicate deforestation from commodity supply chains have focused attention on the agricultural economy of the Southern Amazon. If financial analysts and media pundits are to be believed, the Southern Amazon will soon receive billions of dollars of private and public capital that, hopefully, will transform the business models that have long threatened the Amazon. As usual, the devil will be in the details.

Sustainable intensification: The soy-beef nexus in the Brazilian Amazon

All multi-stakeholder initiatives organised to eliminate deforestation from commodity supply chains include programmes to increase producer productivity.* They are presented as a 'carrot' to farmers and ranchers being

* Multistakeholder initiatives include the Roundtable for Sustainable Palm Oil (RSPO), Round Table for Reasonable Soy (RTRS), Global Roundtable for Sus-

coerced to limit (end) the expansion of their industry via deforestation. The logic is simple: a ten per cent increase in yield can offset a ten per cent reduction in the (future) area under cultivation. This is certainly true at the global scale but less so at local and regional scales. The numbers speak for themselves.

The total soy harvest in Mato Grosso increased from 18 million tonnes in 2008 to more than 35 million tonnes by 2020.[132] Twelve per cent of this increase came from improved agronomic practices (intensification); the rest was due to an expansion of land under cultivation (extensification).* Agribusiness advocates argue that the expansion of cropland (in this instance) was also a form of sustainable intensification because it occurred via the conversion of degraded pastures rather than by expansion into forest. Some assert that law enforcement and market incentives have succeeded in eliminating deforestation from the soybean supply chain.† This happy story, however, has a more nuanced explanation.

The degraded pastures were supplied by ranchers who had accrued a large surplus of under-utilised pasture due to massive deforestation of previous decades. Due to overgrazing, a very large portion (~ 60%) had been degraded.[133] Soil restoration is a significant investment‡ but is much less expensive than clearing forest.§ Soy growers chose growth via pasture conversion because it was the most cost-effective option. Ranchers benefited because they were able to monetise an underperforming asset, either via a sale or by renting their land to a farmer for a determined period of time (~5 years). Those that opt for a lease recover an appreciated land asset with restored soils and renovated pastures.

Approximately five million hectares of pasture were converted to cropland in Mato Grosso between 2008 and 2020; nonetheless, the total area of cultivated pasture remained constant at ~21 million hectares. The

---

tainable Beef (GRSB), Bonsucro; see Chapter 3.

* About ten million hectares of soy were cultivated in Mato Grosso in 2020 with about equal parts from the conversion of pasture and Cerrado savanna. The conversion of Cerrado vegetation has an impact similar to deforestation, and environmental advocates eventually succeeded in incorporating its conservation into sustainability protocols. The moratorium on expanding via Cerrado conversion has increased the demand for pasture and reinforced the economic and political alliances among ranchers and soybean farmers.

† Government sponsored initiatives include the *Plano de Ação para Prevenção e Controle do Desmatamento* (PPCDAm) launched in 2004 by the administration of Luiz Inácio Lula da Silva and in the *Produzir, Conservar e Incluir* (PCI) in Mato Grosso initiated in 2015 as part of Brazil's commitment to the Paris Agreement.

‡ Forest, Cerrado and pasture soils all require agricultural lime (calcium carbonate) to resolve acidity and aluminum toxicity; applications of macro (NPK) and micro-nutrients optimise plant health and productivity.

§ Clearing forest costs ~ $US 1,000 more per hectare compared to the conversion of pasture.

conversion of pasture was offset by new deforestation on the forest frontier and within forest remnants on consolidated landscapes. Simultaneously, the cattle herd expanded from 26 to 32 million head, which translates into an improvement of the mean stocking rate from 1.3 head per hectare to 1.5 head per hectare.[134] Grazing management is only one aspect of beef productivity and the industry also invested in genetics, animal health and nutrition, which has further increased the productivity of its supply chain (see Chapter 3)

Both the beef and soy industries have expanded their production by intensification: soy farmers have increased yields and expanded onto pasture while ranchers have increased stocking rates and improved animal health. Claims that they have avoided deforestation are inaccurate, however, because intensive cropping displaced cattle ranching in an industry that continues to expand via deforestation. In the vernacular of natural resource economics, this is called indirect land-use change, while carbon accountants refer to it as leakage.[135] Environmental advocates label it as greenwash.[136]

Eventually, all the pastureland suitable for annual crops, estimated at about ten million hectares, will be occupied by farmers. Mato Grosso's ranchers will need to double stocking-rates to maintain current levels of beef production if they hope to avoid future deforestation. They will probably attain that level of productivity;* however, other factors will influence whether they expand their spatial footprint. As mentioned previously, the appreciation of land is an integral part of a rancher's business model. Intensification tends to improve profit margins, which provides producers with more capital and, like businessmen everywhere, most will use that capital to expand operations.[137] It may be true that the supply and demand for commodities is a zero-sum equation at the global scale, but it is certainly not true at the local or regional scale

Meat-packing companies and commodity traders intend to use ESG finance to eliminate deforestation from their supply chain. Perhaps. They will use satellite imagery to monitor land use and ear tags embedded with block-chain-coded chips to document the origin of a cow.[138] It is not clear, however, how technology can resolve the issue of indirect land-use change or detect cattlemen who trade calves via informal markets. Investors should pay close attention to the Key Performance Indicators (KPI) used to evaluate whether their creditors meet ESG criteria – or not.

Irrigation: A problematic intensification strategy

The agro-industrial farms of Mato Grosso are among the most efficient on the planet; they benefit from abundant rainfall and a long rainy season,

* This is feasible because ranchers in São Paulo and Paraná average about 3.5 head per hectare: Source: Arantes et al. (2018).

which allows them to plant and harvest two crops per year. They do not operate without risk, however. Both crop cycles fail during periodic droughts (1985, 1991, 1993, 2009, 2016),* and the second harvest is often constrained by limited soil moisture at the end of the rainy season. The recent decline in mean annual precipitation, the consequence of climate change and deforestation, manifests largely in a delay in the onset of the rainy season.[139] Farmers must wait until it rains to seed their crops, a delay that reverberates through the subsequent first harvest (typically soy) and the sowing of the second crop (typically maize) and its eventual harvest, which is known as the *safrinha*.† The gradual (non-tipping point) decline in precipitation has increased the probability of poor maize yield.[140]

Some producers are using centre-pivot irrigation systems to manage the risk of dryland farming; the number has grown from fifty in 2000 to more than a thousand in 2021.‡ The original motivation may be to alleviate drought stress during the *safrinha* (May, June, July), but precision water management can increase yields during both crop cycles by ensuring optimum soil moisture during key stages of plant development (seedling, flowering and grain filling). Once the investment is made, producers use the equipment throughout the year. Some are planting a third crop.[141]

Irrigation circles were observed on approximately 1.5 per cent (150,000 hectares)[142] of the area under intensive cultivation in 2020 (ten million hectares).[143] Most producers are pumping water directly from rivers or from small impoundments on upstream watercourses. Expansion has been most notable on the headwaters of the Teles Pires (Tapajós) and Río das Mortes (Araguaia), followed by the Juruena (Tapajós) and Alto Xingu.

Irrigation systems are regulated by the state environmental authority *Secretaria de Estado de Meio Ambiente do Mato Grosso* (SEMA) in coordination with a state water council (*Conselho Estadual de Recursos Hídricos de Mato Grosso* – CEHIDRO) and basin-specific governance committees. Current guidelines stipulate that surface water removals should be limited to seventy per cent of minimum waterflows (Q_{95})§ and no individual stakeholder can take more twenty per cent of that total.[144] Current levels of take-off are well within those guidelines, but the ongoing rate of expansion (ten per cent annually) will eventually outstrip surface water supplies. Long before

* Basin-scale droughts are triggered when cyclical dipole systems in the Atlantic and Pacific Oceans coincide to suppress atmospheric waterflows from the Atlantic Ocean into the Amazon (see Ch. 10).

† The Portuguese term *safrinha* translates as 'small harvest' because it is small in comparison to the first crop cycle.

‡ The largest number are located in the municipalities of Primavera do Leste (193), Sorriso (176) and Novo Ubiratã (84).

§ The minimum waterflow is defined as Q_{95}, a water flow metric (m^3/s) that was exceeded 95% of the time within the specific flow record; it is roughly equivalent to a dry season flow in a permanent watercourse.

Source: Google Earth

Satellite images reveal that farmers are extracting surface water directly from watercourses and small impoundments in the headwaters of the Tapajos and Xingu watersheds. Some are using solar energy to power their systems, which might make them eligible for green bonds and other forms of ESG investment.

that happens, producers will start exploiting the groundwater resources of the Parecis Aquifer, a massive reservoir in the sandstone rock formations that overlay the Amazon Craton (Figure 4.14).*

Information on both surface and groundwater resources was used by the national water agency (*Agência Nacional de Águas e Saneamento Básico* – ANA) when preparing the *Atlas Irrigação*, a national planning document that has mapped the nation's irrigation potential.[145] According to that document, Mato Grosso has the water resources necessary to irrigate 3.9 million hectares of cropland, an area equivalent to ~40% of the total area under cultivation in 2020.[146] Of that total, about 500,000 hectares would depend upon groundwater resources, while the remainder would be extracted from the region's rivers (Figure 4.15).

---

* The *Sistema Aquífero do Parecis* (SAP) has a horizontal surface area of between 200,000 to 300,000 km^2. It is composed of two subsystems: a confined aquifer (static level: 15 m; mean flow: 23 m^3/hour; mean thickness: 80 m) and non-confined aquifer (static level: 17 m; mean flow: 50 m^3/hour; mean thickness: 93). Source: Silva (2013).

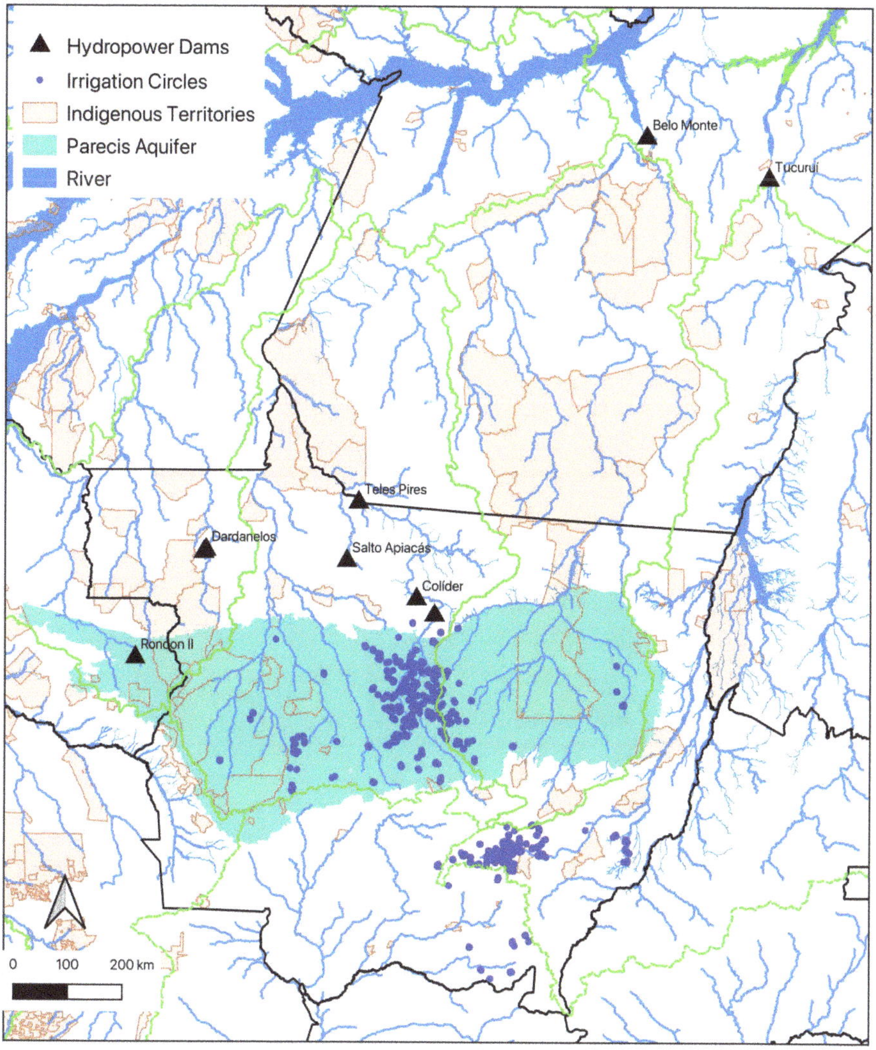

Figure 4.14: The use of centre-pivot irrigation systems has increased from less than twenty in 2010 to more than 1,000 in 2020. The systems use surface water extracted from the headwaters of the Tapajos, Xingu and Araguaia basins. The expansion of these systems risks the hydraulic regimes that power four hydropower facilities with a combined capacity of 2.0 GW of electricity generation and key ecosystem services that sustain the indigenous fisheries of the middle Tapajos River.

Data sources: ANA (2021), RAISG (2021) and Pereira et al. (2014).

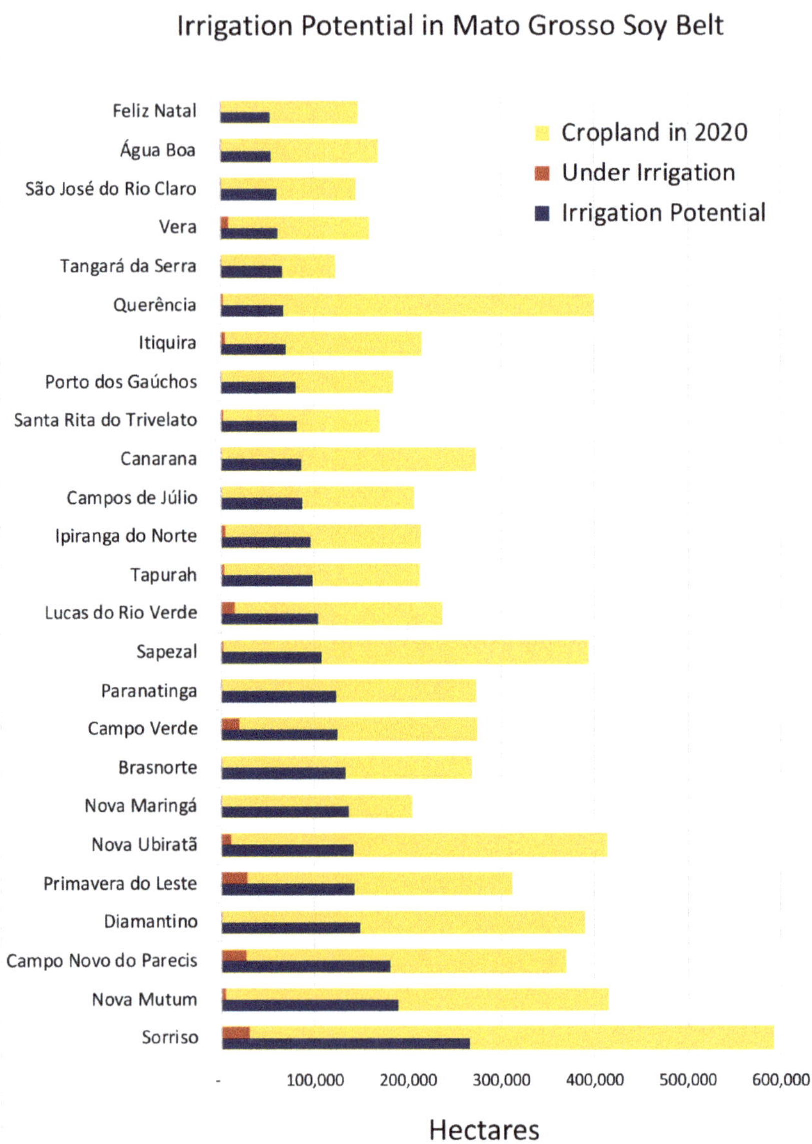

Figure 4.15: Although irrigation has been growing at 10% annually since 2010, it still represents only a small fraction of both the projected potential and the current crop area. According to projections by the national water agency (ANA), the combined surface and groundwater resources could support 3.9 million hectares of irrigation agriculture — 40% of the current area under intensive cultivation in 2020.

Data sources: ANA (2021) and MapBiomas (2021).

Ironically, irrigation at that scale would partially compensate for the decline in rainfall caused by deforestation because it would enhance evapotranspiration and convection over agrarian landscapes. Simultaneously, however, industrial-scale irrigation would disrupt the seasonal waterflows that support the biodiversity and ecosystem function on the upper Tapajós, Xingu and Araguaia watersheds.[147] Climate change and deforestation have already caused a significant decline in precipitation over the Southern Amazon;[148] nonetheless, waterflows in those rivers have maintained historical levels because of an increase in runoff from deforested landscapes.[149] As irrigation systems multiply, these waterflow subsidies will disappear and, if the Southern Amazon passes a deforestation-induced climatological tipping point, the impacts on downstream ecosystems and communities will be catastrophic.

If (when) that occurs, agribusiness will scramble to install irrigation systems. If history is any guide, they will use all the available surface water and overexploit the Parecis Aquifer – unless regulatory authorities act pre-emptively to limit irrigation. That may be difficult, however, because the current 35-year span of the current permit system is based on historical climate data that overestimates future water resource availability. The current regulatory framework has been questioned by hydrologists, who contend surface water abstractions should be limited to twenty per cent of the total water volume during any specific temporal period.[150] Removals from the aquifer are more complicated to measure and regulate because they are based on balancing the recharge rate with the rate of extraction.* The recharge rate in the future is unknown.

As surface water abstractions in the Tapajós basin increase, they will threaten the economic viability of four hydropower facilities.† Simultaneously, the riparian habitats and Munduruku communities on the middle Tapajós will suffer from reduced water flows, particularly if dam operators mitigate reduced waterflows by retaining a greater share of water. On the Xingu, the river corridor is protected by the *Parque Indígena do Xingu* (PIX), home to sixteen different ethnic tribes (see Chapter 11), but the headwaters are located entirely on private land. The much-debated provisions of the Mato Grosso ZEE (see above) could potentially limit the expansion of centre-pivot systems, because it expressly identifies landscapes that are important for the management of the water resources of the Amazon.‡

---

* Water reserves have been estimated at between 2.89×10^{12} to 1.13×10^{13} m^3; hydrologists estimate that between 8 and 12% of the total aquifer could be sustainably exploited (extraction = recharge). Source: Silva (2013).

† Sinop, Colider, Teles Pires, São Manoel; see Ch. 2.

‡ Category 1.2 (*Agricultura e Pecuaria*) 'areas of hydric importance as headwaters and recharge zones for the Amazon, Tocantins-Araguaia and Paraguai hydrographic regions.' Source: *Dispõe sobre o Zoneamento Socioeconômico Ecológico do Estado de Mato Grosso* – ZSEE/MT, http://seplag.mt.gov.br/index.php?pg=ver&id=6304&c=117&sub=true

As of October 2021, there was no evidence that green bonds were financing irrigation agriculture in the Brazilian Amazon. This situation will change in the near future because irrigation projects are eligible under ESG standards[151] and figure prominently in discussions about climate change adaptation.[152] Solar energy powers irrigation systems at multiple locations in Mato Grosso, and the ongoing expansion of the maize-based biofuel industry is being financed by green bonds.* Assertions that these operations are compliant with ESG criteria must, eventually, be reconciled with their long-term impact on water flows on aquatic ecosystems of the Tapajós, Xingu and Araguaia rivers and the livelihoods of indigenous communities on those rivers.

Other crops for which irrigation technology is employed at scale are rice in Tocantins (100,000 hectares), coffee in Rondônia (43,000 hectares) and oil palm in Pará (25,000 hectares). Irrigation technology has been installed in an unknown area, perhaps as large as 10,000 hectares, as part of a new business model to cultivate açaí in plantations (see below).[153] Centre-pivot irrigation systems have been introduced on the alluvial plain of Santa Cruz but they have not been widely embraced, despite the considerable drought risk that characterises that region.[154]

Conservation agriculture and agroforestry

There are three fundamental rules of financial planning: (1) save continuously, (2) invest in a diversified portfolio of assets and (3) exercise patience via a long-term strategy. This common-sense advice is at the heart of Conservation Agriculture (CA), a land-management philosophy that seeks to reconcile the technologies of modern agriculture with the time-worn practices of organic farming. These include multi-crop systems that minimise risk from weather, pests and markets, and the spatial and temporal rotation of crops. When integrated, these practices will increase soil organic matter (carbon), which improves the water holding capacity and the nutrient status of soils. Agroforestry systems are particularly advantageous because deep-rooted perennials contribute to evapotranspiration, which supports regional rainfall, while individual farmers benefit by reducing energy and labour costs, as well as locking in a long-term revenue stream.

Agribusiness is not unsympathetic to common-sense advice, and most farmers have diversified their choice of crops and adopted minimum

* *FS Bioenergia*, the largest maize biofuel processor in Brazil, issued $US 550 million in green bonds in 2020 to finance the construction of ethanol refineries in municipalities that are also leading in the expansion of centre-pivot systems: Primavera do Leste, Sorriso, Novo Mutum, Lucas do Rio Verde and Campos Novo de Parecis. Source: https://www.fs.agr.br/

© Tarcisio Schnaider/Shutterstock.com

© Vinicius Abe/Shutterstock.com

Land-use intensification (land sparing) is a strategy embraced by corporate farms because it usually maximises profitability on a per hectare basis. Top: Tree plantations provide an attractive return when calculated over several decades, while supporting landscape-scale evapotranspiration essential for maintaining regional rainfall regimes. Bottom: Minimum tillage techniques reduce soil erosion, increase soil organic matter and improve water-use efficiency.

tillage technologies.* Nonetheless, they almost invariably choose industrial commodities (soy, maize, sorghum, sunflower, cotton) and genetically engineered varieties designed for use with herbicides. Plantations are almost always composed of exotic species (eucalyptus, pine or gmelina). A few corporate entities have allocated a portion of their land to an integrated production model known as ILFP (*Integração Lavoura-Pecuária-Floresta),* a type of industrial agroforestry that seeks to optimise the benefits from three major production systems (row crops, livestock, tree farms). Nonetheless, the overwhelming majority of large-scale farmers are enamoured with (addicted to) the financial returns from monoculture, and they are not likely to change their business models.

Ranchers might be more prone to changing land-management practices because they have a surplus of under-utilised land that has suffered from poor management, as evidenced by their use of joint ventures with farmers as a strategy to restore degraded pasture. Nonetheless, cattlemen (and women) belong to a conservative cultural tradition that is notoriously resistant to change. They will adopt new technologies but only after there is a clear demonstration of economic benefit – preferably one they can observe on a neighbour's landholding. A solid majority are in violation of the Forest Code,† and many have made a legal commitment to come into compliance via a mechanism known as a TAC (see above).[155] Most have not followed through on these commitments because of weak enforcement mechanisms, but that may change if future ESG finance obligations force reforms onto beef supply chains.

Conservation agriculture, agroforestry and reforestation are key components of *Agricultura de Baixo Carbono* (Table 4.9), an innovative finance programme managed by Banco Nacional de Desenvolvimento Econômico e Social (BNDES). The bank has a long and unfortunate history of funding infrastructure projects in the Amazon (see Chapter 2), but it has the financial power to influence development, at least with the corporate sector. In 2021, BNDES announced it would float green bonds on international capital markets in collaboration with the Interamerican Development Bank

* A central tenet of conservation agriculture that avoids the mechanical manipulation (ploughing) of top soil. Variants include no-till (*plantio direto*), minimum till (the use of discs but not ploughs) and strip till. The goal is to maximise straw on the soil surface to avoid erosion and slow the decomposition of roots to foster the formation of humus.

† According to Soares-Filho et al. (2014), the forest liabilities in the Amazon Biome are estimated at 899,000 hectares of APPs (Permanent Protection Areas) and 7.2 million hectares of RL (legal reserve), but 56% of the RL liabilities do not need to be restored and can be compensated through mechanisms such as the Environmental Reserve Quota (Law No. 12,651/2012) or via commercial plantations.

Table 4.9: The loan programmes managed via the Agricultura de Baixo Carbono programme of the Banco Nacional de Desenvolvimento Econômico e Social.

Loan Type	Business Model / Investment Projects
ABC Recuperação	Restoration of degraded pastures
ABC Orgânico	Organic agricultural production practices
ABC Plantio Direto	Direct planting systems (no-till or minimum-till systems)
ABC Integração	Crop-livestock, crop-forest, livestock-forest or crop-livestock-forest integration systems and agroforestry systems
ABC Florestas	Implantation and management of commercial forests, including those destined for industrial use or the production of charcoal
ABC Ambiental	Regularisation of rural properties in accordance with Forest Code, particularly the restoration of *RL* and *APP*; restoration of degraded areas; and sustainable forest management
ABC Tratamento de Dejetos	Manure and waste treatment systems from animal production for energy generation and composting to reduce methane emissions
ABC Dendê	Zero-deforestation oil palm plantations, particularly those established on degraded pastures
ABC Fixação	The cultivation of species that support the biological fixation of nitrogen
ABC Manejo dos Solos	Soil conservation, including the correction of soil acidity and fertility using agricultural lime ($CaCO_3$)
ABC Bioinsumos.	Biological control pest management and the transformation of biofertilizer

Source: BNDES: https://www.bndes.gov.br/wps/portal/site/home/financiamento/produto/programa-abc

(IDB); some of those resources would be used to fund projects dedicated to ABC-like investments.[156]

An opportunity for smallholders

Smallholders should be more willing to diversify their production systems and adopt practices that increase resilience. Mitigating risk is integral to their livelihoods because crop failure can lead to hunger and bankruptcy. Smallholders exist in all parts of the Pan Amazon, including within jurisdictions dominated by large landholders (Figure 4.16). As such, improving the sustainability of smallholders would yield multiple benefits, ranging from the stabilisation of the regional climates to ameliorating the inequality that defines the rural economy.

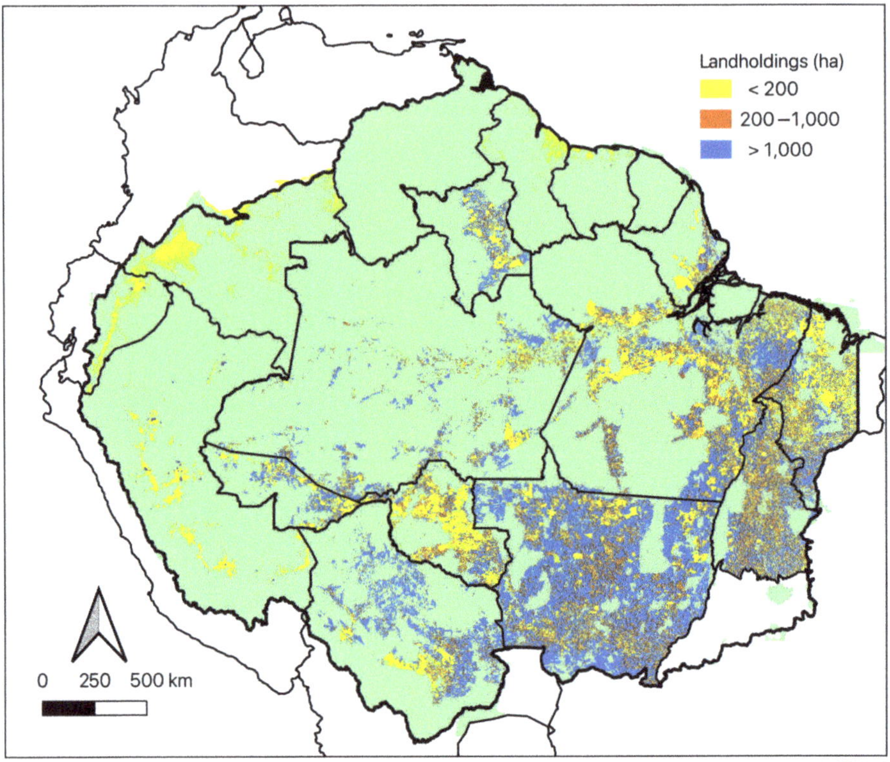

Figure 4.16: Policies that help smallholders invest in tree-based production systems could restore evapotranspiration on strategically located landscapes in the Pan Amazon. See Annex 4.1 to 4.16 for jurisdiction-specific graphics, maps and sources.

The potential is greatest in Peru and Ecuador, where small farmers occupy more than ninety per cent of previously deforested landscapes. Most grow basic foodstuffs for household consumption and for sale to domestic consumers, as well as large number who cultivate coffee and cacao for international markets, including a subastantial minority that receive a premium for adopting organic practices. Oil palm is expanding because it provides a steady stream of income on a monthly basis, while the concept of zero-deforestation palm oil is gaining currency within producer associations (see Chapter 3). Livestock operations are primitive, but producers are adept at adopting new technology, as evidenced by the ongoing expansion of aquaculture (see Chapter 8). Recruiting the small farmers of the Andean piedmont to pursue climate-friendly production has a good probability of success, because it aligns with their own experiences, traditions and aspirations.

Agroforestry is a land sharing production model where tree crops are grown in combination or rotation with annual crops; when combined with forest conservation, these systems tend to be more resilient to drought, pests and market volatility while providing greater ecosystem services when compared to land intensification schemes.

Bolivia is similarly well-positioned to implement policies that benefit small family farms, which occupy ~35% of the agricultural landscapes created by deforestation. Like their peers in Peru and Ecuador, Bolivian smallholders are accomplished farmers engaged in commercial agriculture and open to innovation. Many have been enticed into the soy monoculture model, but they will respond to other options if they are economically competitive (see Chapter 3). The situation is more complicated in Colombia, where rural peasants have been coopted by drug cartels, land grabbers and cattle ranchers. Most would welcome a less onerous livelihood, but that will require peace and the establishment of the rule of law.[157]

It will be challenging to engage the smallholders of Brazil because their land has been captured by the Brazilian beef industry. These producers are more accurately described as small ranchers rather than small farmers. Their avocation for livestock also explains the relatively low proportion of secondary forest on their properties (Figure 4.17). Tropical farmers have

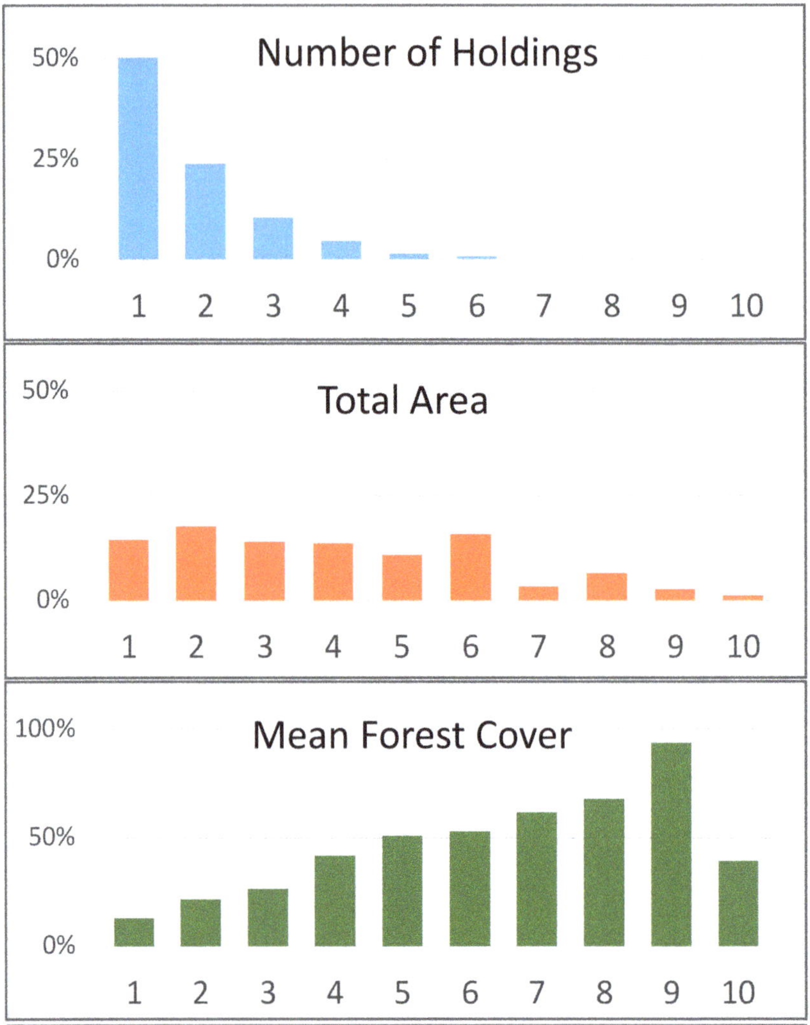

Size classes (ha)

1	< 50	6	1,000—5,000
2	50—100	7	5,000—10,000
3	100—200	8	10,000—50,000
4	200—500	9	50,000—100,000
5	500—1,000	10	> 100,000

Figure 4.17: Rondônia has the largest population of smallholders in the Brazilian Amazon. As a group, they also have the lowest forest cover of any landholder. See Annex 4.10. Data sources: IMAFLORA (2019) and MapBiomas (2021).

Source: Google Earth

The proportion of remnant forest on smallholdings (blue polygons) in Brazil varies greatly. Top: In Teixeirópolis (RO), degraded pastures cover ~90% of the area and watercourses lack forest corridors. Middle: In Altamira (PA), remnant forest is more abundant and pastures are less overgrazed, but there is little evidence of agroforestry systems. Bottom: In São Luis (RR), native forest predominates and forest fallow is prominent (a), but ongoing deforestation continues to reduce forest cover.

forest fallows but ranchers just convert everything to pasture. They also tend to consume remnant forest over time. Municipalities dominated by small ranchers in Acre, Rondônia and Pará are all characterised by a superabundance of pastures and an almost total absence of other production systems.

Fortunately, there are exceptions that show a different pathway for smallholders in Brazil. Municipalities near urban areas are a major source of basic foodstuffs and tropical fruits.* Region-wide, this production represents about eighty per cent of non-beef revenues, far greater than the value reported for cash crops that are grown in regions with specific programmes to support producers: coffee (Rondônia), cacao (along BR-230 in Pará), oil palm (northeast Pará) and black pepper (more widely in Pará). As in the Andean Amazon, aquaculture could revitalise the smallholder sector; however, it requires a significant capital investment and know-how that is different than traditional livestock systems. Small ranchers cannot shift to aquaculture without extension assistance and access to credit.

The potential to revitalise smallholder production could benefit from the expanding market for *açaí*, the most valuable food commodity in Amazonian Brazil (after soy and maize). Most of the current harvest originates from intensely managed natural populations located within communal territories.† Global demand will soon outstrip the capacity of natural populations and, eventually their consumers will pressure for changes in supply chains currently reliant on child labour and the over-exploitation of natural populations.[158] When this happens, the *açaí* industry will shift to cultivated plantations.‡ Fortunately, EMBRAPA has developed a technological package for cultivating *açaí* on upland landscapes using irrigation technology, and middle-class farmers near Belem have been cultivating the palm for more than a decade. The transition to cultivation, which is inevitable, could revitalise smallholder landscapes across the Central Amazon. The production model could also be exported to the high rainfall areas in the Andean Amazon.

---

* Cassava, rice, beans, bananas, pineapple, citrus watermelons, passion fruit, papaya; excludes soy, maize, sorghum, cotton and oil palm, as well as industrial production of rice and sugar cane. Source: SIDRA – Sistema IBGE de Recuperação Automática (2021) Produção Agrícola Municipal: https://sidra.ibge.gov.br/pesquisa/pam/tabelas.

† *Projeto de Assentamento Agroextrativista* (PAE), Projeto de Desenvolvimento Sustentável (PDS); *Reserva Extrativa* (RESEX), *Reserva de Desenvolvimento Sustentável* (RDS).

‡ In 2019, gross revenues were reported as $US 900 million but are projected to reach $US 2.5 billion by 2025. North America is the largest market (greater apparently than Brazil), while the fastest growing market is Asia Pacific where consumers from China are renowned for their avocation for health food. Source: Market data forecast (April 2021), https://www.marketdataforecast.com/market-reports/acai-berry-market

The success of *açaí* has highlighted the potential of other palm fruits with unique nutritional value or an innate capacity to produce large volumes of vegetable oil. Most have nascent markets based on the exploitation of natural populations but are also candidates for domestication and incorporation into agroforestry systems (see Chapter 8). In the western Amazon, this includes *Oenocarpus bataua*[*] and *Mauritia flexuosa,*[†] which, like *açaí,* are adapted to high rainfall areas and marsh habitats. In the Southern Amazon, *Acrocomia aculeata,*[‡] a savanna species adapted to upland soils could easily be integrated into beef production operations for large and small ranchers (see Text Box 4.2).

Text Box 4.2: The Macaúba Palm

Acrocomia aculeta is a single-stem palm native to the dry forest and Cerrado savannas of the Southern Amazon; it is also a common constituent of secondary vegetation and cultivated pastures. The fruit and seeds are rich in fats and have been consumed by indigenous people for millennia. The species has potential as an industrial crop because its fruits and seeds can be processed into edible vegetable oils or used as a feedstock for the manufacture of consumer goods and biofuels. Pilot projects have been underway in Brazil for more than two decades, including one rated 'best in the world' by the Forest Investment Program (FIP) of The World Bank. That project successfully integrated macaúba's cultivation into cattle ranches as a 'silvopastoral' system (a type of agroforestry). EMBRAPA estimates that Brazil has more than 32 million hectares of degraded pastures that are appropriate for the cultivation of macaúba. If it were cultivated only in degraded pastures, with iron-clad proof that its cultivation did not stimulate indirect land-use change, it would qualify as a sustainable aviation fuel (SAF). In 2030, the projected demand for SAF has been valued at $US 30 billion, but the total addressable market for aviation fuel in 2020 was $300 billion. Presumably, the ranchers of the southern Amazon could be convinced to embrace a zero-deforestation production strategy for a global market that is complimentary to the production of beef.

* Common names: *Ungurahui, Bataua; Majo* (Bolivia); *Palma seje* (Venezuela); *Patauá* (Portugués). Source: d'Eeckenbrugge and Ferla (2000).

† Common names: *aguaje* (Perú), *buritĭ* (Brasil), o *morete* (Ecuador), *palma de moriche* (Venezuela, Bolivia, Colombia), *palma real* (Bolivia). Source: d'Eeckenbrugge and Ferla (2000).

‡ Common names: *macauba* (Brazil, Argentina Paraguay), *corozo* (Venezuela, Colomvia), *totaí* (Bolivia), macaw palm (English). Source: d'Eeckenbrugge and Ferla (2000).

© Mirian Goulart Nogueira / Shutterstock.com

© Raoni Silva, INOCAS CC BY 4.0

© Raoni Silva, INOCAS CC BY 4.0

The Macauba palm. (See Text Box 4.2)

Reforestation and Restoration

One of the benefits of agroforestry and plantation forestry is the ability of tree crops to capture and store carbon in their above-ground biomass. Although they must be renovated (cut down and replanted) at approximately twenty-year intervals, agroforest and plantation landscapes can sequester ~20% of the carbon stored in a natural forest.* Even more carbon can be captured via the restoration of natural habitat; there are ~10 million hectares of land that should never have been cleared because it was legally protected by the Forest Code.[159] Presumably, this land will eventually be reforested, and hundreds of initiatives are underway to facilitate that outcome.† It is not, however, an inexpensive proposition.

Brazilian foresters have developed models that reflect the cost and benefits of different reforestation strategies. Passive restoration approaches, which rely on natural ecological succession, are less expensive and function well for lands that retain a certain level of vegetative cover (shade and soil organic matter). Active approaches, which employ soil amendments, nurseries, weed control and periodic culling, are more expensive; however, they allow the landowner to manipulate tree populations to favour native hardwoods (silviculture) and obtain a comfortable financial return – if the landholder can afford to wait three decades (Figure 4.18). Investments in reforestation and restoration must be protected from fire and grazing, particularly during the early years of their establishment. This is an additional cost but also a long-term commitment that may not accompany all passive approaches focusing on natural systems.

According to hydrological models, tropical landscapes need to retain about sixty per cent of their canopy cover to maintain the atmospheric recycling that supports historical levels of precipitation. This would require the reforestation of approximately fifteen million hectares in the most heavily denuded municipalities of the Southern Amazon. A reforestation programme of this magnitude would require at least a decade to implement, if not longer, and demand a total budget of between $US 20 billion (passive) to $US 100 billion (active).

---

* The proportion depends on time scales and natural disturbance. Primary forest in the Amazon varies between 100 and 300 tons C per hectare, while plantation forests harvested at 20-year intervals have a mean carbon density of about 50 tons C per hectare.

† *Aliança pela restauração na Amazônia* has identified 2,773 forest restoration initiatives in the Brazilian Amazon, totaling 113,500 hectares. The largest number were agroforestry systems (59%), but ecological restoration covered the largest spatial areas. Civil society organisations are leading the way, but companies account for 52% of the total area under restoration. See: https://aliancaamazonia.org.br/

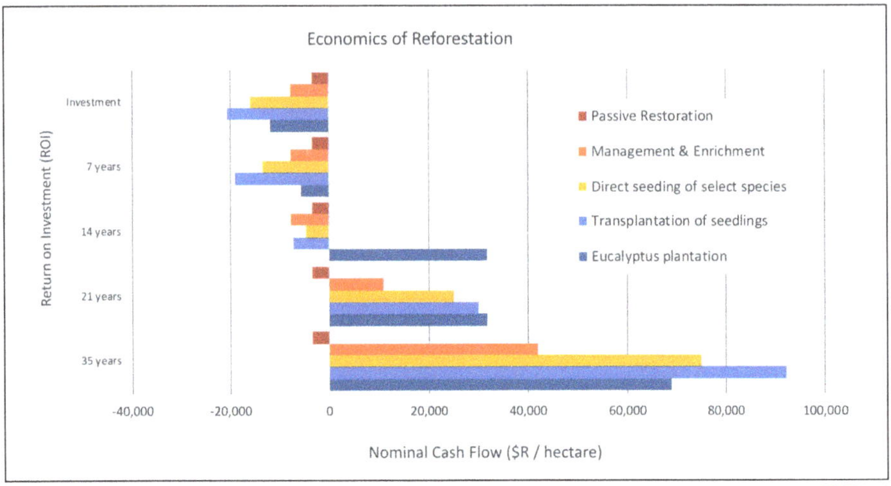

Figure 4.18: The financial return from reforestation and plantation forestry depends on the capital outlay associated with different management options and the value of native hardwood timber. Passive restoration is the least expensive, but silvicultural interventions dramatically improve the return on investment if the landholder can wait decades to monetise the investment. Eucalyptus plantations are culled at seven years and are harvested and replanted every fourteen years.

Data source: Instituto Escolha (2021)

Large-scale ranchers and farmers can access the ESG finance necessary to realise an investment of this magnitude, but they will only do it if they are forced to comply with the Forest Code. Most will use passive methods wherever possible, particularly on land that is off-limits to any future economic exploitation (*Áreas de Proteção Permanente*). They will use active protocols on landscapes that allow them to recover their costs (*Reserva Legal*) and many will opt for cultivating commercial (exotic) species in monocultural plantations, which is allowed in some instances (see Chapter 7). This type of forest investment will be eligible for loans that have been underwritten by green bonds or, perhaps, via direct equity investments for investors with an appetite for an illiquid long-term asset.

Smallholders are not likely to attract investment in reforestation from private capital markets. However, local and regional jurisdictions could access climate finance via carbon markets and use those funds to subsidise programmes that target this constituency. Environmental advocates tend to favour reforestation schemes and offer landholders a modest stipend as a form of payment for ecosystem services.* Regardless of the fate of climate finance linked to carbon markets, however, agroforestry systems will be more popular among small ranchers and farmers in need of a reliable source of income.

Agroforestry systems can cost between $US 1,500 to $US 2,000 per hectare but would require a parallel investment in logistical systems to collect and process a diffuse supply chain scattered across thousands of smallholdings, particularly for a production system based on vegetable oil subject to spoilage (see Text Box 4.2). Investments to create a tree-based production system on approximately half of the denuded landscapes of central Rondônia would require at least $US 10 billion, which translates into ~$US 100,000 for each small farm in Rondônia.

Rebalancing the Tipping Point in the Southern Amazon

The threat of climate change has highlighted the importance of the Amazon Rainforest in the global carbon cycle, while underscoring the fragility of its atmospheric water recycling system. Deforestation risks pushing that system past a 'tipping point', which would trigger a collapse in precipitation across the South American continent. Ominously, some climate models project the Southern Amazon could pass this critical threshold even if the region's inhabitants agree to end all future deforestation.

If a loss of forest cover tips the atmospheric scale towards drought, then an increase in tree cover should rebalance the ecological fulcrum. Support for reforestation is universal, but it is actually more difficult than stopping deforestation. Clearing a forest generates revenues over the short term, while restoring a forest is extraordinarily expensive and inherently slow. Moreover, the word 'reforestation' means different things to different people: an ecologist uses the term to describe the restoration of a [quasi] natural ecosystem, while some foresters use it describe commercial tree plantations. Both concepts are valid and both must be harnessed to rebalance the tipping point in the Southern Amazon.

Amazonian societies have never asked for a handout, but their representatives have stated, repeatedly, the need for economic incentives to reward forest conservation and reforestation. Environmental economists have long predicted that carbon markets would provide those incentives, but they have failed, repeatedly, to materialise. Carbon offsets are being

* The agreement announced at COP26 should expand demand for corporations seeking nature-based carbon offsets; unfortunately, it did not resolve the arcane accounting issues that characterise REDD+ projects. Fortunately, it is supportive of jurisdictional approaches that are being developed and tested in the Pan Amazon. Source: C. Streck, (2021) What Does the Article 6 Rulebook mean for REDD+? https://www.ecosystemmarketplace.com/articles/what-does-the-article-6-rulebook-mean-for-redd/

promoted again following the agreements at COP26 in 2021, and they may succeed finally in ending the modern era of deforestation. Unfortunately, subsidies based on offsets are unlikely to change the economic logic that constrains tree planting at the scale and speed necessary to rebalance the tipping point.

Amazonian producers, large and small, grow commodities for global and national markets. They are not likely to abandon their conventional production systems for reforestation projects that are overly reliant on regulatory subsidies, particularly if they take two or more decades to provide substantive revenues. The farmers and ranchers of the Southern Amazon might, however, use climate finance to invest in tree-based systems to produce green commodities that provide solid returns over the medium-term. Commodity markets have driven the deforestation of the Amazon. Markets for green commodities can drive its reforestation.

What defines a green commodity? A green commodity must be absolutely and verifiably carbon negative. It cannot rely on a carbon offset to reach neutrality. Its production must actually sequester carbon. There can be no leakage or indirect land-use change. If it originates in the Amazon, its production must respond to the inequality that defines land ownership and provide governance mechanisms to prevent cheating and ensure transparency within its supply chain.

The key to rebalancing the tipping point in the Southern Amazon is to discover business models that provide landholders with an economic return that is demonstrably superior to conventional production systems. The goal, simple in concept but difficult to implement, is to make planting trees more profitable than clearing forest. Tree-based production systems established on the previously deforested landscapes in the Southern Amazon can meet that criterion.

Annex 4.1: Acre

In Acre, relatively large forest holdings (registered in the SNCR) and claims (registered in the CAR) jointly represent the largest category of land tenure. Most inhabitants in INCRA PA-type (agrarian) settlements have registered their landholdings in the CAR database as individual landholdings.

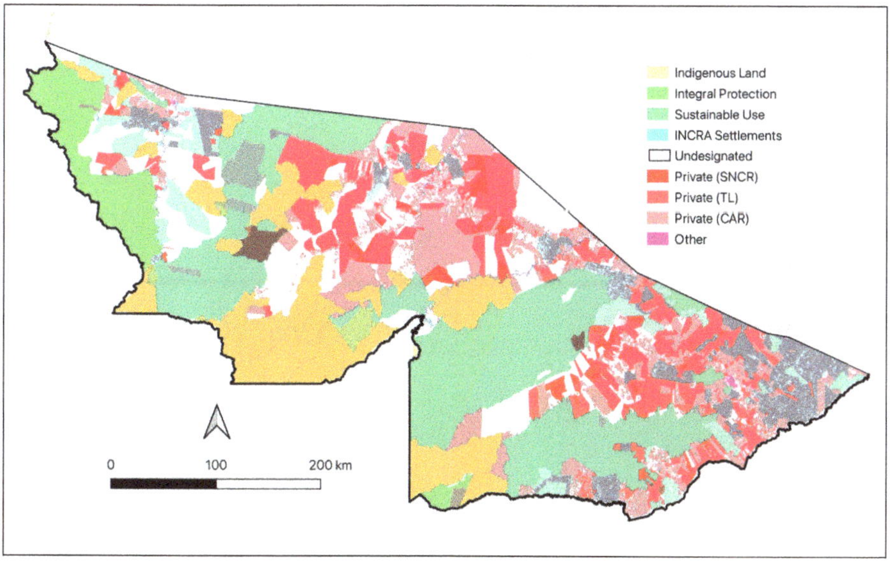

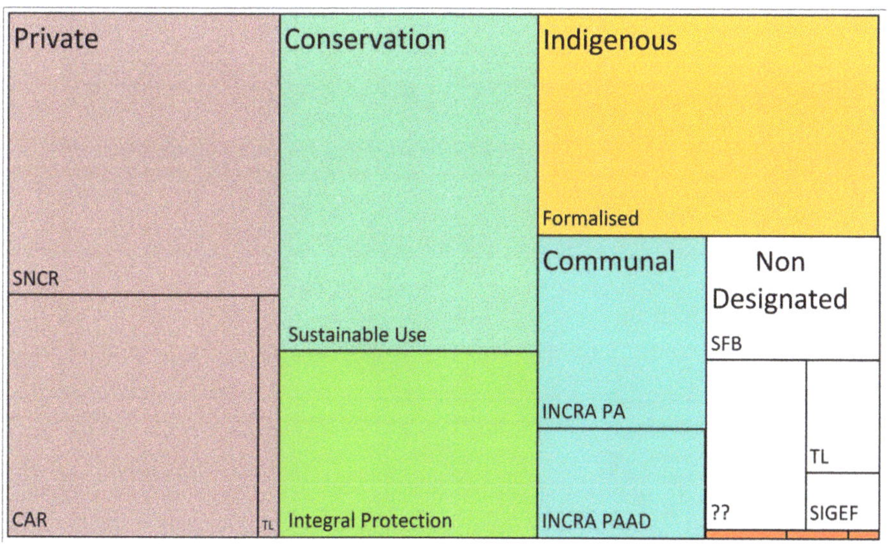

Data sources: INCRA (2020) and IMAFLORA (2019).

Annex 4.2: Amapá

Amapá is the only state in the Brazilian Amazon where Conservation Units incorporate more forest than Indigenous Land. The Terra Legal (TL) programme registered 1.6 million hectares of landholdings, more than five times that previously registered in the SNCR; approximately 43% are located within the Floresta Estadual do Amapá, a sustainable use protected area.

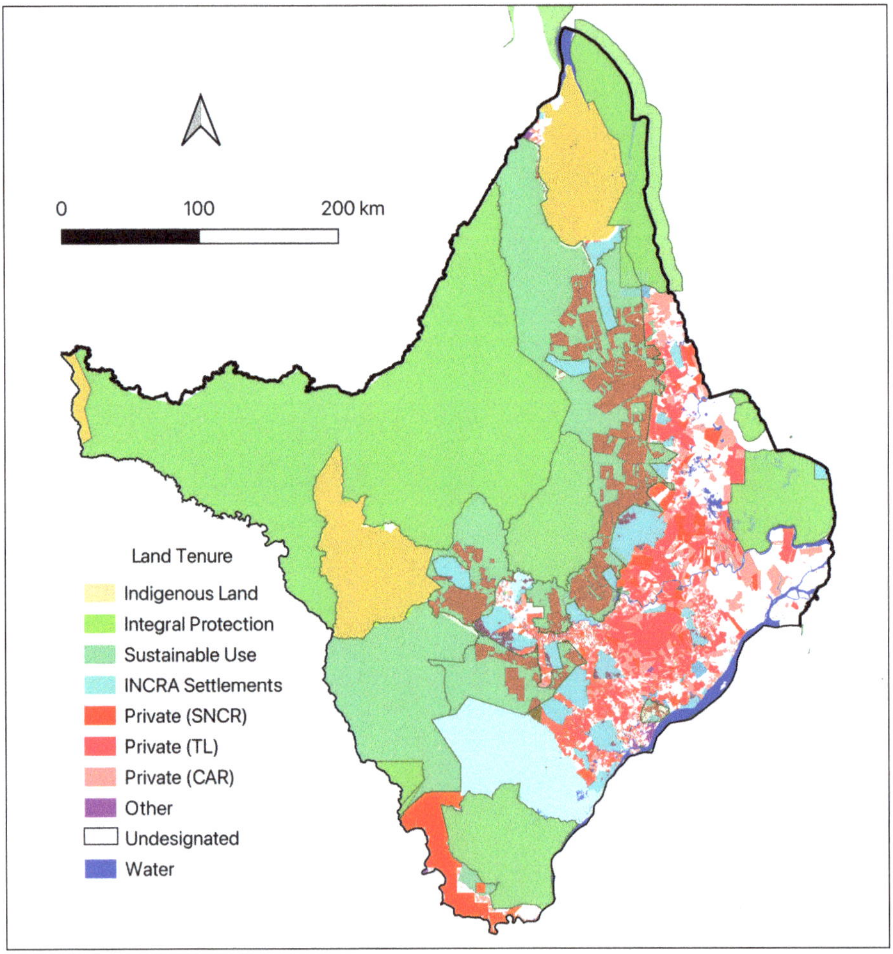

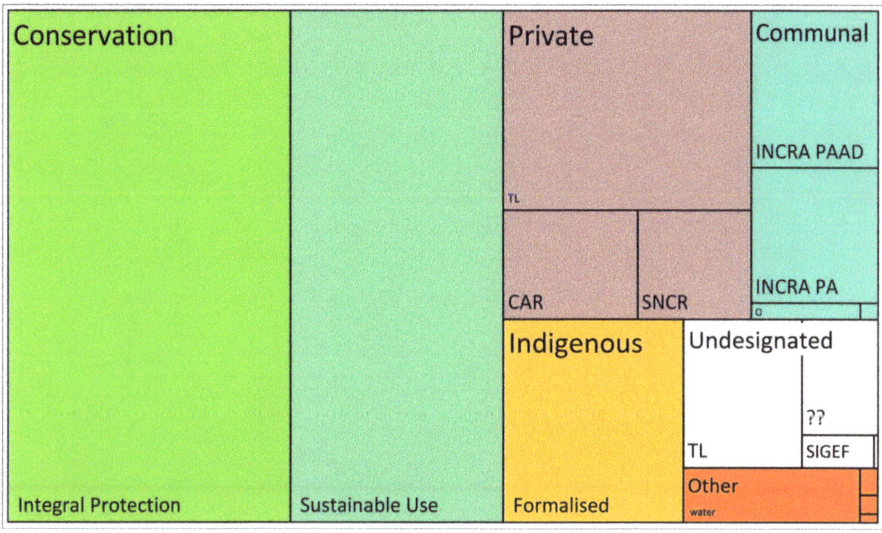

Amapá Land Tenure
Data sources: INCRA (2020) and IMAFLORA (2019).

Annex 4.3: Amazonas

In Amazonas state, large forest estates represent the greatest area of private holdings and claims in both the certified (SNCR) and environmental (CAR) cadasters. Most INCRA projects are PAAD-type settlements similar to ICMBio conservation units (RESEX, RDS, FLONA), managed for the sustainable use of forest and aquatic resources. Approximately 31% (37.2 million hectares) has yet to be allocated into a specific land use category.

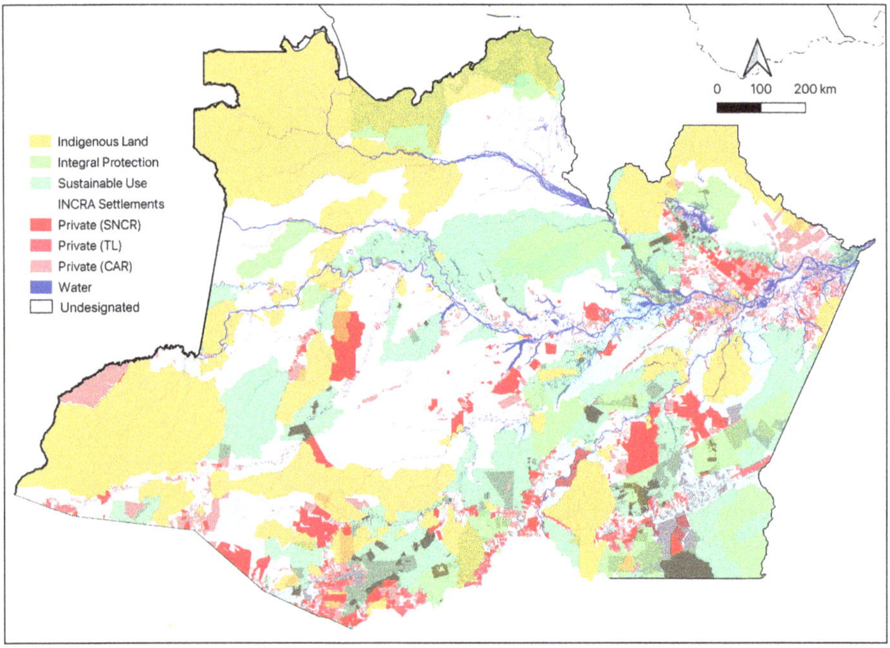

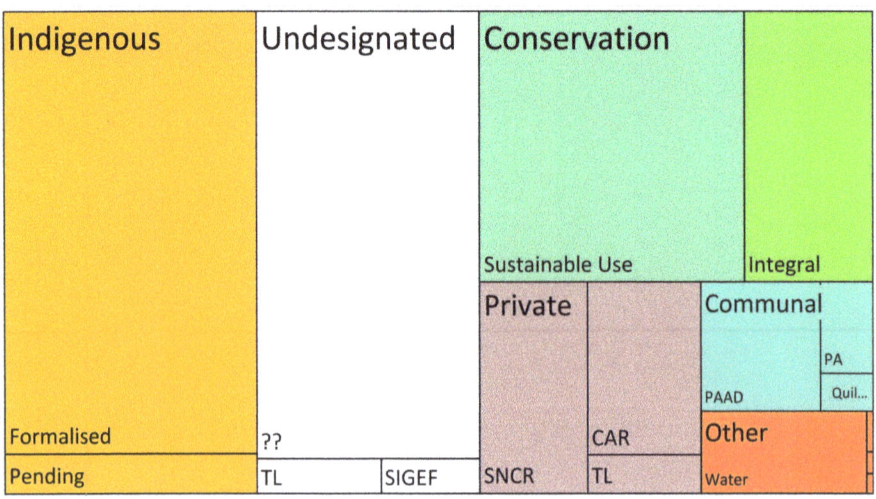

Data sources: INCRA (2020) and IMAFLORA (2019).

Annex 4.4: Maranhão

The limited number of conservation units and indigenous territories in Maranhão are surrounded by agrarian landscapes. The greatest concentration of certified landholdings are located in the intensive agricultural landscapes of the Cerrado Biome in the southern sector of the state.

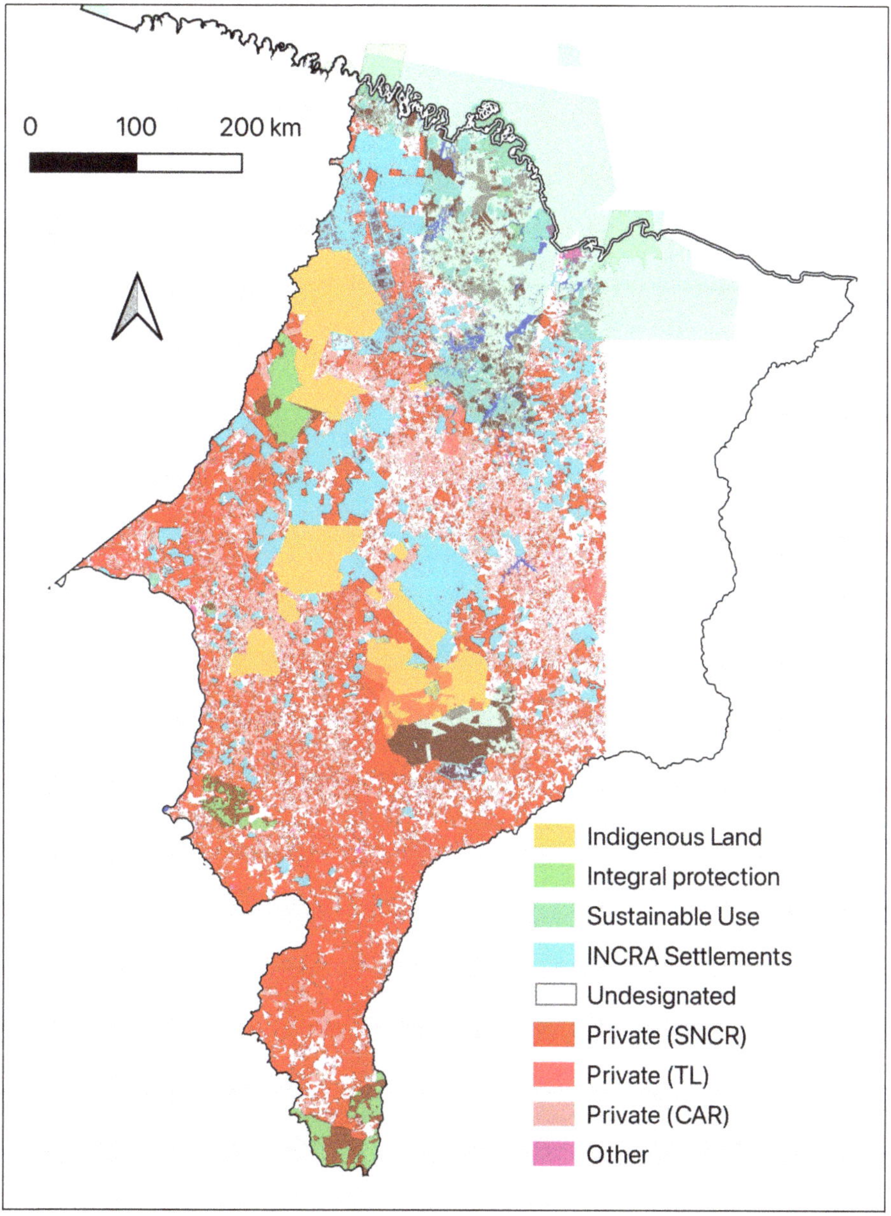

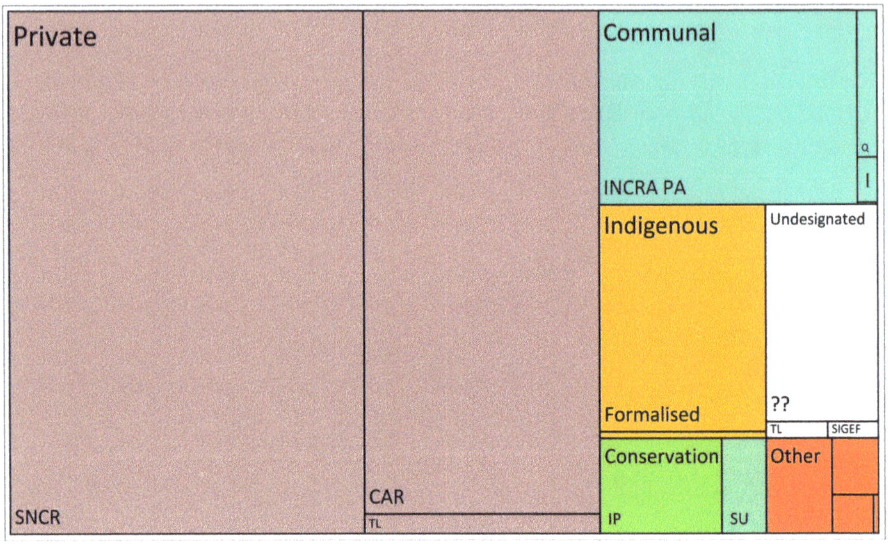

Maranhão Land Tenure.
Data sources: INCRA (2020) and IMAFLORA (2019).

Annex 4.5: Mato Grosso

The tenure regularisation process is well advanced in Mato Grosso. Landholdings certified by the SNCR cover a greater area than those registered only in the CAR. The state is notable for its scarcity of conservation units and the importance of indigenous territories for the long-term conservation of natural forest; approximately 53% of remnant natural forest is located within privately held landholdings. Most of the undesignated land is probably claimed or occupied by individuals who have not registered their holdings in either the CAR or the SNCR.

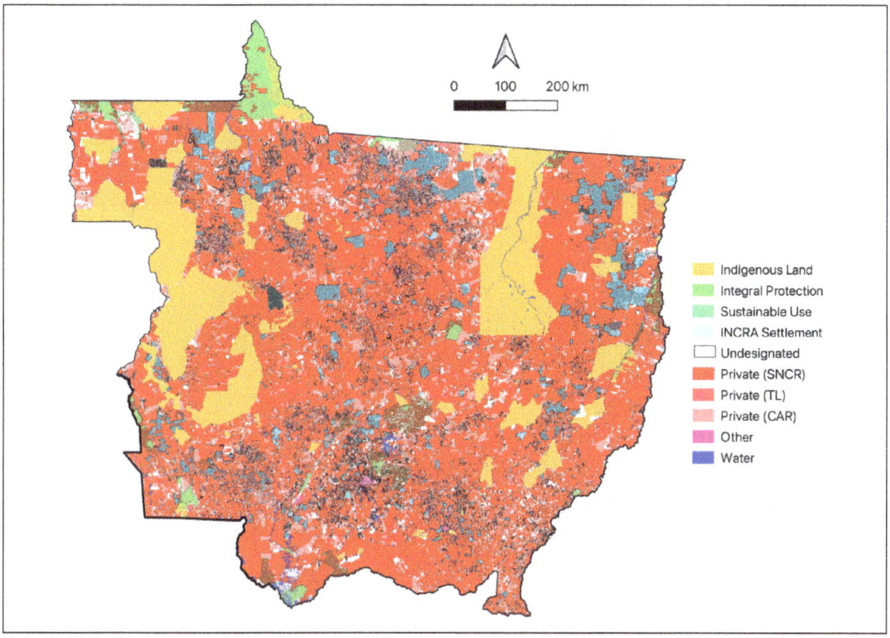

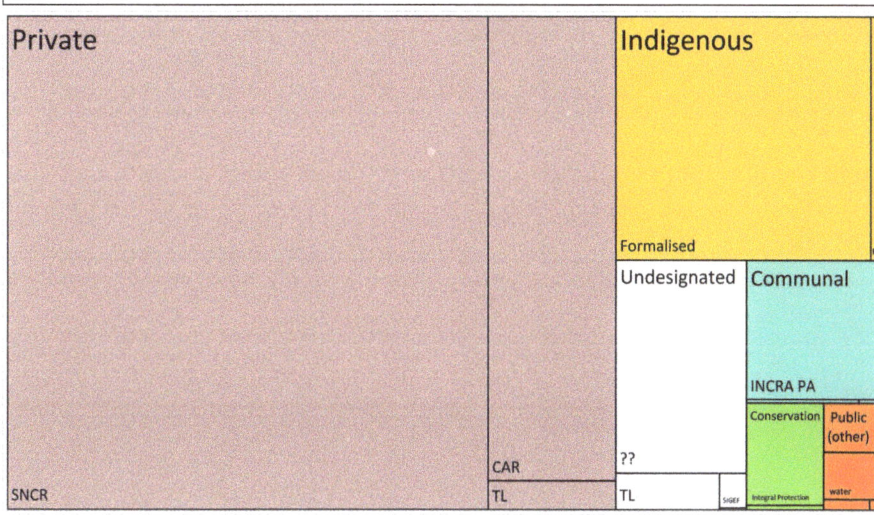

Data sources: INCRA (2020) and IMAFLORA (2019).

Annex 4.6: Pará

There is a large gap in the title regularisation process in Pará due to: an abundance of smallholdings; the presence of numerous medium- to large-scale cattle ranches claiming lands within Sustainable Use Conservation Units; and the dubious claims of land grabbers and settlers on public land on frontier landscapes zoned for settlement.

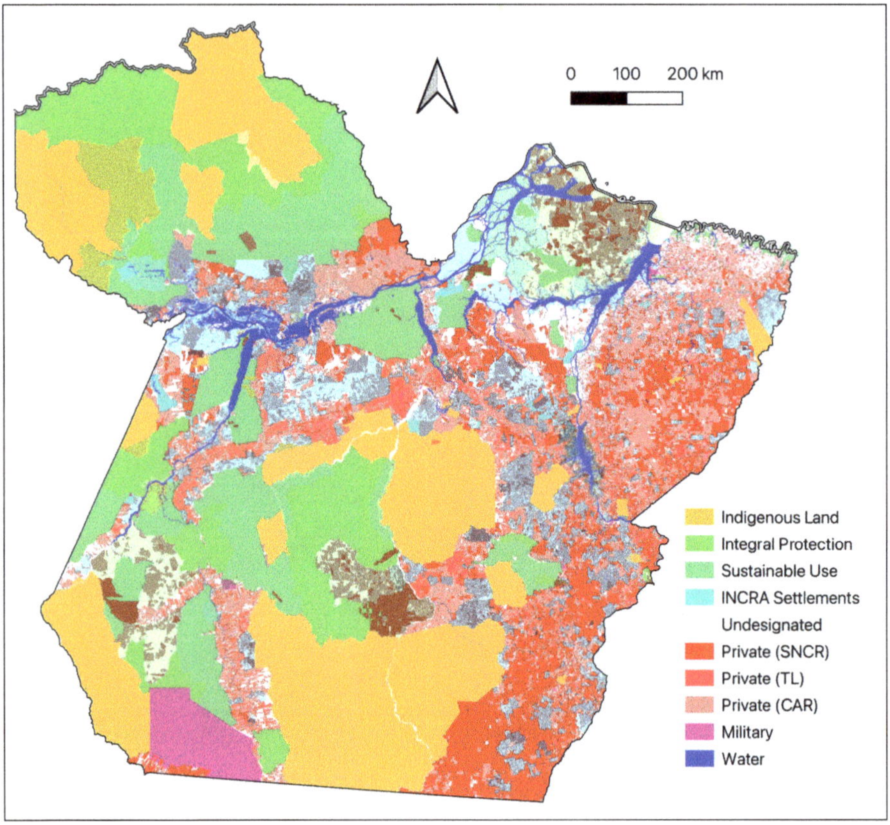

Annex 4.6: Pará

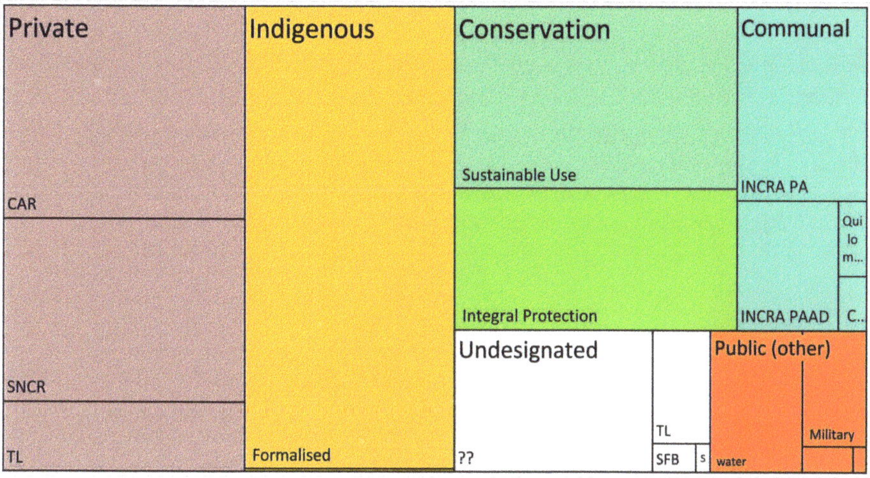

Pará Land Tenure

Data sources: INCRA (2020) and IMAFLORA (2019).

Annex 4.7: Rondônia

In Rondônia, landholdings registered in the SNCR are concentrated in areas dominated by large to medium-sized estates. Most smallholdings have not been regularised, although the vast majority are registered within the CAR, as are the individual holdings within the INCRA PA-type settlements. The most conflictive landscapes are in the North where settlers are invading sustainable-use conservation units; undesignated public lands are occupied by individuals with plausible claims to ownership.

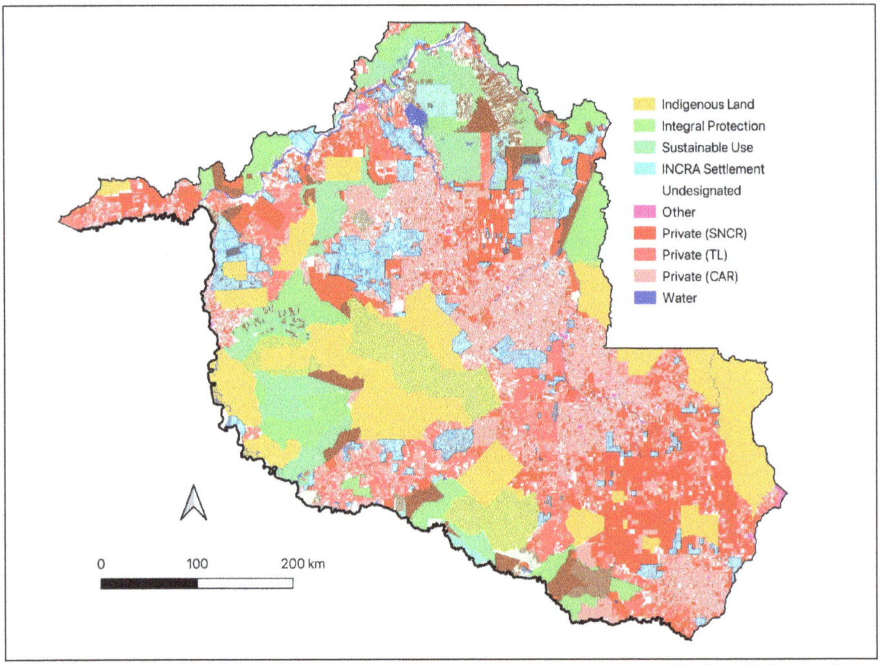

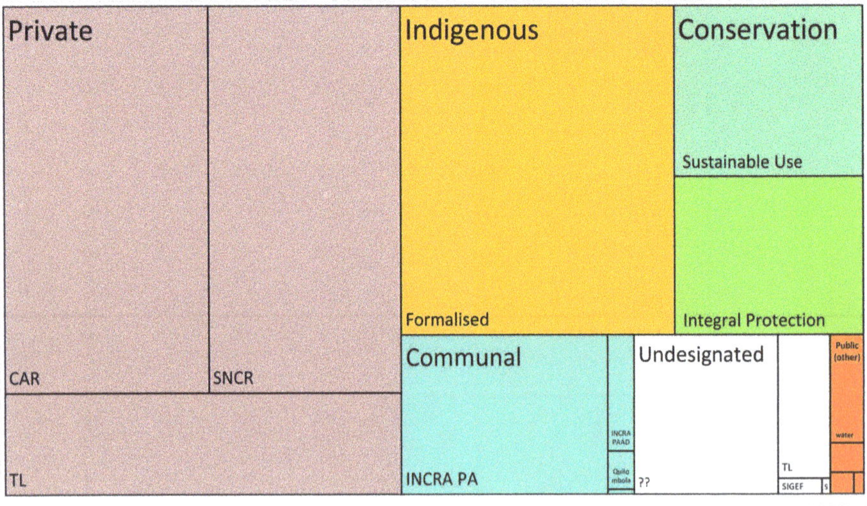

Data sources: INCRA (2020) and IMAFLORA (2019).

Annex 4.8: Roraima

Indigenous Territories are the largest land tenure category in Roraima. The regularisation process has advanced for larger holdings on the savanna landscapes but less so in the smallholder landscapes in the southern municipalities, where inhabitants of INCRA PA-type settlements are also registering their holdings in the CAR system. The holdings registered by the Terra Legal (TL) process are largely forest properties with low levels of deforestation.

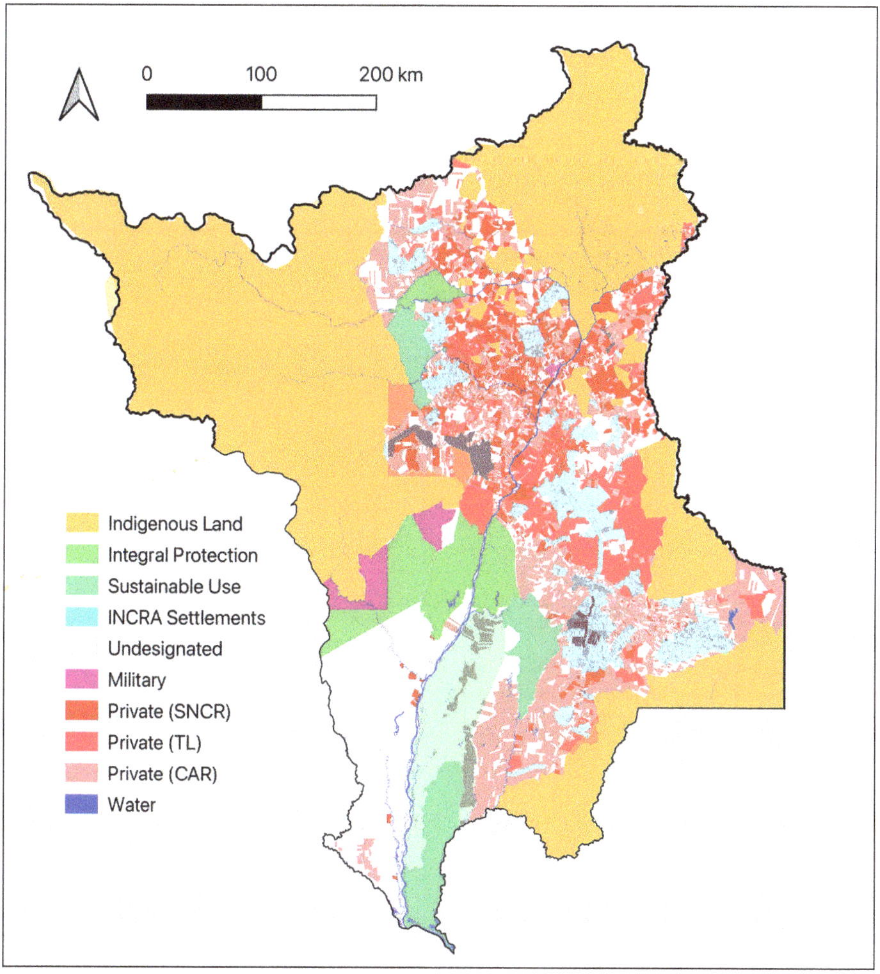

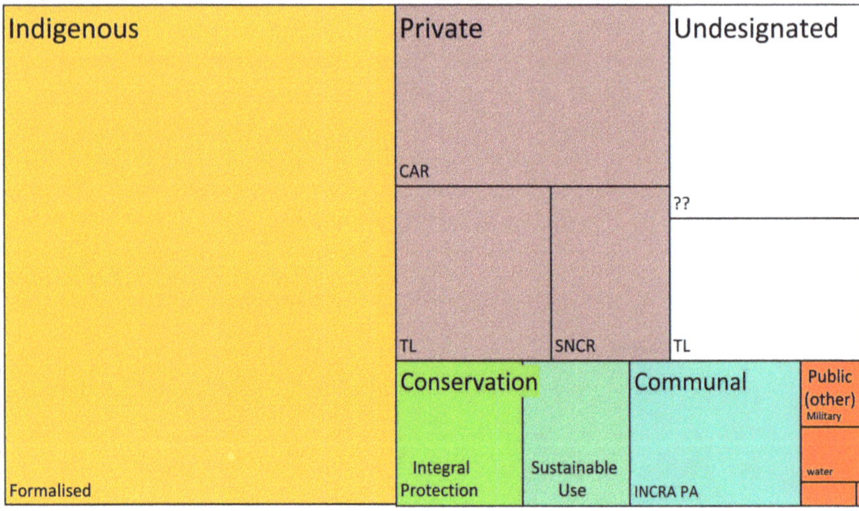

Roraima Land Tenure

Data sources: INCRA (2020) and IMAFLORA (2019).

Annex 4.9: Tocantins

Private landholdings registered within the SNCR are the largest category in Tocantins, in part due to the predominance of medium- to large-scale estates on Cerrado landscapes that were settled in the 1960s and 1970s. Most (all) of the conservation units have inholdings within their boundaries.

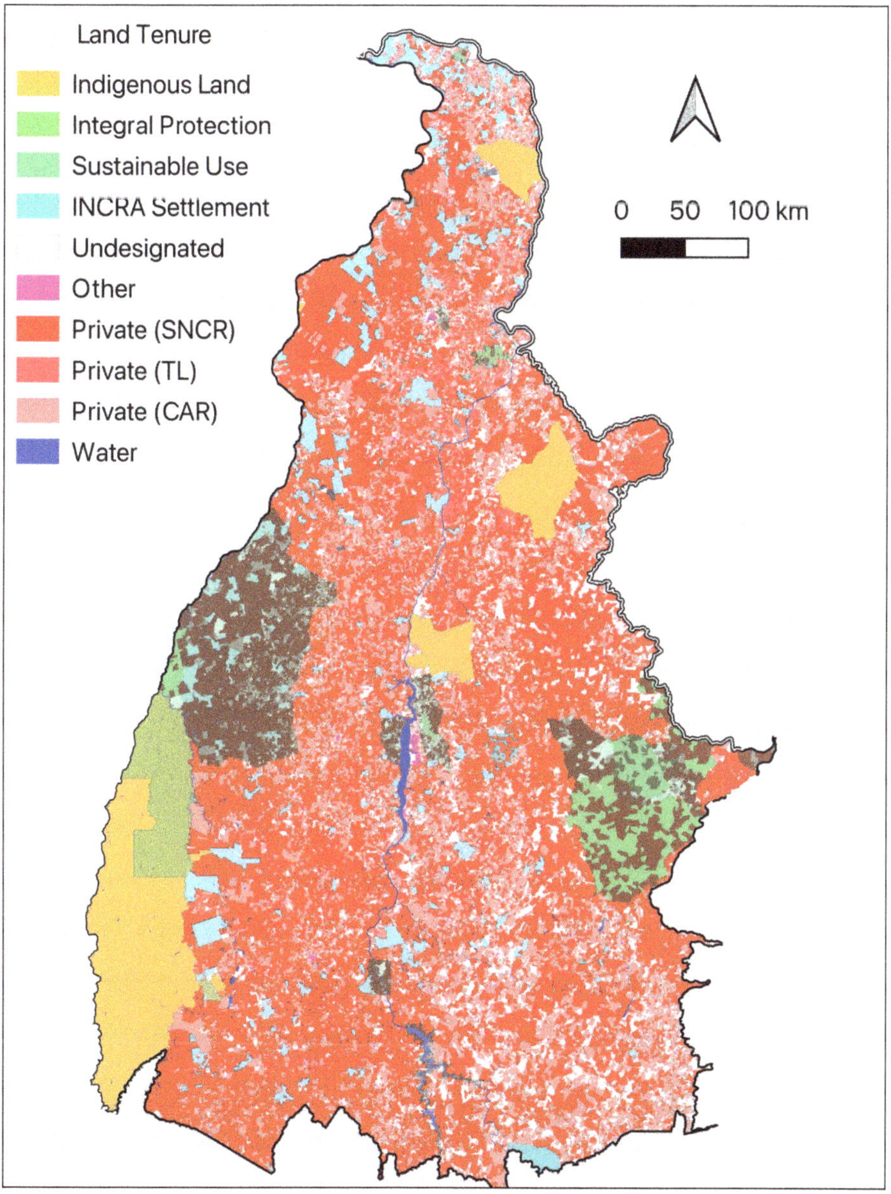

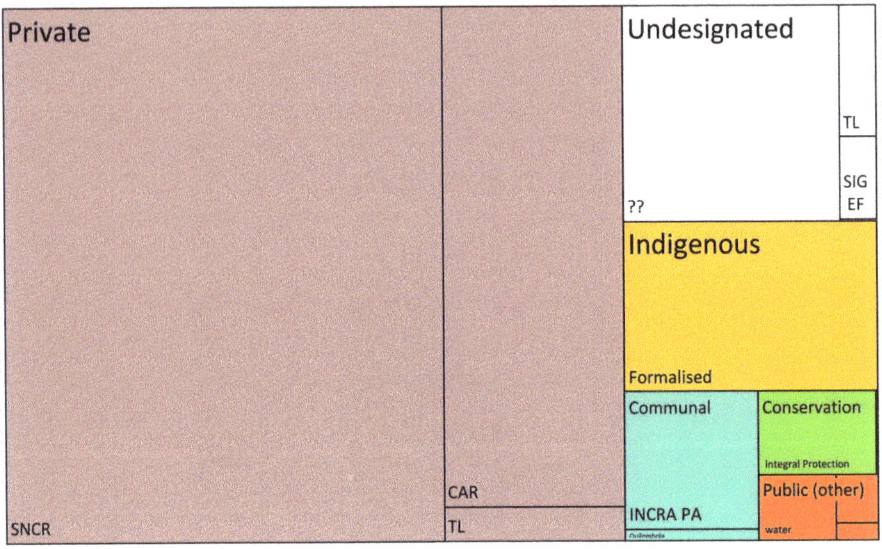

Tocantins Land Tenure

Data sources: INCRA (2020) and IMAFLORA (2019).

Annex 4.10: Land Tenure Metrics for Private Landholdings in the Brazilian Amazon

Inequality in land tenure is evident when comparing the number (left column) and surface area (middle column); only Rondônia and Acre have relatively egalitarian patterns of land tenure. Smallholdings retain much less remnant forest when compared to medium and large-scale estates (right column); the disparity is particularly noticeable in Rondônia, Pará, Acre and Mato Grosso.

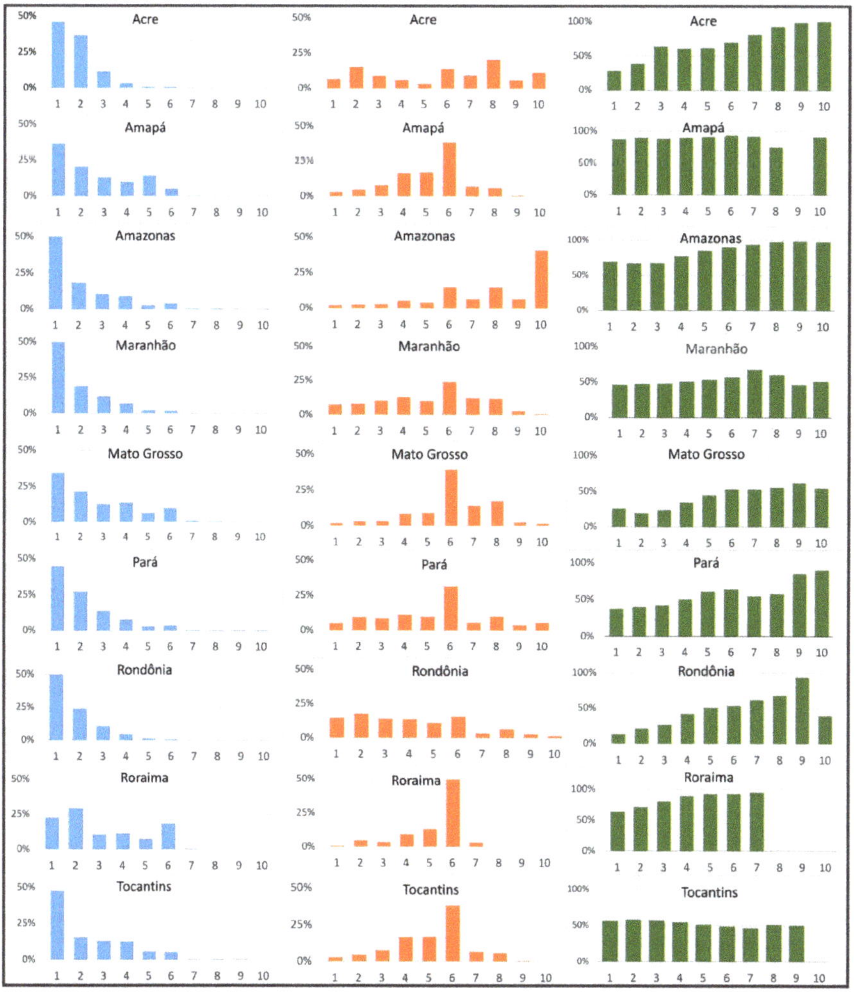

Size classes (ha)			
1	< 50	6	1,000—5,000
2	50—100	7	5,000—10,000
3	100—200	8	10,000—50,000
4	200—500	9	50,000—100,000
5	500—1,000	10	> 100,000

Data sources: IMAFLORA (2019) and MapBiomas (2021).

Annex 4.11: Bolivia

In Bolivia, the land tenure review process (saneamiento) has prioritised smallholders and communal landholdings in the northern part of the country (Castañeros). Recent land grants to settler syndicates with residents that self-identify as Interculturales will probably be redistributed as smallholdings. A backlog of medium and large-scale land-holdings is awaiting review, particularly on the savanna landscapes in the Department of Beni. Undesignated lands include forest blocks that were granted as 30-year timber concessions in the late 1990s; most have been clawed back by the state and are viewed by many as a land bank for eventual distribution and settlement.

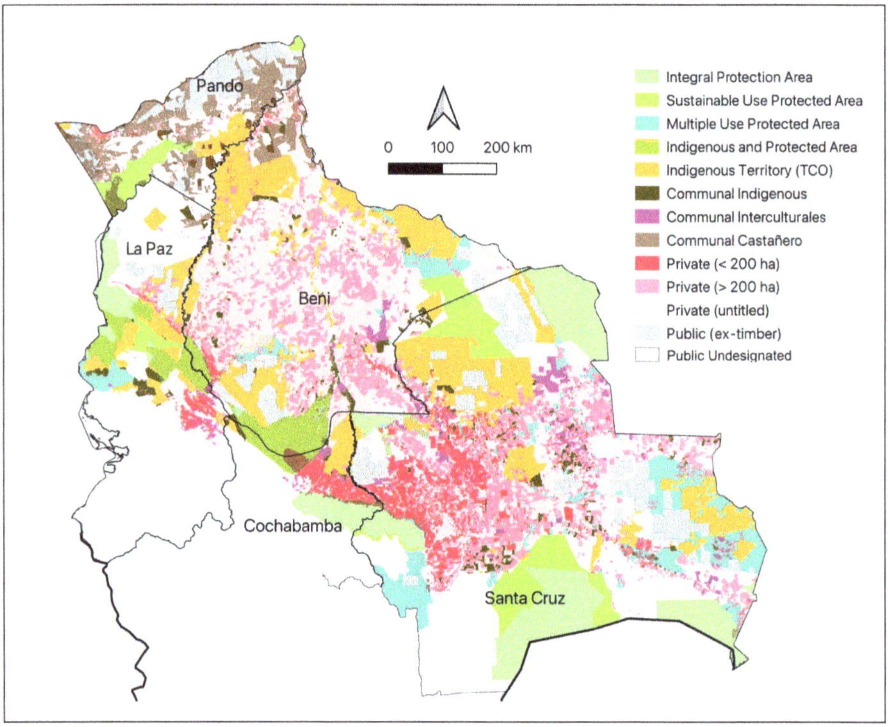

Annex 4.11: Bolivia

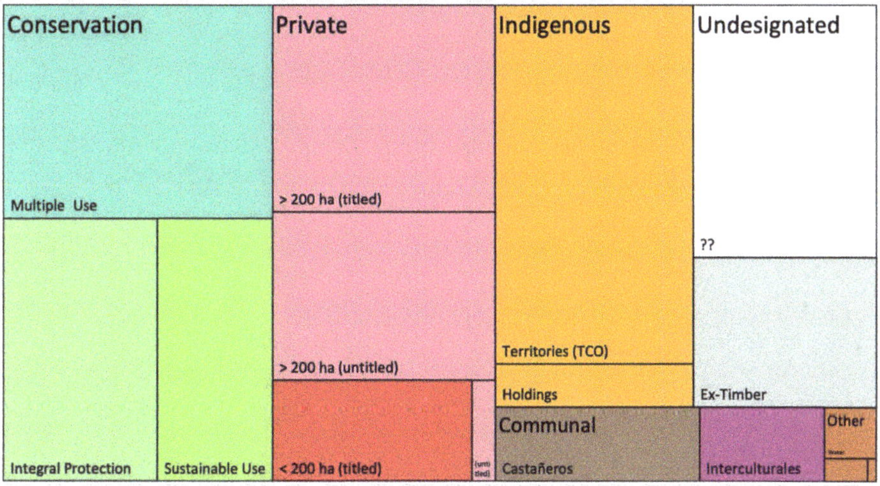

Bolivia Land Tenure

Data sources: Colque et al. (2016) and INRA (2018).

Annex 4.12: Peru

The distribution of landholdings in the Peruvian Amazonian is the most egalitarian in the Pan Amazon, but less than 25% have passed through the title review process. The country has the second largest area of undesignated public lands, a source of contention and competition among the timber sector, settlers, indigenous groups and ribereña communities.

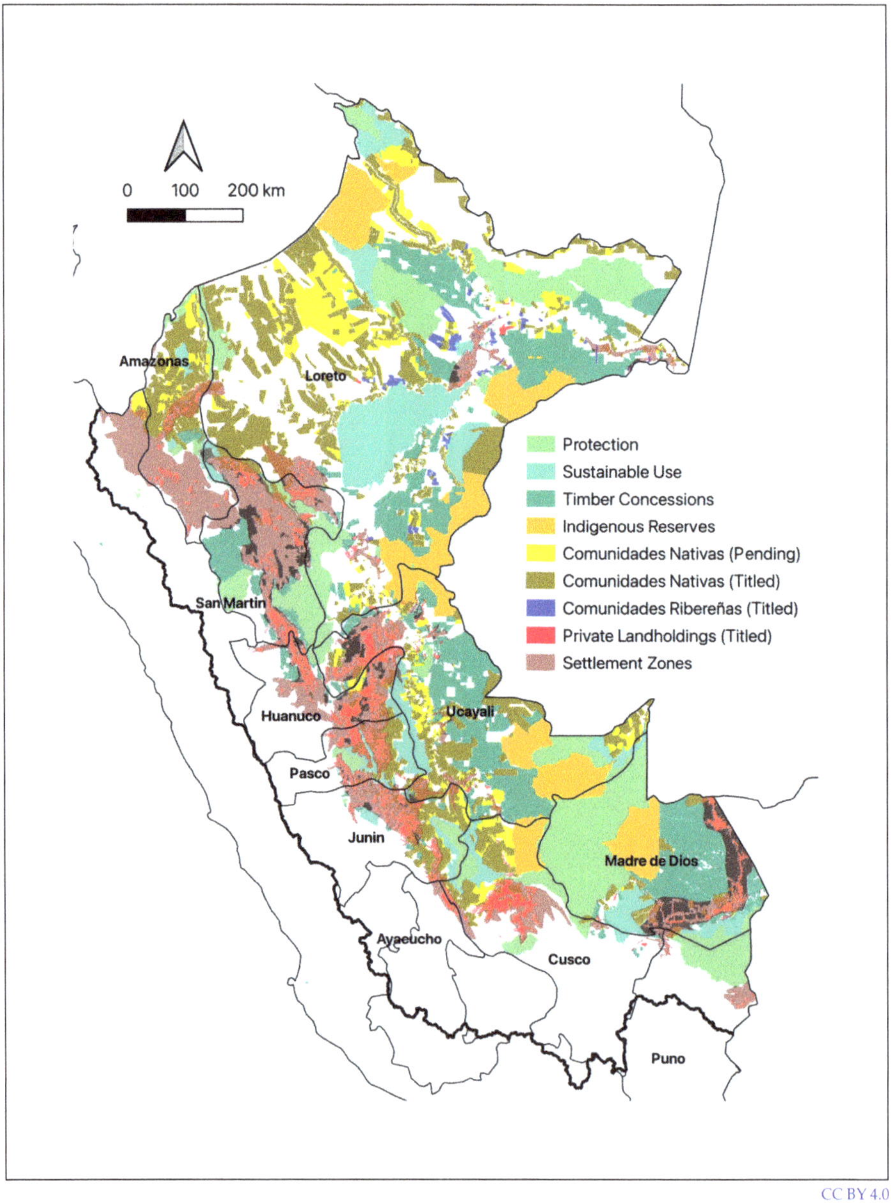

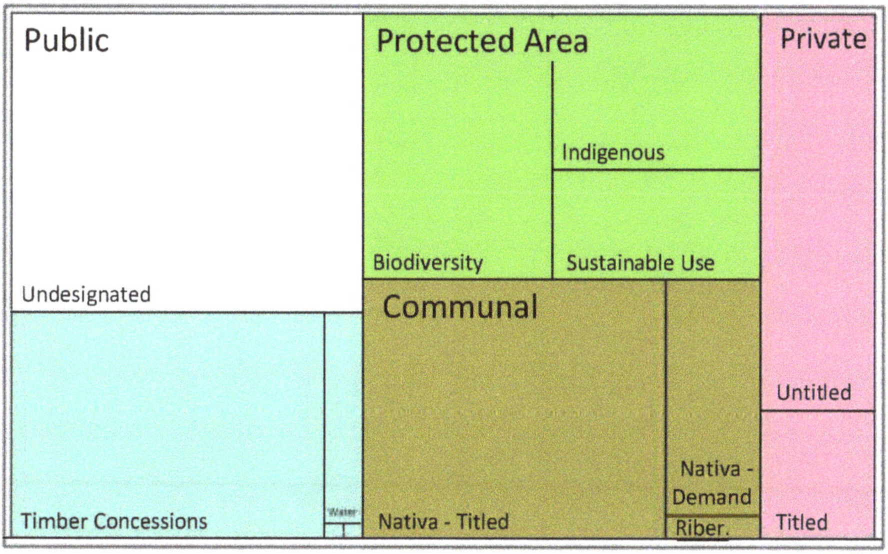

Peru Land Tenure

Data sources: IBC (2021), SICAR (2020) and RAISG (2021).

Annex 4.13: Ecuador

Ecuador has essentially closed its Amazonian provinces to new settlement, but agriculture and deforestation continues to expand along the margins of the existing agrarian landscapes. Most landholders hold a title, but the title review process has only advanced through three of 47 municipalities (cantones). There are two types of indigenous territories: landholdings surrounded by private properties and large reserves that span 'wilderness' landscapes.

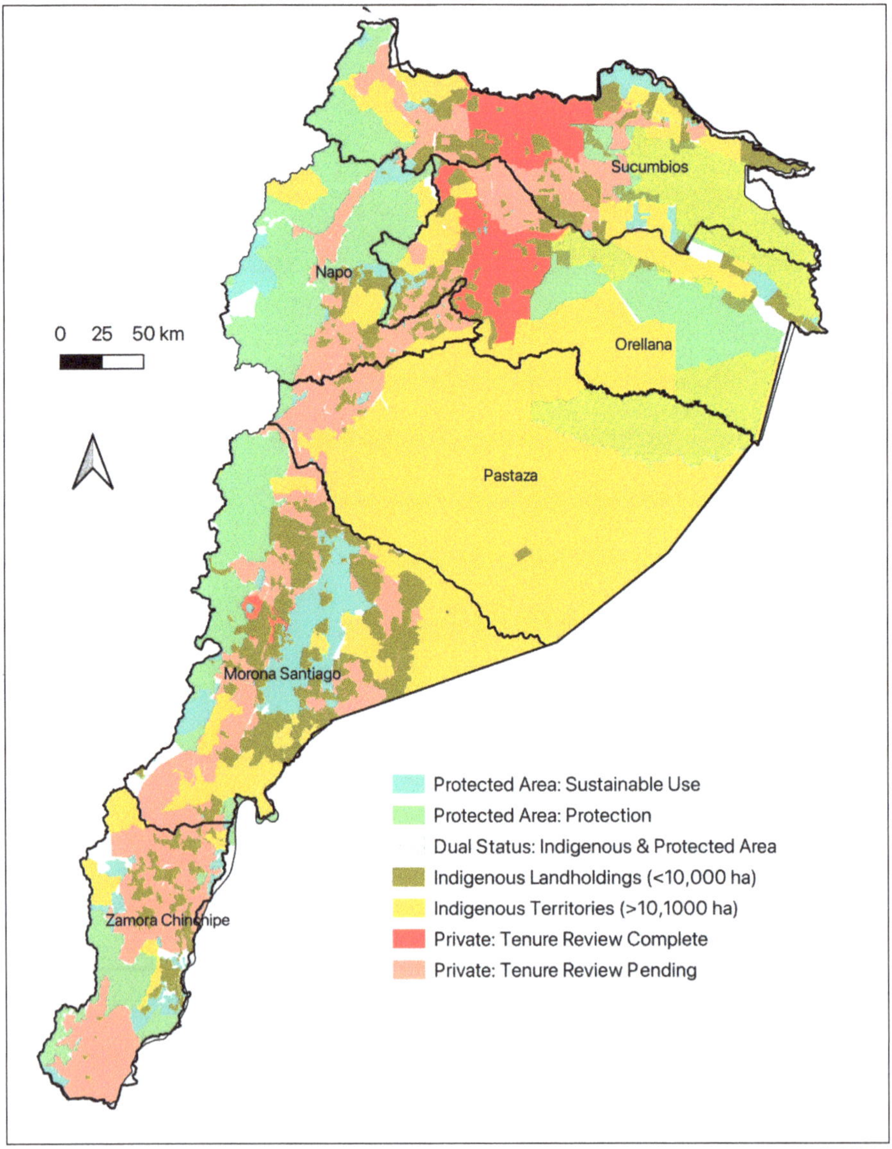

Annex 4.13: Ecuador

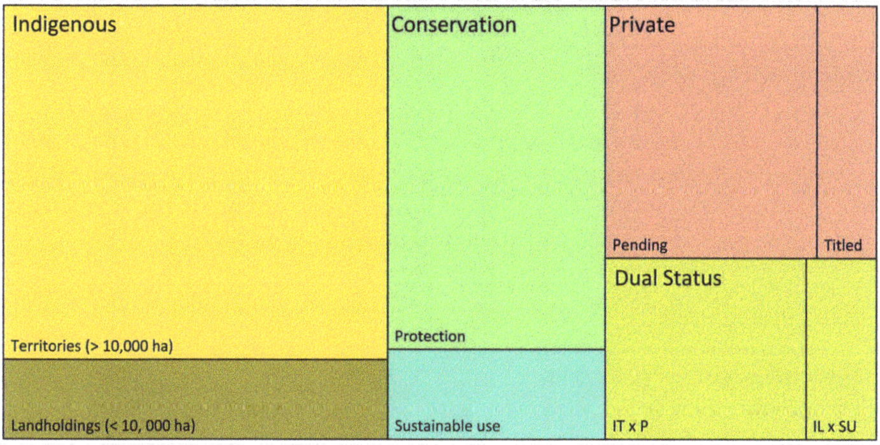

Ecuador Land Tenure
Data source: RAISG (2021).

Annex 4.14: Colombia

Most of the Colombian Amazon region was declared a 'Forest Reserve' in 1969 but only areas that have been reallocated into a specific indigenous reserve or protected area enjoy a (limited) level of protection from settlers and land grabbers. There is an active forest frontier that extends along the Andean piedmont through Putumayo, Caquetá and Guaviare Departments. The number and size of land holdings is unknown because there is no organised land registry at any scale.

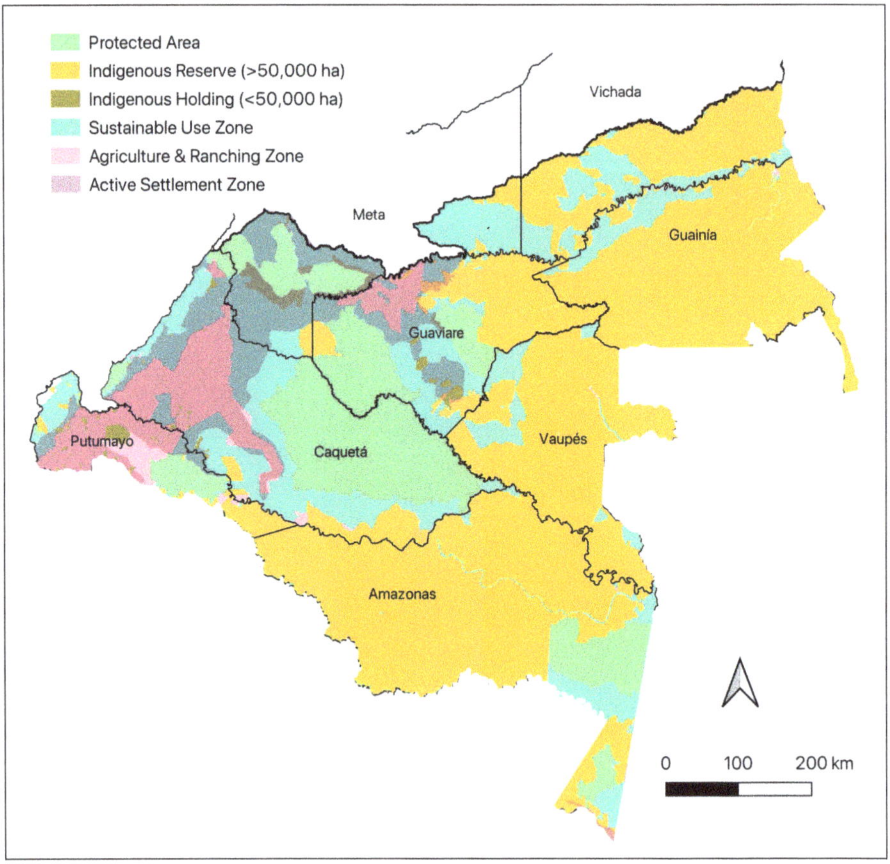

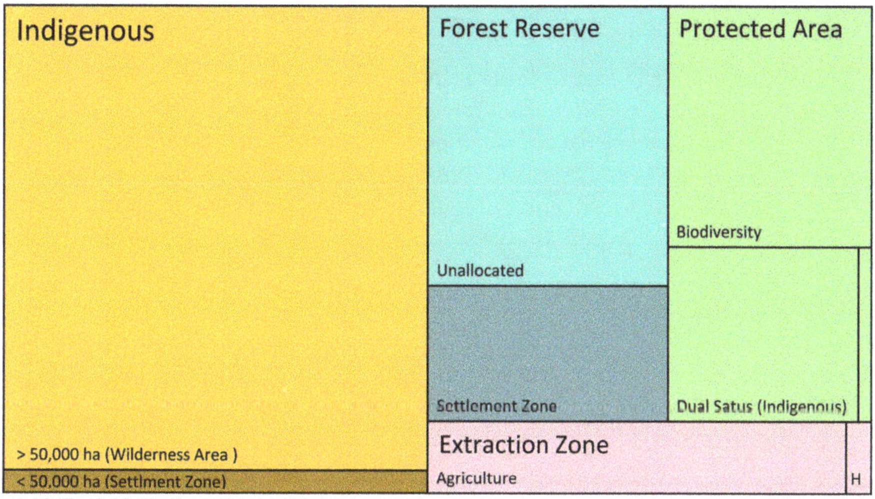

Colombia Land Tenure

Data sources: SINCHI (2016) and SIAT-AC (2020).

Annex 4.15: Venezuela

Venezuela was a pioneer in establishing a protected area system, whose Natural Monuments category provides protection to dozens of tepuye mountains. The process to formalise the territorial rights of indigenous communities halted following the death of Hugo Chavez in 2014. There is no agricultural frontier but wildcat miners roam widely across the region. The country has the third largest component of undesignated public lands in the Pan Amazon.

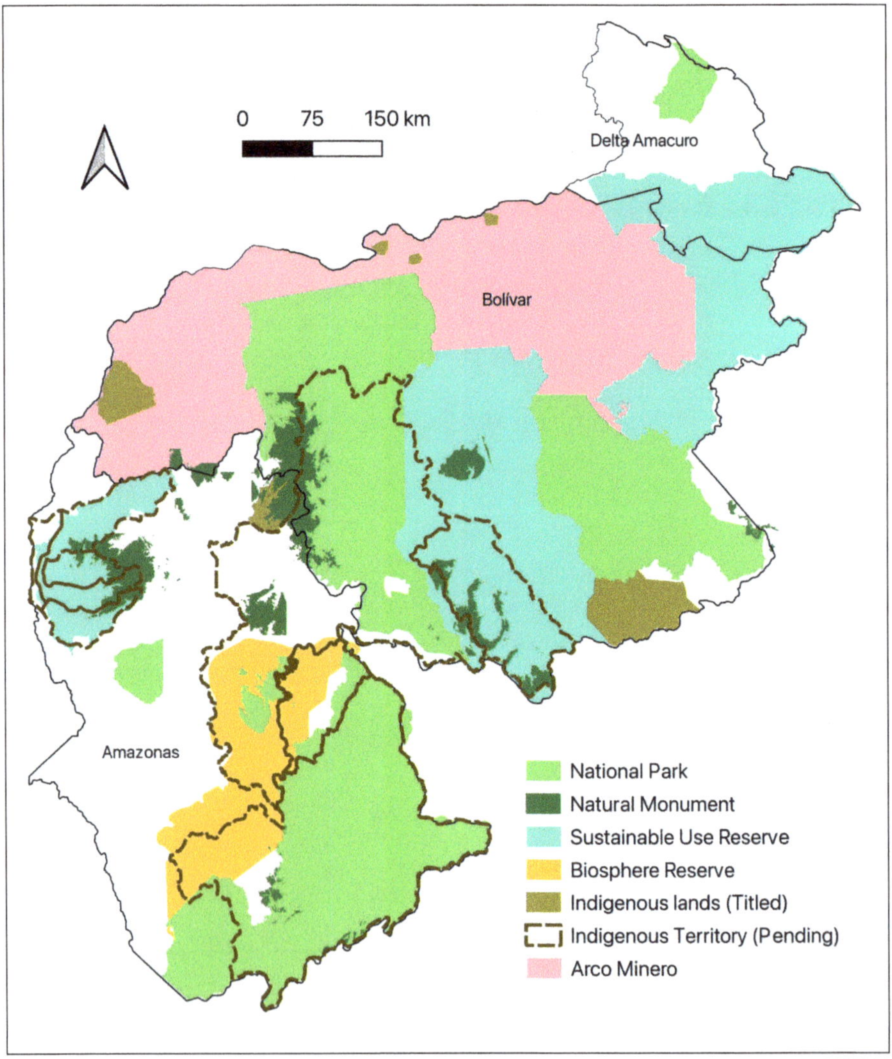

Annex 4.15: Venezuela

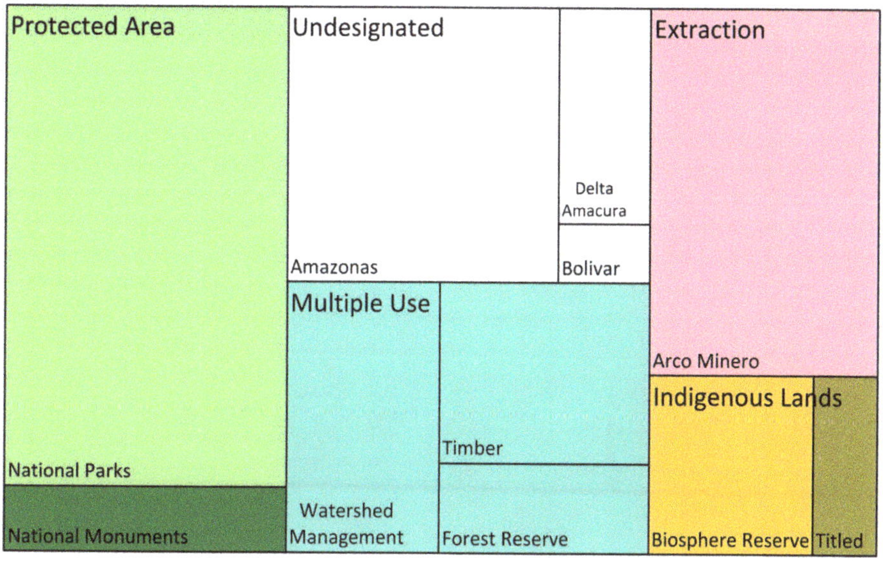

Venezuela Land Tenure

Data sources: MPPEA (2020) and RAISG (2021).

Annex 4.16: The Guianas

Guyana and Suriname have similar patterns of land tenure, with private landholdings (freeholds and leaseholds) on the coast and current timber and mining concessions located in the northern half of their territories. Guyana has recognised the territorial rights of individual villages, but as of January 2022, Suriname had yet to formally recognise the territorial rights of its indigenous citizens. The southern regions of both countries (combined) encompass about 13.6 million hectares of undesignated public forest. France has allocated all the territory in French Guiana to integral protection or sustainable use, while recognising the use-rights of its indigenous people.

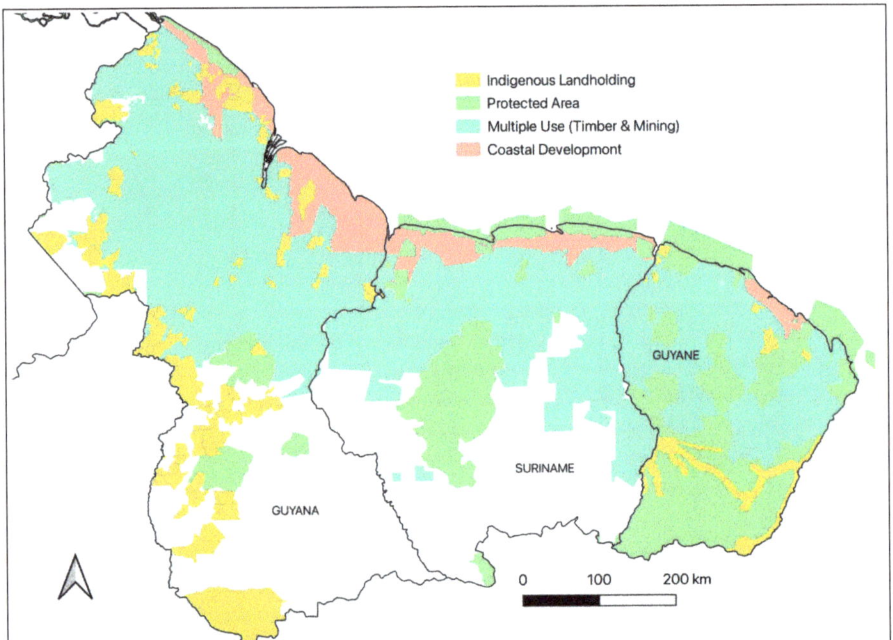

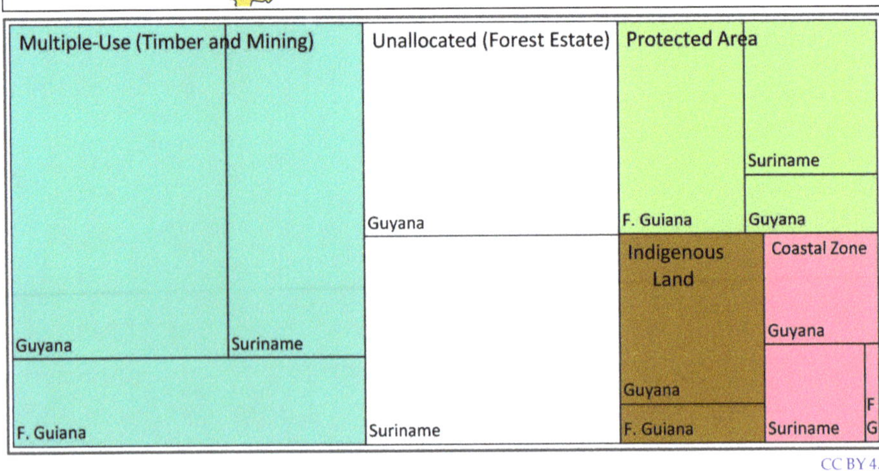

Data sources: RAISG (2021), The Guyana Forest Commission (2020) and Government of Suriname (2018).

Bibliography

Alencar, A., C. Pereira, I. Castro, A. Cardoso … and R. Novaes. 2016. *Desmatamento nos assentamentos da Amazônia: Histórico, tendências e oportunidades.* IPAM, Brasília, DF. https://ipam.org.br/wp-content/uploads/2016/02/Desmatamento-nos-Assentamentos-da-Amaz%C3%B4nia.pdf

ANA – *Agência Nacional de Águas e Saneamento Básico (Brasil).* 2021. *Atlas irrigação: uso da água na agricultura irrigada* / Agência Nacional de Águas e Saneamento Básico. 2nd ed. Brasília: ANA. https://portal1.snirh.gov.br/ana/apps/storymaps/stories/a874e62f27544c6a986da1702a911c6b

Andrea, M.C.D.S., K.J. Boote, P.C. Sentelhas and T.L. Romanelli. 2018. 'Variability and limitations of maize production in Brazil: Potential yield, water-limited yield and yield gaps'. *Agricultural Systems* **165**: 264–273.

Arantes, A.E., V.R.D.M. Couto, E.E. Sano and L.G. Ferreira. 2018. 'Livestock intensification potential in Brazil based on agricultural census and satellite data analysis'. *Pesquisa Agropecuária Brasileira* **53** (9): 1053–60.

Arias, M.E., F. Farinosi, E. Lee, A. Livino, J. Briscoe and P.R. Moorcroft. 2020. 'Impacts of climate change and deforestation on hydropower planning in the Brazilian Amazon'. *Nature Sustainability* **3** (6): 430–36.

Assunção, J. and R. Rocha. 2016. *Rural Settlements and Deforestation in the Amazon.* Climate Policy Initiative. https://climatepolicyinitiative.org/wp-content/uploads/2017/02/Rural_Settlements_and_Deforestation_in_the_Amazon_Working_Paper_CPI.pdf

Azevedo, A.A., R. Rajão, M.A. Costa, M.C. Stabile … and R. Pacheco. 2017. 'Limits of Brazil's Forest Code as a means to end illegal deforestation'. *Proceedings of the National Academy of Sciences* **114** (29): 7653–58.

Azevedo-Ramos, C., P. Moutinho, V.L.D.S. Arruda, M.C. Stabile … and J.P. Ribeiro. 2020. 'Lawless land in no man's land: The undesignated public forests in the Brazilian Amazon'. *Land Use Policy* **99**: 104863.

Bahia, R.B.C., M.A. Martins Neto, M.S.C. Barbosa and A.J. Pedreira. 2007. 'Análise da evolução tectonossedimentar da Bacia dos Parecis através de métodos potenciais'. *Revista Brasileira de Geociências* **37** (4): 639–649. https://rigeo.cprm.gov.br/jspui/bitstream/doc/578/1/art_bahia_2007.pdf

Barthel, K., V. Cespedes, B. Salazar, R. Torres and M. Varón. 2016. Land and rural development policy reforms in Colombia: The path to peace. USAID/Colombia Land and Rural Development Program (LRDP): https://www.globalcommunities.org/publications/2016-Colombia-Rural-Development-Policy.pdf

Berry, A. 2006. *Has Colombia Finally Found an Agrarian Reform that Works. Human Development in the Era of Globalization: Essays in Honor of Keith B. Griffin.* Portland OR: Book News Inc.

Botero Garcia, R. 2021. *A deforestación reciente en la Amazonia Colombiana: Consideraciones para su Análisis, Repensar el future de America Latina y el caribe.* Alternativas parac la transfomación social- ecologica: https://fcds.org.co/publicaciones/?sf_paged=2

Brito, B. and P. Barreto. 2011. *A regularização fundiária avançou na Amazônia? Os dois anos do programa Terra Legal.* Belém: Imazon. https://imazon.org.br/publicacoes/1404-2/

Cabezas, J.E.P. 2017. 'La planificación nacional en Ecuador: Planes de desarrollo y ordenamiento territorial, y el sistema de seguimiento y evaluación SIGAD/National planning in Ecuador: Development and territorial planning plans, and the SIGAD monitoring and evaluation system'. *Ciencia Unemi* **9** (21): 168–79.

Chiavari, J., C.L. Lopes and J.N. de Araujo. *Panorama dos direitos de propriedade no Brasil rural*. Relatório. Rio de Janeiro: Climate Policy Initiative, 2021.

Chiavari, J., C.L. Lopes, D. Marques, L. Antonaccio and N. Braga. 2016. *Panorama dos direitos de propriedade no Brasil rural: Legislação, gestão fundiária e código florestal*. Rio de Janeiro: Climate Policy Initiative. Núcleo de Políticas Climáticas da PUC-Rio.

Cochrane, T.A., O. Rosales and T.J. Killeen. 2007. *Agua, gas y agroindustria: gestión sostenible de agua para riego agrícola en Santa Cruz, Bolivia*. La Paz: Conservation International. https://ir.canterbury.ac.nz/bitstream/handle/10092/184/12595702_Main.pdf?sequence=1

Colque, G., E. Tinta and E. Sanjinés. 2016. *Segunda Reforma Agraria: Una historia que incomoda*. La Paz: Fundación Tierra.

Corral, L.R. and C.E.M. Olea. 2020. 'What drives take-up in land regularization: Ecuador's rural land regularization and administration program, Sigtierras'. *Journal of Economics, Race, and Policy* **3** (1): 60–75.

Dammert, J.L., C. Cárdenas and E. Canziani. 2012. Potenciales impactos ambientales y sociales del establecimiento de cultivos de palma aceitera en el Departamento de Loreto: https://repositorio.spda.org.pe/bitstream/20.500.12823/265/1/Potenciales_Impactos_Ambientales_2012.pdf

Dourojeanni M. 2013. Loreto Sostenible 2021, Derecho Ambiente y Recursos Naturales (DAR), Lima, Peru: https://dar.org.pe/archivos/publicacion/lsostenible2021_p2.pdf

Dourojeanni, M. 2017. *Belaúnde en la Amazonía*. Centro Amazónico de Antropología y Aplicación Práctica (CAAAP): https://www.caaap.org.pe/2017/06/12/belaunde-en-la-amazonia-por-marc-j-dourojeanni/

Dourojeanni, M. 2016. 'Aprovechamiento del barbecho forestal en áreas de agricultura migratoria en ia amazonía peruana'. *Revista forestal del Perú* **14** (2): http://cedinfor.lamolina.edu.pe/Articulos_RFP/Vol14_2_87_(20)/vol14_no2_art2.pdf

Durwin, H. 2018. Land Administration in Guyana, Natural Resources for Sustainable Rural Development: Access and Governance: http://www.fao.org/3/ca4407es/ca4407es.pdf

d'Eeckenbrugge, G.C. and D.L. Ferla. 2000. *Fruits from America: An Ethnobotanical Inventory*. Cali, Colombia: International Plant Genetic Resources Institute /CIRAD-FLHOR.

Farthing, L., 2019. 'An opportunity squandered? Elites, social movements, and the government of Evo Morales'. *Latin American Perspectives* **46** (1): 212–29.

Feder, E. 1965. 'Land reform under the alliance for progress'. *Journal of Farm Economics* **47** (3): 652–668.

Ferreira, A.B.H. 1986. *Novo dicionário da língua portuguesa. 2ª edição*. Rio de Janeiro: Nova Fronteira. p. 1372.

Fort, R. 2007. Property rights after market liberalization reforms: Land titling and investments in rural Peru. Ph.D. Thesis. Wageningen: Wageningen Academic Publisher. http://www.grade.org.pe/publicaciones/buscar/PETT

Fórum Mato-Grossense da Agropecuária. 2021. Proposta ZSEE / MT – 2018: https://www.boamidia.com.br/wp-content/uploads/2021/02/Apresentacao_Deputados-1.pdf

Freitas, F.L.M., V. Guidotti, G. Sparovek and C. Hamamura. 2018. 'Malha fundiária do Brasil, v.1812'. In *Atlas - A Geografia da Agropecuária Brasileira*: https://www.imaflora.org/atlasagropecuario

García Yapur, F.L., L.A.G. Orellana and M.S. Romero. 2014. *MAS legalmente, IPSP legítimamente. Ciudadanía y devenir Estado de los campesinos indígenas en Bolivia.* https://bitacorraintercultural.org/mas-legalmente-ipsp-legitimamente-ciudadania-y-devenir-estado-de-los-campesinos-en-bolivia/

Girardi, E.P. 2015. 'Uma leitura da questão agrária em Mato Grosso Confins'. *Revue franco-brésilienne de géographie/Revista franco-brasilera de Geografia.* https://journals.openedition.org/confins/10446

Gondard, P. and H. Mazurek. 2001. '30 años de reforma agraria y colonización en el Ecuador (1964-1994) – Dinámicas espaciales'. *Estudios de Geografía* **10**: 15–40 http://horizon.documentation.ird.fr/exl-doc/pleins_textes/pleins_textes_7/carton01/010026095.pdf

Government of Suriname. 16 Mar. 2018. Onze Natuur op 1 – natuurwetgeving: http://www.gov.sr/themas/milieu-en-omgeving/project-onze-natuur-op-1-natuurwet.aspx

Guyana Forest Commission. 2020. Forest resources Allocation Map of Guyana. Forest resources Information Unit: https://forestry.gov.gy/wp-content/uploads/2021/11/Forest-Allocation-Map-2021-copy-scaled.jpg

Guereña, A. 2016. *Unearthed: Land, Power and Inequality in Latin America.* Oxfam International.

Gutiérrez, R.G. and L. Quevedo. 2008. *El sistema de concesiones forestales en Bolivia.* CADEFOR: https://docplayer.es/81287039-El-sistema-de-concesiones-forestales-en-bolivia.html

Haghtalab, N., N. Moore, B.P. Heerspink and D.W. Hyndman. 2020. 'Evaluating spatial patterns in precipitation trends across the Amazon basin driven by land cover and global scale forcings'. *Theoretical and Applied Climatology* **140**: 1–17.

Hecht, S. and A. Cochburn. 2010. *The Fate of the Forest.* Chicago: University of Chicago Press.

Hecht, S.B. 1985. 'Environment, development and politics: Capital accumulation and the livestock sector in eastern Amazonia'. *World Development* **13** (6): 663–684.

Hoekstra, A.Y. 2014. 'Sustainable, efficient, and equitable water use: The three pillars under wise freshwater allocation'. *Wiley Interdisciplinary Reviews. Water* **1** (1): 31–40.

Harten, S. 2011. *The Rise of Evo Morales and the MAS.* London and New York: Zed Books.

IBC – Instituto del Bien Común. 2021. Sistema de Información sobre Comunidades Nativas de la Amazonía Peruana (SICNA): https://ibcperu.org/en/servicios/sicna/

IIAP – Instituto de Investigaciones de la Amazonía Peruana. 2010. ¿Qué es la Zonificación Ecológica y Económica? http://terra.iiap.gob.pe/index.html#queeszee

IMAFLORA. 2019. Atlas Agropecuária - A geografia da agropecuária brasileira: http://atlasagropecuario.imaflora.org/

INCRA – Instituto Nacional de Colonización y Reforma Agraria. 2017 (15 Dec). Incra nos Estados – Informações gerais sobre os assentamentos da Reforma Agrária: http://painel.incra.gov.br/sistemas/index.php

INCRA – Instituto Nacional De Colonização e Reforma Agrária. 2019. Preços referenciais de terras e imóveis rurais: https://www.gov.br/incra/pt-br/assuntos/governanca-fundiaria/relatorio-de-analise-de-mercados-de-terras

INCRA – Instituto Nacional De Colonização e Reforma Agrária. 2020. Acervo Fundiario: http://acervofundiario.incra.gov.br/acervo/acv.php

INRA – Instituto Nacional de Reforma Agraria. 2018. Estado de saneamento, 2018: https://www.inra.gob.bo/

Instituto Escolha. 2021. Quantoé? Plantar Floresta: http://quantoefloresta.escolhas.org/

Kaimowitz, D., G. Thiele and P. Pacheco. 1999. 'The effects of structural adjustment on deforestation and forest degradation in lowland Bolivia'. *World Development* **27** (3): 505–20.

Kohlhepp, G., 2018. 'O Programa Piloto Internacional de Proteção das Florestas Tropicais no Brasil (1993–2008)'. *Revista Nera* **42**: 309–31.

Lambin, E.F. and P. Meyfroidt. 2011. 'Global land use change, economic globalization, and the looming land scarcity'. *Proceedings of the National Academy of Sciences* **108** (9): 3465–72.

Laskos, A.A., A.A. Cazella. and P.B.M. Rebollar. 2016. 'O Sistema Nacional de Cadastro Rural: História, limitações atuais e perspectivas para a conservação ambiental e segurança fundiária'. *Desenvolvimento e Meio Ambiente* **36**: https://revistas.ufpr.br/made/article/view/39124

Lathuillière, M.J., M.T. Coe, A. Castanho, J. Graesser and M.S. Johnson. 2018. 'Evaluating water use for agricultural intensification in Southern Amazonia using the Water Footprint Sustainability Assessment'. *Water* **10** (4): 349.

Lemel, H. 2001. *Patterns of Tenure Insecurity in Guyana*, Working Paper No. 43, Land Tenure Center, University of Wisconsin–Madison: https://ageconsearch.umn.edu/bitstream/12801/1/ltcwp43.pdf

Lourenço, A. 2009. Regularização Fundiária e Desenvolvimento na Amazônia, *Interesse Nacional* **2** (6): 29–41. https://interessenacional.com.br/wp-content/uploads/2021/12/IN-06.pdf

Lovejoy, T.E. and C. Nobre. 2018. 'Amazon tipping point'. *Science Advances* **4** (2): eaaat22340. https://doi.org/10.1126/sciadv.aat2340

MapBiomas- Mapeamento Anual do Uso e Cobertura da Terra no Brasil. 2021. Cobertura E Transições Municípios: https://mapbiomas.org/estatisticas

Marengo, J.A. and J.C. Espinoza. 2016. 'Extreme seasonal droughts and floods in Amazonia: Causes, trends and impacts'. *International Journal of Climatology* **36** (3): 1033–1050.

McInddewar, L. and T. Reis. Aug. 2016. *Land Tenure and the Effect of the Terra Legal Program on Deforestation*. Technical Report: https://doi.org/10.13140/RG.2.2.30811.16169

MMA – Ministério do Meio Ambiente. 2009. Mapa Integrado dos Zoneamentos Ecológico Econômicos dos Estado da Amazônia Legal: Situação atual. Secretaria de Extrativismo e Desenvolvimento Rural Sustentável, Departamento de Zoneamento Territorial – DZT: https://antigo.mma.gov.br/estruturas/225/_imagens/mapa_estados_integrados_225.jpg

MMA – Ministério do Meio Ambiente. 2016. O zoneamento ecológico-econômico, na Amazônia Legal, Trilhando o caminho do future, Sectratari de Recuros Hidricos e Amnbiente Urbano. Departamento de Zoneamento Territiorial, Brasilia, MMA: https://antigo.mma.gov.br/images/arquivo/80253/ZEE_amazonia_legal.pdf

MPPEA – Ministerio de Poder Popular para Ecosocialismo y Agua. 2020. Las Áreas Bajo Régimen de Administración Especial (ABRAE): http://sigta.minec.gob.ve/sigta/abrae.php https://i.pinimg.com/originals/58/86/db/5886db0a462ac24ae-773592c0dc19395.jpg

Morais, H.M.A.D.O. 2017. *Regularização fundiária rural na Amazônia Legal: Uma análise da Lei nº 13.465 de 11 de julho de 2017*. Universidade de Brasília – UnB: https://www.bdm.unb.br/handle/10483/18853

Pacheco, P. 2009. 'Agrarian reform in the Brazilian Amazon: Its implications for land distribution and deforestation'. *World Development* 37 (8): 1337–1347.

Pacheco, P. and J.H. Benatti 2012. 'Land appropriation under changing environmental governance in the Amazon: The cases of lowland Bolivia and the State of Para, Brazil'. In *International Conference on Global Land Grabbing II,* organised by the Land Deals Politics Initiative (LDPI), Department of Development Sociology, Cornell University, Ithaca.

Pallares. G. 2021. 'Lessons learned from a decade of REDD+ in Guyana'. *Forest News*. https://forestsnews.cifor.org/71458/lessons-learned-from-a-decade-of-redd-in-guyana?fnl=en

Pedraza Isaza, D. 2016. *Colombia's Old Agrarian Reforms Challenge a New Rural Reform to Finally Benefit the Rural Poor. Agrarian, Food and Environmental Studies (AFES)*. International Institute of Social Studies: http://hdl.handle.net/2105/37259

Peña, R.E.G. and M.I.S. Viera. 2013. 'Las ABRAE versus las áreas protegidas en Venezuela'. *Revista COPÉRNICO* **8** (19): 27–39. http://abrae85.blogspot.com/

Pereira, L.A.C., C.C.A. Cajazeiras, T.L.F.D. Paula, M.A. Freitas … and C.J.B.D. Aguiar. 2014. *Síntese da hidrogeologia da Bacia dos Parecis*. PRM – Serviço Geológico do Brasil.

Pinto, L.F.G., V.G. Faria, G. Sparovek, B.P. Reydon … and T. Carvalho. 2020. 'Quem são os poucos donos das terras agrícolas no Brasil-O mapa da desigualdade'. *Sustentabilidade Em Debate* **10**: 1–21.

del Prado, N. 2015. *Support Sound Land Use Planning in Suriname: A Review of the Legal and Institutional Framework for LUP.* Workshop land use planning: http://www.wwfguianas.org/news/publications/workshop_land_use_planning_march_17th_2015/

RAISG – Rede Amazônica de Informação Socioambiental. 2021. Dados Cartográficos: https://www.amazoniasocioambiental.org/pt-br/mapas/#download

Rattis, L., P.M. Brando, M.N. Macedo et al. 2021. 'Climatic limit for agriculture in Brazil'. *Nat. Clim. Chang.* https://doi.org/10.1038/s41558-021-01214-3

Rausch, L. 2014. 'Convergent agrarian frontiers in the settlement of Mato Grosso, Brazil'. *Historical Geography* **42**: 276–297.

Richani, N. 2013. *Systems of Violence: The Political Economy of War and Peace in Colombia*. New York: SUNY Press.

Rodrigues Alvarez, G. 2018. *Sinergia Público-Privada por la Soberanía Alimentaria De Bolivia, Santa Cruz – Bolivia,* Comercio Exterior, N° 267, Instituto Boliviano de Comercio Exterior (IBCE): https://ibce.org.bo/images/publicaciones/ce-267-UCAB-Hacia-la-soberania-alimentaria-en-Bolivia.pdf

Rojas Calizaya, J.C. and A. Anzaldo Garcia. 2020. 'El nuevo PLUS del Beni, excluye a los actores y sus diversas visiones de desarrollo y atenta contra la Amazonía boliviana'. *Mundos Rurales* **15** (1): 87–104. https://cipca.org.bo/publicaciones-e-investigaciones

Sanchez, B. and J. Rosales. 2008. Una revision del plan de manejo de la cuenca del Rio Caroní Venezuela, desde una perspectiva de la valoracion del recurso hidrico. Conference: VI Congreso Ibérico de Planeación y gestión del agua, Vitoria, Spain: https://www.researchgate.net/publication/281101043_UNA_REVISION_DEL_PLAN_DE_MANEJO_DE_LA_CUENCA_DEL_RIO_CARONI_VENEZUELA_DESDE_UNA_PERSPECTIVA_DE_LA_VALORACION_DEL_RECURSO_HIDRICO.

Schönenberg, R., K. Hartberger, C. Schumann, J.H. Benatti and L. Cunha Fischer. 2015. 'What comes after deforestation control? Learning from three attempts of land-use planning in southern Amazonia'. *GAIA – Ecological Perspectives for Science and Society* **24** (2): 119–27.

SEMAS- Secretaria de Estado de Meio Ambiente e Sustentabilidade (2012) Informações Gerais sobre o Zoneamento Ecológico-Econômico: https://www.semas.pa.gov.br/diretorias/digeo/zee/

SEPLAG – Secretaria de Estado de Planejamento e Gestão. 2018. *Dispõe sobre o Zoneamento Socioeconômico Ecológico do Estado de Mato Grosso—ZSEE/MT, e dá outras providências*: http://seplag.mt.gov.br/index.php?pg=ver&id=6304&c=117&sub=true

SEPLAN – Secretaria de Estado de Planejamento de Mato Grosso. 2018. *Zoneamento Socioeconômico Ecológico De Mato Grosso, Superintendência de Estudos Socioeconômicos e Geográficos*: http://seplag.mt.gov.br/index.php?pg=ver&id=6304&c=117&sub=true

SFB – Serviço Florestal Brasileiro. 2020. SICAR, Sistema de Cadastro Ambiental Rural: https://www.car.gov.br/publico/municipios/downloads

SIAGEO Amazônia. 2021. Sistema Interativo de Análise Geoespacial da Amazônia Legal, Empresa Brasileira de Pesquisa Agropecuária, Ministério da Agricultura, Pecuária e Abastecimento: http://www.amazonia.cnptia.embrapa.br/publicacoes_macrozee.html

SIAT-AC – Sistema de Información Ambiental Territorial de la Amazonia Colombiana. 2020. Dinámica de Paisajes Agropecuarios: https://siatac.co/paisaje-agropecuario/

SICAR – Sistema Catastral para Predios Rurales. 2020. Mapa Geo Rural del Ministerio de Agricultura y Riego, Ministerio de Desarrollo Agrario y Riego: http://georural.minagri.gob.pe/sicar/

SIDRA – Sistema IBGE de Recuperação Automática. 2021. Número de estabelecimentos agropecuários, Instituto Brasileiro de Geografia e Estatística: https://sidra.ibge.gov.br/tabela/6780

SINCHI – Instituto Amazónico de Investigaciones Científicas Sinchi. 2016. *Síntesis general de la zonificación ambiental y ordenamiento de la Reserva Forestal de la Amazonia, creada mediante la Ley 2ª de 1959, en la región amazónica Colombiana*. Grupo de Gestión de Información Ambiental y Zonificación del Territorio Amazonia Colombiana – GIAZT. Bogotá: https://siatac.co/resultados-zonificacion-ambiental-de-ley-segunda-de-1959/

Silva, T.M.G.D., 2013. *Caracterização do Sistema Aquífero Parecis na região centro-norte do Estado de Mato Grosso: Subsídios para a gestão dos recursos hídricos subterrâneos. Dissertação de Mestrado, Instituto de Geociências, Universidade de Brasília*: https://repositorio.unb.br/bitstream/10482/13677/1/2013_TalitaMenezesGomesSilva.pdf

Soares-Filho, B., R. Rajão, M. Macedo, A. Carneiro … and A. Alencar. 2014. 'Cracking Brazil's forest code'. *Science* **344** (6182): 363–64.

Soliz. T. 2015. *Cumbre agropecuaria 'Sembrando Bolivia', resultados, ecos y primeros pasos hacia su implementación*. Centro de Investigación y Promoción del Campesinado, La Paz, Bolivia: https://cipca.org.bo/docs/publications/es/8_cumbre-agropecuaria-sembrando-bolivia-resultados-ecos-y-primeros-pasos-a-su-implementacion-1.pdf

Souza, M.L. and A.S. Alencar. 2020. *Assentamentos sustentáveis na amazônia: Agricultura familiar e sustentabilidade ambiental na maior floresta tropical do mundo*. Instituto de Pesquisa Ambiental da Amazônia, Brasília, pp. 176. https://ipam.org.br/wp-content/uploads/2020/05/IPAM-Livro-Projeto-Assentamentos-Sustentaveis-na-Amazo%CC%82nia.pdf

Sparovek, G., B.P. Reydon, L.F.G. Pinto, V. Faria … and G.P. Siqueira. 2019. 'Who owns Brazilian lands?' *Land Use Policy* **87**: 104062.

Steinwig, T., G. Thoumi and B. Tomais. 2017. Grupo Palmas: First Peruvian NDPE Policy Creates Business Opportunities but Strands Land, Chain Reaction Research: https://chainreactionresearch.com/commodities/palm-oil/

Struiken, H. and C. Healy. 2003. Suriname: the challenge of formulating land policy. *Land in the Caribbean: Issues of Policy, Administration and Management in the English-Speaking Caribbean*, pp. 315–44: http://terrainstitute.org/pdf/landincarrib.pdf#page=321

Szlafsztein, C.F., A.A. Azevedo and A. Alencar. 2016. *Amazônia em Pauta No 6: Análise da implementação do Zoneamento Ecológico-Econômico (ZEE) sobre o uso e a ocupação do solo na Amazônia Brasileira*. Brasília: Ipam. http://www.academia.edu/29403221/Boletim_Amaz%C3%B4nia_em_Pauta_6_-_An%C3%A1lise_da_implementa%C3%A7%C3%A3o_do_zoneamento_Ecol%C3%B3gico-Econ%C3%B4mico_ZEE_sobre_o_uso_e_a_ocupa%C3%A7%C3%A3o_do_solo_na_Amaz%C3%B4nia_brasileira

Velasco, F. 1979. *Reforma agraria y movimiento campesino indígena de la sierra*. Quito: FJ Conejo. https://biblio.flacsoandes.edu.ec/libros/digital/52614.pdf

Wasserstrom, R. and D. Southgate. 2013. 'Deforestación, reforma agraria y desarrollo petrolero en Ecuador, 1964–1994'. *Natural Resources* **4**: 34–44. http://www.scirp.org/journal/nr

Yulán Morán, M. 18 May 2017. *Ley y reglamento de tierras: La clausura de la redistribución*: https://lalineadefuego.info/2017/05/18/ley-y-reglamento-de-tierras-la-clausura-de-la-redistribucion-por-milton-yulan-moran/

Zegada, A. 10 Feb. 2019. *Casi 90% de tierras fue para campesinos e interculturales*. The Land Portal Foundation: https://landportal.org/es/blog-post/2021/02/casi-90-de-tierras-fue-para-campesinos-e-interculturales

Notes to Chapter 4

1. Pacheco, P. 2009. 'Agrarian reform in the Brazilian Amazon: Its implications for land distribution and deforestation'. *World Development* 37 (8): 1337–1347; Rausch, L. 2014. 'Convergent agrarian frontiers in the settlement of Mato Grosso, Brazil'. *Historical Geography* **42**: 276–97.
2. Sierra Prael, Y. 9 Apr. 2021. Menonitas en Perú: La historia oculta de la entrega de bosques en Masisea: https://es.mongabay.com/2021/04/menonitas-peru-historia-entrega-bosques-masisea/ ; Mongabay.com 31 May 2021. Colombia: Menonitas deforestan un territorio ancestral en el Meta, La Liga Contra el Silencio, Mongabay Latam, Rutas del Conflicto: https://es.mongabay.com/2021/05/colombia-menonitas-deforestan-un-territorio-ancestral-en-el-meta/
3. CPT – *Comissão Pastoral da Terra 2021. Massacres No Campo.* https://www.cptnacional.org.br/noticias/acervo/massacres-no-campo
4. Ibid.
5. CPT – *Comissão Pastoral da Terra 2021. Massacres No Campo.* https://www.cptnacional.org.br/downlods/category/5-assassinatos
6. Hanbury, S. 14 Dec. 2019. Murders of indigenous leaders in Brazilian Amazon hits highest level in two decades. Mongabay: https://news.mongabay.com/2019/12/murders-of-indigenous-leaders-in-brazil-amazon-hit-highest-level-in-two-decades/ ; CPT – *Comissão Pastoral da Terra 2021. Massacres No Campo.*
7. Carillo, J., C. Cárdenas, Y. Atamain, G. Santos and R. Zapta. 22 Apr. 2020. Morir por la tierra: catorce indígenas asesinados en la Amazonía desde el 2013. *Ojo Público*: https://ojo-publico.com/1779/morir-por-la-tierra-indigenas-asesinados-en-la-amazonia
8. Hallazi, L. 16 Mar. 2021. Qué hay detrás de los asesinatos de líderes indígenas en la Amazonía peruana? *El País*: https://elpais.com/planeta-futuro/2021-03-16/que-hay-detras-de-los-asesinatos-de-lideres-indigenas-en-la-amazonia-peruana.html
9. Jones, K. and M.F. Ramirez. 1 Sept. 2021. Deforestación en la Amazonía colombiana: delimitación del problema. *InSight Crime:* https://es.insightcrime.org/investigaciones/deforestacion-amazonia-colombiana-delimitacion-problema/
10. INCRA – Instituto Nacional De Colonização E Reforma Agrária. 2019. *Relatório de Análise de Mercados de Terras.* https://www.gov.br/incra/pt-br/assuntos/governanca-fundiaria/relatorio-de-analise-de-mercados-de-terras
11. ISU – Iowa State University. 2021. *2020 ISU Land Value Survey Results.* https://www.card.iastate.edu/farmland/isu-survey/2020/
12. Freitas, F.L.M., V. Guidotti, G. Sparovek and C. Hamamura. 2018. 'Malha fundiária do Brasil, v.1812'. In *Atlas - A Geografia da Agropecuária Brasileira*, 2018: www.imaflora.org/atlasagropecuario
13. Feder, E. 1965. 'Land reform under the alliance for progress'. *Journal of Farm Economics* **47** (3): 652–68.
14. Assunção, J. and R. Rocha. 2016. *Rural Settlements and Deforestation in the Amazon.* Climate Policy Initiative. https://climatepolicyinitiative.org/wp-content/uploads/2017/02/Rural_Settlements_and_Deforestation_in_the_Amazon_Working_Paper_CPI.pdf

15. MST – Movimento dos Trabalhadores Rurais Sem Terra. 28 June 2021. *What is the MST?* https://www.mstbrazil.org/content/what-mst

16. INCRA – Instituto Nacional de Colonización y Reforma Agraria. 15 Dec. 2017. Incra nos Estados – Informações gerais sobre os assentamentos da Reforma Agrária. http://painel.incra.gov.br/sistemas/index.php

17. Lourenço, A. 2009. *Regularização Fundiária e Desenvolvimento na Amazônia*. Diretor de Amazônia da Secretaria de Assuntos Estratégicos da Presidência da República. http://interessenacional.com.br/2009/07/01/regularizacao-fundiaria-e-desenvolvimento-na-amazonia/

18. Alencar, A., C. Pereira, I. Castro, A. Cardoso… and R. Novaes. 2016). *Desmatamento nos assentamentos da Amazônia: Histórico, tendências e oportunidades*. IPAM, Brasília, DF. https://ipam.org.br/wp-content/uploads/2016/02/Desmatamento-nos-Assentamentos-da-Amaz%C3%B4nia.pdf

19. Lourenço 2009.

20. Chiavari, J., C.L. Lopes and J.N. de Araujo. *Panorama dos Direitos de Propriedade no Brasil Rural*. Relatório. Rio de Janeiro: Climate Policy Initiative, 2021.

21. Souza, M.L. and A.S. Alencar. 2020. *Assentamentos sustentáveis na amazônia: Agricultura familiar e sustentabilidade ambiental na maior floresta tropical do mundo*. Instituto de Pesquisa Ambiental da Amazônia, Brasília, pp. 176. https://ipam.org.br/wp-content/uploads/2020/05/IPAM-Livro-Projeto-Assentamentos-Sustentaveis-na-Amazo%CC%82nia.pdf

22. Fundo Amazonia. 28 June 2021. *Portfolio de Projetos*. http://www.fundoamazonia.gov.br/pt/carteira-de-projetos/

23. Alencar, A. et al. 2016. *Desmatamento nos Assentamentos da Amazônia: Histórico, Tendências e Oportunidades*. IPAM, Brasília, DF: http://ipam.org.br/wp-content/uploads/2016/02/Desmatamento-nos-Assentamentos-da-Amaz%C3%B4nia.pdf

24. CPI – Comissão Parlamentar de Inquérito. 2003. *Relatório da Comissão Parlamentar de Inquérito destinada a investigar a ocupação ae terras públicas na região Amazônica*: https://www2.camara.leg.br/atividade-legislativa/comissoes/comissoes-temporarias/parlamentar-de-inquerito/51-legislatura/cpiamazo/relatoriofinal.pdf

25. Hecht, S. and A. Cochburn. 2010. *The Fate of the Forest*. Chicago: University of Chicago Press; Hecht, S.B. 1985. 'Environment, development and politics: Capital accumulation and the livestock sector in eastern Amazonia'. *World Development* **13** (6): 663–684.

26. Peixoto, F. 23 July 2009. *Linha do tempo: Entenda como ocorreu a ocupação da Amazônia da BBC Brasil em Brasília*: https://www.bbc.com/portuguese/noticias/2009/07/090722_amazonia_timeline_fbdt

27. Laskos, A.A., A.A. Cazella. and P.B.M. Rebollar. 2016. 'O Sistema Nacional de Cadastro Rural: História, limitações atuais e perspectivas para a conservação ambiental e segurança fundiária'. *Desenvolvimento e Meio Ambiente* **36**: https://revistas.ufpr.br/made/article/view/39124

28. Greenpeace. 2004. *Grilagem de terras na Amazônia: Negócio bilionário ameaça a floresta e populações tradicionais*: https://greenpeace.org.br/amazonia/pdf/grilagem.pdf

29. *Revista Veja*. 13 Jan. 1999. 'O maior latifundiário do mundo'; see description of article at https://uc.socioambiental.org/noticia/42072

30. CPI – Comissão Parlamentar de Inquérito. 2003. *Relatório da Comissão Parlamentar de Inquérito destinada a investigar a ocupação ae terras públicas na Região Amazônica.*
31. Ibid.
32. Brito, B. and P. Barreto. 2011. *A regularização fundiária avançou na Amazônia? Os dois anos do programa Terra Legal.* Belém: Imazon. https://imazon.org.br/publicacoes/1404-2/
33. INCRA – Instituto Nacional de Colonización y Reforma Agraria. 21 June 2021. *Acervo Fundiária do incra.* https://acervofundiario.incra.gov.br/acervo/acv.php
34. Sparovek, G., B.P. Reydon, L.F.G. Pinto, V. Faria ... and G.P. Siqueira. 2019. 'Who owns Brazilian lands?' *Land Use Policy* **87**: 104062.
35. Chiavari, J., C.L. Lopes, D. Marques, L. Antonaccio and N. Braga. 2016. *Panorama dos direitos de propriedade no brasil rural: Legislação, gestão fundiária e código florestal.* Rio de Janeiro: Climate Policy Initiative. Núcleo de Políticas Climáticas da PUC-Rio.
36. CNIR – Cadastro Nacional de Imóveis Rurais. 30 June 2021. *Perguntas e respostas –* CNIR. Receota Federal, Ministério da Economia: http://receita.economia.gov.br/orientacao/tributaria/cadastros/portal-CNIR
37. McInddewar, L. and T. Reis. August 2016. *Land Tenure and the Effect of the Terra Legal Program on Deforestation.* Technical Report. https://doi.org/10.13140/RG.2.2.30811.16169
38. SERFAL – Programa Terra Legal. 2019. *Secretaria Especial de Agricultura Familiar e do Desenvolvimento Agrário –* SEAD: https://dados.gov.br/dataset/serfal-programa-terra-legal
39. SIDRA Sistema IBGE de Recuperação Automática. 2021. *Censo Agropecuário, Tabela 6780 – Número de estabelecimentos agropecuários:* https://sidra.ibge.gov.br/tabela/6780
40. SFB – Servicio Forestal Brasilero. 8 Oct. 2021. *Números do Cadastro Ambiental Rural.* http://www.florestal.gov.br/numeros-do-car
41. LEI N° 11.952, DE 25 DE JUNHO DE 2009.
42. Morais, H.M.A.D.O. 2017. *Regularização fundiária rural na Amazônia Legal: Uma análise da Lei n° 13.465 de 11 de julho de 2017.* https://www.bdm.unb.br/handle/10483/18853
43. https://cimi.org.br/2020/06/pl-da-grilagem-ainda-representa-ameaca-a-sociobiodiversidade/
44. https://www.oeco.org.br/noticias/senado-vota-pl-que-regulariza-a-grilagem-de-terras-publicas-ate-dezembro-de-2014/ ; Wenzel, F. 15 Apr. 2021. Bills before Brazil Congress slammed for rewarding Amazon land grabbers. Mongabay: https://news.mongabay.com/2021/04/bills-before-brazil-congress-slammed-for-rewarding-amazon-land-grabbers/
45. Pacheco, P. and J.H. Benatti 2012. 'Land appropriation under changing environmental governance in the Amazon: The cases of lowland Bolivia and the State of Para, Brazil'. In *International Conference on Global Land Grabbing II,* organised by the Land Deals Politics Initiative (LDPI), Department of Development Sociology, Cornell University, Ithaca.

46. Kaimowitz, D., G. Thiele and P. Pacheco. 1999. 'The effects of structural adjustment on deforestation and forest degradation in lowland Bolivia'. *World Development* **27** (3): 505–20.
47. Manfred Ledermann, pers. comm. July 2021.
48. Gutiérrez, R.G. and L. Quevedo. 2008. *El sistema de concesiones forestales en Bolivia.* CADEFOR: https://docplayer.es/81287039-El-sistema-de-concesiones-forestales-en-bolivia.html
49. FOBOMADE – Foro Boliviano sobre Medio Ambiente y Desarrollo. 2011. *Evo y los empresarios pactan una 'alianza productiva' para alimentar al pueblo.* Boletín N. 89 del Servicio de Noticias Ambientales (SENA) del Fobomade: http://www.rallt.org/PAISES/LATINOAMERICA/BOLIVIA/boli28.htm
50. CENDA – Centro de Comunicación y Desarrollo Andino. 2014. *El TIPNIS símbolo de la defensa de los Derechos Territoriales y de la Madre Tierra.* Tierra, Territorio y Derechos Colectivos: https://cenda.org/especial-tipnis
51. Fundación Tierra. 4 July 2021. *Gonzalo Colque: Se está incubando un conflicto muy grande por la tierra en Santa Cruz*: http://www.ftierra.org/index.php/tema/tierra-territorio/987-gonzalo-colque-se-esta-incubando-un-conflicto-muy-grande-por-la-tierra-en-santa-cruz
52. Ibid.
53. Ibid.
54. SERVINDI. 15 Mar. 2015. *Abogado León: Dirigente de Cidob manejó Bs 1.000 millones.* https://www.servindi.org/actualidad/125302
55. Rodrigues Alvarez, G. 2018. *Sinergia Público-Privada por la Soberanía Alimentaria De Bolivia, Santa Cruz – Bolivia,* Comercio Exterior, N° 267, Instituto Boliviano de Comercio Exterior (IBCE): https://ibce.org.bo/images/publicaciones/ce-267-UCAB-Hacia-la-soberania-alimentaria-en-Bolivia.pdf
56. BIC – Bank Information Center. 2020. *Bolivia Rural Land Regularization and Titling Program*: https://bankinformationcenter.org/en-us/project/bolivia-rural-land-regularization-and-titling-program/
57. IDB – InterAmerican Development Bank. 2020. *PMR Operational Report.* BO-L1113: https://www.iadb.org/projects/document/EZSHARE-522669563-33366?project=BO-L1113
58. MINAG. 13 Dec. 2017. *El proceso de reforma agraria, Objetivos de la reforma agraria.* Ministerio de Agricultura y Riego: https://www.midagri.gob.pe/portal/objetivos/70-marco-legal/titulacion-agraria-en-el-peru/413-el-proceso-de-reforma-agraria.
59. Belaunde Terry, F. 1985. Peru 1985, *Mensaje a al Nación President de la Republica Fernando Belaunde Terry*: https://fernandobelaundeterry.com.pe/libros/2-segundo-gobierno/2-mensaje-presidencial-impresa/MENSAJEALCONGRESODELPRESIDENTEDELAREPUBLICAFERNANDOBELAUNDETERRY1985.pdf
60. Dourojeanni, M. 2017. *Belaúnde en la Amazonía.* Centro Amazónico de Antropología y Aplicación Práctica (CAAAP): https://www.caaap.org.pe/2017/06/12/belaunde-en-la-amazonia-por-marc-j-dourojeanni/
61. Dourojeanni, M. 2016. 'Aprovechamiento del barbecho forestal en áreas de agricultura migratoria en ia amazonía peruana'. *Revista forestal del Perú* **14** (2): http://cedinfor.lamolina.edu.pe/Articulos_RFP/Vol14_2_87_(20)/vol14_no2_art2.pdf

62. Geobosques. 27 Aug 2021. *Plataforma de monitoreo de cambios sobre la cobertura de los bosques.* http://geobosques.minam.gob.pe/geobosque/view/perdida.php
63. INEI – Instituto Nacional de Estadística e Informática. 2001. *Estimaciones y Proyecciones de Población, 1950–2050*: https://www.inei.gob.pe/media/MenuRecursivo/publicaciones_digitales/Est/Lib0466/Libro.pdf
64. http://www.minem.gob.pe/minem/archivos/Fusi%C3%83%C2%B3n%20COFOPRI-PETT.pdf
65. http://www.regionlalibertad.gob.pe/noticias/regionales/9337-gobierno-regional-asume-catastro-rural-y-titulacion-de-tierras-de-cofopri
66. IDB – Interamerican Development Bank. 2014. *Proyecto de Catastro, Titulación y Registro De Tierras Rurales en el Perú.* Tercera Etapa (Pe-L1026), Perfil De Proyecto: http://idbdocs.iadb.org/wsdocs/getdocument.aspx?docnum=39045138
67. IBC – Instituto de Bien Comun. 2021. *Poblado Indigena, spatial database.* https://ibcperu.org/en/
68. INEI – Instituto Nacional de Estadísticas e Informáticas. 2017. *III Censo de Comunidades Nativas.* https://www.inei.gob.pe/media/MenuRecursivo/publicaciones_digitales/Est/Lib1598/TOMO_01.pdf
69. Baldovino S. 2016. Situación Legal de La tenencia de Tierras Rurales en el Perú, Sociedad Peruana de Derecho Ambiental: https://biblioteca.spda.org.pe/biblioteca/catalogo/_data/20170109155049_Tenencia%20de%20Tierras%20-Silvana%20Baldovino%20-%20Libro%20completo.pdf
70. IDB – Interamerican Development Bank 2014. *Evaluación Comparativa: Proyecto de Regularización y Administración de Tierras. Perú, Estudio de caso #4. Anexo Técnico 6.* Oficina de Evaluación y Supervisión: https://issuu.com/idb_publications/docs/technicalnotes_es_84377
71. MINAGRI – Ministerio de Desarrollo Agrario y Riego. 2016. *MINAGRI asume administración de la información nacional referida a saneamiento y titulación de predios rurales, y de comunidades campesinas y nativas*: https://www.midagri.gob.pe/portal/present-catastro-rural
72. http://gobernanzadelatierra.org.pe/2019/02/22/ptrt3-el-proyecto-termina-en-el-ano-2021-y-no-hay-avance/ ; IIPDRS –Instituto para el Desarrollo Rural de Sudamérica. 2014. *Perú: Conversatorio 'Titulación Rural y Agricultura Familiar'*: http://www.sudamericarural.org/noticias-peru/que-pasa/2941-peru-conversatorio-titulacion-rural-y-agricultura-familiar
73. Dammert, J.L., C. Cárdenas and E. Canziani. 2012. Potenciales impactos ambientales y sociales del establecimiento de cultivos de palma aceitera en el Departamento de Loreto: https://repositorio.spda.org.pe/bitstream/20.500.12823/265/1/Potenciales_Impactos_Ambientales_2012.pdf
74. Steinwig, T., G. Thoumi and B. Tomais. 2017. Grupo Palmas: First Peruvian NDPE Policy Creates Business Opportunities but Strands Land, Chain Reaction Research: https://chainreactionresearch.com/commodities/palm-oil/
75. Sierra Praeli., Y. 6 Nov. 2020. Peru prosecutors probe Amazon deforestation linked to Mennonite communities. Mongabay: https://news.mongabay.com/2020/11/peru-prosecutors-probe-amazon-deforestation-linked-to-mennonite-communities/
76. Ibid.

77. Velasco, F. 1979. *Reforma agraria y movimiento campesino indígena de la sierra*. Quito: FJ Conejo. https://biblio.flacsoandes.edu.ec/libros/digital/52614.pdf
78. Wasserstrom, R. and D. Southgate. 2013. 'Deforestación, reforma agraria y desarrollo petrolero en Ecuador, 1964–1994'. *Natural Resources* **4**: 34–44: http://www.scirp.org/journal/nr
79. Gondard, P. and H. Mazurek. 2001. '30 años de reforma agraria y colonización en el Ecuador (1964-1994) - Dinámicas espaciales'. *Estudios de Geografía* **10**: 15–40 http://horizon.documentation.ird.fr/exl-doc/pleins_textes/pleins_textes_7/carton01/010026095.pdf
80. Ibid.
81. *El Universo*. 1 Nov. 2010. Conflictos de tierra pasarán del INDA a Secretaría a fin de mes: http://www.eluniverso.com/2010/11/01/1/1355/conflictos-tierra-pasaran-inda-secretaria-fin-mes.html
82. Yulán Morán, M. 18 May 2017. *Ley y reglamento de tierras: La clausura de la redistribución*. https://lalineadefuego.info/2017/05/18/ley-y-reglamento-de-tierras-la-clausura-de-la-redistribucion-por-milton-yulan-moran/
83. CONAIE – Confederación de Nacionalidades Indígenas del Ecuador. 2016. Conaie, Red Agraria Y Demás Organizaciones Del Sector Campesino Realizarán Cumbre Agraria En 2016 Y Reiteran Su Rechazo A La Ley De Tierras: https://conaie.org/tag/ley-de-tierras/
84. SIGTIERRAS Ecuador. 26 Oct. 2017. http://www.sigtierras.gob.ec/area-de-intervencion-del-proyecto-de-catastro-sigtierras/
85. *El Universo*. 11 Nov. 2020. Sí, pero: 'El Ecuador tiene más de 850.000 unidades de producción agropecuaria que tienen un enorme potencial para generar empleo': https://www.eluniverso.com/noticias/2020/11/11/nota/8045320/si-ecuador-tiene-mas-850000-unidades-produccion-agropecuaria-que/
86. IDB – Interamerican Development Bank. 27 Mar. 2013. *Titulación de tierras para impulsar el acceso al crédito agropecuario en Ecuador*: https://www.iadb.org/es/noticias/articulos/2013-03-27/titulacion-de-tierras-y-acceso-a-credito-en-ecuador,10342.html
87. MAG – Ministerio de Agroicultura y Ganaderia. 2017. *SIGTIERRAS, Catastro Rural en el Ecuador*: https://www.scribd.com/document/472152488/Sigtierras-CatastroRural
88. Corral, L.R. and C.E.M. Olea. 2020. 'What drives take-up in land regularization: Ecuador's rural land regularization and administration program, Sigtierras'. *Journal of Economics, Race, and Policy* **3** (1): 60–75.
89. Pedraza Isaza, D. 2016. *Colombia's Old Agrarian Reforms Challenge a New Rural Reform to Finally Benefit the Rural Poor. Agrarian, Food and Environmental Studies (AFES)*. International Institute of Social Studies: http://hdl.handle.net/2105/37259
90. Berry, A. 2006. *Has Colombia Finally Found an Agrarian Reform that Works. Human Development in the Era of Globalization: Essays in Honor of Keith B. Griffin*. Portland OR: Book News Inc.
91. IDMC – Internal Displacement Monitoring Center. 31 Oct. 2017. Colombia: http://www.internal-displacement.org/countries/colombia

92. Richani, N. 2013. *Systems of Violence: The Political Economy of War and Peace in Colombia*. New York: SUNY Press.
93. OXFAM. 2017. Radiografía de la desigualdad. Lo que nos dice el último censo agropecuario sobre la distribución de la tierra en Colombia: https://www-cdn.oxfam.org/s3fs-public/file_attachments/radiografia_de_la_desigualdad.pdf
94. Barthel, K., V. Cespedes, B. Salazar, R. Torres and M. Varón. 2016. Land and rural development policy reforms in Colombia: The path to peace. USAID/Colombia Land and Rural Development Program (LRDP): https://www.globalcommunities.org/publications/2016-Colombia-Rural-Development-Policy.pdf
95. Botero Garcia, R. 2021. *A deforestación reciente en la Amazonia Colombiana: Consideraciones para su Análisis, Repensar el future de America Latina y el caribe*. Alternativas parac la transfomación social- ecologica: https://fcds.org.co/publicaciones/?sf_paged=2
96. Struiken, H. and C. Healy. 2003. Suriname: the challenge of formulating land policy. *Land in the Caribbean: Issues of Policy, Administration and Management in the English-Speaking Caribbean*, pp. 315–44: http://terrainstitute.org/pdf/landincarib.pdf#page=321
97. Lemel, H. 2001. *Patterns of Tenure Insecurity in Guyana*, Working Paper No. 43, Land Tenure Center, University of Wisconsin–Madison: https://ageconsearch.umn.edu/bitstream/12801/1/ltcwp43.pdf
98. Durwin, H. 2018. Land Administration in Guyana, Natural Resources for Sustainable Rural Development: Access and Governance: http://www.fao.org/3/ca4407es/ca4407es.pdf
99. MINAM – Ministerio do Meio Ambiente 2016. O Zoneamento ecológico – económico na Amazônia Legal, Trilhando o caminho do Futuro: https://antigo.mma.gov.br/images/arquivo/80253/ZEE_amazonia_legal.pdf
100. Szlafsztein, C.F., A.A. Azevedo and A. Alencar. 2016. *Amazônia em Pauta No 6: Análise da implementação do Zoneamento Ecológico-Econômico (ZEE) sobre o uso e a ocupação do solo na Amazônia Brasileira*. Brasília: Ipam. http://www.academia.edu/29403221/Boletim_Amaz%C3%B4nia_em_Pauta_6_-_An%C3%A1lise_da_implementa%C3%A7%C3%A3o_do_zoneamento_Ecol%C3%B3gico-Econ%C3%B4mico_ZEE_sobre_o_uso_e_a_ocupa%C3%A7%C3%A3o_do_solo_na_Amaz%C3%B4nia_brasileira
101. SEMAS-PA. 7 Mar. 2018. Zoneamento Ecológico-Econômico do Pará, Secretaria do Meio Ambiente e Sustentabilidade: https://www.semas.pa.gov.br/servicos/zee__trashed/futuros-zees/
102. JusBrasil. 2012. *Comissão Nacional rejeita Zoneamento de Mato Grosso*, Ministério Público do Estado do Mato Grosso: https://mp-mt.jusbrasil.com.br/noticias/3075748/comissao-nacional-rejeita-zoneamento-de-mato-grosso
103. Aprosoja. 2021. *Zoneamento Socioeconômico e Ecológico (ZSEE) de 2018*. https://www.youtube.com/watch?v=zjidSry9XF4
104. ALMT – Assembleia Legislativa do Estado de Mato Grosso (6 May 2021) *ALMT instala comissão especial para debater Zoneamento Socioeconômico Ecológico*; https://www.al.mt.gov.br/midia/album/comissao-especial-do-zoneamento-socioeconomico-ecologico-zdee/visualizar

105. CORDECRUZ – Corporación Regional de Desarrollo de Santa Cruz. 1995. *Plan de Uso del Suelo PLUS, Planificación y Proyectos*. División de Estudios Básicos y Planificación.

106. FCBC – Fundación para la Conservación del Bosque Chiquitano. 12 Mar. 2018. *Gestión Integral Del Territorio Y Áreas Protegidas*: http://fcbc.org.bo/getionterritorial/

107. Soliz, T. 2015. *Cumbre agropecuaria 'Sembrando Bolivia', resultados, ecos y primeros pasos hacia su implementación*. Centro de Investigación y Promoción del Campesinado, La Paz, Bolivia: https://cipca.org.bo/docs/publications/es/8_cumbre-agropecuaria-sembrando-bolivia-resultados-ecos-y-primeros-pasos-a-su-implementacion-1.pdf

108. GAB – Gobierno Autonomo del Beni 2019. *Plan de Uso del Suelo del Departamen to del beni*: https://cedib.org/wp-content/uploads/2019/12/plan-de-uso-de-suelo-beni-2019.pdf

109. CEJIS – Centro de Estudios Jurídicos e Investigación Social. 2020. *Análisis Socioambiental, Plan de Uso del Suelo 2019 en Territorios Indígenas del Departamento del Beni*: https://www.cejis.org/wp-content/uploads/2020/12/cartilla_PLUS_2019_f.pdf

110. Rojas Calizaya, J.C. and A. Anzaldo Garcia. 2020. 'El nuevo PLUS del Beni, excluye a los actores y sus diversas visiones de desarrollo y atenta contra la Amazonía boliviana'. *Mundos Rurales* **15** (1): 87–104. https://cipca.org.bo/publicaciones-e-investigaciones

111. MINAM – Ministerio del Ambiente – Perú. 2019. *Conociendo la Zonificación Ecológica y Económica*: https://sinia.minam.gob.pe/documentos/conociendo-zonificacion-ecologica-economica-zee

112. IIAP – Instituto de Investigaciones de la Amazonia Peruana. 2011. Meso ZEE de la provincia de Satipo 2010. *Zonificación Ecológica y Económica de la provincia de Satipo*: http://terra.iiap.gob.pe/assets/files/meso/08_zee_satipo/Propuesta_ZEE_Satipo.pdf

113. IIAP – Instituto de Investigaciones de la Amazonia Peruana. 2011. Propuesta de Zonificación Ecológica Económica de la Cuenca del Río Aguaytia: http://terra.iiap.gob.pe/mesozee.html

114. MINAM – Ministerio del Ambiente – Perú. 2015. *Orientaciones básicas sobre el Ordenamiento Territorial en el Perú / Dirección General de Ordenamiento Territorial*. 2nd ed. Lima: Ministerio del Ambiente. https://www.minam.gob.pe/ordenamiento-territorial/wp-content/uploads/sites/129/2017/02/Orientaciones-basicas-OT-1.pdf

115. MINAM – Ministerio del Ambiente. Oct 2021. Registro Nacional de Procesos de ZEE: https://www.minam.gob.pe/ordenamientoterritorial/registro-nacional-de-procesos-de-ot/

116. SENPLADES – Secretaría Nacional de Planificación y Desarrollo. 2014. *Lineamientos y directrices para la planificación y ordenamiento territorial SENPLADES* 1st ed. Quito. http://sni.gob.ec/planes-de-desarrollo-y-ordenamiento-territorial

117. Cabezas, J.E.P. 2017. 'La planificación nacional en Ecuador: Planes de desarrollo y ordenamiento territorial, y el sistema de seguimiento y evaluación SIGAD/National planning in Ecuador: Development and territorial planning plans, and the SIGAD monitoring and evaluation system'. *Ciencia Unemi* **9** (21): 168–79.

118. PROAmazonía. 20 Oct. 2021. Amazonian integral forest conservation and sustainable production program: https://www.proamazonia.org/en/
119. ST-CTEA – Secretaría Técnica de la Circunscripción Territorial Especial Amazónica. 20 Oct 2021. *Fondo Común*: https://www.secretariadelamazonia.gob.ec
120. GDC – Gobernación de Caquetá. 2017. *Directrices de Ordenamiento Territorial para el departamento del Caquetá y sus entidades territoriales*. CORPOAMAZONIA and Gobernación de Caquetá and The Nature Conservancy, Bogotá: http://www.caqueta.gov.co/planes/directrices-ordenamiento-territorial
121. Pallares, G. 2021. 'Lessons learned from a decade of REDD+ in Guyana'. *Forest News*. https://forestsnews.cifor.org/71458/lessons-learned-from-a-decade-of-redd-in-guyana?fnl=en
122. del Prado, N. 2015. *Support Sound Land Use Planning in Suriname: A Review of the Legal and Institutional Framework for LUP*. Workshop land use planning: http://www.wwfguianas.org/news/publications/workshop_land_use_planning_march_17th_2015/
123. GoS – Government of Suriname. 16 Mar. 2018. Onze Natuur op 1 – natuurwetgeving: http://www.gov.sr/themas/milieu-en-omgeving/project-onze-natuur-op-1-natuurwet.aspx
124. Peña, R.E.G. and M.I.S. Viera. 2013. 'Las ABRAE versus las áreas protegidas en Venezuela'. *Revista COPÉRNICO* **8** (19): 27–39. http://abrae85.blogspot.com/
125. EDELCA – Electrificación del Caroní C.A. 2004. *Estudio Plan Maestro de la Cuenca Del Río Caroní, Corporación Venezolana de Guayana*: https://www.pilcomayo.net/media/uploads/biblioteca/libro_1116_LG-130.pdf
126. Sanchez, B. and J. Rosales. 2008. Una revision del plan de manejo de la cuenca del Rio Caroní Venezuela, desde una perspectiva de la valoracion del recurso hidrico. Conference: VI Congreso Ibérico de Planeación y gestión del agua, Vitoria España: https://www.researchgate.net/publication/281101043_UNA_REVISION_DEL_PLAN_DE_MANEJO_DE_LA_CUENCA_DEL_RIO_CARONI_VENEZUELA_DESDE_UNA_PERSPECTIVA_DE_LA_VALORACION_DEL_RECURSO_HIDRICO.
127. MINEA – Ministerio del Poder Popular para Ecosocialismo y Aguas. 31 May 2016. *Plan de Ordenamiento y Reglamento de Uso*: http://www.minec.gob.ve/inician-consulta-publica-de-los-planes-de-ordenamiento-del-rio-caroni/
128. Azevedo-Ramos, C., P. Moutinho, V.L.D.S. Arruda, M.C. Stabile and … J.P. Ribeiro. 2020. 'Lawless land in no man's land: The undesignated public forests in the Brazilian Amazon'. *Land Use Policy* **99**: 104863 ; L.F.G. Pinto, V.G. Faria, G. Sparovek, B.P. Reydon … and T. Carvalho. 2020. 'Quem são os poucos donos das terras agrícolas no Brasil-O mapa da desigualdade'. *Sustentabilidade Em Debate*: 10: 1–21.

129. Servicio Nacional Forestal y de Fauna Silvestre – SERFOR. 2021. Cuenta de Bosques del Peru, Lima, Servicio Nacional Forestal y de Fauna Silvestre (SERFOR) and Instituto Nacional de Estadística e Informática (INEI): https://www.inei.gob.pe/media/MenuRecursivo/publicaciones_digitales/Est/Lib1811/libro.pdf ; SINCHI – Instituto Amazónico de Investigaciones Científicas Sinchi. 2016. Síntesis general de la zonificación ambiental y ordenamiento de la Reserva Forestal de la Amazonia, creada mediante la Ley 2ª de 1959, en la región amazónica colombiana. Grupo de Gestión de Información Ambiental y Zonificación del Territorio: Amazonia Colombiana – GIAZT. Bogotá: https://siatac.co/resultados-zonificacion-ambiental-de-ley-segunda-de-1959/ ; Freitas, F.L.M., V. Guidotti, G. Sparovek and C. Hamamura. 2018. 'Nota técnica: Malha fundiária do Brasil, v.1812'. In *Atlas – A Geografia da Agropecuária Brasileira*: https/:/www.imaflora.org/atlasagropecuario ; Colque, G., E. Tinta and E. Sanjinés. 2016. *Segunda Reforma Agraria: Una historia que incomoda*. La Paz: Fundación Tierra; SIGTA – Sistema de Información para la Gestión Territorial del Ambiente. 2021. Las Áreas Bajo Régimen de Administración Especial (ABRAE), Minesterio de Poder Popular para Ecosocialismo y Agua: https://i.pinimg.com/originals/58/86/db/5886db0a462ac24ae773592c0dc19395.jpg ; GFC – Guyana Forest Commission. 2020. Forest resources Allocation Map of Guyana. Forest resources Information Unit: https://forestry.gov.gy/wp-content/uploads/2021/11/Forest-Allocation-Map-2021-copy-scaled.jpg ; SBB – Stitching voor Bosbeher en Bostoesicht. 2020. Oversicht Uitgegeven Houtkaprechten. Ministerie voor Ruimtelijke Ordening, Grond- en Bosbeheer (RGB): https://sbbsur.com/wp-content/uploads/2021/05/bosbouwlegger_a0_04-februari-2020-.pdf
130. Lovejoy, T.E. and C. Nobre. 2018. 'Amazon tipping point'. *Science Advances* **4** (2): eaaat22340. https://doi.org/10.1126/sciadv.aat2340
131. Rattis, L., P.M. Brando, M.N. Macedo et al. 2021. 'Climatic limit for agriculture in Brazil'. *Nat. Clim. Chang.* https://doi.org/10.1038/s41558-021-01214-3
132. SIDRA – Sistema IBGE de Recuperação Automática. 2021. Produção Agrícola Municipal: https://sidra.ibge.gov.br/tabela/289
133. Projeto MapBiomas – Mapeamento Anual de Cobertura e Uso da Terra do Brasil Coleção 6, https://mapbiomas-br-site.s3.amazonaws.com/Fact_Sheet_PASTAGEM_13.10.2021_ok_ALTA.pdf
134. SIDRA 2021: https://sidra.ibge.gov.br/tabela/289
135. Lambin, E.F. and P. Meyfroidt. 2011. 'Global land use change, economic globalization, and the looming land scarcity'. *Proceedings of the National Academy of Sciences* **108** (9): 3465–72.
136. Watts J. 1 Nov. 2021). Do not trust Brazil's 'greenwashing' promises, say Amazon activists. *The Guardian*: https://www.theguardian.com/environment/2021/nov/01/do-not-trust-brazils-greenwashing-promises-say-amazon-activists
137. Lambin and Meyfroidt. 2011.
138. Samora, R. 2020. Brazil's JBS vows to monitor deforestation through whole cattle supply chain. Reuters: https://www.reuters.com/article/us-jbs-amazon/brazils-jbs-vows-to-monitor-deforestation-through-whole-cattle-supply-chain-idUSKCN26E20I

139. Marengo, J.A. and J.C. Espinoza. 2016. 'Extreme seasonal droughts and floods in Amazonia: Causes, trends and impacts'. *International Journal of Climatology* **36** (3): 1033–1050.
140. Andrea, M.C.D.S., K.J. Boote, P.C. Sentelhas and T.L. Romanelli. 2018. 'Variability and limitations of maize production in Brazil: Potential yield, water-limited yield and yield gaps'. *Agricultural Systems* **165**: 264–273.
141. Pacheco, V. 19 Sept. 2020. Sistema de Irrigação permite tres safras. Vida Rural, Youtube: https://www.youtube.com/watch?v=YrTRQ0x0xx8
142. ANA – *Agência Nacional de Águas e Saneamento Básico (Brasil)*. 2021. *Atlas irrigação: uso da água na agricultura irrigada* / Agência Nacional de Águas e Saneamento Básico. 2nd ed. Brasília: ANA. https://portal1.snirh.gov.br/ana/apps/storymaps/stories/a874e62f27544c6a986da1702a911c6b
143. MapBiomas – Mapeamento Anual do Uso e Cobertura da Terra no Brasil. 2021. *Cobertura e Transições Municípios*: https://mapbiomas.org/estatisticas
144. SEMA – Secretaria de Estado de Medio Ambiente. 3 Nov. 2021. *Outorga De Direito De Uso Da Água No Estado De Mato* Grosso: http://www.sema.mt.gov.br/site/index.php/outorga
145. ANA – *Agência Nacional de Águas e Saneamento Básico (Brasil). Atlas irrigação: uso da água na agricultura irrigada.*
146. MapBiomas — Mapeamento Anual do Uso e Cobertura da Terra no Brasil. 2021.
147. Lathuillière, M.J., M.T. Coe, A. Castanho, J. Graesser and M.S. Johnson. 2018. 'Evaluating water use for agricultural intensification in Southern Amazonia using the Water Footprint Sustainability Assessment'. *Water* **10** (4): 349.
148. Haghtalab, N., N. Moore, B.P. Heerspink and D.W. Hyndman. 2020. 'Evaluating spatial patterns in precipitation trends across the Amazon basin driven by land cover and global scale forcings'. *Theoretical and Applied Climatology* **140**: 1–17.
149. Arias, M.E., F. Farinosi, E. Lee, A. Livino, J. Briscoe and P.R. Moorcroft. 2020. 'Impacts of climate change and deforestation on hydropower planning in the Brazilian Amazon'. *Nature Sustainability* **3** (6): 430–36.
150. Hoekstra, A.Y. 2014. 'Sustainable, efficient, and equitable water use: The three pillars under wise freshwater allocation'. Wiley Interdisciplinary Reviews. *Water* **1** (1): 31–40.
151. CBI – Climate Bond Initiative. 2021. Agriculture Criteria Climate Bonds Standard & Certification Scheme, Agriculture Criteria – Climate Bonds Initiative: https://www.climatebonds.net/standard/agriculture
152. MAPA – Ministério da Agricultura, Pecuária e Abastecimento. 2021. *Estratégias de adaptação às mudanças do clima dos sistemas agropecuários brasileiros: MAPA/SENAR, 2021*: https://www.gov.br/agricultura/pt-br/arquivos/abc_final.pdf
153. SEDAP – Secretaria de Estado de Desenvolvimento Agropecuário e da Pesca. 2016. *Programa de Desenvol Vimento da Cadeia Produtiva do Acaí No Estado do Pará*: http://www.sedap.pa.gov.br/sites/defauln/files/PROGRAMA_PRO_ACAI.pdf
154. Cochrane, T.A., O. Rosales and T.J. Killeen. 2007. *Agua, gas y agroindustria: gestión sostenible de agua para riego agrícola en Santa Cruz, Bolivia.* La Paz: Conservation International. https://ir.canterbury.ac.nz/bitstream/handle/10092/184/12595702_Main.pdf?sequence=1

155. Azevedo, A.A., R. Rajão, M.A. Costa, M.C. Stabile … and R. Pacheco. 2017. 'Limits of Brazil's Forest Code as a means to end illegal deforestation'. *Proceedings of the National Academy of Sciences* **114** (29): 7653–58.

156. BNDES – Banco Nacional de Desenvolvimento Econômico e Social. 1 Apr. 2021. *BNDES cria nova estrutura para emissão de bônus verdes, sociais e sustentáveis, com o apoio do BID*: https://www.bndes.gov.br/wps/portal/site/home/imprensa/noticias/conteudo/bndes-cria-nova-estrutura-para-emissao-de-bonus-verdes-sociais-e-sustentaveis

157. Schmidt, S. and D. Durán. 22 Nov. 2021. Colombia is pitting two vulnerable groups against each other. At stake is the Amazon. *The Washington Post*: https://www.washingtonpost.com/world/2021/11/21/colombia-peace-accords-land-titles/

158. McCoy, T. 2021. Small children are climbing 60-foot trees to harvest your açaí. *The Washington Post*: https://www.washingtonpost.com/world/2021/11/28/brazil-acai-child-labor/

159. Soares-Filho, B., R. Rajão, M. Macedo, A. Carneiro … and A. Alencar. 2014. 'Cracking Brazil's forest code'. *Science* **344** (6182): 363–64.

www.ingramcontent.com/pod-product-compliance
Lightning Source LLC
LaVergne TN
LVHW050906210926
848482LV00012B/359